Mission-Critical and Safety-Critical Systems Handbook

Mission-Critical and Safety-Critical Systems Handbook

Design and Development for Embedded Applications

Edited by
Kim Fowler
Technical Consultant, Sharfus Draid, Inc.

AMSTERDAM • BOSTON • HEIDELBERG • LONDON
NEW YORK • OXFORD • PARIS • SAN DIEGO
SAN FRANCISCO • SINGAPORE • SYDNEY • TOKYO

Newnes is an imprint of Elsevier

ELSEVIER

Newnes

Newnes is an imprint of Elsevier
30 Corporate Drive, Suite 400, Burlington, MA 01803, USA
Linacre House, Jordan Hill, Oxford OX2 8DP, UK

Notices
Knowledge and best practice in this field are constantly changing. As new research and experience broaden our understanding, changes in research methods, professional practices, or medical treatment may become necessary.

Practitioners and researchers must always rely on their own experience and knowledge in evaluating and using any information, methods, compounds, or experiments described herein. In using such information or methods they should be mindful of their own safety and the safety of others, including parties for whom they have a professional responsibility.

To the fullest extent of the law, neither the Publisher nor the authors, contributors, or editors, assume any liability for any injury and/or damage to persons or property as a matter of products liability, negligence or otherwise, or from any use or operation of any methods, products, instructions, or ideas contained in the material herein.

Library of Congress Cataloging-in-Publication Data
Mission-critical and safety-critical systems handbook : design and development for embedded applications / edited by Kim Fowler ; technical consultant, Sharfus Draid, Inc.
     p. cm.
     Includes bibliographical references and index.
     ISBN 978-0-7506-8567-2 (alk. paper)
     1. Reliability (Engineering) 2. Fault tolerance (Engineering) 3. Embedded computer systems–
Design and construction. 4. Safety factor in engineering. I. Fowler, Kim.
     TA169.M574 2010
     620'.00452–dc22   2009038373

British Library Cataloguing-in-Publication Data
A catalogue record for this book is available from the British Library.

For information on all Newnes publications
visit our Web site at www.elsevierdirect.com

08 09 10 10 9 8 7 6 5 4 3 2 1

Printed in the United States of America

Contents

About the Editor

Kim Fowler has spent 27 years in the design, development, and project management of medical, military, and satellite equipment. He cofounded Stimsoft, a medical products company, in 1998 and sold it in 2003; he has also worked for JHU/APL and Ixthos. Kim currently consults in technical development for both commercial companies and government agencies; his focus is on engineering processes in designing and developing products and systems.

Kim is the Executive Vice President of the IEEE Instrumentation & Measurement Society for 2009. He spent 9 years as Editor-in-Chief of the award-winning *IEEE Instrumentation & Measurement* magazine and writes its "Tried and True" column. Kim is adjunct professor for the Johns Hopkins University Engineering Professional Program and lectures internationally on systems engineering and developing real-time embedded systems. Kim has written *Electronic Instrument Design: Architecting for the Life Cycle* (New York: Oxford University Press; 1996) and *What Every Engineer Should Know About Developing Real-Time Embedded Products* (Boca Raton, FL: CRC Press; 2008). He has published over 50 articles in engineering journals and proceedings, and has 17 patents—granted, pending, or disclosed.

About the Contributors

Brian Biersach is the founder of and senior biomedical engineer for Medical Equipment Compliance Associates, LLC., which handles the evaluation and testing of medical equipment to the IEC/UL/CSA/EN 60601-1 and other medical standards. He holds a BS degree in biomedical engineering from the Milwaukee School of Engineering, and a BA degree in economics from the University of Wisconsin-Madison. Brian is also a U.S. expert for AAMI on IEC/ISO Working Groups for international medical standards writing committees. Prior to founding MECA, Brian worked for Underwriters Laboratories, Inc. (UL), as a medical device reviewer and project engineer, a primary instructor for UL's medical seminars and workshops, and an accredited FDA 510(k) reviewer, under the Food and Drug Administration Third Party Review Program.

Timothy Cathcart is consultant and owner of ISQ Systems in Rhode Island. He has been involved in military systems engineering for many years.

Jeff Geisler has 25 years of experience in embedded software development with firms large and small. Trained as an analytical chemist, he decided early in his career that he would rather build instruments than tend them. He joined a start-up as a scientist, and then moved into developing the software for their laser-based spectroscope used to measure respiratory gases and inhaled agents during anesthesia. He then joined the leading firm in portable medical x-ray fluoroscopy as a software team leader. Currently he is software architect at another start-up, developing the software for a next-generation blood pump that assists the pumping action of native hearts in patients with congestive heart failure.

Jeff has lectured at multiple conferences on software development for medical devices, approaches to safety critical software, and software testing. He lives in Salt Lake City, Utah, with his wife. His current non-work enthusiasm is smart phone software development.

Chris Hersman works at the Johns Hopkins University/Applied Physics Laboratory (JHU/APL) and is currently the mission system engineer for NASA's New Horizons mission to Pluto. Since the mission proposal in 2001, he has been responsible for the spacecraft system engineering. After the successful launch of New Horizons in January 2006, his role was expanded to include other mission-related technical responsibilities. He also serves as assistant chief engineer of the Space Department at JHU/APL.

After receiving an MS in electrical engineering from the Ohio State University, Chris began his career in the space industry in 1990 at JHU/APL. Over the past 19 years he has worked on a number of Earth-orbiting and interplanetary missions, including the Midcourse Space Experiment, TIMED, OrbView-3, and Near Earth Asteroid Rendezvous, as well as New Horizons.

Jeremi Peck is the Vice President and senior biomedical engineer of Medical Equipment Compliance Associates, LLC (MECA), which provides medical companies with certification and regulatory compliance services. He holds a BS degree in biomedical engineering. Prior to working for MECA, Jeremi was a medical device services engineer and reviewer for Underwriters Laboratories, Inc. (UL). He was an instructor for UL's medical seminars and workshops, an accredited FDA 510(k) reviewer under the FDA Third Party Review Program, a technical expert for healthcare electrical systems per NFPA 70 and NFPA 99, and a technical expert for the European Medical Device Directive (for CE marking).

Jeffrey M. Sieracki is an "extremely applied mathematician" with over 25 years experience in computer-based systems engineering for industry, Department of Defense, and academia. He holds degrees in physics and applied mathematics from Johns Hopkins University, and a doctorate in applied mathematics and scientific computing from the University of Maryland. He presently directs SR2 Group, LLC., a private R&D firm with which he has worked on diverse applications, including medical devices, optical electronics, pollution control, RADAR, acoustics, terrain mapping, and signal and image processing. Prior to the SR2 Group, Sieracki co-founded Stimsoft, Inc., a medical device company focused on neural stimulation for pain control. Prior to Stimsoft, he consulted for industry and university clients on numerous projects in both mission-critical, real-time systems, and scientific computing. Sieracki holds or has pending 15 U.S. patents, has published in journals, including the *Proceedings of the National Academy of Science*, and maintains an adjunct appointment with the University of Maryland as a visiting research scientist.

David Tyler is Vice President of Advanced Automation Corporation in Rome, New York. He has been involved in military systems engineering for many years.

Best Practices in Mission-Assured, Mission-Critical, and Safety-Critical Systems

Kim Fowler

1. Roadmap to This Book

The material in this book presents best practices in developing mission-assured, mission-critical, and safety-critical systems for medical devices, avionics, military equipment, and spacecraft subsystems. It is a disparate grouping. The good news is that there are threads of commonality in best practices among these systems. The acronyms may vary but best practices translate fairly easily from one area to another. The following are selected areas where best practices in one market often translate well into other markets:

- Defined processes, procedures, and quality assurance (QA) programs

- Collecting and managing requirements and setting priorities for their implementation

- Experimentation and prototyping

- Risk assessment—technological feasibility and business feasibility

- Rigorous software development processes

- Review and inspection

- Documentation

- Test planning, verification and validation, and acceptance testing

1.1. Systems Engineering

The material in this book takes a high-level architectural approach. In essence, it is systems engineering. Whether medical, avionics, military, or space, all deal with multiple

Doi:10.1016/B978-0-7506-8567-2.00001-9

1

disciplines, including (but not limited to) software, electronic hardware, mechanical subsystems, and operations. The systems engineering approach pulls these disciplines together into a framework that helps elucidate interactions between disciplines. Basic definitions that you will need for the remainder of the book follow:

System: A combination of elements or parts forming a complex or unitary whole; composed of components, attributes, and relationships. Typically these elements within a system form definable inputs, processing, and outputs. The interrelated components work together toward a common objective [1].

Systems engineering: An "engineering discipline whose responsibility is creating and executing an interdisciplinary process to ensure that the customer and stakeholder's needs are satisfied in a high quality, trustworthy, cost efficient, and schedule compliant manner throughout a system's entire life cycle. This process is usually comprised of the following seven tasks: **S**tate the problem, **I**nvestigate alternatives, **M**odel the system, **I**ntegrate, **L**aunch the system, **A**ssess performance, and **R**e-evaluate.... [The] [s]ystems [e]ngineering [p]rocess is not sequential. The functions are performed in a parallel and iterative manner" [2]. Brian Mar states that systems engineers should adhere to the following basic core concepts:

- Understand the whole problem before you try to solve it

- Translate the problem into measurable requirements

- Examine all feasible alternatives before selecting a solution

- Make sure you consider the total system life cycle

- Test the total system before delivering it

- Document everything [2]

Everyone on the team should exercise the discipline of systems engineering. The leader of the effort, aside from the program manager, should be a systems engineer or a systems architect. The difference between engineering and architecting is that engineering is deductive work based on hard science, while architecting is inductive work that tends to be likened to art [3]. Both are important in systems engineering.

Mission assurance: "Mission [a]ssurance is a full life-cycle engineering process to identify and mitigate design, production, test, and field support deficiencies of mission success" [4].

"Mission [a]ssurance includes the disciplined application of system engineering, risk management, quality, and management principles to achieve success of a design, development, testing, deployment, and operations process. Mission [a]ssurance's ideal is achieving 100% customer success every time. Mission [a]ssurance reaches across the enterprise, supply base, business partners, and customer base to enable customer success" [5].

"Mission-[a]ssured [d]esign is aimed at creating a robust, manufacturable system—one that does exactly what it is intended to do (its mission) despite the presence of variations, stresses, and uncertainties that can result in mission failure" [6].

Mission critical: "The term mission critical ... refers to any factor (equipment, process, procedure, software, etc.) ... [that] is crucial to the successful completion of an entire project. It may also refer to a project the success of which is vital to the mission of the organization which attempts it" [7].

Safety-critical systems: "A computer, electronic or electromechanical system whose failure may cause injury or death to human beings.... [Examples include] an aircraft or nuclear power station control system. Common tools used in the design of safety-critical systems are redundancy and formal methods" [8].

Quality: The degree for which the sum total of product characteristics fulfill all of the requirements of customers.

Process: A group of interrelated activities and resources that transforms inputs into outputs, often described by a block or flow diagram of events.

Procedure: Specific implementation of the process for a single, focused area of concern; typically step-by-step instructions.

Validation: The confirmation that the design, function, and operation of the final product satisfies the customer's intent.

Verification: The objective tests of metrics that show that the final product meets the quantitative requirements.

1.2. Important Issues

Five basic issues affect every development:

- Integrity—Development requires an integrated approach.
- Interfaces—Most important actions occur at interfaces.
- Humanity—All problems have a human origin.

- Iteration—All levels of development should have feedback to revise direction and design.

- Multidimensions—There are no silver bullets for completing a project.

Integrity: An integrated approach requires integrity, which is defined as the "seamless whole." This requires a "big-picture view" of how the parts fit into the whole. The project lead, in particular, needs to maintain this vision of development and needs to communicate it to the team. Doing so also requires a balanced view of concerns and a willingness to listen to differing opinions from people in different disciplines. This will build trust and a more effective team. A colleague and friend of mine, Jeff Sieracki, occasionally will comment on the need to understand the integration of different disciplines within a project's development. Jeff will say somewhat wryly, "If it's not in my field of expertise, then it must be trivial." Jeff's point is that this colloquial attitude needs to be jettisoned if a project is to truly succeed.

Interfaces: All, or nearly all, important actions occur at interfaces. Think about it, most interesting things happen where two different domains meet. Wave propagation, for example, doesn't get interesting until it encounters an obstacle; then diffraction, refraction, and reflection take place. Semiconductor operations occur at the boundary of different materials. Even the interface between human operations and machine functions is interesting and important. (The one place this view weakens is in software— but even then mistakes like unprotected underflow or overflow or illegal operations are a sort of "boundary.")

One part of successful development is to understand the various types of interfaces (e.g., electrical, material, mechanical, software, signal flow, environmental, human–machine interactions). This requires defining the boundaries and subsystems and then using these to partition the development.

Humanity: All problems have a human origin. Even the very best designs will fail eventually or outlive their usefulness. Human design cannot account for all possibilities—circumstances that were completely unexpected or unknown, multiple and simultaneous interactions that lead to failure, and even human abuse from inappropriate, stupid, or malicious operations. Beyond these concerns, the finite lifetime will limit normal use over time. Failure, furthermore, can have other causes, such as business climate, socioeconomic factors, or politics, that defy technical solution.

Iteration: Development needs a closed-loop, feedback form of operation that includes planning, execution, and review. Discussion of this is addressed later in the section covering the PERRU (plan, execute, review, report, and update) concept.

Multidimensions: Every interesting problem that needs solution requires the involvement of diverse disciplines. There are no "silver bullets." Simply no tool or method or process fits all operations. Don't look for one. Any "silver bullet" will narrow your focus and you will overlook important issues and concerns.

1.3. Material Covered

The material in this book covers best practices, guidelines, government regulations, and standards in developing mission assurance with mission-critical and safety-critical equipment. With these concerns, the book addresses the following disciplines within product and system development:

- Systems

- Operations research

- Design and development

- Software

- Electronic hardware

- Mechanics

- Human interface

- Testing

- Fielding

Areas that are important but not addressed in detail in this book are information technology (IT), manufacturing, distribution, marketing, logistics and support, legal, and business and accounting.

Medical devices, military equipment, and spacecraft instruments share common concerns in many areas, even though at first blush they appear to be very different and operate in very different environments. Common concerns include mission assurance, reliability, availability, vibration, shock, rigorous development, and extensive testing. In particular, mission assurance with mission-critical and safety-critical systems does not and should not rely on any single discipline to ensure performance.

The main difference between areas is that emphasis may change in usage, environment, software rigor, change or update frequency, dependability, and maintenance. Table 1.1 gives examples of some differences in emphasis; these examples only point to the majority of situations, obviously exceptions exist.

Table 1.1: Comparison of Differences of Emphases by Application Type

	Medical Devices	Military Equipment	Spacecraft Instruments
Human interface	Implanted: none External: simple and very clear	Can be fairly complex requiring considerable training of users	Instrument: none GSE: can be complex but only a very few "gurus" will run it; much less effort needed to build intuitive interfaces; interface needs to allow extensive control of system
Environment	Implanted: body temperature, minimal shock, low vibration, corrosive fluids External: office ambient, 0 to +40°C, shock of 1-m drop to concrete floor, minor vibration	COTS: office ambient, 0 to +40°C, shock of 1-m drop to concrete floor, minor vibration Field: –40 to +80°C, wide variety of shock and vibrations	Instrument: –40 to +80°C, shock and vibration of launch GSE: office ambient, 0 to +40°C, shock of 1-m drop to concrete floor, minor vibration
Software development rigor	Class III devices that have life-sustaining operations require certified compilers and RTOS, and very rigorous development and review by FDA	COTS: none Field: rigorous development, depends on review and oversight by customer, some guidelines and standards	Instrument: rigorous development, depends on review and oversight by customer, some guidelines and standards GSE: only review required by customer
Update frequency and type	None, once approved by FDA; requires completely new approval cycle for changes and updates	Depends on architecture of equipment; line replaceable units (LRUs) allow changes quite readily	Instrument: usually none unless minor patches transmitted up GSE: as required
Certification	FDA approval	None unless UL or CE required by customer	None
Reliability	Implanted: a top priority External: some importance	Important, particularly for field applications where repair facilities are not available	Instrument: a top priority GSE: no requirement for COTS usually
Availability, diagnostics, and repair	Not important—devices replaced, not repaired in patient or field	Very important for repair	Instrument: none GSE: no requirement for COTS usually
Safety	Highest priority	Usually a matter of training, not of design requirements	Instrument: none, except for assembly GSE: none

Table 1.1: Cont'd

	Medical Devices	**Military Equipment**	**Spacecraft Instruments**
Maintenance	Either none or personnel trained to perform replenishment and logistics	Very important in both design requirements and in training of operating personnel	Instrument: none GSE: none
Disposal	FDA requirement for supplier to track, record, and dispose	As required by customer or by international standards (i.e., RoHS and WEEE)	Instrument: none GSE: RoHS and WEEE

Notes: Specific conditions and parameters are for examples only. Considerations for spacecraft instruments pertain to unmanned spacecraft only.
GSE, ground support equipment; RoHS, restriction of use of certain hazardous substances, which requires the phase-out of lead, mercury, cadmium, chromium, halogen, bromide, and some fire retardants; WEEE, waste from electrical and electronic equipment, which requires the treatment, recovery, and recycling of electric and electronic waste.
Source: © 2009 by Kim Fowler. Used with permission. All rights reserved.

2. Best Practices

2.1. What and Why?

Best practices are processes, procedures, methods, and techniques that help development of good-quality products. These best practices have to do with what works well in conceiving, planning, architecting, analyzing, designing, building, and testing mission-critical and safety-critical systems.

Best practices, properly applied, will improve the probability of success; they will not guarantee success. Interestingly, best practices in one field or market arena often have analogs in other fields or markets.

Since the 1970s while attempting to achieve "better, faster, cheaper" development of quality products, companies in most industries have acceded that good quality is a necessary ingredient to profitability and even survivability. Best practices include aspects of implementing and operating a quality system. A quality management system (QMS) should blanket a company, which means that it affects every area as a philosophy that guides operations. It should never be an "added management structure" for looking over people's shoulders. The most important realization of a QMS is that it becomes an individual responsibility, as well as a corporate mandate.

2.2. Rationale

A primary goal of a QMS is mission assurance—delivering the right product to the right people in the right application and operating as expected. Toward that goal, consistent application of the QMS across the enterprise is important, which means instituting

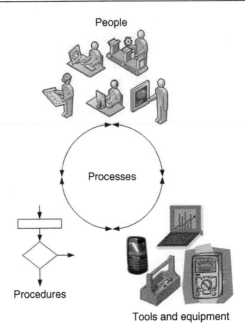

Figure 1.1: Processes link people, tools, and procedures to support consistent application of effort toward quality across the enterprise. (© 2008–2009 by Kim Fowler. Used with permission. All rights reserved.)

appropriate processes that involve people, tools of industry and business, and procedures (Fig. 1.1). The most important part of QMS is the people. The wrong people—incompetent or unmotivated or worse—will undermine and destroy all other efforts. From this point on, the book will assume that the right people are already in place. The right people can make anything work; good processes tying together people, tools, and procedures, will help them be more effective and efficient.

2.3. Standards and Guidelines for a QMS

2.3.1. QMS Components

The QMS has the following requirements: It identifies processes, controls processes, documents the effort through record keeping, and controls the records. Regardless of their formats, all quality systems should have the following primary components:

- Implementation

- Management and staff responsibility

- Resource management

- Product (or service) realization

- Measurement, analysis, and improvement

2.3.2. Responsibility within QMS

Usually a quality manager will establish policy to cover the following concerns:

- Customer focus

- Quality policy

- Planning

- Authority

- Communications

- Review

These areas tend to be the most visible; staff and customers often associate these with the QMS. Ultimately a quality system and its best practices within a company should be everyone's responsibility. Table 1.2 lists the people for establishing the QMS and leading the infusion of best practices into your company.

2.3.3. Resource Management within QMS

Resource managers within the QMS must deal with the following components:

- Human competence, awareness, and training

- Communications and data flow between customer, business areas, and staff

- Infrastructure, including work space, utilities, and equipment

- Environment, including cleanliness, safety, and sound levels

Table 1.2: People Who Can Establish the QMS and Implement Best Practices within a Company

Company Department/Activity	Responsible Leader
Business	CEO, CFO, company president, or business VP
Engineering	CTO, engineering VP, or an engineering manager
Product and quality assurance	CEO, quality manager, or quality lead
Manufacturing and production	Manufacturing VP or manufacturing manager
Personnel	Human resources manager
Infrastructure	CEO, COO, or plant manager

2.3.4. Product Realization within QMS

Product realization within the QMS includes the following components:

- Planning

- Customer input and requirements

- Design and development

- Purchasing

- Production and manufacturing

- Service and support

- Control of measuring equipment

2.3.5. Development and Certification of the QMS

Four types of quality processes are prominent in many industries: ISO 9001, AS9100, Six Sigma, and CMMI. Choosing one depends on its fit to the industry and the understanding of the particular QMS by both employees and customers. Brief descriptions of each type follow.

ISO 9001

ISO is the acronym for International Organization for Standardization, which is based in Geneva, Switzerland. ISO 9001 provides a quality management system defined as, "a series of components logically linked together that provides measures and controls to manage and improve products" [1]. To be ISO certified, a company needs to purchase the standard and then be regularly audited.

ISO 9001 comprises two basic programs—one provides structure and the other provides for improvement. The structure program accomplishes the following:

- Identifies processes and documents them

- Describes sequences and interactions

- Ensures resources to run the quality management system

The improvement program provides the following:

- Measures performance

- Judges effectiveness

- Improves the quality management system [9]

ISO 9001 requires an eight-section QMS. The most prominent are the last five. Section 4 is about documentation, particularly the quality manual. Section 5 concerns management responsibility, and Section 6, resource management. Section 7 is about product realization, and Section 8, measurement, analysis, and improvement.

AS9100

AS9100 is defined specifically by and for the aerospace industry [10]. It was developed from AS-9000 (1997), which adopted the 20 elements of ISO-9001-1994, and aligned them with the needs of the air transport industry. The AS9100 incorporates additional provisions to cover civil and military aviation and aerospace industry standard requirements more comprehensively. It also covers the needs of suppliers, regulatory bodies, and customers of the aerospace industry.

AS9100 has the following objectives:

- Reduce defects in the supplier chain

- Reduce costs

- Continuously improve quality

- Enhance customer satisfaction

- Reduce the person-hours spent by an organization on quality management activities, primarily by replacing the various customer requirements and specifications with a single industry standard that's acceptable to everyone

Key features of AS9100 include:

- Quality improvement

- Variation control or management for key product characteristics

- Production and service provisions unique to the aerospace industry such as part accountability, foreign object detection, production documentation, part identification, and part traceability

- Process/tooling change control and management

- Supply-chain quality control, which covers the control of purchasing and acceptance processes

- Design and development control

- Product configuration control management

- Product quality, reliability, and safety control

- Continual improvement

Some AS9100 definitions are as follows:

- Product (AS9100 service)—Activity that adds value for a customer. In aerospace this might be transporting cargo.

- Product (AS9100 software)—Computer program that adds value for a customer.

- Product (AS9100 hardware)—Physical entity that adds value for a customer. A hardware product might be a mechanical engine or motor or it might be an electronic circuit board or it might be a mechanism or a structural enclosure.

- Product (AS9100 processed material)—Substance requiring replenishing that adds value for a customer. One example is lubricant.

- Authority (AS9100)—National aviation authority having jurisdiction over activities affecting the airframe.

- Key characteristic (AS9100)—Features of a material, process, or part whose variation has a significant influence on product life, performance, service life, or manufacturability.

Six Sigma

The Six Sigma quality system originated in high-volume production and manufacturing. Its main goal is to identify and measure variances. Six Sigma does not guarantee quality but provides expectations of program performance based on customer satisfaction [11].

CMMI

CMMI is the acronym for Capability Maturity Model Integration. It grew out of development guidelines for software through the Software Engineering Institute at Carnegie Mellon University. The guidelines have unlimited distribution rights and can be downloaded from www.sei.cmu.edu.

CMMI has five levels of maturity that represent stages and capability of a company with its processes. Table 1.3 lists and briefly describes the levels of maturity.

CMMI has 22 process areas (PAs); each PA is a cluster of related best practices to satisfy goals to improve processes. Table 1.4 lists the PAs along with the maturity level for a staged representation.

Table 1.3: Maturity Levels in CMMI for Staged Representation

Level	Name	Comments
1	Initial	Success depends on competence and heroics.
2	Managed	Processes are planned and executed according to policy.
3	Defined	Processes are tailored to each project and described more rigorously.
4	Quantitatively managed	Quantitative objectives are measured and demonstrate predictable performance.
5	Optimizing	Continually improves processes based on quantitative understanding.

Table 1.4: General Areas and PAs for Stage Representation of CMMI

Areas	Maturity Level
Requirements management Project planning Project monitoring and control Supplier agreement management Measurement and analysis Process and product quality assurance Configuration management	2
Requirements development Technical solution Product integration Verification Validation Organizational training Integrated project management Risk management Integrated teaming Integrated supplier management Decision analysis and resolution Organizational environment for integration	3
Organizational process performance Quantitative project management	4
Organizational innovation and deployment Causal analysis and resolution	5

Selected CMMI definitions follow:

- Process area (PA)—Cluster of related practices in an area that, when implemented collectively, satisfy a set of goals considered important for making improvement in that area.

- Generic goal (GG)—Goal that applies to multiple PAs and describes the characteristics to institutionalize processes; it is a required model component and is used in appraisals to help determine whether a PA is satisfied.

- Specific goal (SG)—Goal that has unique characteristics and satisfies a specific PA; it is a required model component and is used in appraisals to help determine whether a PA is satisfied.

- Generic practice (GP)—Describes an activity that is important for achieving the associated generic goal; it is an expected model component.

- Specific practice (SP)—Description of an activity that is important for achieving the associated specific goal; it is an expected model component.

Comparison of ISO 9001 versus CMMI

ISO 9001 tends to be more of a "pass-or-fail" sort of effort. You describe a company's processes and then show how the company meets and performs these processes. It does not provide as many guidelines or details to incorporate into processes.

CMMI is more "granular" or has higher resolution than ISO 9001. CMMI provides more detailed guidelines than ISO 9001. The intent of CMMI is to focus on process improvement through assessment and maturity levels. Table 1.5 provides a comparison between ISO 9001 and CMMI.

Table 1.5: ISO 9001 versus CMMI

	ISO 9001	CMMI
Characteristics	Stereotyped; "pass-fail" test	More flexible; focus on assessment and improvement
Advantages	Easily adaptable to manufacturing Customers understand	Adapts well to software Easy-to-get standard
Disadvantages	Less direction for improvement Cost to undergo audits	Learning curve for the 22 PAs Must train customers to understand

CMMI Framework   CMMI attempts to prevent sole focus on a specific business area. It transcends disciplines and includes software, hardware, mechanical, logistics, service, and maintenance. CMMI incorporates best practices for both development and maintenance. Watts Humphrey outlines five ideas that inspired the goals of CMMI:

- Planning, tracking, and schedule management
- Requirements definition and configuration control
- Process assessment
- Quality measurement and continuous improvement
- Evolutionary improvement [12]

Table 1.6 gives the stages of implementation and the PAs implemented in each stage. Appendix A contains a description of each PA.

2.3.6. Training for QMS

Implementing a quality system and best practices within a company requires training all staff and personnel. The goal is to make the quality system useful and to make the training efficient and appropriate. Consequently, training should occur at several stages and levels. For example, a company might do the following:

- Short, weekly session of about 20 minutes
- Monthly session of about 1 hour
- Several days every 3 months to evaluate the processes, suggest changes, and train for more detail in the current stage or prepare for the next stage

The most important concern is for training to be regular, appropriate, effective, and inclusive.

2.3.7. Measurement, Analysis, and Improvement within QMS

Measurement, analysis, and improvement within the QMS have the following goals and components:

- Ensure product, service, and QMS uniformity
- Customer satisfaction
- Internal audits
- Monitor processes
- Control of nonconforming product

Table 1.6: CMMI PAs Established at Each Stage of Implementation

Level	Components Implemented	PA	CMMI Level
1	Work order (WO), Problem report corrective action (PRCA), requirements development	Project planning	2
		Requirements development	3
		Requirements management	2
		Measurement and analysis	2
		Process and product quality assurance	2
2	Engineering processes and checklists, configuration management, document templates	Project monitoring and control	2
		Technical solution	3
		Supplier agreement management	2
		Configuration management	2
		Organizational training	3
		Integrated project management	3
3	Engineering updates, manufacturing updates, vendor qualification and management	Product integration	3
		Verification	3
		Validation	3
		Risk management	3
		Decision analysis and resolution	3
		Organizational environment for integration	3
4	R&D, project processes integrated	Integrated teaming	3
		Integrated supplier management	3
		Organizational process performance	4
		Quantitative project management	4
5	Business processes integrated	Organizational innovation and deployment	5
		Causal analysis and resolution	5

- Analyze data

- Improvement (continuous operation, corrective action, preventive action)

Measurements, analyses, and updates improve the quality of the process and procedures. The planning block establishes the metrics for measurement; these metrics are a part of the requirements. Every requirement should have a metric or measurable/observable quantity to allow later verification and validation.

Once the measurements are made, the data are analyzed for methods of improvement, which happens during the review stage or block. The updates then implement the improvements that derive from the review.

The review and report are both qualitative and quantitative. Learning derives from the update activities.

3. Project Management and Systems Engineering

There is considerable overlap between project management and systems engineering—both are leadership positions and both must deal with many different people and disciplines while maintaining budget, schedule, and output quality. The difference between project management and systems engineering is small; project management is more business oriented while systems engineering is more technical.

For smaller projects, typically 20 or fewer people, one person usually occupies both positions and leads the projects. For large-scale system development, such as a vehicle or facility design, the team may have multiple levels of both project managers and systems engineers. For these large developments, a systems architect may also join the team and assist system engineers and team leaders with the technical architecture of the system.

3.1. Project Management

Managing a project has several component activities. Project management establishes the contract for the project and selects the type of development. It determines the team and subcontractors who will complete the project. Project management sets the schedule and budget. It controls the project through task or work orders and then monitors the project through the completion of task/work orders, various reviews, and the control board. Finally, project management closes out the project.

Verzuh states, "The ultimate challenge for project managers is to meet the cost, schedule, and quality of the project without damage to the people. That means the project ends with high morale, great relationships with customers, and vendors [who] can't wait to work with you on the next project" [13].

3.1.1. Definitions

Project: Development work that has a clear beginning and end and is performed once to produce something unique. This work is different from ongoing operations such as manufacturing [13].

Project management: Verzuh claims that project management is "art informed by science." He lists five characteristics of a successful project management:

1. Agreement among the project team, customers, and management on the goals of the project.

2. A plan that shows an overall path and clear responsibilities and will be used to measure progress during the project.

3. Constant, effective communications among everyone involved in the project.

4. A controlled scope.

5. Management support [13].

3.1.2. Inputs and Outputs

The inputs to project management are the customer's expressed intent and requirements, project constraints, market conditions, standards, guidelines, and the contract. Outputs are descriptive documents, reviews and reports, and physical product.

3.1.3. Operations

In between the inputs and outputs are the operations. Operations include the following areas:

- Concept and market
- People and disciplines
- Architecting, architecture, and process flows
- Scheduling
- Documentation
- Requirements and standards
- Analyses
- Design tradeoffs
- Test and integration
- Manufacturing
- Support and service

Project management has the following functions:

Definition—Determine the purpose, goals, and constraints for the project, and set the rules and guidelines.

Enlist people—Get a sponsor and determine the stakeholders for the life cycle of the project.

Communicate—Prepare and deliver the statement of work, the responsibility matrix, the charter of authority, and the communication plan.

Planning—Estimate effort and cost, prepare the budget and schedule, define the risk.

Control—Measure progress, manage risk, ensure that corrective actions take place, hold reviews, and maintain communications [13].

3.1.4. Best Practices—Scheduling

The project manager leads planning and scheduling but, in general, planning and scheduling are shared responsibilities with the team. It takes a committed team to plan well, work the plan, stick to the plan, and learn from their mistakes. Conversely, failing to plan properly often results in failed schedules.

You can schedule a project several different ways; the most typical are top-down scheduling and bottom-up scheduling. Top-down scheduling sets delivery dates and milestones first and then determines the development that can fit into the periods between dates. Bottom-up scheduling starts with the work break-down structure, which contains the basic details of everything that needs to be done, and builds up to the dates when stages can be finished. While top-down planning has the overall perspective needed to complete a project, it can over-constrain a project by planning to do too much in too short a time. While bottom-up planning can be very detailed, inappropriately estimated margins and tolerances in timing can build inefficiencies into the process and extend the schedule too far. Use both methods in combination to develop an appropriate schedule. Start with a top-down plan and then see where a bottom-up plan meets it. If they are misaligned, iterate the exercise by relaxing constraints in top-down and tightening tolerances that may be too loose in bottom-up. This kind of exercise can help you identify and remove the pitfalls of either method.

When planning and scheduling, you must first understand the activities that make up project development—the work break-down structure is a great tool here. Remember to include meetings, desk work, lab work, communications, documentation, travel, debugging, testing, fabrication, support, installation, and training, along with the obvious activities of design, production, and big customer reviews.

Think carefully about these things. A simple meeting of several people requires more time than that used to sit around the table and talk. It includes preparation of notes and slides, communicating and scheduling, and even the very simple act of walking to and from the location. Simple and short telephone calls can fill a day. I have found that calling a vendor and inquiring about the price and delivery on a single, generic, stocked item will average about 8 minutes. Simple activities can eat up time!

Finally, build in a schedule margin for "surprises." Study where problems have occurred in previous projects and find correlations between the activities and the results. Once you better understand the trouble spots, then you can apportion margin to those activities.

3.2. Systems Engineering

3.2.1. Definitions

Systems engineering: An interdisciplinary approach encompassing the entire technical effort to evolve and verify an integrated and life-cycle balanced set of system, people, product, and process solutions that satisfy customer needs. Systems engineering encompasses the following:

- Technical efforts related to the development, manufacturing, verification, deployment, operations, support, disposal of, and user training for, system products and processes

- Definition and management of the system configuration

- Translation of the system definition into work break-down structures

- Development of information for management decision making [14]

Work break-down structure (WBS): A tool for breaking down a project into its component tasks to initiate, execute, and complete the development of the project. Each task is an elemental, self-contained effort that has clear inputs and outputs and a description of the work. The WBS helps provide a detailed outline of the project scope, monitors progress, contributes to the estimation of cost and schedule, and designates the project team [13]. For military projects, MIL-STD-499B described WBS as "[a] product-oriented family tree composed of hardware, software, services, data, and facilities which result from systems engineering efforts during the development and production of a defense materiel item, and which completely defines the program. Displays and defines the product(s) to be developed or produced, and relates the elements of work to be accomplished to each other and to the end product" [14].

Summary tasks: A higher-level description of a group of tasks that completes a subset of the project or its process. Summary tasks may contain lower-level (more detailed) summary tasks or work packages or both.

Work packages: The most detailed, elemental task that clearly defines a single effort, such as analyzing a simple circuit design or performing a code review of a single software module. Often a work package will be completed in less than 80 hours of effort.

3.2.2. Inputs and Outputs

Inputs and output for systems engineering are very similar to those in project management. The focus tends to be more technical in nature and the requirement metrics more objectively measurable.

3.2.3. Operations

Operations for systems engineering cover the same areas listed above in the discussion of project management. The technical detail is greater for systems engineering and more time is spent on developing the architecture, and confirming, verifying, and validating the work product with the requirements. Systems engineering also focuses on the interactions of components and subsystems within the system. Often the systems engineer will develop the WBS, or a portion of it, for use by the program manager.

3.2.4. Best Practices—Requirements

Requirements form a list of descriptions that define and constrain system design. They specify system behavior and operational parameters, provide a reference for modifications, and characterize potential responses to unexpected situations. A requirement may have one of several different forms, "from a high-level abstract statement of a service or of a system constraint to a detailed mathematical functional specification" [15]. Oshana writes, "A good set of requirements has the following characteristics:

- *Correct*. Meets the need.

- *Unambiguous*. Only one possible interpretation.

- *Complete*. Covers all the requirements.

- *Consistent*. No conflicts between requirements.

- *Ranked for importance*.

- *Verifiable*. A test case can be written.

- *Traceable*. Referring to requirements is easy.

- *Modifiable*. Adding new requirements is easy [15].

Establishing the requirements for a project has several stages, each with associated best practices. The stages follow:

- Gathering

- Establishing

- Applying

- Refining and updating

Gathering: You need direct contact with customers and potential customers to gather requirements. The problem is to understand what they really need. (Note the term "need," not "want"—there is a significant difference. People often will tell you what they want but don't always know the extent of their true need.) Your visits to customers are very important; these meetings succeed when discussions between designers and users have prepared templates of guided scripts. A variation on this theme is the use of focus groups, which provide a controlled environment and trained facilitators; focus groups can elicit ideas and directions from potential customers that may not have occurred to you.

Another primary area for gathering requirements is to find and understand the necessary standards and regulations within the particular market. These can often be found online.

Establishing: Once you have begun gathering information for a potential new project, then you can begin developing requirements. Realize that requirements are the codification of intent, which is the desire that a product have particular characteristics and an appropriate function and performance. The trick is to translate intent into requirements, which define and set measurable quantities for later testing, verification, and validation.

Applying: Requirements direct both design and evaluation. The requirements define the metrics of the design. Evaluating the metrics indicates project progress; appropriate requirements will provide a good and complete picture of the progress.

Refining and updating: No list of requirements is ever complete or correct in the beginning. Requirements need to be analyzed through various means: simulation, trade studies, risk analysis, laboratory bench tests, field trials, and discussions with clients and customers. Later as the project develops, constraints and contradictory requirements will appear—they always do. The list of requirements must be refined and updated to represent the current reality. Prepare for requirements to be flexible and plan for cycles of updates to the requirements list.

3.2.5. Best Practices—Efficient Development of Requirements

The process of developing requirements is an inexact science. Unforeseen problems and unpredictable interactions often arise because the requirements are either too sparse or too complex. Sparse requirements leave out important concerns leading to unexpected problems. Complex requirements can "overspecify" a product leading to unforeseen interactions that cause problems. Reinertsen reports that many projects fail because requirements are too complete, too stable, and too accurate. These types of requirements shift attention from the few, critical specifications to the many, unimportant ones [16].

Here are some ways to establish useful requirements:

- Write the catalog page for the product first. If a feature warrants mention, then it is important. Any feature that is left out is deemed less important. This will help you prioritize requirements [16].

- "Brief specifications almost always force higher levels of customer contact" [16]. More customer contact leads to a deeper understanding of customer needs, quicker resolution of interpretations, and faster reaction to market shifts.

- High-quality requirements make for concise specifications. The converse is not true—short specifications do not necessarily lead to good quality specifications [17].

- Allow flexibility in the requirements and make development a "closed-loop system." Reinertsen claims that initial accuracy is not as important as assuming that the specifications have inaccuracies and then using feedback to detect and correct them [16].

- Pazemenas argues for shortening "the fuzzy front end" [18]. Identify new product opportunities and improvements in technology often—quarterly evaluations may be necessary—so that the fuzzy front end does not drag out from a long learning curve.

Some companies are using or moving toward model-based specifications for subsystems to capture requirements. Model-based specifications use a model of the device or system with mathematical constructs, such as sets and sequences, and define system operations by how they modify the system state. Use cases and UML are an example of model-based approaches; these can reveal interactions between the system and external inputs and can force you to consider obscure sequences [15].

Another potential best practice in developing requirements is testdDriven development. "Test-driven development (TDD) ... uses short development iterations based on pre-written test cases that define desired improvements or new functions. Each iteration produces code [or requirements] necessary to pass that iteration's tests" [19].

TDD originated as a software design technique but serves well for system requirements as well (Fig. 1.2). "We employed TDD to test core behavior and used the best practices of object-oriented design to isolate areas of uncertainty. Work on requirements continued in parallel, reducing the uncertainty in the process. After six months of development, we had only two or three bugs that required more than a few minutes of debugging. The total time spent debugging was only about a week over the six-month project!" [20]

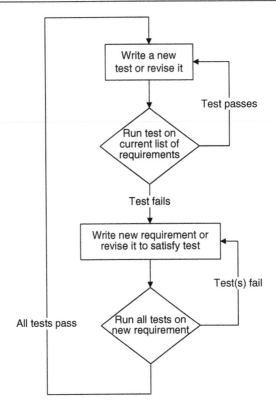

Figure 1.2: Generalized flow for developing requirements through TDD. (© 2009 by Kim Fowler. Used with permission. All rights reserved.)

3.3. Mission Assurance

Mission assurance is the goal of project management and systems engineering. There are a number of ways to approach mission assurance. Aerospace Corporation has a fine description of how the company strives for mission assurance in the Fall 2007 issue of *Crosslink Magazine* [21].

Aerospace Corporation's *Mission Assurance Guide* explains six core mission assurance processes:

1. **Requirements analysis and validation** is a review of formally specified user requirements and their consistency with formally or informally stated user needs and expectations.

2. **Design assurance** is a set of planning, analysis, and inspection activities to assess whether the evolving designs can produce a system that will perform as intended overall operating conditions throughout its design life.

3. **Manufacturing assurance** ensures [19] that the planned manufacturing processes are repeatable and reliable and can produce the system as designed.

4. **Integration, testing, and evaluation** verify that assembled components meet requirements individually and as part of the finished system.

5. **Operational readiness** encompasses all activities required to transport, receive, accept, store, handle, deploy, configure, field-test, and operate launch and space vehicles and supporting ground systems.

6. **Mission assurance reviews and audits** are forms of independent technical analysis that facilitate understanding of the interfaces and composite performance of a system while synchronizing government and contractor expectations. The three types are technical reviews, audits, and readiness reviews [21].

Seven mission assurance disciplines support the six core mission assurance processes identified above in Aerospace Corporation's *Mission Assurance Guide*. These supporting disciplines "provide the more technically oriented underpinning of mission assurance application and include engineering methodologies specifically geared toward system design validation and product verification" [21]. These seven mission assurance disciplines follow:

1. **Risk management** is a structured approach to identifying and evaluating risk and risk-control measures and communicating mission threats to program stakeholders.

2. **Reliability engineering** includes the development and validation of probabilistic system reliability requirements and design trade-off studies in the early program phases. During design development and production, it assists with the determination of component failure rates and development of probabilistic reliability models, analysis of failure modes and effects, identification and control of critical items, application of worst-case and parts-stress analyses, and analysis of accelerated-life test data.

3. **Configuration management** seeks to control the technical hardware and software baselines of a program—the requirements, specifications, designs, interfaces, data, and supporting documentation. The goal is to ensure that the functional, allocated, developmental, test, and product baselines are consistent, accurate, and repeatable and that any changes to those baselines will be recorded and will maintain the same accuracy, consistency, and repeatability.

4. **Parts, materials, and processes (PMP) engineering** seeks to provide a standardized set of qualified components from which to build a reliable product at a reasonable cost and risk.

5. **Quality assurance** is the engineering and management discipline intended to ensure that a product meets the specified performance parameters. A well-defined and properly

implemented quality assurance program instills confidence that all quality requirements have been met through control of operations, processes, procedures, testing, and inspection.

6. **System safety assurance** applies engineering and management principles and techniques to control system hazards within the constraints of operational effectiveness, schedule, and cost.

7. **Software assurance** seeks to ensure that system software will meet performance requirements and user expectations and will be dependable, maintainable, and applicable to the user's operational environment" [21].

4. Process Flows for Developing Products

4.1. Plan, Execute, Review, Report, and Update (PERRU)

Every process and procedure should have a set of activities that is both consistent and thorough; each process and procedure should provide for self-evaluation. They should have a structure that can be summarized in the acronym PERRU for plan, execute, review, report, and update. Figure 1.3 illustrates this structure and data flow; it also provides some of the

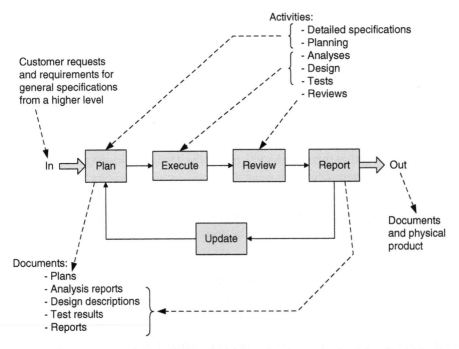

Figure 1.3: Basic, self-similar unit of operation that suits any level of detail within a project. (© 2008–2009 by Kim Fowler. Used with permission. All rights reserved.)

types of activities and types of documents generated during a project. This structure has an interesting self-similarity principle that exists at every level of process, from high- to low-level processes and finely detailed procedures, within project and product development.

4.2. Development Processes

A development process provides the flow of events and procedures that should occur to develop a product. The necessary characteristics of a process follow:

- Addresses the field and market of the product

- Covers all the activities needed to complete product development

- Useful and does not force "busy work" or overburden with extraneous efforts

- Consistent with company goals

- Flexible for the constraints and demands of particular projects

4.3. Processes vs. Procedures

The difference between a process and a procedure is a qualitative description, as well as diversity in details. A process contains the overarching principles that drive best practices, and tends to be general. A procedure is a specific implementation of the process for a single, focused area of concern.

4.4. General Process Models

A process model indicates the activities and their general phasing. Process models typically comprise a visual view of those activities. The activities include concept, design, evaluation, integration, acceptance sign-off, production, and operation.

Figure 1.4 shows a simple representation of the V-model of development. The figure outlines six primary phases, and shows how verification and validation connect between various activities and across phases. (Recall that verification is an objective set of tests to confirm that the product meets the metrics of the requirements, while validation means that the product in a more subjective perspective meets the original intent.)

A limitation of the V-model is that it seems to imply that the requirements are complete in the conceptual or preliminary stage. This does not occur in most, if not all, product development. Requirements, design, and evaluation often iterate several times before final integration and acceptance. This iterative situation leads to spiral development as a defined discipline.

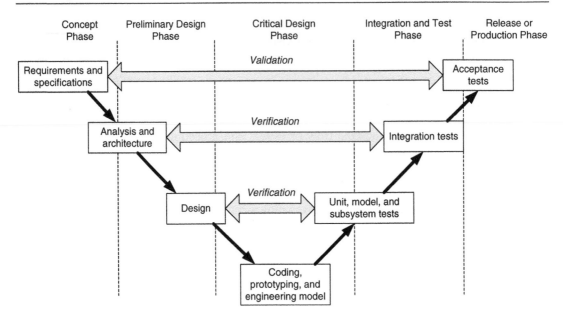

Figure 1.4: Simplified V-model of development that shows how it fits development phases.
(© 2008–2009 by Kim Fowler. Used with permission. All rights reserved.)

Figure 1.5 shows spiral development, which adds components, modules, and subsystems, in succession, to a system. Spiral development thoroughly tests the system for functionality and meeting the appropriate set of requirements after each module addition; this is one form of verification. An important aspect of spiral development is that system functionality builds in stages. Each stage plans a new set of implemented features, tests them, evaluates them, and then updates the system configuration to include them.

4.5. An Example of Phases, Processes, and Procedures

Figure 1.6 shows an example of implementing a V-model process; it requires the procedures, checklists, and all forms of documentation to be generated for a project that develops a new product. The figure displays the first level of detail in the task summaries for each phase.

Diving into the next level of detail, consider as an example the critical design phase, the phase that refines and completes the architecture and requirements for a new product. It also contributes a number of final analyses. A critical design review completes this phase. Figure 1.7 gives the process flow for the critical design phase. Figure 1.8 illustrates an example of one potential work package (an elemental or kernel task) within the critical design phase.

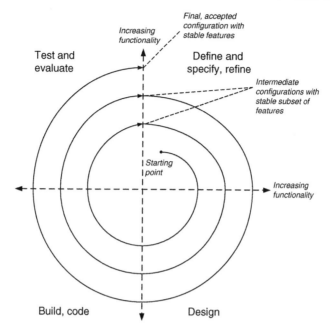

Figure 1.5: Simplified spiral model of development. A system develops in stages, and each stage passes through four separate development phases. (© 2008–2009 by Kim Fowler. Used with permission. All rights reserved.)

The primary *responsibilities* in the critical design phase might be held by the following individuals or groups:

- Hardware (H/W) design: Electronic engineer(s)

- Software (H/W) design: Software engineer(s) or developer(s)

- Mechanical design: Mechanical engineer(s)

- Manufacturing design: Manufacturing or production engineer(s)

- Product architecture: System architect

- Requirements analysis: System engineer(s)

- Control board: Program manager, system architect, system engineer(s)

Other company employees may contribute to the architecture and requirements. All involved in the project might participate in reviews. Employees outside the project can sometimes provide unbiased considerations. Subcontractors and customers often participate in reviews.

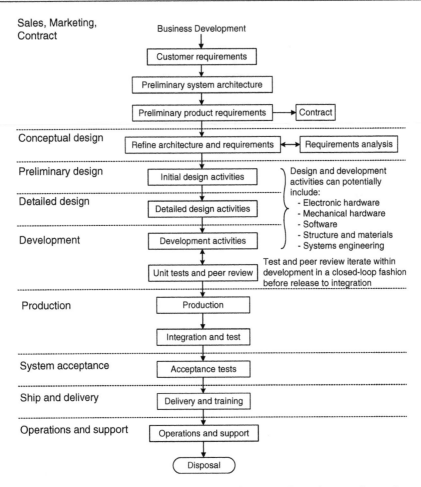

Figure 1.6: An overview of a high-level process framework and procedures for product development, production, and delivery. This is the first level of detail for task summaries. (© 2008–2009 by Kim Fowler. Used with permission. All rights reserved.)

Inputs include the requirements, the state of work (SOW) and work orders prepared by the project manager. The *outputs* include analyses (e.g., power, reliability, environmental, worst-case stress), Product Requirements Document, Interface Control Documents, development plans, manufacturing plan, test plans, and Engineering Change Notices, among many other documents.

The Critical Design Phase has unpredictable *durations* that depend on customer requirements and your company's loading with other projects. For small projects and subsystems, this phase can often take from several weeks to several months in duration. For large systems, this phase can be measured in years.

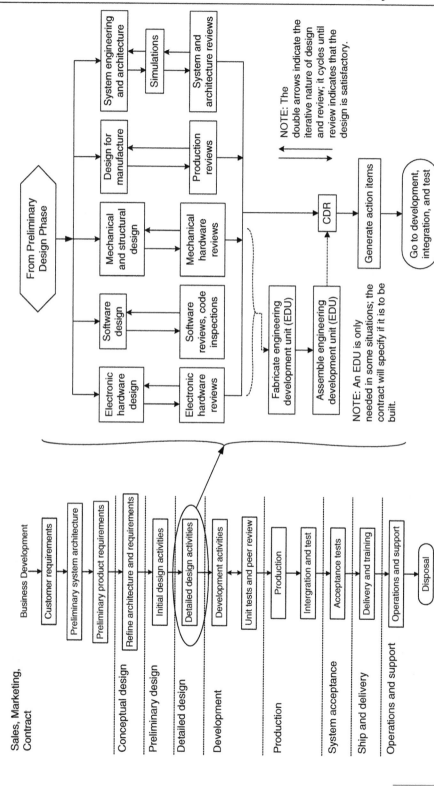

Figure 1.7: Process for the critical design phase. This is the second level of detail for task summaries. (© 2009 by Kim Fowler. Used with permission. All rights reserved.)

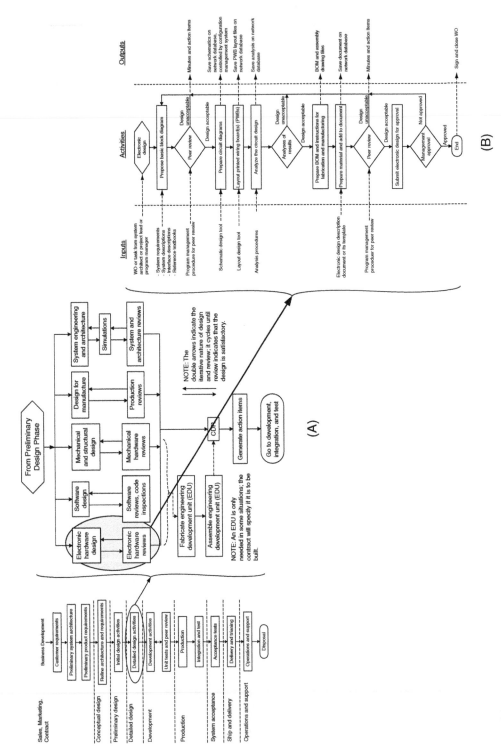

Figure 1.8: An example of a procedure for electronic circuit design during the critical design phase. This is the third level of detail and approaches a single task. (A) Location of successively more detailed levels. (B) Example of a work package or elemental, kernel task. (© 2009 by Kim Fowler. Used with permission. All rights reserved.)

5. Standards

"These lists are of necessity incomplete. There are many standards-setting organisations and standards that are specific to particular industries, types of product, national markets, etc., and changes are frequent" [22].

5.1. General Standards Organizations

- International Organization for Standardization (ISO) (http://www.iso.ch).

- International Electrotechnical Commission (IEC) (http://www.iec.ch). The electrical/electronic section of ISO.

- U.S. National Institute of Standards and Technology (NIST) (http://www.nist.org). Coordinates the production of U.S. standards.

- European Committee for Standardisation (CEN). European standards are prefixed EN (Euro Norm).

- European Committee for Electrotechnical Standardization (CENELEC).

- European Commission (EC). The EC issues directives related to compliance with standards.

- British Standards Institution (BSI) (http://www.bsi.org.uk). The BSI coordinates the production of UK standards. British standards are prefixed BS.

- The UK Accreditation Service (UKAS) (http://www.ukas.com) is responsible for the accreditation of laboratories offering measurement and calibration services. UKAS also provides accreditation for ISO9000 certification organizations.

- German Association for Electronics and Information Technology (VDE) (http://www.vde.de). VDE produces standards for electronics and IT, provides accreditation services, and operates test laboratories. The VDE mark is affixed to equipment that complies with relevant VDE standards.

"All other industrialised countries have national standards organisations, which contribute to international standardisation work. However, only a few maintain standards that differ significantly from international or other major national standards" [22].

5.2. Industry-Based Standards Organizations

5.2.1. Mechanical and Systems

- American Society of Mechanical Engineers (ASME) (http://www.asme.com).

- American Society of Testing and Materials (ASTM) (http:www.astm.com).

5.2.2. Electrical and Electronic

- Institute of Electrical and Electronic Engineers (IEEE) (http://www.ieee.org). (The IEEE has 37 societies that generate standards and organize seminars on a variety of aspects of system development.)

- Telcordia Technologies (was Bellcore). Telcordia standards are applied in the telecommunications industries worldwide (http://www.telcordia.com).

5.3. Military Standards Organizations

- U.S. Department of Defense—Military standards, handbooks, and specifications (MIL-STDs, MIL-HDBKs, MIL-SPECs).

- North Atlantic Treaty Organization—NATO Standards (STANAGS).

- UK Ministry of Defence—Defence Standards (DEFSTANs).

5.4. Aviation and Aerospace Standards Organizations

- International Civil Aviation Organization (ICAO).

- Air Radio Inc. (ARINC). Produces standards for aviation electronic systems (avionics).

- National Aeronautics and Space Administration (NASA).

- Federal Aviation Administration (FAA) (USA).

- Civil Aviation Administration (CAA) (UK).

- European Space Agency (ESA).

6. Potential Procedures, Checklists, and Documents

A company building mission-assured, mission-critical, or safety-critical systems has six primary areas that contribute to the operation of the business. These areas are:

- Business

- Marketing and sales

- Personnel

- Infrastructure

- Manufacturing and production

- Engineering

Each area has a number of procedures that are to be followed within its purview.

Business deals with company organization, governance, relationships (e.g., suppliers, bankers, tax accountants), legal concerns, government regulations, facilities, and office equipment, among other things. Marketing and sales deal with initial contact and customer relations, order fulfillment, billing, and service support. Personnel and human resources deal with hiring, benefits, training and development, personnel files, and termination. Infrastructure has to do with equipment and systems and facilities to carry out the company business; it is often split among several groups. Manufacturing and production deals with inventory, fabrication, assembly, and distribution; sometimes it provides service support as well.

In this book, we are only focusing on the development engineering aspects. Following is a partial list of major concerns for which engineering has responsibility. (Clearly other disciplines exist; this listing gives an indication of the types of issues.)

1. Project flowcharts

2. Project tracking—Engineering concerns

   - Visibility

   - Accountability

   - Configuration management

   - Version control

   - Database implementation

   - Engineering change notice

   - Feedback

   - Improvement

   - Closeout

   - Archiving

   - Outsourcing

   - Qualification of vendors

   - Templates—Plans, documents, reports, action item, design documents, test results

3. General engineering processes

- Requirements

    Input—Customer, company, standards

    Analysis

    Producing metrics

    Modification and change

    Compliance matrix

- Analysis—Business feasibility

- Design

- Reviews

    Peer review

    Control board review

    Design review—CoDR, PDR, CDR, PRR

- Development

- Test

- Integration

- Acceptance

- Delivery

4. Electronic and electrical engineering

- Tools

- Tool certification

- Design

    Schematics

    Circuit board layout

    Assembly

- Analysis

    Signal integrity

      Electromagnetic compatibility (EMC)

      Power consumption

      Thermal

      Radiation

      Shock and vibration

      Humidity

      Corrosion

      Worst-case stress

5. Software engineering

   - Tools
   - Tool certification
   - Style guide
   - Source listing
   - Compilation and transfer to hardware

6. Mechanical engineering

   - Tools
   - Tool certification
   - Design

           Schematics

           Mechanical layout

           Assembly

   - Analysis

           Mass, size, and volume

           Materials

           Structural model

           Thermal

Radiation

Shock and vibration

Humidity

Corrosion

Worst-case stress

7. Systems engineering

- Tools

- Tool certification

- Design

   Schematics

   Mechanical layout

- Analysis

   Event tree analysis (ETA)

   Failure mode effects analysis (FMEA)

   Fault tree analysis (FTA)

   Reliability

   Radiation

   Interface inspections

   Integration coverage—functional, physical, environmental

8. Metrology

- Qualification of vendors

- Calibration

- Certification of measurements

- Archiving

9. Service support

7. Review of Procedures and Processes

7.1. Difference between Procedures and Processes

The difference between processes and procedures is a qualitative description, as well as a difference in the detail. A *process* contains the overarching principles that drive best practices. It tends to be general. *Procedures* are specific implementations of the process for a single focused area of concern; often they are step-by-step instructions.

7.2. Why Review Procedures and Processes?

Procedures and processes characterize every system of business or technical development. These systems are necessarily limited due to flaws, gaps, or changing circumstances. Nevertheless, a system should have an avenue for audit, self-examination, correction, change, and update.

Reviews of procedures and processes are a quality assurance function. These reviews may be directed by either external or internal sources. Customers, audit agencies, and consultants may conduct external reviews with the agreement of your company. Employees of your company and consultants might conduct internal reviews.

7.3. Types of Review

Design and Peer Review: Effective review of the project should have a stereotyped format including agenda, checklists, and minutes. An outline example of review minutes follows:

- Date of the review

- Review agenda

- Who attended

- Who presented the design

- Lead and independent reviewers

- What major decisions were made

- What action items were generated, their due dates, and who is responsible for each.

Reviews should generate action items to ensure that identified issues are addressed. All action items should be tracked in a database. Each action item should have the following fields:

- Unique identifier number

- Status (open, closed, in work, in sign-off)

- Date opened

- Brief summary

- Response summary

- Requestor

- Assignee

- Due date

Design Reviews: For larger programs and projects, the design review should have independent reviewers who are not directly associated with the project. The review committee for each formal review should consist of at least four members plus a designated chairperson; none of which should be members of the project team. I have to say this rarely happens—it is the ideal but most companies do not have the people, time, or resources to devote to independent review.

Most companies organize design reviews with members of the project team presenting and reviewing. Sometimes they will ask a customer or client to attend and critique. This form of review is still effective.

Regardless of the format, you should send a review package to each reviewer about two weeks ahead of the review. The review package should contain a copy of all the slides to be presented at the review along with appropriate background material.

Code Inspections and System Reviews: Software code walkthroughs are a legitimate form of review as well. They are a form of static testing, but highly effective. These forms of peer review are an excellent way to encourage proper designs and good development processes. For some types of products, such as medical devices, they are an important part of the formal development. You should still have procedures, which can be simple and straightforward, for recording notes or minutes and then maintain a database of action items.

This same sort of review would be good for early system integration. If done regularly by team members, these reviews would reveal more problems earlier and give you a better chance at ironing them out sooner.

7.4. Frequency of Review

External reviews of procedures and processes for quality assurance should occur at least once a year. Your company might have a frequency goal of two or more external reviews each year—but reviews do take time.

Internal reviews of procedures and processes for quality assurance should occur at least twice a year. Your company might have a frequency goal of four internal reviews per year.

7.5. Review Content

Reviews of procedures and processes for quality assurance should address the following questions:

- What processes and procedures are being followed?

- What processes and procedures are not being followed?

- What processes and procedures are not sufficient?

- What processes and procedures need updating?

- What processes and procedures need revision?

7.6. Course of Action, Changes, and Updates Following Review

Reviews of procedures and processes for quality assurance will proceed according to the following general course of action:

- Agreement to conduct an audit within your company or between your company and an outside party.

- Schedule the audit.

- Conduct the audit.

- Review the findings. In general, answer the questions immediately above.

- Perform a general review of the procedures and processes.

- Combine the findings of the audit and general review and present to your company's control board.

- Implement adjustments, corrections, and updates to your company's procedures and processes.

7.7. Review Responsibilities

Customers, audit agencies, and consultants may conduct external reviews. Employees of your company and consultants may conduct internal reviews. Management determines when the reviews and audits may take place. Your company's control board is responsible for implementing the revisions or updates that derive from a review.

8. Configuration Management

8.1. Rationale for Configuration Management

Every business or technical development system needs configuration management because problems, mistakes, and changing circumstances occur frequently, if not continuously. Configuration management helps reduce the consequences of these problems from becoming too disruptive by maintaining an understanding of current system state.

8.2. Configuration Management Coverage

Configuration management monitors the state of just about everything in a project. Here are some examples of areas within configuration management:

- Software

- Hardware

- Products

- Identification

- Database and network

8.3. Records Responsibility

Your company's control board is responsible for implementing configuration management. The project manager then may designate who will manage the configuration of a particular project.

8.4. System and Location

Maintain the configuration management system in a database on a server located on the premises. The data on the server should be backed up regularly to an off-site storage system. Back up the remote site once a day if possible but once a week at the very longest. Should the system fail or be put out of commission, the data can be retrieved and restored on a replacement system.

8.5. Version Control

A centerpiece of that configuration management is the version control system. Version control is software that resides on the system server and archives project data so that it may be traced from inception to the end of the project. A number of commercial software packages implement reasonable version control systems.

Version control has four different functions:

- Performs the library operations of checking files and documents in and out of the system in an unambiguous fashion. Prevents two or more people checking in the same file, each with different modifications, and having one version overwrite and destroy the other modifications. Can merge modifications so that all changes are recorded.

- Uniquely assigns identification to each file so that it can be traced, tracked, followed, and found.

- Records all changes to each file and the list of versions of each file. This allows someone to "unroll" the changes and return to a previous version, if so desired.

- Time-tags and date-tags every file and each revision to any file.

8.6. Design Repository

One particular portion of a project's configuration management is the design repository. This repository contains the design descriptions and instructions for production. It is a particular "folder" or "file drawer" in the project's storage area that contains these files of the design descriptions and instructions. Version control, as mentioned, controls document storage into and retrieval from the design repository.

8.7. File Structure

Each project has its own separate storage area. Here is an example of partitioning, in which the files can segregate into the following areas, but these are at the discretion of the program manager or project lead:

- Design repository

    Architecture—Requirements, descriptions

    Software—Source listings, requirements, descriptions

    Hardware—Schematics, requirements, descriptions

    Sensors and optics—Schematics, requirements, descriptions

- Business operations

    POs

    Work orders (WOs)

        Contract, request for proposal (RFP), rough-order-of-magnitude (ROM) quotes, SOW

- Products

        Manufacturing and fabrication instructions

        In-process routers (IPRs) for production

        Subcontractor reports

        Identification

- Documents

        Plans

        Descriptions

        Test results

        Checklists

- Records

        Minutes—Peer reviews, design reviews (CoDR, PDR, CDR, MRR, TRR, PSRR), control board meetings, failure review board meetings

        Logs—Problem reports/corrective actions (PRCAs), engineering change requests (ECRs), engineering change notices (ECNs)

        Test results—Bench tests, unit tests, integration, acceptance tests

        Communications—Memos, email messages, notes, letters, customer audits

8.8. Obsolete Documents

When documents or company procedures or processes change, revise, or become subsumed, the original must be removed from the active templates folder. They should be placed in a folder or location marked "obsolete." Each obsolete document, procedure, and process should have a watermark of "obsolete" inserted on each page when they are removed from circulation.

Another concern is for paper copy of documents being obsolete when compared to the current version in the design repository. A good practice to put a disclaimer in the header or footer of each page to the effect that "This paper copy is for reference only—please consult the company database repository for the current version."

8.9. Training for Use of the System

A quality assurance team or management-designated personnel should train staff in the use of the configuration management system. Training should cover the basic operations, some of which are listed in the following:

- Overview of the configuration system and version control

- Entering the system

    Where to find it on the server database

    User ID and password

    Navigation through system

- File transfers

    Entering a new file into the system

    Retrieving a file from the system

    Returning a file to the system

- Reviewing file archives

- How the system is set up and run

- Backups

- Restoration

9. Documentation

9.1. Rationale for Documentation

Documentation generally serves three purposes:

- Record the specifics of development (the "who, how, and why") that include, but are not limited to, engineering notebooks, software source listings, schematics, and test reports. These records help when modifications, upgrades, fixes, and recalls occur.

- Account for progress (the "what, when, and where") toward satisfying requirements and provides an audit trail of the development. These types of documents include memos, but are not limited to, meeting notes and minutes, review action items, and project plans. An appropriate plan for documentation supports rigorous testing, validation, and verification.

- Instruct users and owners on the extent of functionality of your product. Instructions include, but are not limited to, user manuals, DVDs with instructions, websites, and labels and warnings that present concise instructions [23].

9.2. Coverage and Responsibility for Documentation

Documentation addresses every aspect of business; it explains the "who, what, when, where, why, and how" of a project. For your company to develop a product, there are two primary types: company documentation and project documentation. The appendices at the end of this chapter provide examples of both for a smaller project, such as a self-contained piece of equipment (not a large structure or vehicle such as an aircraft, which is a domain unto itself).

The project manager designates who will manage the documentation and the document plan. Typically everyone on a project team contributes specific documents that represent their area of work.

9.3. Types of Documentation

Many types of documents exist. Each set is project-type specific. The appendices at the end of this chapter provide examples of types of documents for a smaller project. Figure 1.9 illustrates several of the major documents that direct development for a smaller project. The set of documents for a project should answer the "who, what, when, where, why, and how" of product development. Here is a general outline of the types of documents that are part of most products:

- Plans (who, what, where, when, why)

- Design documents (how, why)—Source listings, schematics, engineering notebooks

- Reviews and reports (who, what, where, when, why)

- Instructions (what)—User manuals, brochures, training materials, maintenance and repair guides

The appendices at the end of this chapter provide example outlines of some documents for a smaller project.

9.4. Best Practices for Documentation

First, understand the audience for each document and then tailor the document to that audience. That means an appropriate level of detail and reading comprehension with an

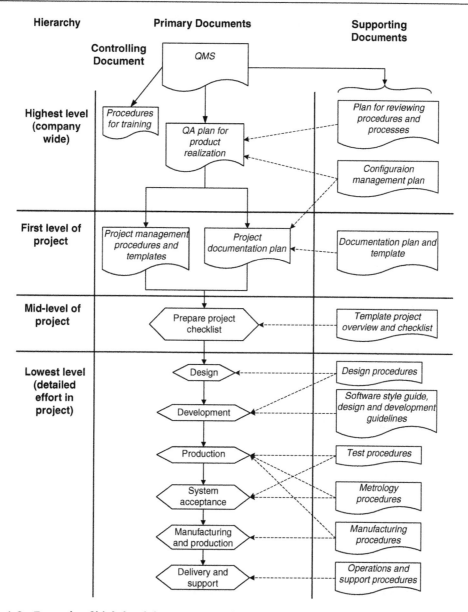

Figure 1.9: Example of high-level document tree for direct development of a smaller project, such as a self-contained piece of equipment; not recommended for a large structure or vehicle such as an aircraft. (© 2009 by Kim Fowler. Used with permission. All rights reserved.)

intuitive format or instructional flow in the layout of text and graphics. Every document should have attributes in the following order of priority:

1. Correct

2. Complete

3. Consistent

4. Clear

5. Concise

Good documents should share similar characteristics that cover the following points:

- Basic content

  What: Prominently note the model number and version

  Why: Clearly explain the need or theory

  How: Detail the operation

- Clear organization

  Introduction: Purpose, Scope, Definitions, Acronyms, and Abbreviations, References

  Overview, with rationale

  Description, report, or procedures

  Appendices

- Modular format

- Clearly illustrated figures and tables

- Detailed schematics

- References to source listings and other technical records

- Table of contents, index, and cross-references

Three heuristics about documentation that I have learned follow:

- Documentation is integral to every product.

- While good documentation cannot help a poor product, poor and inadequate documentation can destroy a good product.

- While some good products have poor documentation, I have *never* found a poor product with good documentation [23].

References

[1] Blanchard BS, Fabrycky WJ. Systems engineering and analysis. 4th ed. Upper Saddle River, NJ: Pearson Prentice; 2006.

[2] International Council on Systems Engineering. A consensus of the INCOSE Fellows. Available at www.incose.org/practice/fellowsconsensus.aspx.

[3] Maier MW, Rechtin E. The art of systems architecting. 2nd ed. Boca Raton, FL: CRC; 2002.

[4] Mission assurance. Available at: en.wikipedia.org/wiki/Mission_assurance; 2009.

[5] Mission assurance—government needs to manage risk differently in face of new operating realities. Booz Allen Hamilton Inc.; March 8. Available at: www.boozallen.com/consulting-services/services_article/659316; 2004.

[6] Gordon P. Ground zero for mission assurance, PRTM. Available at: www.prtm.com/uploadedFiles/Strategic_Viewpoint/Articles/Article_Content/PRTM_Ground_Zero.pdf; 2006.

[7] Mission critical. Available at: http://en.wikipedia.org/wiki/Mission_critical; 2009.

[8] Farlex. The free dictionary. Safety-critical system. Available at:encyclopedia2.thefreedictionary.com/safety-critical+system; 2009.

[9] Sedlak JR. Quality management system aerospace requirements: differences between ISO 9001:2000 and AS9100B, Akron, OH: Smithers Quality Assessments, Inc; August. Available at: www.isots16949.com/AS9100B-Differences-from-ISO9001.pdf; 2006.

[10] AS-9100 Standard—QMS. Available at: www.siliconfareast.com/as9100.htm.

[11] Penn L. CMMI and Six Sigma. In: Chrissis MB, Konrad M, Shrum S, editors. CMMI: guidelines for process integration and product improvement. 2nd ed. Addison-Wesley: Upper Saddle River, NJ; 2007, pp. 5–8.

[12] Watts H. CMMI: history and direction. In: Chrissis MB, Konrad M, Shrum S, editors. CMMI: guidelines for process integration and product improvement. 2nd ed. Addison-Wesley: Upper Saddle River, NJ; 2007, pp. 5–8.

[13] Verzuh E. The fast forward MBA in project management. 2nd ed. New York; 2005, p. 3, 7, 8, 20, 21, 113, 114.

[14] MIL-STD-499B (draft). Systems engineering; May 6, 1994.

[15] Oshana R. DSP software development techniques for embedded and real-time systems. Burlington, MA; 2006, p. 507–509.

[16] Reinertsen D. In search of the perfect product specification. IEEE Instrumentation Measurement Mag 2000;3(2):28–31.

[17] Fowler K. What every engineer should know about developing real-time embedded products. Boca Raton, FL: CRC; 2008, p. 32.

[18] Pazemenas V. Rapid development for medical products. IEEE Instrumentation Measurement Mag 2000;3(2):32–37.

[19] Test-driven development. Available at: en.wikipedia.org/wiki/Test-driven_development; 2009.

[20] Grenning J. Applying test driven development to embedded software. IEEE Instrumentation Measurement Mag 2007 (December):21.

[21] Guarro S. The mission assurance guide: system validation and verification to achieve mission success, Crosslink Mag 2007;8(2). Available at: www.aero.org/publications/crosslink/fall2007/03.html.

[22] O'Connor P. Test engineering: a concise guide to cost-effective design, development and manufacture. New York: Wiley; [Reliability, testing and related regulations and standards. Available at: www.pat-oconnor.co.uk/relteststds.htm.]; 2001.

[23] Fowler KR. Electronic product development: architecting for the life cycle. 2nd ed. New York: Oxford University Press; [forthcoming].

Appendix A: Example Document Outlines

The following document outlines cover three primary areas that contribute to realizing a smaller product, such as a self-contained piece of equipment (not for a large structure or vehicle such as an aircraft). The areas include:

- Infrastructure

- Manufacturing and production

- Engineering

Example document outlines described here can lead to document templates that suggest elements to include in each document.

Work Order (WO)

A WO system can be implemented on the company's server and file system. The WO should have the following components:

- Defines a task

- Assigns the task

- Assigns an expected completion date

- Provides instructions for the task

- Attaches a procedure

- Attaches templates of documents and records to complete

- Attaches references (if needed)

A WO has an author, typically the project manager; unique ID number; project title; and comment box for the assignee to use, if needed.

Minutes

Minutes from reviews should have the following components:

- Title

- Date

- Place
- Attendees
- Points of discussion and concern
- Action items
- Assignees to each action item
- Expected date of completion of each action item

Problem Report/Corrective Action (PRCA)

PRCAs should have the following components:

- Title
- ID number
- Date
- Place
- Description of problem
- Assignee to address the concern
- Expected date of completion of concern

Engineering Change Request (ECR)

ECRs should have the following components:

- Title
- ID number
- Date
- Place
- Description of engineering change request
- Assignee to address the request
- Expected date of completion of addressing the request

Engineering Change Notice (ECN)

ECNs should contain the following components:

- Title

- ID number

- Date

- Place

- Description of engineering change

- Assignee to address the change

- Expected date of completion

Project Management Plan (PMP)

The PMP is the centerpiece of planning and documentation for a project, as it aims to cover the "who, what, when, and where" of project development. Sometimes this material is split into two separate documents—a systems engineering plan (SEP) and a systems engineering management plan (SEMP). The structure of a PMP should follow this basic outline:

1. Introduction

    1.1. Purpose

    1.2. Scope

    1.3. Definitions, Acronyms, and Abbreviations

    1.4. References

    1.5. Overview

2. Project Overview

    2.1. Project Purpose, Scope, and Objectives

    2.2. Assumptions and Constraints

    2.3. Project Deliverables

    2.4. Evolution of the Project Plan

3. Project Organization

    3.1.   Program Structure

    3.2.   Organizational Structures

    3.3.   External Interfaces and Organizations

    3.4.   Roles and Responsibilities—Cite responsibility matrix

4. Project Estimates

    4.1.   Phase Plan—Very standardized for most companies and specifies which of three different types of development might be followed: custom, tailored-COTS (commercial off-the-shelf), or build-to-print

    4.2.   Iteration Objectives

    4.3.   Releases

    4.4.   Project Schedule

    4.5.   Project Resources

    4.6.   Budget

5. Iteration Plans—These might fold into the individual development plans for hardware, software, or mechanics, and not require a separate section in the PMP

6. Management Plan—This might fold into the responsibility matrix and not need a separate section in the PMP

    6.1.   Cite the responsibility matrix

        6.1.1. Program Manager

        6.1.2. Project Lead Engineer

        6.1.3. Systems Engineer

        6.1.4. Hardware Engineering

        6.1.5. Software Engineering

        6.1.6. Mechanical, Packaging, and Thermal Engineering

        6.1.7. Fabrication or Manufacturing or Production Engineering

        6.1.8. Parts Quality Assurance

    6.2.   Explain lines of authority

7. Project Monitoring and Control

   7.1. Configuration Management Plan—Can cite Configuration Management Plan if already a separate document

   7.2. Project Procedures Checklist—List containing all the procedures to be called out during development

   7.3. Document Checklist—List containing all the documents to be called out during development; may already reside in a Document Plan as a separate document

   7.4. Requirements Management Plan

   7.5. Schedule Control Plan

   7.6. Budget Control Plan

   7.7. Quality Control Plan

   7.8. Product Acceptance Plan

   7.9. Approval, Distribution, and Archiving Plan—Can cite Configuration Management Plan

8. System Specification and Performance Verification

   8.1. Requirements and Flow-down

   8.2. Technical Performance Standards

   8.3. Interface Definition and Control (ICDs)

      8.3.1. Electrical

      8.3.2. Mechanical

      8.3.3. Optics or sensor

      8.3.4. Data or software

      8.3.5. Support equipment

      8.3.6. Miscellaneous interfaces

   8.4. Configuration Management and Change Tracking

   8.5. System Validation

   8.6. Performance Verification

8.7. Technical Performance Trending

8.8. System-Level Design Guidelines

9. Reviews—Can cite Configuration Management Plan if already a separate document

9.1. Peer Reviews

9.2. Formal Reviews—Audits, design reviews (CoDR, PDR, CDR, MRR, TRR, PSRR)

9.3. Action Item Management

10. Documentation Plan—Can cite Documentation Plan or a checklist derived from it if already a separate document

11. Development Plans

11.1. System Architecture

11.2. Software

11.3. Electronics

11.4. Mechanical or Packaging

11.5. Sensor or Optics

11.6. Support Equipment

12. Close-out Plan

13. Supporting Process Plans—All of these may not be necessary for a given project

13.1. Test Plan—Often a separate document

13.1.1. EMC/EMI Test Plan

13.1.2. Shock and Vibration Test Plan

13.1.3. Thermal-Vacuum Test Plan

13.1.4. Extreme Environment Test Plan

13.2. Problem Resolution Plan—Might be sufficient to cite company's Quality Assurance Plan

13.3. Infrastructure Plan

14. Glossary

15. Technical Appendices

Interface Control Documents (ICDs)

ICDs specify the interface between subsystems or subsystems and the system to which they connect. Topic areas that ICDs may cover include the following:

- Electrical

- Mechanical

- Optics or sensor

- Data or software

- Support equipment

- Miscellaneous interfaces

Contents includes the following:

1. Author of the ICD, date of creation, record of revisions and dates

2. Scope

3. List of inputs and outputs

4. Subsystems or systems that connect to the particular subsystem

5. Parameters covered and allowed variances on the parameters

6. Requirements

Development Plans

Development plans cover many engineering aspects, such as the following: Architecture Development Plan, Electronic System Development Plan, Software System Development Plan, Mechanical System Development Plan, Support Equipment Development Plan, and Special (e.g., optics) Development Plan(s). Each of these has a standard format:

1. Introduction

    1.1. Purpose

    1.2. Scope

    1.3. Definitions, Acronyms, and Abbreviations

    1.4. References

2. Overview (of particular development area, such as hardware, software, mechanical)

    2.1. Project Purpose, Scope, and Objectives

    2.2. Assumptions and Constraints

    2.3. Deliverables

3. Management and Staffing (of particular development area, such as hardware, software, mechanical)—Cite responsibility matrix in PMP

4. Design Inputs

    4.1. Customer Requirements and Requests

    4.2. Standards and Practices

    4.3. Company Guidelines

    4.4. Appropriate ICD

    4.5. Architectural Interactions with Other Areas

5. Design Outputs

    5.1. Requirements

        5.1.1. Listing

        5.1.2. Compliance matrix

    5.2. Documents—Cite PMP and Documentation Plan

    5.3. Schematics or source listings

    5.4. Analyses

    5.5. Test results

    5.6. Demonstrations

    5.7. Reviews—peer, design

6. Tools, Techniques, and Methodologies

7. Contributions to Risk Management

Requirements

Requirements group under three primary headings: Operations, Functions, and Performance.

Operational Requirements

Operational requirements pertain to the scenarios within which the product will operate. Operational requirements usually cover the following concerns:

- Mission profiles

- Infrastructure needed

- Logistics and maintenance

- Responsible party for generating the requirements

- Environment

    Shock

    Vibration

    Corrosion

    Humidity

    Temperature range

    Radiation

Functional Requirements

Functional requirements pertain to the physical plant and situations within which the product will reside. Functional requirements usually cover the following concerns:

- Interfaces (ICDs)—Human, mechanical, electrical, software, special (e.g., optics)

- Mechanical—Size, shape, weight, volume, density

- Electrical—Power sources, distribution

- Responsible party for generating the requirements

Performance Requirements

Performance requirements pertain to the metrics and parameters that describe the product's capability. Performance requirements usually cover the following concerns:

- Responsible party for generating the requirements

- Sensor parameters

    Measurand

    Speed of transduction—samples per second

   Span

   Full-scale output

   Linearity—%, SNR

   Threshold

   Resolution—ENOB

   Accuracy—SNR

   Precision

   Sensitivity—%

   Hysteresis

   Specificity

   Noise—SNR, % budget

   Stability

- Data throughput

   Bytes or samples per second

   Data transmission protocol

   Data storage

   Control

- Operation

   Electrical—Power consumption, efficiency, signal integrity

   Mechanical—Strength, motion required

   Structural—Capability to withstand environments in mission profiles

   Optical

- Calibration

- Dependability

   Reliability

   Maintainability

Testability

Fault tolerance

Longevity

- Power consumption

- Dissipation and cooling

- Electromagnetic compatibility (EMC)

    Conducted susceptibility

    Radiated susceptibility

    Conducted interference

    Radiated interference

Risk Management Plan

This plan strives to identify problem areas before they become large problems, and categorize their effects, or criticality. A Risk Management Plan usually covers the following concerns:

- Identify areas of risk

    Business

    Operations

    Technical disciplines—software, electronics, mechanical, optics, materials

    Security

- Estimate criticality of each area

- Determine primary risk elements

- Responsible party for activity and document

    Fault tree analysis (FTA)

    Failure modes and effects analysis (FMEA)

    Event tree analysis (ETA)

    Margin management

Configuration Management Plan

The Configuration Management Plan strives to provide consistent control over the status of the project and state of the developing product. See the section on configuration management earlier in this chapter for more information.

Documentation Plan

A project's Documentation Plan strives to provide consistency among all the documents generated in a project. Using the checklist in this document might be sufficient to apply to a project. See the section on documentation earlier in this chapter for more information.

Analysis Reports

Multiple analyses attempt to understand the feasibility of building a product and the tradeoffs that need to be made before realizing the actual product. Examples of reports and their potential topics that can derive from design procedures follow:

- System feasibility

     ETA, FTA, FMECA (specify at a subsystem or system level)

     Technical feasibility (specify important issues and likelihood of success)

- Business case: cost vs. benefit
- Technical tradeoffs

     Architecture (examples follow)

  - Distributed vs. centralized
  - Modular vs. custom monolithic
  - Loose vs. tight coupling
  - Hardware vs. software
  - Electronic actuation vs. mechanical linkage
  - Testability
  - Manufacturability
  - Human interface

- Buy vs. build

Electronic (examples follow)

- Types of processors

- Power

- Types of signal transmission

- Signal conditioning

Software (examples follow)

- Languages

- Tool suites and compilers

- Module size

Mechanical

Structure

Materials

Production

Maintenance

Design Description

A design description for a product provides detailed coverage, description, and instructions. The level of detail needs to be sufficient to allow it to be either replicated in production and manufacturing or available for future modifications. A design description might have the following format:

1. Introduction

    1.1. Purpose

    1.2. Scope

    1.3. Definitions, Acronyms, and Abbreviations

    1.4. References

    1.5. Overview

2. Overview—Should provide rationale so that the intent of the design may be understood in the event that future revisions or modifications are needed

3. Description

    3.1. Function—Includes the logical intent of the function

    3.2. Operation—Includes the logic of operation

    3.3. List of inputs and outputs

    3.4. Subsystems or systems that connect to the particular unit

    3.5. Partitioning of modules

    3.6. Requirements compliance

4. Appendices

Topic areas that design descriptions may cover include the following:

- Architecture

- Electrical

- Mechanical

- Optics or sensor

- Data or software

- Support equipment

- Miscellaneous systems

Test Plan

Tests are critical to confirming operation and function of a product. The Test Plan for a product gives detailed description and instructions for all the tests necessary or requirement within a project. A Test Plan performs two basic functions: verification and validation.

- "The purpose of [v]erification (VER) is to ensure that selected work products meet their specified requirements" (pp. 579–595). This is an objective measure of how well the product meets the stated requirements.

- "The purpose of [v]alidation (VAL) is to demonstrate that a product or product component fulfills its intended use when placed in its intended environment" (pp. 565–578). This statement is a more subjective evaluation of how well the product meets its desired intent.

(The cited definitions derive from Mary Beth Chrissis, Mike Konrad, and Sandy Shrum, titled, *CMMI: Guidelines for Process Integration and Product Improvement*, 2nd ed.)

A test plan can have the following content:

- Verification (primarily for subsystems)

    Code inspections

    Unit tests

    Electronic tests

    Mechanical tests

    EMI/EMC tests

    Radiation tests

    Highly accelerated stress screening

    Manufacturing inspections

    Manufacturing tests

    Functional tests for system operational behavior

- Validation (primarily for the system)

    Field or prototype tests

    Integration tests

    - Shock

    - Vibration

    - Thermal-vacuum

    - Condensation and corrosion

    - EMI/EMC

    - Functional behavior

    Acceptance test for customer signoff (may be part of integration tests)

- Integration Plan (may be part of Test Plan) comprises a combination of verification and validation. Some or all of the previously stated tests might help fulfill the integration plan.

- Test Results format—All test descriptions should cover the "who, what, when, where, and how" to ensure correct instructions for testing.

  Who—Author of test, responsible party to run the tests

  What—Detailed description of the type of tests and their functions

  When—Detailed instructions for when the tests should be run in the development cycle of the project

  Where—Detailed instructions for the facilities to run the tests

  How—Detailed instructions for running the tests

Operation Plan

This plan is more important for a product that requires significant involvement from the vendor or contractor after delivery. The Operation Plan, which instructs support and maintenance issues after delivery, should address the following content:

- Installation
- Training
- Logistics
- Maintenance and repair
- Disposal

Metrology Concerns and Procedures

The list of concerns for which metrology has responsibility are listed below. Major concerns follow.

- Outsourcing of calibration
- Qualification of vendors to test and calibrate
- Calibration
- Certification of measurements
- Archiving

Appendix B: Program Management Documents for Project Development

Doc. #	Business and Contract Documents	Pre	Ph0	Ph1	Ph2	Ph3	Ph4A	Ph4B	Ph4C	Ph5	Responsibility
Xxx	Catalog	Δ									Marketing
Xxx	Marketing brochures	F									Marketing
Xxx	Preliminary product description and requirements	F									Marketing
	Company procedures										
Xxx	Company introduction and operations	Δ									Senior management
Xxx	Quality assurance plan	Δ									Senior management
Xxx	Product assurance plan	Δ									Senior management
Xxx	Company infrastructure plan	Δ									Senior management
Xxx	Parts, materials, and processes plan	Δ									Senior management
Xxx	Program-approved parts list	Δ									Senior management
Xxx	Program-approved materials and processes list	Δ									Senior management
xxx	Production and manufacturing plan	Δ									Senior management
xxx	Printed wiring board procurement plan	Δ									Senior management
xxx	Vendor qualification test plan	Δ									Senior management
xxx	Contamination control plan	Δ									Senior management
xxx	Prohibited materials verification plan	Δ									Senior management

Doc. #	Business and Contract Documents	Pre	Ph0	Ph1	Ph2	Ph3	Ph4A	Ph4B	Ph4C	Ph5	Responsibility
xxx	Documentation plan	Δ									Senior management
xxx	Metrology and calibration plan	Δ									Systems engineer

Doc. #	Business and Contract Documents	Pre	Ph0	Ph1	Ph2	Ph3	Ph4A	Ph4B	Ph4C	Ph5	Responsibility
001	Customer requirements inputs		F	Δ	Δ	Δ	Δ	Δ	Δ		PM, lead, systems engineer
002	Rough-order-of-magnitude (ROM) quotes		F								PM
003	Quote		F								Senior management
004	Quote review report		F								Senior management
005	Statement of work (SOW)		F	Δ	Δ	Δ	Δ	Δ	Δ		PM
006	Work break-down structure		F	Δ	Δ	Δ	Δ	Δ	Δ		PM
007	Financial plan		F	Δ	Δ	Δ	Δ	Δ	Δ		PM
008	Schedule deadlines		F	Δ	Δ	Δ	Δ	Δ	Δ		PM
009	Business case		F	Δ	Δ	Δ	Δ	Δ	Δ	Δ	Senior management
010	Contract		F	Δ	Δ	Δ	Δ	Δ	Δ		Senior management
011	Contract reviews		U	U	U	U	U	U	U		Senior management
012	Purchase order (PO)		F				Δ	Δ	Δ		Senior management
013	PO review report		U				U	U	U		Senior management

(Continues)

Appendix B: Cont'd

Doc. #	Project Documents/Databases	Pre	Ph0	Ph1	Ph2	Ph3	Ph4A	Ph4B	Ph4C	Ph5	Responsibility
020	Product management plan (PMP)		I	F	△	△	△	△	△		PM
021	Budget		I	F	△	△	△	△	△		PM
022	Schedule		I	F	△	△	△	△	△		PM
023	Architecture development plan		I	F	△	△	△	△	△		PM
024	Software development plan		I	F	△	△	△	△	△		PM
025	Hardware electronics development plan		I	F	△	△	△	△	△		PM
026	Mechanical development plan		I	F	△	△	△	△	△		PM
027	Support equipment development plan		I	F	△	△	△	△	△		PM
028	Safety and mission assurance plan		I	F	△	△	△	△	△		Systems engineer
029	Product acceptance plan		I	F	△	△	△	△	△		PM
030	Operation support plan—policies, processes, standards		I	F	△	△	△	△	△		PM
031	Analysis plan—feasibility, requirements		I	F	△	△	△	△	△		Systems engineer
032	Records, logs of performance and checklists—vendor, engineering change request/engineering change notices (ECR/ECNs), problem report/correction actions (PRCAs), minutes from meetings, action items from meetings, work orders (WOs) completed		I	F	△		△	△	△		Systems engineer
033	Monitoring vendors		I	F	△	△	△	△	△		PM
034	Configuration management plan		I	F	△	△	△	△	△		PM

Doc. #	Project Documents/Databases	Pre	Ph0	Ph1	Ph2	Ph3	Ph4A	Ph4B	Ph4C	Ph5	Responsibility
035	Signature list			F	△	△	△	△	△		PM
036	Data storage—version control, data retention, archiving			F	△	△	△	△	△		PM
037	Problem reporting, corrective action plan			F	△	△	△	△	△		Systems engineer
038	Engineering change request plan			F	△	△	△	△	△		Systems engineer
039	Software style guide			F	△	△	△	△	△		Software engineer
040	Documentation plan—listing and flow-down of documents		I	F	△	△	△	△	△		PM
050	Programmatic risk management plan—database, risk severity tables, risk watch list, monitor technical risk		I	F	△	△	△	△			PM
	Analysis		I	F	△	△	△	△	△		PM
060	Business feasibility		I	F	△	△	△	△	△		
061	Business requirements		I	F	△	△	△	△	△		
070–099	*Customer-specified documents*		*I*	*D*	*D*	*D*	*D*	*D*	*F*		*PM*
Doc. #	**Project Documents/Databases**	**Pre**	**Ph0**	**Ph1**	**Ph2**	**Ph3**	**Ph4A**	**Ph4B**	**Ph4C**	**Ph5**	**Responsibility**
	Monitor plans and documents			up	up	up	up	up	up	up	PM
	Monitor logs			up	up	up	up	up	up	up	PM
	Monitor phase design review reports										PM
	Kickoff review		F								PM
	Conceptual design review (CoDR)			F							PM
	Preliminary design review (PDR)				F						PM

(Continues)

Appendix B: Cont'd

Doc. #	Project Documents/Databases	Pre	Ph0	Ph1	Ph2	Ph3	Ph4A	Ph4B	Ph4C	Ph5	Responsibility
	Critical design review (CDR)					F					PM
	Test readiness review (TRR)						F				PM
	Manufacturing readiness review (MRR)							F			PM
	Pre-ship readiness review (PRR)								F		PM
	Customer sign-off									F	PM

Doc. #	Project Documents/Databases	Pre	Ph0	Ph1	Ph2	Ph3	Ph4A	Ph4B	Ph4C	Ph5	Responsibility
	Review and monitor design outputs		U	U	U	U	U	U	U	U	PM
	Individual review reports		U			U	U	U	U	U	PM

Doc. #	Project Documents/Databases	Pre	Ph0	Ph1	Ph2	Ph3	Ph4A	Ph4B	Ph4C	Ph5	Responsibility
	Monitor manufacturing, fabrication, and assembly								F		PM
	Monitor test results								F		PM
	Certify logs									F	PM

Doc. #	Business and Delivery Documents	Pre	Ph0	Ph1	Ph2	Ph3	Ph4A	Ph4B	Ph4C	Ph5	Responsibility
900	Monitor acceptance test results									F	PM
901	Monitor delivery checklists									F	PM
902	Customer sign-off									F	PM, senior management
903	Quality assurance support documents									F	PM
904	QA plans									F	Senior management

Doc. #	Business and Delivery Documents	Pre	Ph0	Ph1	Ph2	Ph3	Ph4A	Ph4B	Ph4C	Ph5	Responsibility
905	QA audits									F	Senior management
906	Marketing updates for catalog and brochures									F	PM
907	Review debrief report									F	PM
908	Assess business case									F	PM, senior management
909	Monitor updates to documents (if needed)									F	PM
910	Monitor support (if contracted)									F	PM

Key:

U—Unique, standalone report in final form; generally a snapshot of a single item or issue.

up—Update log(s).

I—Initial development. A full outline of the document has been established. Writing of some sections has begun.

D—Complete draft—The document is completely written and undergoing review. A very small number of TBDs can remain, but these are limited to specific pieces of information, not entire sections or subsections.

F—Released final version or completed initial release.

Δ—Updates to released version, re-released with changes.

Pre—Presales phase.

Ph0—Sales phase.

Ph1—Concept phase.

Ph2—Preliminary design phase.

Ph3—Critical design phase.

Ph4A—Development phase.

Ph4B—Test and integration phase.

Ph4C—Acceptance test phase.

Ph5—Production phase.

PM—Program manager.

Doc. #—Example of a three-digit suffix (yyy) added to a company's document numbering system to distinguish the documents within a project.

Appendix C: Technical Project Documents for Project Development

Doc. #	Business and Contract Documents	Pre	Ph0	Ph1	Ph2	Ph3	Ph4A	Ph4B	Ph4C	Ph5	Responsibility
100	Preliminary architecture		F								PM, Lead, systems engineer
101	Operational concept document (OCD)		F								PM, Lead, systems engineer
102	Preliminary product requirements document (PPRD)		F								PM, Lead, systems engineer
103	Preliminary interface control documents (PICDs)		F								PM, Lead, systems engineer
104	Electrical		F								PM, Lead, systems engineer
105	Optics		F								PM, Lead, systems engineer
106	Mechanical		F								PM, Lead, systems engineer
107	Data or software		F								PM, Lead, systems engineer
108	Support equipment		F								PM, Lead, systems engineer

Doc. #	Project Documents/Databases	Pre	Ph0	Ph1	Ph2	Ph3	Ph4A	Ph4B	Ph4C	Ph5	Responsibility
110	Technical risk management plan—database, severity tables, watch list		I	F	△	△	△	△	△		Systems engineer
120–199	Customer-specified documents			I	D	D	D	D	F		PM or systems engineer
200	Analysis			I	D	F	△	△	△		Systems engineer
201	Technical feasibility			I	D	F	△	△	△		Systems engineer
202	Technical requirements			I	D	F	△	△	△		Systems engineer
203	Power consumption			I	D	F	△	△	△		Electrical lead
204	Thermal dissipation			I	D	F	△	△	△		Electrical or mechanical lead
205	Environmental—shock, vibration, thermal-vac			I	D	F	△	△	△		Mechanical lead
206	Risk and hazard analysis			I	D	F	△	△	△		Systems engineer
207	Fault tree analysis (FTA)			I	D	F	△	△	△		Systems engineer
208	Failure modes effects analysis (FMEA)			I	D	F	△	△	△		Systems engineer
209	Event tree analysis (ETA)			I	D	F	△	△	△		Systems engineer
210	Radiation hardness and survivability analysis			I	D	F	△	△	△		Systems engineer
211	Structural model			I	D	F	△	△	△		Mechanical lead
212	Reliability and dependability			I	D	F	△	△	△		Systems engineer

(Continues)

Appendix C: Cont'd

Doc. #	Project Documents/Databases	Pre	Ph0	Ph1	Ph2	Ph3	Ph4A	Ph4B	Ph4C	Ph5	Responsibility
213	Worst-case stress analysis			I	D	F	△	△	△		Electrical or mechanical lead
220	Product requirements document (PRD) and compliance matrix			I	D	F	△	△	△		Systems engineer
221	Software requirements document			I	D	F	△	△	△		Software lead
222	Electronic hardware requirements document			I	D	F	△	△	△		Electrical lead
223	Mechanical requirements document			I	D	F	△	△	△		Mechanical lead
224	Optics and sensor requirements document			I	D	F	△	△	△		Sensor lead
225	Support equipment requirements document			I	D	F	△	△	△		Support equipment lead
230	Interface control documents (ICDs)			I	D	F	△	△	△		Systems engineer
231	Electrical			I	D	F	△	△	△		Electrical lead
232	Mechanical			I	D	F	△	△	△		Mechanical lead
233	Optics or sensor			I	D	F	△	△	△		Sensor lead
234	Data or software			I	D	F	△	△	△		Software lead
235	Support equipment			I	D	F	△	△	△		Support equipment lead

Doc. #	Project Documents/ Databases	Pre	Ph0	Ph1	Ph2	Ph3	Ph4A	Ph4B	Ph4C	Ph5	Responsibility
236	Miscellaneous interfaces			I	D	F	△	△	△		
240	Product test plan			I	D	F	△	△	△		Lead engineer
241	Software test plan			I	D	F	△	△	△		Software lead
242	Electronic hardware test plan			I	D	F	△	△	△		Electrical lead
243	Mechanical test plan			I	D	F	△	△	△		Mechanical lead
244	Optics test plan			I	D	F	△	△	△		Sensor lead
245	Support equipment test plan			I	D	F	△	△	△		Support equipment lead
246	System verification & validation (V&V)					I	D	F	△		System or test engineer
247	EMC/EMI test plan					I	D	F	△		Electronic lead or test engineer
248	Environmental test plan—shock, vibe, therm-vac					I	D	F	△		Electronic lead or test engineer
249	Radiation test plan					I	D	F	△		Electronic lead or test engineer
250	Highly accelerated stress test (HAST)—burn-in					I	D	F	△		System, lead engineer
251	Qualification test procedure					I	D	F	△	△	System, lead engineer
252	Acceptance test plan					I	D	F	△	△	System, lead engineer
253	Calibration plan					I	D	F	△	△	System, lead engineer

(Continues)

Appendix C: Cont'd

Doc. #	Project Documents/Databases	Pre	Ph0	Ph1	Ph2	Ph3	Ph4A	Ph4B	Ph4C	Ph5	Responsibility
	Engineering request/change notices (ECRS/ECNSs)			U	U	U	U	U	U	U	Lead engineer
	Review reports										
260	Systems engineering and architecture				U	U	U	U			System, lead engineer
261	Software			U	U	U	U	U			Software lead
262	Electronic hardware			U	U	U	U	U			Electronic lead
263	(FPGA)			U	U	U	U	U			Electronic lead
264	Mechanical			U	U	U	U	U			Mechanical lead
265	Support equipment			U	U	U	U	U			GSE engineer
270	Phase design review reports, close-out action items			F	F	F		F	F		Lead engineer
271	Action item database			F	Δ	Δ	Δ	Δ	Δ		Lead engineer
280–299	*Customer-specified documents*			*I*	*D*	*D*	*D*	*D*	*F*		*Systems engineer*
	Description documents										
300	Architecture				I	D	F	Δ	Δ		Lead engineer
301	Software				I	D	F	Δ	Δ		Software lead
302	Electronic hardware				I	D	F	Δ	Δ		Electronic lead
303	FPGA				I	D	F	Δ	Δ		Electronic lead
304	Mechanical				I	D	F	Δ	Δ		Mechanical lead
305	Optics/sensors				I	D	F	Δ	Δ		Optical lead

Doc. #	Project Documents/Databases	Pre	Ph0	Ph1	Ph2	Ph3	Ph4A	Ph4B	Ph4C	Ph5	Responsibility
306	Support equipment				I	D	F	△	△		Support equipment engineer
310	Design outputs										
311	Software source code and listings				I	D	F	△	△		Software lead
312	Electronic schematics and notes				I	D	F	△	△		Electronic lead
313	FPGA source code listings, schematics, notes				I	D	F	△	△		Electronic lead
314	Mechanical schematics and notes				I	D	F	△	△		Mechanical lead
315	Optics/sensor				I	D	F	△	△		Optical lead
316	Support equipment source code listings, schematics, notes				I	D	F	△	△		Support equipment engineer
317	Users manual				I	D	D	D	F		Lead, systems, software engineer
318	Maintenance manual				I	D	D	D	F		Lead, systems, software engineer
319	Training manual—operational rules or constraints				I	D	D	D	F		Lead, systems, software engineer

(Continues)

Appendix C: Cont'd

Doc. #	Project Documents/Databases	Pre	Ph0	Ph1	Ph2	Ph3	Ph4A	Ph4B	Ph4C	Ph5	Responsibility
320	Problem reporting corrective actions (PRCAs)				U	U	U	U	U	U	Systems engineer
370–399	*Customer-specified documents*				I	D	D	D	F		*Systems engineer*
400	Manufacturing plan						D	F	Δ		Manufacturing lead
401	Vendor evaluation and qualification					I	D	F	Δ		Manufacturing lead
402	Vendor data					I	D	F	Δ		Manufacturing lead
403	Bill of materials (BOM)					I	D	F	Δ		Manufacturing lead
404	Fabrication and assembly instructions					I	D	F	Δ		Manufacturing lead
405	Inspection reports					I	D	F	Δ		Manufacturing lead
406	Travelers					I	D	F	Δ		Manufacturing lead
407	Parts control plan					I	D	F	Δ		Manufacturing lead
408	Parts inventory list					I	D	F	Δ		Manufacturing lead
409	Circuit fabrication and assembly plan					I	D	F	Δ		Manufacturing lead

Doc. #	Project Documents/Databases	Pre	Ph0	Ph1	Ph2	Ph3	Ph4A	Ph4B	Ph4C	Ph5	Responsibility
410	Mechanical fabrication and assembly plan					I	D	F	Δ		Manufacturing lead
411	Support equipment fabrication and assembly plan					I	D	F	Δ		Manufacturing or support equipment lead
412	System assembly plan					I	D	F	Δ		Manufacturing lead
420	Test procedures					I	D	F	Δ		Systems or test engineer
450	Integration procedures					I	D	F	Δ		Systems or test engineer
470–499	*Customer-specified documents*					*I*	*D*	*D*	*F*		*Systems or manufacturing lead*
500	Design outputs—components and assemblies										
501	Jigs and test assemblies						F				Manufacturing lead
502	Subsystems						F				Manufacturing lead
503	Final product								F		Manufacturing lead
	Manufacturing fabrication, and assembly checklists										
510	Circuit fabrication and assembly						F				Manufacturing lead

(Continues)

Appendix C: Cont'd

Doc. #	Project Documents/ Databases	Pre	Ph0	Ph1	Ph2	Ph3	Ph4A	Ph4B	Ph4C	Ph5	Responsibility
511	Mechanical fabrication and assembly						F				Manufacturing lead
512	Support equipment fabrication and assembly						F				Manufacturing or support equipment lead
513	System assembly						F				Manufacturing lead
514	Travelers						F				Technicians
515	Inspection reports						F				Technicians
516	Vendor reports						F				Manufacturing lead
517	ECNs—summary									F	Systems engineer
	Test results										
530	Software unit test/ verification results						F	△	△		Software lead
531	Hardware subsystem test/verification results						F	△	△		Electronic lead
532	Support equipment subsystem test/ verification results						F	△	△		Support equipment lead
533	Integration and test (I&T) results							F	△		Lead engineer
534	System test results							F	△		Systems engineer

Doc. #	Project Documents/Databases	Pre	Ph0	Ph1	Ph2	Ph3	Ph4A	Ph4B	Ph4C	Ph5	Responsibility
535	Test results—EMI and EMC								F		Test technicians
536	Test results—environmental								F		Test technicians
537	Test results—radiation								F		Test technicians
538	Stress tests								F		Test technicians
539	Burn-in								F		Test technicians
540	Acceptance test results									F	Lead engineer
	Logs										
550	Equipment operational power cycles						U	U	U	U	Test technicians
551	Connector mate-demate cycles						U	U	U	U	Test technicians
552	Calibration						U	U	U	U	Test technicians
553	Clean room maintenance						U	U	U	U	Test technicians
554	Training						U	U	U	U	Test technicians
560–599	*Customer-specified documents*						I	D	F		*Systems engineer*
600	Delivery checklists									F	Systems engineer
601	Product data package									F	Manufacturing lead
602	Digital photographic images									F	Manufacturing lead

(Continues)

Appendix C: Cont'd

Doc. #	Project Documents/ Databases	Pre	Ph0	Ph1	Ph2	Ph3	Ph4A	Ph4B	Ph4C	Ph5	Responsibility
603	Documents									F	Manufacturing lead
640	Debrief report									F	Systems engineer
650	Updates to documents (if needed)									F	Lead engineers
660	Support summary (if contracted)									F	Lead engineers
700	Applications notes (if needed)									F	Lead engineers
800	White papers (if needed)									F	Lead engineers

Key:
U–Unique, standalone report in final form; generally a snapshot of a single item or issue.
I–Initial development. A full outline of the document has been established. Writing of some sections has begun.
D–Complete draft. The document is completely written and is undergoing review. A very small number of TBDs can remain, but these are limited to specific pieces of information, not entire sections or subsections.
F–Released final version or completed initial release.
Δ–Updates to released version re-released with changes.
Pre–Pre-sales phase.
Ph0–Sales phase.
Ph1–Concept phase.
Ph2–Preliminary design phase.
Ph3–Critical design phase.
Ph4A–Development phase.
Ph4B–Test and integration phase.
Ph4C–Acceptance test phase.
Ph5–Production phase.
PM–Program manager.
Doc. #–Example of a three-digit suffix (yyy) added to a company's document numbering system to distinguish different documents within a project.

CHAPTER 2

Failsafe Software Design:
Embedded Programming in a Fail-Certain World

Jeffrey M. Sieracki

1. Software Matters

Ask yourself why you trust your microwave oven. Think about that as your daughter stares in the window and watches her milk warm. There in your kitchen is a ubiquitous lifestyle tool with components that could, in seconds, do severe harm to you or your family. Yet careful design of the physical cavity, the door, and various electronic and software interlocks render it as benign as a cereal spoon in most circumstances and far safer than the ordinary kitchen range next to it.

Meanwhile, tired and grumpy in the morning, you push a few buttons trusting utterly that your tea water or frozen sausage will heat up to your liking. A few beeps, the numbers glow, the magnetron energizes, and breakfast is served. The buttons always respond, the timer always counts down to zero, the door interlock always knows when you are in a hurry and safely disengages the innards. Somebody cares.

In my kitchen, on the wall across the room, hangs another piece of modern machinery: A two-line telephone by one of those late 1990 companies formerly known as AT&T. It's lovely to look at, includes autodialing, conferencing, speaker phone, and two-line caller ID. The large LCD display is bordered by four "soft" buttons for menu control and—herein lies the rub—a red "new call" LED for each of the two lines.

As calls come in, the caller ID system dutifully decodes the name and number of the caller, recording this information together with the time, date, and incoming line, and the embedded software lights the "new call" LED for the appropriate line. Each call is marked "new" on the text screen until I review it once by sequencing through the list.

Curiously, sometimes as I sequence back through the list until I have seen the last new call, the LED steadfastly remains illuminated. If I sequence back through again, I see the software has clearly removed the "new" designation from each listed call; yet even if I delete the calls one by one, erasing the entire caller-ID list, the LED cheerfully continues to

DOI: 10.1016/B978-0-7506-8567-2.00002-0

announce "new call." If I pull the power to the phone, its clever brain keeps track of the situation and my Hal 9000 wall phone instantly boots back up glowing as if to announce, "I'm sorry Dave, but I can't allow you to turn off that LED." In designing this phone, somebody didn't care quite so much.

It *is* possible to turn off the LED by a special sequence of strokes that clears memory, losing all record of all calls. In other words, the "soft reset" works. It didn't take me long to think of that. In fact, the first time I witnessed my persistent "new call" light, I recognized the failure mode as an old friend and began tinkering to see which software pathways had been addressed and which had not. I have not only seen such an error before, I've created one. The circumstances were far less benign.

This chapter is about good software design for mission- and safety-critical systems. We will review simple principles and practices that help ensure systems work well and keep working. Our guiding mantra is *Assume Fallibility*.

We are generous in this assumption; don't just blame your sales team or the sketchy documentation from your system architect. Assume fallibility in the hardware you are controlling, in the sensors you rely upon, in the next programmer that picks up your code, in your communications channels, in the idiot-proof end users, and, ego to the wind, in yourself. We will draw frequently from real-world examples to reinforce these points.

This chapter is *not* about the myriad guidance and regulating documents available for safety-sensitive and mission-critical software design. Certainly to those working in specific fields, documents such as DO-178b (avionics), MIL-STD-498 and -882 (military), and 21 CFR 820 (medical devices), will eventually be of keen interest. We shall instead consider some key unifying principles behind such guidance documents and attempt to motivate further interest in what sometimes appears to be tedious recitations of burdensome process.

Modern software engineers are often heavily schooled in how to employ good process and design strategies to find and avoid bugs. But in contrast to more physical engineering fields—electrical, mechanical, or civil—there is little emphasis placed on designing and compensating for failure modes. Where other fields employ redundancy and experience-driven safety factors as a matter of course, software developers are often tying their own hands in a tradition of minimalism and false assumptions.

This chapter is laid out in sections, each building on the previous. Section 2 is about process, and how and why it can work to your favor in engineering reliable critical systems. Section 3 delves into important core principles and patterns that arise repeatedly at all levels of failsafe design and implementation. Section 4 discusses the user interface and the key role it may play in the success of a mission—or safety—critical system. Section 5 addresses some of the design patterns in the context of the common practice of creating one's own real-time operating software. Finally, section 6 considers an extended "what-if" scenario in which we examine a failsafe hardware design and consider how it might be usefully mimicked in software.

The recommendations and principles presented here were developed over many long years of both book learning and hands-on, hard-knocks experience. The topics will be of interest to applications software engineers moving into mission-critical/safety-sensitive systems work, as well as to hardware engineers transitioning to software work. They may also be of interest to those already in the field, either looking for new ways to think about their work or looking for material to help their manager understand why solid safety- and mission-critical software doesn't often come cheap.

2. The Essence of Process

The development of engineered systems, and of software in particular, has been characterized from the start by evolving descriptions of *process*. The classic waterfall process, first defined in the 1950s, lays out a feed-forward effort in which requirements feed into design, design to implementation, implementation into test and verification, and so on down to delivery. The heart of the idea, sometimes referred to as the *unmodified* waterfall model, is illustrated in Fig. 2.1(A). Why "unmodified"? Because in practice, the waterfall model is nearly always modified.

The waterfall model is just an attempt to sketch on paper how engineers interact in sequential stages to produce a product. Many practitioners would add pre-process steps, such as a *concept* phase, and post-process steps, such as *maintenance*, as shown in Fig. 2.1(B). Many would further break down the stages into sub-blocks: *design* might become *analysis*, *design*, and *test design*, while *implementation* might become *module implementation*, *module test*, and *integration*. One example of a more refined form of the waterfall is shown in Fig. 2.1(C).

The added degrees of detail are attempts to capture a very complicated reality. Any engineer with real project experience could sit all day adding new boxes to the diagram—and probably remain unsatisfied. In fact, a true waterfall process is almost never used.

The exceptions are those rare cases where there is no time for, no money for, or simply no possibility of changing one's mind. A one-off product, for example, might be sent out the door, ready or not, on an absolute limited budget; likewise, a satellite in orbit is not easy to reengineer once the final test data (launch!) are recorded.

Yet, even in those less common cases, there will almost certainly be some feedback between development stages—a prototype implemented gives information that changes the design, a test failed that changes the implementation—these small elements of feedback occur all the time. Even the termination of the process in "maintenance" is artificial since, nearly always, the results of any given project will feed into the next-generation product.

The closed-loop nature of most actual development leads to the spiral model of the development process, which simply recognizes the natural cycling between *requirements definition*, *analysis*, *development*, and *testing*. Various authors when describing the spiral call

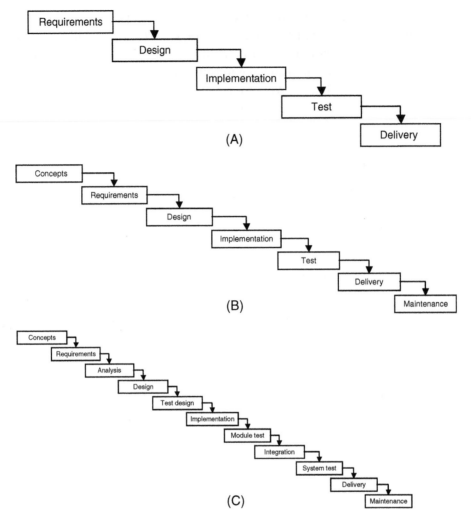

Figure 2.1: (A, B, C) The waterfall process diagram is a good metaphor for thinking about development, but does not capture the complicated reality of most engineering projects.

out different elements of the process, but the important concept is that good engineering is very seldom linear. Rather, good engineering is usually an iterative process of refining concepts and designs to converge on a solid result.

Attempts to better codify the development process have led to a host of ideas, and libraries of books on the topic. In addition to learning more about the waterfall and spiral methods, the avid reader may enjoy perusing the Internet for keywords like "agile," "extreme," "iterative," "RUP," and "RAD," and seeing where those links take them. Attempts to firmly and definitively separate one process model from another or to quantify their respective failings fill nearly as many volumes as descriptions of the methods themselves.

At heart, the *process* chosen is often much less critical to success than the simple existence of a process. In fact, most guidance documents come to a similar conclusion. Well-known quality certifications, such as ISO 9001, do not specify a particular process; rather, they specify the existence of a process and consistent supporting documentation. The development process itself remains up to the organization. ISO 9000 is a body of quality standards, not an engineering guideline, but for purposes of applying process to mission- and safety-critical systems, thinking in terms of quality management is not far off the mark.

The purpose of defining a development process from a critical system perspective is to provide a clear framework for *provability* of the system. In the next section, provability is discussed as one of the fundamental principles of failsafe software design. In essence, the goal is to demonstrate that the system will perform as intended.

We typically speak about meeting requirements and specifications in terms of *verification* and *validation*. In general usage, *verification* refers to the process of ensuring that the project elements meet specifications, while *validation* considers whether the project satisfies its intended purpose. The common use of the two together, typically shortened to V & V, considers that the line between one and the other is sometimes fuzzy.

With this in mind, let us consider one more process model that captures concepts in a different way: the *V-model*, illustrated in Fig. 2.2(A). As drawn, the process clearly tends to flow from left to right, beginning in project definition, traversing implementation, integration, and test, and ending in operations and maintenance. However, the time arrow is only intended to indicate general flow, since information gained during test may easily feed back to a new iteration of coding or design. The V-model is not so concerned with a firm sequence of steps; one may use spiral, iterative, waterfall, or any other sequencing ideas to move the development team from requirements to delivery.

Rather than sequencing, the V-model is largely focusing on testability. The most important links of the diagram, from our perspective, are the *verification-and-validation* (V&V) flow arrows bridging the gap of the V. Definitions including requirements and specifications, analysis and architecture, and detailed design are set at stages on the descending leg on the left and are matched to corresponding tests, including system tests, integration testing, and module test at each ascending stage on the right.

System tests will naturally include testing of failure modes. We will see in the course of this chapter that encapsulating mission- and safety-critical elements of the design in ways that they can be isolated, implemented, and tested independently, goes a long way toward provability of those elements. Encapsulation lies at the heart of the *design-for-test* concept as it applies to software.

Iterative, stepwise, documented testing allows the engineer to build the critical elements of the system on a rock solid foundation. In particular, risk assessment and hazard analysis

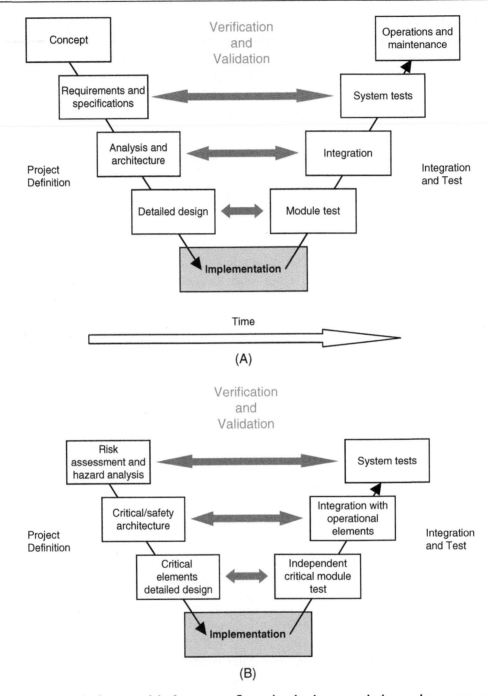

Figure 2.2: (A, B) The V-model of process reflects the ties between design and test aspects of a project. It can be adapted as a very good metaphor for understanding the process of developing, testing, and integrating safety- and mission-critical project pieces.

(RA/HA) become integral elements in the requirements and, subsequently, in the testing. Figure 2.2(B) illustrates how mission- and safety-critical elements can, and should, be called out, independently implemented, and independently verified. This critical-element subset of the V-model neatly captures how process relates to failsafe software design.

Any process model of software development is, at best, an approximation of reality, and the V-model is no better in that respect. The extreme emphasis on testing certainly will not appeal to applications programmers driven by tight deadlines rather than by hard mission- or safety-critical requirements. Rigid adherence will often be overkill for less demanding aspects of a project, such as general operations software. However, the V-model is an excellent graphic guide to best practices of traceability and provability for critical systems.

3. Three Principles for Design and Coding

3.1. What Does It Mean to Be Failsafe?

Having emphasized the importance of V&V in critical-system development, let's take a step back and remember that testing isn't everything. One problem with overemphasizing the design and test philosophy and relying on a third-party verification process is that if we design our tests to only test what we designed, then we are potentially blinding ourselves to unforeseen interaction of system elements.

Understanding what it means to be failsafe means understanding what can fail. We need to allow for unexpected failures both inside and outside the box. In a tip of the hat to Gödel's classic logic proof, we need to recognize that we are nearly certain to experience failure modes in the field that were not realistically possible to anticipate in our design reviews.

Defining your embedded system as "the box," consider first the outside-the-box elements. As the saying goes, nothing is idiot proof, given a sufficiently talented idiot. Your users are probably not idiots, in most cases, but they don't necessarily think like you do. It is almost a certainty that your use cases and failure charts will not capture all the possible variations of input the system will ever see. Nor can you completely control the environments in which your system will be used—this is particularly true when the system has a long life cycle, such as medical or military projects that are likely to persist as legacy systems long after your primary design life.

Example 2.1. Outside-the-Box Testing

Sometimes what is obvious to the designer is not so obvious to the user. A medical test system required us to design a graphical interface that allowed patients to indicate their pain or other sensations by drawing on diagrams of their body. We very carefully thought out the

---

(Continues)

Example 2.1.   Cont'd

interaction: The user circled regions of interest with a virtual pen. Each time the pen was lifted, a clever algorithm closed the curve and reduced the drawing to a shaded area of the body. This provided instant feedback to the patient and was very straightforward to anyone familiar with pen and paper. With a quick introduction by the medical practitioner, the patient was generally off and running.

Intriguingly, about 10% of the users found a second mode of interaction. Rather than circle the body region they wished to indicate, these users instead shaded it in, crayon style. The algorithm behaved predictably in response to this unplanned entry mode, providing positive feedback to the users just like a pen on paper so they continued to use it. This was a triumph of lucky design in some ways, but a big problem in others.

Data entered by shading produced vastly more closed-curve objects in a given drawing than data entered by simple circling. Our data analysis routines stepped through these closed-curve objects in a way that rendered the shaded data extremely slow to evaluate. So these 10% of our users required vastly disproportionate analysis time. We eventually wrote subroutines to combine overlapping closed curves into minimal convex hulls for purposes of both data compression and analysis speed—compensating for our own unplanned features.

The users had stepped beyond the designer's intent and found another way of successfully using the system, with unforeseen consequences. They were unforeseen because the use case variation was completely outside the design scope. It just happened to work. No amount of planned testing was likely to reveal the issue.

In this very same system a bored alpha tester also took us quite literally outside the box. The young gentleman, a university student and research intern, was asked to test our system between his other tasks. He wasn't given specific test instructions, just asked to try out different functions and let us know if anything broke. It wasn't long until he succeeded.

Asked to demonstrate for us, he proceeded to circle the entire screen several times, lift his pen, and then make one more dot somewhere else on the screen. The system crashed instantly. In fact, he had triggered a rather unlikely buffer overrun condition that we promptly addressed. This was completely outside any rational use case since there was no reason any user would ever need to circle the entire screen, let alone more than once. It took a bored lad scribbling at random to find our failure mode.

What could a bored lad achieve by punching buttons at random on your embedded control device? What unintended user interactions might be supported, even reinforced by your user interface? You will not predict them all. Keep that in mind when contemplating the value of cross checks and critical element redundancies.

Consider next the inside-the-box elements. Today's embedded processors sport an abundance of power and memory resources and encourage increasingly complex embedded code: complex code that is less and less provable. We now need to triage our complex systems and isolate critical elements from ordinary operations code. Even so, code for critical-system elements often interact with less reliable subsystems. Consider, for example, the now common situation of running your front-end operations and user interface on consumer-grade Windows. Even in the best of circumstances, bugs in large software systems can be insidious, with sleeper failure modes lying in wait for just the right combination of inputs or just the right intermittent hardware glitch. What about the future? If your code is subject to future modification, how resilient is it to bugs introduced later? How much testing will need to be repeated?

We simply cannot rely on high-level system testing of complex mission- and safety-critical designs. We need to fight the battle on the ground, at the foundation level of our design and coding. We begin with three strategic principles:

- Readability

- Redundancy

- Provability

The first of these principles, *readability*, will be familiar in some form or other to every software engineer who has cracked a textbook in the last 30 years. The details of the idea have evolved substantially and coding styles have gone in and out of fashion, but the concept remains and certain fundamentals persist, including clear blocking of code, transparent naming of variables, functions, and properties, extensive commenting: All of these niceties make it easier for your colleagues to read and understand your code and for you to remember what you were doing when the boss asks you to modify the code a year from now.

Example 2.2. Human Readable Is Human Recoverable

All software engineers go through a phase where they enjoy maximizing efficiency. It's amazing how much information you can pack into a binary file, coding bits and bytes in tight structures. It's also amazing how useless that information can become if corresponding "reader" software isn't available. If the binary data files become corrupted, then one can only hope the original engineer took the time to write recovery functions into the reader software. Otherwise, the hour is truly dark.

There are instances where tight data are key. Perhaps there are system storage limitations. Perhaps the data are naturally structured, such as in image or audio files. There are also instances where an embedded system is fundamentally designed to process information for compression or transmission—think mobile phones.

There are, however, many, many instances where data are being logged or recorded for testing purposes on typical modern systems with typical modern storage capacities. In such cases the additional overhead of writing an ASCII field name with the data or in writing ASCII coded

---

(Continues)

Example 2.2.   Cont'd

text is often trivial compared to the benefits of having a human readable output quickly available to review in any handy text editor.

The Linux world has long recognized the usefulness of ASCII control files and logs. Ironically, after years of increasingly obscure file formatting, even Windows now incorporates ASCII-coded configuration and log files.

In my work in experimental test systems and embedded design, it becomes impossible to count the occasions where an ASCII log file, or an ASCII coded header to a binary data file, has saved the day. Cases include testing, debugging, crash recovery, and a few substantial projects in legacy data recovery where systems have fallen into disuse.

Not long ago I received a phone call from a researcher asking my help on understanding data that we collected with a custom test system almost 20 years ago. The system no longer exists, and even the hardware it ran on is impossible to reproduce—but this gentleman had file after file of archived data from our work. I walked him through the data file headers, written in long-form text, which fully documented the experimental setup and each data field written in the file below. Needless to say, he was rather pleased.

Documentation gets lost. Code gets lost. Processors and operating systems change. Even storage media evolves. If your data are not stored in a well-documented and broadly applied standard format and your requirements are not truly data intensive by modern standards, consider keeping it human readable wherever possible.

Readability maximizes transparency, facilitating verification and bug fixes. In the realm of applications programming, these are *good practices*. In the realm of mission- and safety-sensitive software, these are *critical practices*. When the failure of your code could result in the loss of a quarter-billion-dollar aircraft and human lives, clarity counts.

As with process, it is not my intent to espouse a particular style convention. Depending on your background, the tools you work in, and your organization's standards, your preferred style may vary quite a bit. However, general experience suggests that using as long and as descriptive a name as your own convention and software tools permit often goes a long way toward success. When code reads like a sentence, it becomes much more readable and much easier to verify. A friend who programs for a major airline-industry software provider relates the story of getting slightly embarrassing attention over her innocent choice of the polymorphic method name "isTurnedOn()" for checking the status of system objects. However, one may rest assured that the code reviewers had no trouble in understanding and verifying the logic during the walk-through.

Keep in mind that meaningful naming conventions also require consistent spelling. In an age when modern programming languages are reintroducing untyped, undeclared, and runtime allocated variables, it pays to get a little old school when it comes to your critical system elements. When writing code, turn on compiler flags that require explicit declarations and use code evaluation tools that help you find orphan spellings.

Programming standards have evolved through the introduction of functions and subroutines in software pre-history, to the organizational rules of *structured programming*, to the current *object-oriented* vogue. Modern embedded programmers may code in languages from assembly to Java these days, so there is not a one-size-fits-all solution. However, adapting a consistent strategy and complying with your own standards is the first step in any good practice manual.

Side Bar: Is OO the Enemy of Transparency

To the experienced embedded software engineer, there are advantages and tradeoffs to every tool. Some occasions call for hand-optimized assembly code, while others afford the luxury of the latest UML graphic design tools. Having lived through both the structured programming and object-oriented (OO) heydays, it is interesting to note that not all the purported gains of OO are blessings to the mission-critical programmer. In particular, inheritance and polymorphism can be sticky concepts since, in practical situations, progressive development of software or reuse of older code often entails modifications of root classes, potentially triggering an extensive re-verification process. If someone's safety is on the line, every test pass may need to be deep and exhaustive.

OO gurus will take exception. In fact, the same aspects of OO that allow one to define inheritance will aid in tracing dependencies and defining test plans. It is certainly possible to adhere to good OO design patterns and build systems where properties of OO work to the mission-critical programmer's advantage. In the author's experience, however, this is seldom achieved, since evolving hardware choices and other system considerations combine with delivery time pressures to spoil design perfection. Particularly in complicated event-driven object code where user input and time-cued events continually modify the system state, it becomes remarkably easy to lose track of which methods or call-backs handle which hardware checks, resets, and state updates under which circumstances. Independence of the critical safety code as espoused in the chapter can help mitigate those risks.

OO programming does have advantages. In particular, OO allows easy abstraction of real-world elements so that the code interaction model is more intuitively understandable. The "Hardware as Software" exercise in this chapter (see section 6) is intentionally sketched using an old-school, structured subroutine model. It would be much more transparent to sketch a UML diagram where an object like the "PrimaryInterlockSwitch" had a property "Closed." However, while the OO model clarifies intent, it deliberately hides functionality. Inside the "PrimaryInterlockSwitch" object, someone would eventually have to write and verify a subroutine that acted on changes in the property "Closed," and the hardware control aspects of that subroutine code would look rather similar to the subroutine presented. The UML object model must be linked to the hardware eventually.

OO may have a natural appeal to hardware engineers moving into software engineering. OO programming can be thought of analogously to component-based hardware design. In many ways, each object class is like a discrete device, with external interactions governed entirely

---

(Continues)

Side Bar: Cont'd

by fixed public "properties" and "methods" in the manner of the documented electrical connections on a chip. The internal manner of function is not of concern to the user; only the published behavior is important.

Relying on the hidden embedded functionality, however, dictates thorough component module testing. It also dictates isolation and tracking of critical functions—a recurrent theme of this chapter.

This brings us to another point: a key positive feature that both structured and OO programming styles share is encapsulation. Use it! As discussed in the text, encapsulation helps enable both readability and provability. However, the existence of encapsulation is not sufficient—think very carefully at the design stage about how to employ encapsulation. If safety and critical elements of the system are scattered across the system classes, the design may be a recipe for disaster. Complex interactions between classes can conceal unanticipated failure modes. Extensive, repeated testing may become necessary at each software change to verify critical system behaviors. Because of lack of transparency, verification can become a matter of probabilities rather than certainty.

How does one overcome the transparency issues that can arise in OO programming? First and foremost, by applying the principles of redundancy and provability to the safety- and mission-critical aspects of the system. Burying critical safety checks in your operational classes is both unnecessary and poor practice. Instead, encapsulate critical software aspects in independent classes that can be verified independently of other operations code.

Yes, one may very well incorporate secondary safety and status checks in operational code that are redundant with the independent safety modules; however, by clearly calling out critical hazard mitigations in the object design, the system will become readily verifiable and inherently lower risk.

The second principle is *redundancy*. This concept may be natural to the mechanical engineer designing bridges, but it is seldom second nature to the applications software engineer who relies habitually on computer hardware and operating systems to perform as advertised. It has been said that the best engineers are lazy—that is to say, we innovate intensively to save work in the long run. For many, the elegance of software somehow lies in the art of creating minimal code to achieve maximal effect. It may seem counterintuitive to save work by allowing more than one piece of code to perform the same task. It's not.

In fact we save work, in the long run, because by carefully employing redundancy we are enabling the last of the three principles: provability. In particular, we enhance provability by reducing the required testing load.

Consider what happens if you introduce a change in the operational code that mistakenly removes or inactivates a sequential safety check. Consider what happens if someone else comes along and mucks up your code with their own ideas—perhaps simply because they didn't fully understand the dependencies. No one is perfect and any practicing engineer can probably recount a few such experiences.

At another level, consider that software is inherently reliant upon the hardware that supports it. Consider how completely you want to rely on your internal state variables accurately reflecting the real-world environment if an erroneous hardware setting could permanently injure an innocent user. Hardware engineers certainly don't take their hardware's continued normal operation on faith; likewise, critical system software engineers should not take either the hardware or their own software on faith.

We address these issues by clearly calling out and enabling mission-critical and safety elements in separate, verifiable code. If the disabled in-line safety check were backed up by a redundant, independent-watchdog safety check, a potentially life threatening bug can be reduced to merely an important future software patch. If the system is able to use independent sensors to verify reality, rather than relying on internal state variables, entire classes of failure modes can be detected and avoided.

Working field software should not even trust itself. In space, military, medical, and heavy-industry environments, well-designed embedded systems always start up with a thorough self-check. Is the code intact according to checksum tests? Did it load correctly? Are there any stuck bits, registers, or sensors? Space systems, due to ionizing radiation exposure, sometimes include means of continually verifying software, processor, and memory conformity during operation—at times replicating entire embedded subsystems in two or three copies.

Hardware engineers are unashamed of redundancy because they are trained to design to failure modes. Examine the schematic in section 6 (Figure 2.6) and you will find that multiple redundant interlocks consume most of the design. Software engineers are sometimes burdened with an education that suggests if only they follow certain steps they should be able to achieve provable, efficient, bug-free code. That thinking is not wrong, but it is dangerously self-deceiving. Software designers need to fundamentally understand that rushed releases, borrowed code, broken updates, flakey hardware, and honest mistakes are part and parcel to the field. A critically designed highway bridge may, with some luck, carry traffic for years at its design load, but a three-times safety factor over that expected load will truly protect the health and safety of the public in the face of the unpredictable.

Assume fallibility and create failsafe watchdog checks that address not only failure of the system and user error, but also failure of the code. (See Example 2.4.) Just as in building airplanes and bridges, critical-system software must include redundancy: design a complete operational system meeting requirements and, in addition, build in *independent* watchdog and backup elements to protect against critical failures.

Example 2.4. Brain Stimulation Backup

What happens when a little change in code mucks things up in a working system? A failsafe watchdog can really improve your day. In the late 1990s, I was involved in designing software-

---

(Continues)

Example 2.4.   Cont'd

controlled test systems that included electrical brain stimulation. We used the stimulation in conjunction with behavioral tasks to assist surgeons in mapping language and motor skills in the brain before surgery. These systems were noncommercial and we frequently updated our software to incorporate new specialized tests for individual patients under very tight time pressure. Several tightly choreographed computers performed numerous operations, including presenting visual and audio cues to a patient, and recording EEG, speech, and reaction time, all while stimulating across pairs of electrodes laid directly on the brain.

During operation, the general sequence of events was supposed to work like this:

```
    Send trial information to recording computer
    Begin recording
    Turn on electrical stimulation
    Present psychological test cue
    Record patient response
    End psychological test
    Turn off electrical stimulation
    End Recording
    Store data
Repeat for each trial until done
```

Caveats abounded. Chief among them was that electrical stimulation could under no circumstances remain active for longer than 5 seconds for safety reasons. The length of the psychological test could vary and patient response time was up to the subject, meaning that the timing and even the sequence of events were never completely predictable. The system incorporated several levels of software cross-checks and, of course, emergency hardware disconnects.

The hospital monitored human subjects work with an excellent group of independent engineers known as the "clinical engineering" group. Our job was to build it; their job was to try to break it and to make sure, in the worst case, that it didn't hurt anyone.

As our work progressed, the physicians asked us to speed up testing by allowing multiple psychological tests to occur during a single electrical stimulation period. This necessitated a delicate restringing of the control code, and time pressures always meant too little testing. In due course, we demonstrated the updated system to clinical engineering, verifying that we could interrupt stimulation at any point using keyboard inputs and that the multitest sequencing worked properly. After an hour or so, by chance, we discovered a bug: In certain rare cases, where the last scheduled psychological test overlapped the scheduled electrical stimulation endpoint, a flag was miss-set and the state variables lost track of the fact that the stimulator was on.

It was rather stomach-wrenching to have a deeply concerned third party ask, "Why is the stimulation still on?" during this sort of clearance demonstration. However, the clinical engineer immediately checked the electrical stimulator itself and corrected himself: the stimulation was off. In fact, while the computer screen was still indicating "stimulation on" in rather bold and alarming fashion, watchdog timer code that had never before been executed in need, had already detected the 5-second timeout limit and turned the juice off. The system corrected its erroneous display shortly thereafter, verifying reality by checking the stimulation relay status before moving on to the next trial sequence.

Perhaps this reminds you of the "new call" light mentioned in the introduction? It should. (See Example 2.8.) There was no way of directly monitoring stimulation status from the control computer in this case. Instead, a function named something like `ElectricalStimOn()` checked the control relay and reported whether it was open or closed via a hardware register bit. This function did not fail; instead the operations code simply failed to clear the blinking warning from the display because the check-and-update sequence was skipped. The error is akin to the problems with setting internal-state variables to mimic reality; in this case, the presence or absence or a line of text on the display was the "flag" that a human operator saw and it was simply out of synch.

Our clinical system came out ahead of my telephone, with its perpetual "new call" situation, because of strategic redundancy: my software contained independent watchdog safety code and verified reality by forcing things to known conditions at the end of each trial.

The bug was easy to fix once detected. The failsafe watchdog timer not only saved face, it also would have fully protected the patient had one been attached. Ultimately, our third-party reviewer was very favorably impressed with the safe, demonstrable redundancy of the system, especially given the non-commercial nature of the project and the likelihood of frequent future changes.

Independently designed and tested failsafe routines that are not part of the primary control code are key to reliable, demonstrable safety. Put more simply: CYA.

This brings us to the third principle: *provability*. Certain small embedded systems are in fact simple enough to examine and prove exhaustively. These are increasingly rare, and even where they are possible, maintenance and update processes become significant costs when one must re-verify everything.

In general, operational code in modern systems is simply too complex (or too expensive) to test exhaustively. Triage in the form of RA/HA is the key to getting our hands around the problem. We can use the above principles to verify the mission- and safety-critical system aspects to achieve a failsafe product.

Some aspects of provability come about in code validation, others in functional system design. Encapsulation of mission- and safety-critical code not only aids readability, it also enables provability. For example, by designing a watchdog monitoring thread that is independent of my system operations code, I encapsulate a critical aspect of my system. That safety check can be verified in early code reviews and in independent systems testing and thereafter frozen out of most future updates. I then know with some certainty that even if the more intricate operations code is broken in future updates, I have a friendly big brother watching things. Not only is encapsulated code thread independent of the other components, making for a reliable functional design, it is also compact and easy to read and verify without tracking through the full complexities of operations.

Even where code is not multithreaded, a natural extension of this idea applies. Encapsulate the mission- or safety-critical aspect in a standalone function call that can be verified, frozen, and relied upon. For example, `MakeLaserSafe` might be a nice function name for a sequence of events that included testing for lack of energy in the exciter, powering down the laser if necessary, engaging interlocks, and warning the user if anything fails. `TestLaserSafe` might be a nice function name for a routine that checked the exciter, checked the interlocks, and set display warnings accordingly. Once these safety routines were coded, verified, and frozen, the presence of these function calls in any updated operation code would become part of the standard verification procedure.

In single-threaded code designs, even where outside physical events cannot affect hardware status, it is good practice to use these sorts of calls redundantly. In this case, redundancy enables provability. For example, one might call `MakeLaserSafe` followed by `TestLaserSafe` in the operations code as part of ending an ordinary cutting operation. The function `UnlockDoor` might make the very same calls, redundantly, before operating the door solenoid. This would protect against later bugs in which a programmer fails to sequence the commands properly. It would also allow the `UnlockDoor` command to be verified against the RA/HA chart independently of the operations code. Conversely, it would allow operations code updates to be verified without relying on the institutional memory of whether or not `UnlockDoor` makes the system safe.

With modern fast hardware and cheap memory, the only real excuse to avoid such redundancy is ultra tight real-time operating constraints. But generally in such circumstances the entire response-time-critical operating branch becomes a critical component subject to verification and review with each modification.

Another aspect of provability occurs during operation, but must be planned for during design. The idea can be reduced to *verify reality*.

Again, the provability and redundancy ideas complement each other. In the example above, we not only turned off the laser, we made sure it was off, hopefully by means of an independent hardware sensor. Were we to use only software state variables and believe that the laser was off simply because we turned it off, then our user might one day experience a very unpleasant failure mode. Hardware busses fail, registers fail, relays stick. . . . Safety cannot be left to an internal model of the physical system that is not continuously updated and verified.

When software can issue a hardware command but not confirm the result, the situation is often referred to as *open-loop* control. Open-loop control leads to assumptions by its nature; and unverified assumptions are very, very frequently the root cause of high-risk failure modes in safety- and mission-critical systems.

In a safety-critical system, your software control is extremely likely to be backed up by hardware interlocks. In fact, unless it is physically impossible to achieve, it had better be. On the one hand, don't leave home without making the hardware as safe as possible in its own right. On the other hand, do not rely on hardware safeties to compensate for your code failures. Hardware switches and interlocks alone, in some sense, work in big broad strokes, while software can generally act much more nimbly to detect, compensate for, and even predict hardware failures.

3.2. Safety (and Mission) First

For anyone who has wandered the grounds of an industrial lab or manufacturing plant, "safety first" seems like an overworked catch phrase. In fact, in mission- and safety-critical system design, it is an actionable design approach. This approach begins with *risk assessment* and *hazard analysis*.

Hazard analysis is a process used to assess risk. The hazard analysis identifies and quantifies risks and then defines means of controlling or eliminating them. This process occurs in the very early stages of the software development cycle, often before operational features are fully specified.

A common format for such documents is a multicolumn matrix, such as the following.

Failure Mode	Hazard Level	Design Action
(How could it go wrong?)	(How bad could it be?)	(How do we compensate for it?)

Failure modes may be caused by equipment problems, software problems, user actions, environmental situations, hardware or power glitches, or even acts of God. The hazard level typically reflects at least three levels of risk assessment characterizing the significance of the threat. This is sometimes a combination of the possible cost of the failure mode (in safety or mission risk) combined with its likelihood of occurrence. In general, however, one attempts to write down everything bad that could happen since it is often meaningless to sort out likelihood early in the design. The design action is a brief statement of how we intend to control the hazard in our design—always stated in an actionable and traceable way.

The three columns show the essence of the idea. Many practitioners develop their analysis with several more columns, breaking out information and adding additional details specific to the project or to company process. For example, hazard level might be explicitly separated into estimated probability of occurrence and risk level upon occurrence. Design action might be complemented with a column assigning the task to a project group, and each row may include a traceable number or code to be used in tracking.

From the perspective of software, we need to know which of the system-level RA/HA items are actionable. These items should form the basis of a software RA/HA matrix. To that matrix we should add our own additions. These include software-specific risks such as hidden hardware dependencies, potential for state verification, and software bugs that might slip through in initial coding or updates. These are exactly the topics we've discussed above.

Such hidden software risks may not always be on the system engineer's mind, and where they pose significant potential for safety or mission compromise, they should be brought to the overall design group's attention. They should also be documented and tracked within the software group.

We need to recognize that common software failures can be harder to explicitly predict and harder to identify and trace when they occur than common hardware failures (Fig. 2.3). It is incumbent upon the software engineer to recognize the limitations in our craft, to convey the risks to those who need to know, and to design and code in a way that compensates for these recognized, but often nebulous, risks.

In many systems both safety- and mission-critical aspects arise. Sometimes they are related; sometimes they are not. In medical products development, it is common to focus heavily on safety since this is what the Food and Drug Administration and other regulatory agencies will audit and control. But in any embedded software design, there will also be mission-critical aspects that are important to the success and usability of the product. A product may be perfectly safe but ineffective, rendering it a failure in its mission. Why build expensive doorstops? Safe and effective products are our goal, and we achieve this by tracking the balance of mission- and safety-critical aspects of a project.

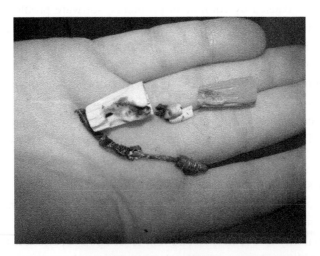

Figure 2.3: Burnt offering: Common software failures can be harder to predict and, when they occur, much harder to identify and trace than common hardware failures. (Photo courtesy of Jeffrey M. Sieracki)

By tracking both aspects in parallel, a critical system designer—or any embedded system engineer—can make clear, traceable choices that increase probability of producing a working, verifiable system that meets design goals.

Consider again the V&V process in Fig. 2.2(B). After we quantify our safety- and mission-critical aspects in an RA/HA chart, we move on to designing the critical architecture components to address this list. We then design code modules to enable this critical backbone architecture. Insofar as is possible, this design will be independent of the operational architecture that we design later to meet the basic functional goals.

After implementation, we assemble our components working bottom up to test and verify. Each descending leg of the V corresponds to an ascending verification step. The critical elements detail design maps to independent testing of those modules, and the critical architecture informs the integration of those modules as well as integration with the operational elements of the system. Finally, the list of critical aspects the original RA/HA chart directly drives set of system-level validations, including both functional tests and RA/HA-specific end-to-end design reviews and code walk-through.

By starting the project with the critical elements rather than general operational aspects of the project, we design to the failure modes. We can then move forward on the feature-driven aspects of the project with some confidence that we have covered ourselves on the critical side. In the author's experience, this safety-first approach often leads to clearer and simpler operational code for two reasons: (1) the operational code can be written directly to the feature list without thinking and rethinking what might go wrong at every step; and (2) the critical design architecture often becomes the backbone of the basic operations architecture with very little additional overhead.

Use cases can also be extended to failure modes. Modern software designers often build up a collection of storylines that helps all stakeholders visualize how the system will interact with users and the environment. These are called use cases. They are generally used to help define user interfaces and operations sequences to enable the system to better serve its application. In practice, their use is typically emphasized in early design stages to help flesh out the look and feel with respect to the basic mission; their importance generally diminishes as development progresses and the actual system elements are available for examination and testing. Application of use causes is often limited to understanding and defining core operations elements. However, they can also serve as a powerful tool for evaluating failure modes and in planning for risk mitigation in safety- and mission-critical systems.

What happens if an alarm sounds in my plant control center or on my aircraft cockpit panel? Is there a defined user checklist? Are there mechanisms by which a user can acknowledge the situation and eliminate the distraction? What if the user simply silences the alarm and moves on—is there contingency code for ensuring that critical alarms cannot be ignored?

Suppose that a primary control fails. Consider the steps by which the user (or the embedded software) will enable a backup system. Is it natural and achievable under expected operating conditions? Are additional steps required to ensure safety during transition?

Consider the myriad insidious software errors discussed earlier. Can the code help a user recognize possible internal failure? At a minimum, we should log inconsistencies in internal states variables and sensors so that they can be reviewed later. Failure of a hardware system to respond as expected could be a hardware error or a software bug. Where detection is possible, it should be addressed automatically to the fullest extent possible.

The failure-mode use-case analysis should assess the risks associated with the problem, whether and how the user will be notified, and how to log the error and notify technicians that something must be investigated. The "check engine" light in your car is a classic example of dealing with low-risk failure modes—by simple means it notifies the user that a non-critical problem exists that cannot quickly be acted upon. The actual problem is logged and a technician can later short a few jumpers to request that the embedded software report the problem in detail.

In some cases, the failure will present a clear and present hazard to safety or to the critical mission of the system. The failure-mode use case should carefully address how the user is notified, and whether and how the system is shut down, the mission aborted, or the situation is otherwise made safe as appropriate.

Example 2.5. Expecting the Unexpected

We give a great deal of emphasis to risk assessment and hazard analysis, as, indeed, we should. These critical paths become part of our design and traceable elements in our test plan. Similarly, requirements documents and specifications lead to our core operational design and operational test plan.

In both commercial and research systems requirements, there are always requirements that will not be documented in advance. These include tacit assumptions or environmental factors that do not become apparent until we get to field releases and beta testing.

In developing a computer-based medical research system designed to be left alone with a patient, we did not fully consider the habits and expectations of the clinical staff. Commercial medical devices are nearly always either permanently installed, equipped with battery backup, or both. Nurses thought nothing of walking in, unplugging the computer, and wheeling it out of the way. In the days before uninterruptible power supplies became cheap and ubiquitous, this was a serious problem. Hours of patient testing could be lost if the system was not shut down properly. Thus "plug-pull tolerance" made its way to the top of the evolving requirements list after a few weeks in the field. By ensuring that the software flushed file buffers early and often and that we tracked recovery information in case of sudden interruption, we prevented loss of data. This meant less wasted patient time and fewer surly exchanges between the research and medical staffs.

To the surprise of the surgeon who directed the lab, the change was fairly quick and easy. The change was accomplished with minor, traceable edits to the working code and one small additional module. There were three changes: (1) The file system was flushed after each minimal unit of meaningful data was collected. This required a single line of code and did not impose any unacceptable delays in operation since it was coordinated with a natural momentary break in the test sequence. (2) An indicator flag file was written before patient testing began and cleared after testing ended. (3) On start-up, the software checked for the flag file and, if present, executed a separate subroutine that checked the last data file for any partially written information and cleaned it up. The software then passed parameters to the existing code to restore the patient interface to the point at which the plug was pulled.

The code was already well encapsulated. The only modification to the critical core control code was the file flush—a single line of code, easily verified. The other modifications were outside the core control code and added to the start-up and wrap-up sequences—again with one or two lines of code, in clear sequence, easily verified. Existing well-tested software modules already worked reliably and major modification and retesting was unnecessary. The new clean-up subroutine was created and verified separately.

Inevitably, your requirements and your code will progress, so encapsulate and generalize safety- and mission-critical aspects. Thinking and designing modularly is efficient at every step, from documentation to code delivery to testing.

In addition to failure-mode use cases, also consider the postmortem in the event of failure. Consider how the technicians can learn what went wrong so that they can repair the system, or convey issues to the designer to fix in the next update.

With respect to good design practices, we can learn a lot from the black boxes used in aircraft and other public transportation. At the hardware level, they follow all of the mission-critical design guidance steps that we are advocating for software. The boxes comprise an independent, encapsulated, and separately tested subsystem. The encapsulated design is limited to a specific, narrow critical mission, and is not subject to revision in conjunction with the rest of the operational system.

Black boxes have become ubiquitous in transportation and are, to some extent, becoming ubiquitous in applications software—in the latter case, they take the form of system event logs. In embedded systems, they have often been limited to simple error-state flags, such as the "check engine" model. With increasingly cheap computing and memory capacity, the safety- and mission-critical designer has little excuse not to include a means to keep detailed track of field problems.

3.3. Verification and Redundancy in the Implementation Process

Relying on encapsulated functionality, as discussed previously, dictates thorough component module testing. These critical paths become part of the design and should be traceable elements in the test plan, just as requirements and specifications documents lead to the core operational design and operational test plan.

To the extent possible, meet specific needs with specific traceable code. In some cases, this may be a watchdog timer running on a separate thread; in others, it may mean encapsulating hardware and sensor interactions; and in still others, it may be as simple as laying out code so that a master loop clearly and unavoidably checks and confirms hardware status on each pass-through.

Ideally, each software mitigation step listed in the RA/HA should be traceable to a specific, encapsulated code segment. This will reduce risk of unforeseen interactions, simplify verification, and increase the likelihood of delivering a robust, operational system.

In practice, not every aspect of every mitigation step can be encapsulated. For example, start-up code needs to make calls to initiate certain modules. Operational sequence code may also need to make specific tests to take advantage of the encapsulated aspects—such as making the laser safe before opening the door. As discussed earlier, however, a certain degree of redundancy can help simplify verification and ensure compliance with required procedure.

Non-encapsulated steps need to be conveyed clearly to code designers and verified as a matter of course in code reviews. Permitting and encouraging strategic code redundancy where safety- and mission-hardware interaction occurs not only aids the reviewer's work, but also provides a safety net. It is certainly possible to achieve Six-Sigma level verification without doing so, but the potential for hidden bugs in complex modern software is so high that the belt-and-suspenders approach of targeted redundancy will almost certainly be a safer and more cost-effective plan.

Targeted redundancy is not synonymous with bloated code. Repeated elements are limited to calls that verify critical hardware status, and the redundancy occurs in that these calls may occur in more than one subroutine that may be part of a sequential code sequence. The hardware functions should be encapsulated and the operational code clean and readable. Old-school software efficiency hounds (the author among them) can take solace in the reality that modern chip designers often must include a high percentage of entirely redundant circuit regions on their silicon dies to compensate for frequent failures in the manufacturing process. A high mission success rate takes precedence, and carefully applied redundancy is ultimately efficient and cost effective.

Version control, test and control plans, and linked verification of the elements in the RA/HA matrix are the means by which safety- and mission-critical elements are traced and ensured in the final product. The process documentation often feels like a burdensome evil when an engineer would rather get down to coding, but it is ultimately a powerful tool for quality assurance. Taken with a positive view, the process can be leveraged by the critical-system engineer to make implementation a very targeted and efficient process.

Other elements of redundancy are becoming standard fare in modern development. A generation ago, programmers tended to be quite curmudgeonly about their peers looking over their shoulders. Today code reviews are ingrained in the development process. By subjecting each function call and line of code to multiple pairs of eyes, the chances of catching and identifying bugs and obscure potential failure modes go up enormously. Code reviews should consider style, insofar as keeping code readable and conforming as necessary to maintain consistency across a work group. However, the emphasis of the review should be on examining whether a particular code segment meets its design purpose and avoids hidden flaws. The best code reviews include multiple team members stepping through the code visually, not only looking for bugs but also challenging each other intellectually with various use cases and code entry conditions to determine whether something was missed.

Adding in safety–or mission–critical crosschecks can become natural. If reviewing encapsulated critical code, make sure that it meets its design parameters, that it is tight and straightforward, and is strictly independent from other systems operation insofar as possible. If reviewing general operational code, have the RA/HA checklist handy, ask risk-associated questions as you go, and make an explicit pass through the checklist once to evaluate that each condition is either addressed or not applicable.

Some shops have introduced more extreme levels of implementation code production redundancy with good effect. These include ongoing multiple programmer integration and even paired "extreme" coding teams that work side by side on every line. The success of these ideas lends more and more credence to the cost effectiveness of carefully applied redundancy when it comes to saving time in achieving a reliable result.

4. The User Interface

On June 22, 2009, a DC metro train at full speed rear-ended a second train that had stopped shy of the station. Six people were killed, and over 50 injured. The failure was not supposed to be possible. Not only are the trains generally operated under automatic control, they also have a separate system that automatically stops a train if it enters a track zone already occupied by another—even when under manual operation. In theory, the system was designed so that if any critical sensor or communications failed, it would "fail safe" by bringing the affected trains to a halt. Clearly something failed out of keeping with anyone's predictions and was either not detected or not acted on by the critical safety systems. The ultimate backup control system—the human operator—also failed to notice or act in time to prevent disaster.

Today even jet aircraft can demonstrably take off, fly, and land themselves without human intervention. For an airline pilot, or a train operator, a boring day at the office is an excellent state of affairs. One reason that systems like this still have human operators is to provide a sanity check on the state of the environment. Our best efforts as designers can never foresee all the possibilities. In spite of all possible engineering finesse, a component will eventually fail or an unplanned situation will eventually arise.

When something does go wrong, letting your user know about it is critical. But just as important is keeping your users informed of the state of things in the ordinary course of operation in a way that is meaningful and actionable. An uninformed user may act unwittingly to cause a problem; conversely a user overwhelmed with information cannot discern and act promptly on important data. Alarms and alerts must also strike a balance. Users can get bored and inattentive in a vacuum, but frequent, low-priority alarms lead to complacency and bad habits.

Numerous transportation and heavy industry safety failures have been attributed to users either inappropriately disabling or sometimes literally fighting with safety systems. This has happened with stick-shaker and other incrementally introduced safety systems in aircraft. In some cases, automated systems can inadvertently hide information. Again, in aircraft, several cases have been documented in which an autopilot system has progressively compensated for a problem, such as flight surface icing, until it either fails suddenly or is manually turned off by a pilot who is instantly thrust into an unexpected critical situation.

Failures have also been attributed to simply having too many features and not enough information. An example of the second sort occurred in the 1990s with an implanted medical device designed to deliver dosed concentrated pain medicine directly to the spinal column. A laptop style controller was used to program the pump system via an RF link. The user interface was seemingly straightforward, with doctors able to simply tab down and adjust a list of operating parameters, such as on–off time, dose rate, and so forth. There are wide variations in delivery rates for different drugs in different circumstances and the device had multiple anticipated uses. For flexibility, the software allowed the users to change the units in which dosing was measured. However, the units were not adjusted as a matter of course and were not on the main operating screen. This led to a situation in which it was possible for an unknowing operator to program a dosage that was off by factors of 10 or 100 from his or her intention—a possibly life-threatening error. In this case "feature creep" had dangerously outpaced hazard analysis.

It's easy to provide information in software, but good user interfaces take hard work. Logical organization of the information and triaging of critical data are extremely important aspects of safety- and mission-critical software design.

Designers of high-performance aircraft were among the first to stumble down this road. Look at the "steam gauge" cockpits in the F-4 Phantom, 1960s jetliners, and most subsequent high-performance aircraft right up until the 1990s. More and more systems were added, each with its own control quadrant and display system. Training on new aircraft became more a process of learning to find and process information than it was of learning to handle the new flight characteristics. First multifunction panels, then critical-information–concentrating heads-up displays, and now modern glass cockpits have all been exercises in clearly using software to *reduce* information and convey knowledge rather than raw data.

Radar and navigation systems designers journeyed down a similar path, adding more and more elements to their display systems to an overload point where buttons literally labeled "de-clutter" were then added to the design. Next-generation designs took a general-purpose engineering approach, giving the user menus by which to select what elements to show or hide on the display. The newest systems are finally taking effective advantage of software by triaging what is displayed by its importance and by situational context (e.g., zoom in when approaching destination). This gives the user "soft" control over the situation, so that they can customize their experience without inadvertently shutting out critical information.

It is impossible to generalize every aspect of critical-system, user interface design. A jet aircraft, a satellite phone, and a nuclear power plant all have very different control and display requirements. However, there are a few critical common factors to all user interface design.

The first of these is efficient information conveyance. We have already touched on several aspects of efficient information conveyance in the examples above. It is important to concisely and clearly convey situational knowledge to the user rather than simply raw data. In some cases, this means designing effective display tools—a graphic speedometer or thermometer conveys information much more rapidly and concisely than a screen full of digital numbers.

In many cases it means combining information to convey situational awareness. A graphical position on a map with an error ring is far easier to process than longitude and latitude or range data from five separate systems. A single indicator that says "laser safe" is sometimes far more functional than a screen full of subsystem status information.

It is critical to work with your target audience to determine what information they need to know in what circumstances, and in what form they would find it most useful. Starting with what is already familiar is a safe bet, but don't fix on preconceived ideas. As designers and engineers, we are full of clever ideas and possibilities. Some of these may, indeed, be the way of the future; all of them need to face the cold light of day in actual user field-testing.

Responsiveness is another significant factor in both user experience and safety- and mission-critical effectiveness. Human reaction time can range from a few hundred milliseconds to over a second, depending on the task, age, and physical condition of the subject. However, we are anticipatory creatures who expect our interactions to be regular and predictable. Poor user interface experiences can degrade safety and increase mission risks.

Consider my cell phone—being a "smart" phone, it has a lot of capabilities and processing capacity. But this capacity is noticeably finite. The limitations become all too evident during boot-up. While the O/S merrily winds up, the phone begins searching for a signal, loading system components, and otherwise verifying its existence. The designers gave some modicum of thought to priority levels, reasoning that when voice communication is your mission, painting pretty pictures can wait. Hence the display is perceptively very low in the queue servicing. Trying to operate most functions during the first minute or two of spool-up is not a good experience—inputs are buffered and multisecond display lag means it takes great care and planning not to get lost in menu navigation. This is simply a hardware limit, but the designers were in one aspect thoughtful about this limitation. While navigation is slow, numbers entered on the keypad are nearly immediately reflected on the display. This means that even while the smart-phone features are bogged down and unusable, I can still make a basic phone call as soon as I have a signal. In this case, the responsiveness has been reasonably tailored to the critical mission: making phone calls.

Consider what information and control inputs are critical to the safety or mission of the system. Make choices that make it possible to meet these needs. This may seem obvious from the perspective of real-time system design: You must service priority elements in a time-deterministic fashion. But it is not always a leap that designers make with respect to the user interface in other circumstances.

Even when the mission is not hard real time per se, the system may include display elements that are time critical. These include warnings and alerts, as well as other safety or mission-critical situational awareness. In addition, general display lags can contribute to safety or mission failures simply through increased user error and frustration.

Flexibility can also be dangerous. As with the medical drug pump example, hidden information and diverse user options can lead to unsafe operation. In adding features, it is important to discern not "what can you do" but "what should you do." What features and display options will enhance the performance without compromising the core mission, compromising safety, or potentially misleading users into inappropriate complacency?

Each option demands its own risk assessment. If the option carries substantial risk and is unnecessary to operations, rethink it. If the option carries substantial risk and is necessary to operations, control access and clearly flag the dangerous operation mode.

For example, the drug pump "units" feature could have been placed in a restricted access menu, requiring a password to change. The units could have been flagged clearly on the usual operating screen, and called to the attention of any operator when not in a standard setting. Even better, the software could have examined other settings, such as the pump type and medication type, to provide a sanity check and flag unusual circumstances.

Consider a train system with automatic proximity braking designed to avoid collision. Clearly there will be circumstances under which the train may need to be operated closer to another train that the system would normally allow—perhaps in joining cars or during testing and maintenance cycles. Access to such an unsafe mode should be controlled, possibly with a physical key system, and warnings of the unsafe mode should be clear and persistent.

This leads us to alarm clarity. Problem notifications should be clear, accurate, and above all actionable. (See Example 2.6.) The user must understand immediately which alarms are advisory and which alarms are critical. Alarm clutter is a serious cause of safety issues. If an alarm becomes routine, it is not an alarm. Users become complacent and may fail to heed the same message when it is critical, or even fail to heed other messages because they are simply conditioned to clear the alarm and ignore salient information.

As mentioned previously, triage is important. Design and RA/HA review should consider how to rank and rate alerts and alarms. Those that are merely warnings can often be flagged clearly but less obtrusively so that users are not conditioned to ignore more significant alerts.

For example, rather than announcing when a non-critical sensor reading is out of range, it might be appropriate to merely highlight the problematic data field in another color. Some successful systems avoid alarm clutter by simply announcing that a non-critical warning exists and leaving it to the user to actively investigate the warning at their convenience by going to another screen.

Determining whether an alert is actionable can play a key role in triaging the event. If the user can do nothing about it, the detection is merely a data point, not an alert. On the other extreme, if the event is so mission- or safety-critical so as to demand instant action, the designer needs to carefully consider whether the system should automatically take action rather than merely begging the user for attention. The most important alerts are those that are safety- or mission-critical and require immediate human action.

Example 2.6. Error Messages Matter

The ubiquitous "check engine" light on the modern automobile has frustrated many an engineer. Why give such incomplete information? The simple answer is to not provide too much information to a user to whom it makes little difference. The errors indicated by your check engine light range from computer errors, to sensor failures, to timing problems and emissions control issues. None of these are correctable by the driver in the short term, nor are they critical enough to demand an immediate change in behavior.

Other "idiot" lights are more specific: "Check oil" demands almost immediate action. "Brakes," "hot," "gen," and "low fuel" demand fairly prompt attention. Give users what they can act on. In aviation, the critical notice is often reduced to "in op," simply red flagging an instrument

---

(Continues)

Example 2.6.   Cont'd

that can no longer be relied upon. The best notices are useful, actionable, and lack hidden assumptions that might cause an inexperienced person to aggravate the situation. Good user interface design anticipates the likely audience.

Give the user information that can be acted upon. Otherwise, give him or her an indication that service is required at their convenience, such as "check engine." Leave it to the technician (or the engineer with a good mechanics manual) to collect the detailed information and correct the subtler problems under controlled conditions.

Software in embedded systems is increasingly reliable. This was not so in the early days of building on-off systems on available hardware such as the IBM PC. In addition to detectable failure modes and code exceptions, these systems would also fail catastrophically because of charming idiosyncratic compiler bugs, operating system errors, or actual hardware burps. Thus it was important to include a catchall exception handler in the code so that the technician could gather information about what happened and try to avoid repeats.

In developing an experimental PC-based medical test system for controlling spinal cord stimulation, I came on just such a situation. The exception handler grew and grew to accommodate and recover gracefully from known issues; nonetheless, when a new error appeared, it needed to be tracked and fixed. From an operational use case perspective, this usually meant shutting down testing and starting over in a controlled fashion. In order to avoid startling or confusing the patient, we did this gracefully with the equivalent of a "check engine" light.

We have all made a few regrettable word choices, and here I must admit to one of mine. Imagine if you will, that you are a patient with an electrical device implanted in your body and you are left alone in your room to work with the computer. You are physically wired to the machine with an RF antenna lead taped to your lower back. You are experiencing strange sensations from the implanted device and probably a bit nervous about the whole situation. Things seemingly are going well when, suddenly, the screen goes blank and then announces, "Fatal error. Please call for assistance now."

In the era of the classic Microsoft Windows "blue screen of death, " we had done them one better. This message occurred exactly once in the field and, fortunately, the patient was computer savvy and very understanding. However, upon consulting with the medical staff, we rapidly changed that particular message to "*system* error" and released a new version.

5. Rolling Your Own

Whether there are tight real-time requirements, good embedded control systems share certain common properties and design aspects. You will benefit from understanding common patterns, and even to rely on an existing legacy systems or commercial RTOS. (See the following side bar.)

Side Bar: RTOS or Not RTOS

An RTOS, or *real-time operating system*, provides standard elements of embedded system operation that facilitate building real-time systems. The RTOS will typically include multitasking with minimal thread switching latency and give developers tools to guarantee deterministic timelines. These advantages can become stars in real-time operating situations.

But one person's "real-time" requirement is another's Sunday walk in the park. Human interaction typically demands multimillisecond response times, with, say, fly-by-wire flight controls requiring a tad tighter time cap than cell-phone display updates. In fact, one often speaks of *hard real time* versus *soft real time* in practical applications. Typically, hard real-time requirements exist when late completion of a task may lead to critical failure or useless results, while soft real-time requirements exist where the system can tolerate latency while degrading gently.

As illustrated by the mobile phone in my pocket, it is often acceptable to delay informational updates on the LCD in favor of boot-up and call processing threads. This does not make me a fan of the designer from a user-experience point of view, but I admit that I would rather have a slow-moving cursor than audio drop-outs and missed calls. In contrast, a fly-by-wire flight control may also degrade gently, at least over the span of a few milliseconds. But pilots operate to some extent by muscle memory and increasingly long or unpredictable dynamic response times will rapidly put the pilot behind the aircraft and degrade flight operations to a high-risk condition. Likewise, in a more down-to-earth example, I would not want my car's anti-lock brake controller waiting for me to adjust the radio.

Car-engine control systems, medical devices such as pacemakers, and systems that include software signal generation are other examples of systems that typically have tight, hard real-time requirements. Not all operations will have the same tight requirements. A good RTOS allows us to define priorities among threads and embed hard deterministic time limits and orders of execution.

So why would you not use an RTOS?

First of all, not all mission- and safety-critical systems have critical real-time aspects. Often, those systems that do have limited, critical time requirements can offload the requirements to separate components—either independent software or hardware.

For example, in working on a spinal-cord stimulation control system, we isolated the timing-critical elements in a separate subsystem. This embedded control device ran simple, tight code that served only the hard real-time aspects of the system, such as inter-pulse timing, energy adjustment, and emergency shutoff. The bulk of the system was soft real-time and resided on the controlling PC running a standard O/S. The O/S provided convenient tools for extensive user interface development, data storage and processing, and provided platform portability. The control system provided safety checks on operational parameters and included mission-critical functions, such as reliable interaction with the patient; however, none of these had hard real-time requirements that were not redundantly backed up by the separate controller subsystem.

This brings us to convenience. If your requirements dictate heavy user-interface development, a small footprint RTOS may not be the best starting point. If your requirements dictate use of

(Continues)

Side Bar: Cont'd

broadly available "standard" platforms for deployment, then an RTOS simply may not be available or practical to install. Licensing and royalty aspects may restrict your company's willingness to invest. RTOSs, like any other third-party components, are subject to changes and discontinuation, thereby adding elements of project risk. Finally, a full-blown RTOS may be overkill for many applications.

Also be aware that multithreading and interrupt-driven operating systems by their nature can introduce a host of other verification issues. For example, so-called *race conditions* may arise in which one process is unexpectedly critically dependent on the sequence or timing of other events. Such issues can occur where hardware responses or user input happen with variable time course on competing threads—they comprise one more reason to emphasize keeping safety-critical system code encapsulated and deterministic.

Some engineers designing smaller embedded systems or working on nonstandard-platforms decide to "roll their own" with respect to the core control code. This is sometimes due to lack of familiarity with the RTOS options; but often it is a carefully considered design choice balancing the system's real-time needs, interaction complexity, investment costs, and learning curves.

Classic time-sharing designs switch between task threads on a regular round-robin basis. When the entire system is yours, time-sharing designs are generally switched on a periodic clock interrupt. Each thread is programmed as if it has sole use of the machine and interaction is based on semaphores and process blocking.

More modern event-driven designs switch threads based on priority, with the occurrence of a higher-priority event preempting a thread in progress. The approach is natural, since many embedded systems respond to real-world dynamics; however, it requires very careful consideration of interrupt interactions, since a process requiring service can make only limited assumptions as to the state of the system or the sequence in which it was triggered.

On very limited processors, it is quite reasonable to simulate aspects of these O/S design philosophies with simple tight loops and switch statements. Consider the flow chart in Fig. 2.4. After a start-up process, the main loop executes continuously until the user shuts it down. Each time through the loop, the process checks the status of a list of system-state variables and acts if they are true. This simple loop is the equivalent of a round-robin list of tasks with equal priority. Setting a flag variable blocks or unblocks a particular task and allows operation. Each of the side-chain processes act as threads that can perform tasks or can set and clear request flags. These system threads interact primarily by setting state and parameter variables that are examined and acted on in the next pass-through.

Provided that no piece of code traps execution, this simple design can actually deliver deterministic real-time behavior. The danger, of course, is that without preemptive interrupts,

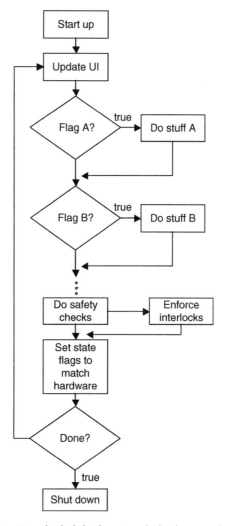

Figure 2.4: A typical tight-loop, switched operating design.

a bug or unexpected condition in the sequential code may lead to lockup. The design also lacks flexibility in task ordering.

We can make this design pattern more flexible with dynamic scheduling. In Fig. 2.5, we replace the fixed switch code order with a tasking process that acts on a sequential list of scheduled events. On each pass, a new event is pulled from the list and the queue trimmed (or a pointer advanced). Switching code processes each active event; this code executes one or more subroutines to handle the event and perform requisite functions. Interaction between task threads can still include semaphore flags; however, these tasks flags can also modify the queue by adding, subtracting, or resequencing upcoming event requests.

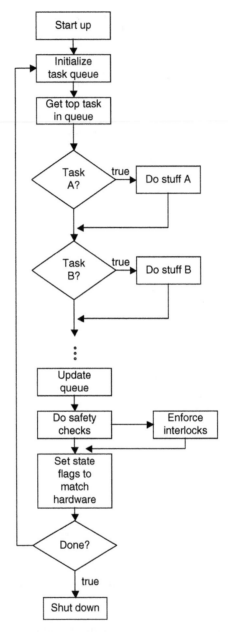

Figure 2.5: A typical tight-loop, queued operating design.

Note that including parameters for the queued tasks in the same data structure is only a small addition, so that the system can vary details of scheduled events. One of these parameters may even be an execution start time, so that events are delayed and left on the queue until a system timer value exceeds the passed parameter. Depending upon implementation, the "update queue" process may be as simple as advancing a pointer or clearing the top item;

however, it might also include a quick sorting operation that tests requested execution times of newly added items and ensures proper priority sequencing.

In practice, the actual switching code will depend on the programming language and the processor: It may be a switch statement, it may be a list of conditional tests, or it may be a jump to an executable memory location. This detail is not important to the design pattern.

Scheduled events will typically include most operational tasks. Interactions can be quite complex: One event may schedule another event, change operational parameters, or remove one or more events from the queue. Note that one can easily schedule and queue some of the same tasks that are in the round-robin test list—taking advantage of the independent encapsulated code to handle both cases.

This brings us directly to the applicability of this design pattern in mission- and safety-critical systems. As shown in the figures, the fixed round-robin event section is a natural place to embed safety checks and interlocks. They are always executed, on each loop, no matter what task is processed from the queue. However, nothing prevents a designer from calling one of these safety checks, interlocks, or other operations within the operational code during a queued task. This permits re-application of heavily tested safety-critical routines that have been established early in the design process. It provides redundancy at the level of code reuse as well as critical crosscheck redundancy at the level of system operation. Memory overhead is minimal since critical code is encapsulated in subroutines. Reviewers can check each task for safety steps and separately confirm that the main loop always contains backups.

If one considers the scheduled events to be ordinary operational processes, it becomes natural to look at fixed round-robin tasks, such as safety checks and hardware state verification, as the equivalent of "interrupts" in a more advanced operating system. These conditions are tested on every loop with finite time latency. If true, they are acted upon. As with true hardware interrupts, good design dictates that these fixed events are quickly executed to minimize delays before returning to ordinary system interaction.

In some cases, rather than executing a long subroutine, a fixed event might instead test a hardware condition and then queue an appropriate action event to process the information through ordinary scheduling. Judgment must be made as to where to encapsulate the safety-critical actions, making sure to include those subroutines in the critical path testing, verification, and freezing process during development.

This section is intended to provide guidance on how one might incorporate fail-safe design patterns, even in very basic, homegrown embedded software. If your embedded system incorporates an O/S that already provides much of the event-handling architecture, the key failsafe ideas of encapsulated independence and redundancy should be retained and adapted to that structure. The choice of whether to rely on such solutions or to turn to a full-blown RTOS that handles deterministic queuing and multithreading is up to the engineer.

6. Hardware as Software: A Thought Exercise in Crossover Thinking

Both hardware and software engineers can benefit from crossover thinking when it comes to failsafe software design. Let us consider the microwave oven as a tangible example. We will start by considering the hazards and failure modes, look at how they are typically addressed in hardware, and then, as an exercise, consider how we would mimic the hardware design in solid traceable software. One would *never fully eliminate hardware interlocks* in a production appliance, simply because hardware can always fail in a manner that is undetectable in software. However, this thought exercise illustrates good practices in design that can be applied where software is required to play a major role in a mission-critical system.

First, let us carry out an abbreviated risk analysis of our basic microwave oven.

Failure Mode	Hazard Level	Design Action
Line power surge	Low	Inline fuse
Internal short	Moderate	Inline fuse
Oven cavity overheat	Moderate	Thermal protection
Magnetron overheat	Moderate	Thermal protection
Energized with door ajar	High	Door interlock switch
Interlock switch failure	High	Interlock monitor switch
Interlock double failure	High	Secondary interlock system Disable control logic Warn user about door ajar

In addition, our usability requirements will include deciding under what conditions to allow the operations control logic to operate, and when to make an error condition apparent to the user so that it may be resolved.

Figure 2.6 illustrates a realistic hardware circuit that addresses the failure modes above. It is somewhat simplified for this discussion, but you will find that ovens in many kitchens comply with this basic model.

In brief, a fuse and two thermal protection devices cover the "low" and "moderate" risk-failure hazards by providing direct interruption of the primary line power to the device. Energizing the magnetron with the door open is an extremely high-risk failure and is addressed with redundant interlock systems.

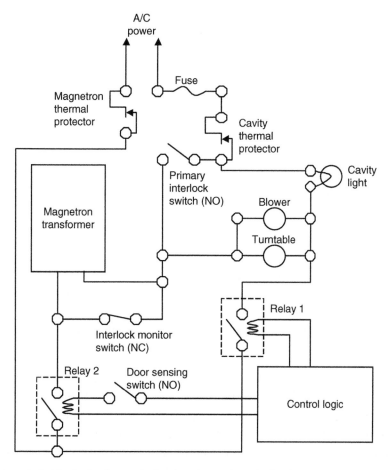

Figure 2.6: Typical schematic for hardware interlocks in a microwave oven.

The primary interlock switch is a mechanical door interlock that is open while the door is ajar. It prevents line power from reaching the magnetron transformer with the door open. Because it is possible that this switch could fail closed (perhaps due to age-related failure or simply dried food wedged in the wrong place), there is a second interlock monitor switch that shorts the transformer while the door is ajar. These two opposite acting physical switches provide primary protection against software errors, wiring shorts, physical abuse, and other significant failure modes.

It is possible, however, unlikely, that both of these switches might fail—one fused shut and the other stuck open. Low probability aside, the results would be very high hazard so the engineer has also incorporated a secondary interlock. This is accomplished with another, physically separate door-sensing switch. The sensing switch informs the control logic unit of the door state. In addition, relay 2, used by the control logic to energize the magnetron under proper conditions, is physically wired through this same switch, providing a final backup to software failure. Note the unabashed redundancy.

In ordinary operation of the oven, the control logic software, sensing the door ajar, disables keypad input so that (1) the start button cannot be activated and (2) the user has feedback that the oven will not function. Additional user feedback is provided by the cavity light, which remains lit until the door is closed. This provides an indication to the user as to why the keypad will not operate; more elaborate software might augment the door light by displaying "door ajar." The error indication is clear and actionable.

There is significant redundancy built into this system. In fact, we could eliminate every mechanical switch but the door sensor and the basic features of the oven would operate correctly. Even though the boss might like to save a little money on switches, it is unlikely that a hardware engineer would be considered wasteful. The liability for a field failure is huge. Likewise the critical system software engineer should not look ill upon a few extra lines of code that add clarity and resilience to the design.

Note the labeling of the components. We consider the direct hardware interlock as the *primary* failsafe protection, and the ordinary operational sensor as the *secondary* system. Many software engineers not versed in mission-critical programming would address this system design in the opposite direction. That is, they would start at ordinary control operations, and then consider detection and notification of error states, and finally consider how to address failures outside their control. This forward approach is dangerous because the programmer cannot help but tacitly assume that his or her code will basically work correctly and that it can and will be fully tested and debugged. Yet forward thinking is entirely natural to the engineer because the control logic and the operational design features are the satisfying part of seeing a project come to fruition—the fun part of software.

Instead we make "safety first" our design approach. We give some real thought to "when all else fails," even before spending much time on the "all else." Then, with basic safety fallbacks in mind and set aside from the operational aspects of the project, we move ahead to the more challenging feature sets.

Let's begin. We will build a basic code structure that mimics our microwave schematic while using principles of testability and encapsulation to isolate and control risks. For purposes of the exercise, we assume that we can replace each circuit-switching device with a software-controlled relay and a sensor. The schematic in Fig. 2.7 shows the idea. Each sensor can

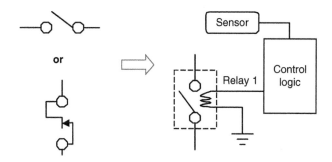

Figure 2.7: Imagine replacing each physical switch with a software-controlled sensor/relay combination.

be read and each relay set directly by control logic via a simple applications programming interface (API).

Such APIs are typical and there is no sense dwelling on the bit-twiddling aspects of this example. More to the point, there should be no tolerance for propagating arcane code lines in a mission-critical project. So we begin by encapsulating our hardware operations in meaningful functions. In pseudo-code, our basic operations for the primary interlock system are:

```
PrimaryInterlockSensorDoorClosed()
    if (primary interlock door detect sensor register) = true then
        return true
    else
        return false
    endif
```

to read the primary interlock door sensor,

```
PrimaryInterlockRelayClosed()
    if (test primary interlock switch state) = (closed) then
        return true
    else
        return false
    endif
```

to check the primary interlock switch relay status,

```
SetPrimaryInterlockSwitch(setstate)
    if setstate = true then
        (set primary interlock switch closed)
        wait (delay)
```

```
            if PrimaryInterlockRelayClosed() = true then
                return true
            else
                return false
            endif
    else
            (set primary interlock switch open)
            wait (delay)
            if PrimaryInterlockRelayClosed() = false then
                return true
            else
                return false
            endif
    endif
```

and to set the primary interlock switch. Similarly we can define other basic operations. These are simply listed in the following:

```
SetInterlockMonitorSwitch(setstate)
        To control the interlock monitor switch

DoorSensorDoorClosed()
        To read the secondary door sensor

SecondaryInterlockRelayClosed()
        To check the door sensor relay switch status

SetSecondaryInterlockSwitch()
        To control the door sensor relay switch

SetMagnetronEnergizeSwitch(setstate)
        To control the Magnetron relay 2
```

The exact naming is not critical, but we emphasize clarity. Here we have used basic camel-back notation to label each function in long form with a meaningful name. Moreover, the name clearly defines what the stated response means, so we need not resort to poring over the documentation every time we call a function. A conditional of the form,

```
    if DoorSensorDoorClosed() then
            Do the right thing
    else
            Make a fuss
    endif
```

is clear at a glance. *Functions* become *methods* in object-oriented languages, and will look a little different. Those who prefer typed variable-naming conventions will create even longer names. Regardless, the meaning should be transparent to you, to your code reviewer, and to whoever is gifted with the task of maintaining your code two years from now.

Each of these base-level routines can be tested in isolation to verify proper operation. For example:

```
PrimaryInterlockDoorTest()
     do
             If PrimaryInterlockSensorDoorClosed() the
                  (report door closed)
             else
                  (report door open)
             endif
     loop
```

This trivial code allows basic testing to ensure that the door sensor is reading the door position correctly. As part of a verification plan, simple programs may be designed to exercise each low-level hardware routine while asking the tester to change the door state and record results. This auxiliary code is the equivalent of a test harness in the hardware world. Again, encapsulation lies at the heart of the "design-for-test" concept.

We can encapsulate the primary interlock system operation as follows:

```
PrimaryInterlockCheck()
     if PrimaryInterlockSensorDoorClosed() then
             switchOk = SetPrimaryInterlockSwitch (true)
             if switchOk = false then
                  trap hardwareerror
             endif
             switchOk = SetInterlockMonitorSwitch (false)
             if switchOk = false then
                  trap hardwareerror
             endif
     else
             switchOk = SetPrimaryInterlockSwitch (false)
             if switchOk = false then
                  trap hardwareerror
             endif
             switchOk = SetInterlockMonitorSwitch (true)
             if switchOk = false then
                  trap hardware error
             endif
     endif
```

The function returns true if the door interlock switch is properly closed, and false if it is not. Physical interlock switches are set appropriately. If physical hardware fails to respond in a verifiable way, the routine does not return at all; rather it executes special-purpose error handling code.

The line "trap hardware error" is shorthand for abandoning ordinary operation in favor of a catchall operation that takes care of detectable hardware failures. This is sometimes called "throwing an exception," developed further in discussion on "Exception Handling" later. In a safety-critical system, this will generally result in an attempt to return everything as quickly as possible to a known safe state.

Every time this function is executed, the system tests the interlock sensor and sets the two control switches appropriate to the door state. Now we ask the question, how often do we execute this? Remember, the door itself is outside your control. Your practical options depend on your O/S and processor. Three options in decreasing order of desirability are considered:

Interrupt Driven: This critical operation is an obvious candidate for a hardware interrupt. If the primary interlock door sensor–state change can trigger a hardware interrupt, then `PrimaryInterlockCheck()` can be executed at each interrupt to change the relay states.

Schedule Driven: If the O/S supports multithread or scheduled execution, then `PrimaryInterlockCheck()` can be executed periodically on an independent thread from the primary operations software. "Periodically" must mean frequently enough to provide protection; hence, in this case, it will likely be on the order of no more than a few milliseconds between loops.

Main Loop: In very basic microprocessors, for simple tasks, one often relies on a tight main loop that tracks and triggers other operations. In this case, we would include the code:

```
do
    PrimaryInterlockCheck()
    query keypad
    ...
    do otherstuff
    ...
loop until shutdown
```

The third method is common in "roll-your-own" real-time designs. We consider this common design pattern further in section 5. It is risky, in that any glitch or bug in the "otherstuff" could freeze or delay the main loop, causing a safety hazard. This choice should be coupled with very careful version control, since any modification to the code that results in a new bug could create a severe safety hazard.

If forced to use method three, one should immediately consider other places to add redundancy. For example, one presumes that somewhere in "otherstuff," there will appear commands to turn on the magnetron; if available code space allows an extra call to `PrimaryInterlockCheck()`, why not be safe?

Independence is key: *Safety critical functions should be designed, tested, and executed independently of control operations code.*

Methods 1 and 2 have the advantage, in that they will better survive code changes and new bugs in the operational code. They are by no means immune from the software failure modes and should also include strategic redundancy. However, with method 3, we are simulating periodic interrupts; hence adding redundant checks at other frequented portions of the control code should be considered, particularly where you expect longer latency. Function over form in this case, as safety is never inelegant. For a personal example, see Example 2.4.

Keep in mind that this is a thought exercise and there is a fourth method, illustrated by the original schematic diagram: analog hardware. No software engineer should consider his or her system beyond fallibility and, in practice, one would not recommend implementing this or any safety-critical system without hardware interlock redundancy. Every critical-system software engineer should study the Therac-25 radiation therapy device failure (see Example 2.8) as inspiration.

Moving down the safety ladder, let us consider the secondary interlock system:

```
SecondaryInterlockCheck()
    if DoorSensorDoorClosed() then
        switchOk = SetSecondaryInterlockSwitch (true)
        if switchOk = false then
            trap hardwareerror
        endif
    else
        switchOk = SetSecondaryInterlockSwitch (false)
        if switchOk = false then
            trap hardwareerror
        endif
    endif
```

As with the first interlock check, it makes sense to run this secondary code frequently in an interrupt-driven fashion or a scheduled thread. If we do not have that option, then, again, redundancy doesn't hurt. For example, we might write simple code to start cooking as follows:

```
TurnOnMagnetron()
    PrimaryInterlockCheck()
    SecondaryInterlockCheck()
```

```
if DoorSensorDoorClosed() then
        SwitchOK = SetMagnetronEnergizeSwitch(true)
        if switchOk = false then
                trap hardwareerror
        endif
        magnetronActive = true
else
        magnetronActive = false
endif
```

As used here, `magnetronActive` is a state variable. This could correspond to a global variable in function-based code, or a property of the Magnetron class object in an OO language like C++. It provides convenience for our control code, allowing us to determine the state where we think that we've placed the Magnetron. In fact, using the principles we have established, it would be better to make `magnetronActive` a function that actually tests the state of the magnetron. In this fashion, there is no way we could mislead ourselves into believing that the magnetron is energized when it is off, or (worse!) that it is off when in fact it is energized. Assuming that our hardware allows us to read back the register that controls relay 2, we can improve our design.

```
MagnetronActive()
        if (test magnetron switch state) = (closed) then
                return true
        else
                return false
        endif
```

Even better would be to get independent secondary information, such as a direct measurement of current to the magnetron or direct detection of microwaves.

Example 2.8. The Therac-25 Nightmare

The Therac-25 radiation therapy debacle has become a standard case study in failure modes of software-based safety-critical systems. Between 1985 and 1987, multiple patients were given radiation doses roughly 100 times higher than intended as a result of an undiscovered software bug. At least six of these overdoses are known to have resulted in severe radiation poisoning and three of the patients eventually died.

This therapy system was feature-rich, with multiple distinct operating modes. In particular, the machine was able to deliver either short-duration direct electron-beam therapy or diffuse x-ray therapy. In order to produce x-rays, the system used the same electron beam at high energy directed against an emitting target. Radiation therapy devices represent an enormous capital investment and this was a clever engineering idea to maximize available treatment options

---

within a single unit. Secondary mechanisms were used to diffuse and collimate the beam to achieve properly controlled dosing for each mode.

The two operation modes required that diverse components be physically moved into and out of the beam path by mechanical actuation. Open-loop control was used to configure the mechanical components. Eventually, a hidden software bug emerged: when certain atypical user-interface entry sequences were made, the unit was temporarily misconfigured, sometimes resulting in a high power beam directed at the patient without being diffused to safe levels.

The Therac-25 software failed to verify hardware conditions. In fact, it has been reported that the open-loop control code had been retained from previous system designs in which hardware interlocks prevented such malfunctions; however, there were no hardware interlocks in the Therac-25. The problem triggered large-scale rethinking of the role of software and software process in safety-critical systems.

Not only was code inappropriately reapplied, the overall architecture was glued together in a way that made contextual verification very difficult. The designers probably relied on past field experience, unaware of having inherited hidden critical dependence on the hardware interlocks.

A chain of design failures contributed to the unsafe system, a list of which reads like a textbook nightmare example. Convenient features were prioritized compared to hazard analysis. The software was not independently reviewed. Borrowed code was reapplied without risk assessment. The hidden bug was triggered by rare keyboard entries, making detection in V&V unlikely and discovery and tracing in the field difficult. The machine detected irregularities in beam energy and shut down, but did so too slowly to protect the patient. Upon shutdown, only obscure alert codes were presented to the operator, offering no actionable information and encouraging complacency.

Each design failure compounded the others and, in not following good practices, the resulting system was not failsafe, user tolerant, nor engineering-flaw robust.

Numerous detailed reviews of the Therac-25 experience have been published, and we strongly encourage the reader to locate a few of them to help drive home the importance of good process and design strategies. A convenient summary with a list of references is available online at *en.wikipedia.org/wiki/Therac-25* and *en.wikipedia.org/wiki/List_of_software_bugs*.

Verify reality: *Avoid mirroring hardware in state variables when the state is directly testable.* In some cases you are at the mercy of the hardware or the physical system. The goal of the mission-critical software engineer, however, should be to ascertain and verify the actual state of the physical hardware insofar as possible so that critical decisions are not made on faulty assumptions.

Following similar patterns, we can write straightforward code to test thermal conditions. For example,

```
MagnetronTempCheck()
      if GetMagnetronTemperature() > 135c then
            SwitchOK = SetMagnetronThermalSwitch(false)
            if switchOk = false then
                  trap hardwareerror
            endif
            return false
      else
            SwitchOK = SetMagnetronThermalSwitch(true)
            if switchOk = false then
                  trap hardwareerror
            endif
            return true
      endif
```

Similarly, we can create a `CavityTempCheck()` function.

Leaving power surges and shorts to the good old hardware fuse, we now have four scheduled tasks to keep track of our safety state. We can bundle them for convenience:

```
SafetyStateCheck()
      PrimaryInterlockCheck()
      SecondaryInterlockCheck()
      MagnetronTempCheck()
      CavityTempCheck()
```

Finally, we can provide an encapsulated crosscheck routine that quickly verifies the state of all hardware and sensors so that the operational code can remain coupled to the state of the hardware and does not rely entirely on state variables.

```
SafeToOperate()
      If    PrimaryInterlockSensorDoorClosed()   and
            DoorSensorDoorClosed()               and
            PrimaryInterlockRelayClosed()        and
            SecondaryInterlockRelayClosed()      and
            GetMagnetronTemperature() < 135      and
            GetCavityTemperature() < 120               then

            return true

      else
            return false
      endif
```

Note what we have achieved. We have clearly named functions that perform our safety checks, bundled outside of the main control code so that they can be individually tested and readily verified by visual inspection in code review. Requirements generated by the hazard analysis are directly traceable to code.

Each function takes care of ensuring safety internally. Certain safety checks may return a status flag to the calling routine for purposes of notifying operating software that it is okay to continue to interact with the user. However, the actual safety of the system is not jeopardized even if these flags are ignored.

Now, with our safety-critical elements bundled, we can turn to implementing the control code. To keep things simple, let us assume this microwave oven has only one button: 60-second heat. Two mission-critical requirements emerge, one of which overlaps with our safety checks. These are:

1. Cook as close as possible to 60 seconds.

2. Interrupt if the door is opened.

Let us assume for maximum programmer hardship that we do not have interrupt or thread-based O/S support. We might code this sequentially as follows for our cooking checks first:

```
CookingChecks()
     timerCnt = timerCnt−1
     if timerCnt = 0 then
          TurnOffMagnetron()
          if MagnetronActive then
               PanicOff()
          endif
     endif
     if DoorSensorDoorClosed() = false then
          TurnOffMagnetron()
          if MagnetronActive then
               PanicOff()
          endif
     endif
```

Note we have put a direct crosscheck on the state of the magnetron. If it fails to go off when requested, we execute a panic routine that attempts to force the system to a known safe state. This is discussed further when we get to exception handling.

Next we code our tiny user interface:

```
ButtonCheck()
      if (button pushed) then
            SoundKeyPressBeep()
            TimerCnt = 60 (time factor)
            SafetyStateCheck()
            If SafeToOperate() then
                  TurnOnMagnetron()
            else
                  (complain to user)
            endif
      endif
```

"Complain to the user" in a simple system could be just a feedback beep that differs from the key-press sound, or could be a flashing error message. If the hardware allows, then longer explanations are more desirable. However, in the event of a dangerous condition out of keeping with expected events—such as a stuck relay or sensor—the designer should seriously consider disabling the interface in a manner that requires technician intervention. This would prevent a frustrated user from ignoring the alarm and possibly escalating the alarm condition into a situation that compromises safety.

Our main operation code is fairly simple:

```
Main ()
      If  SelfTestEmbeddedCodeOK() = false then
            (complain to user)
            end execution
      endif
      If  SelfTestHardwareOK() = false then
            (complain to user)
            end execution
      endif
      SetInitialHardwareState()
      do
            SafetyStateCheck()
            if MagnetronActive then
                  If SafeToOperate() then
                        CookingChecks()
                  else
                        PanicOff()
                  endif
            else
                  if DoorSensorDoorClosed() then
                        CavityLightOff()
                        ButtonCheck()
```

```
        else
              CavityLightOn()
        endif
    endif
loop
```

On cold re-start, the system first established that the software is uncompromised and that each testable hardware component and register responds properly—taking care, of course, not to exercise components during self-tests in any way that could itself compromise safety. Once we verify the system, we set initial hardware states and enter the main execution loop. Since this is an always-ready cooking appliance, there is no need for an explicit shut-down. That will occur either when the plug is pulled or an exception is detected.

We began with the safety-critical elements and then constructed the operational code. Note that we reused hardware status tests both in the safety-critical code and in the operational segments, simplifying implementation. We also make safety check calls at each level of the code, sometimes redundantly in the sense that we make similar checks at different code levels that might, when unwound, result in duplication in the sequence of events.

However, the duplication involves only the re-execution of tiny code segments and will not detract noticeably from response feel. Advantages include easy review of the code, since safety steps are visible at a glance at each level. Failure modes in our RA/HA chart can be mapped directly to independent, encapsulated code during critical safety reviews, while operational code can be verified against design operating sequences without nervously working through possible safety compromises at every software update. The redundancy also provides protection against future bugs introduced in the operational code and even some parts of the critical code segments.

Finally, a note on exception handling: In aviation, pilots have a general mnemonic phrase to keep priorities straight during emergencies. The phrase is, "Aviate, navigate, communicate." The same principles apply to an emergency situation in a critical-system design.

When something goes wrong, priority one is to get the operation of the system under control and into a known safe state. The first step in this example would be to execute a `PanicOff()` operation. In practice, the difference between ordinary "off" routines and the panic routine comes down to assumptions about hardware compliance. The detected failure could represent a transient symptom of an intermittent hardware issue. Rather than simply shutting down the magnetron by the usual means, it is advisable to issue API commands multiple times to each available interlock switch. Do so in a loop that tests for compliance and tests sensors for verification.

Once the hardware is safe, then attempt to sort out the device status and whether it is safe to do anything other than remain shut down. Next, communicate the situation to the user.

If the shut-down loop times out without reaching a verified safe state, take all reasonable measures to otherwise shut down the system. Carefully consider this use case during design of the safety-critical elements: How long should you wait to notify the user? How can you notify the user clearly and effectively to achieve desired action? Generally your software can act far faster than any user can react; however, there are circumstances, such as precautionary personnel evacuation, where it will be advisable to sound the alarm in advance of trying to solve the problem and then communicate details later. Obviously, if active user intervention is required, notification then becomes the safety shut-down step.

Again, this thought exercise is not intended to show useful design patterns. It is not intended to be realistic, as one would never produce a microwave oven or any other dangerous physical device without including additional layers of hardware interlocks. Our purpose is to try to connect safety-first, failsafe redundant hardware engineering practices to similar possible practices in software.

Example 2.9. Telephone Tag

What about that perpetual "new call" light in my kitchen? I've never seen the firmware for my wall phone, but I'd wager a week of Venté Mochas that the error is conceptually similar to that described in Example 2.4. A state variable of some sort (perhaps simply the register that controls the light) is intended to mirror reality. When it gets out of synch with reality, there is no periodic verification, and hence, in contrast to the brain stimulator, there is no automatic means of correcting the problem. Fortunately, this is not a mission-critical function of the phone; it is merely annoying.

A few years of sporadic test data reveal that the stuck "new call" light often occurs when the phone ring is interrupted before an incoming caller ID code is fully processed.

Let us fix these ideas with an example. Consider a block of code triggered by the ringing phone:

```
OnRing()
   ...
setNewCallLight
   start RingerSound Thread
   start GetCallerID Thread
   ...
```

and somewhere in the GetCallerID thread code:

```
getInComingCallerID()
    ...
    do until last byte received
        get data
        accumulate CallIDstring
        if waitTimeOut then
            return "No Caller ID Available"
        endif
        ...
    loop
    ...
    addCallerIDtoCallList(CallIDstring)
    newCallCount = newCallCount + 1
    return CallIDstring
```

Most of the time the phone will ring and a fully processed result will be added to the call list. On this assumption, the engineer probably delegates clearing of the "new call" light to the call review operation: each time a call is viewed in the list, the individual "new" flag is cleared and the newCallCount is decremented. If newCallCount hits zero, then the clearNewCallLight is executed. However, if the newCallCount = 0 already, then clearNewCallLight is never executed. A classic coding error is to use too specific a conditional:

```
...
newCallCount = newCallCount−1
if newCallCount = 0 then
    clearNewCallLight
endif
...
```

The problem, of course, is that if newCallCount = 0 already, then the decrement (depending upon the arithmetic type) either takes the counter negative or wraps to the highest possible integer. With all individual "new" flags cleared, the counter is now misaligned with the actual number of new calls, and the counter light will likely never be cleared no matter how many times we cycle through the review list.

How could this happen? Caller ID is generally transmitted over the beginning of the first ring period; if it is possible to pick up the phone or otherwise interrupt the process between turning on the "new call" light and incrementing the newCallCount, then we will have triggered the intermittent bug.

This is only one concocted scenario and there are ample other possibilities. Perhaps the "new call" light-circuit switch simply intermittently fails to respond and the code never revisits turning off the light once the internal state reflects an off condition. We can be sure that it is not a pure hardware error, such as a stuck gate, since clearing all calls from memory with a "soft reset" reliably switches off the light.

(Continues)

Example 2.9. Cont'd

It does not so much matter what is failing—an interrupted firmware thread, or a slightly sticky bit. Ultimately, the engineer has made an assumption that the state of the light will always remain in synch and has failed to subsequently verify reality in the code.

Defensive programming possibilities abound: forcing the indicator light and program state to synchronize at any regular point in the phone operation could have easily provided a catchall check. Certainly the user expects the new call light to go out when all calls are cleared. Why not make a quick count of "new" call flags in the list as the software is leaving "call review" mode, and simply force the light "off" whenever that count comes up zero? Perhaps that is too time consuming for the tiny embedded processor; then why not at least force the light "off" when the user scrolls back to call number #1 in the review list? If the user has reviewed all the calls in the list, then he or she has obviously seen them all.

This is a benign example of where good coding practices could improve the user experience. Obviously, our curmudgeonly light is not safety sensitive, nor is it mission critical to the operation of the telephone. But it certainly was not the intent of the architect to force phone users to exercise the soft reset—deleting all incoming call records—on a regular basis. Lack of redundancy and lack of provability led to a low-occurrence failure mode that escaped testing. By providing independent secondary code elements as a matter of course one can easily design in low-overhead redundancy that ensures a reliable user experience.

---

7. Conclusions

We discussed three principles that drive all aspects of the critical-system development process that persist through design, implementation, and testing. Our ultimate goal is the last of these: provability. To achieve it we need to carefully apply the other two, and broaden our view with respect to how we think about failure modes, how we think about the fallibility of our system components, and how we think about the fallibility of human designers.

The second principle, redundancy, like the second law of thermodynamics, is perhaps the most significant in obtaining useful results. Strategic redundancy at all stages of the design process and in the code architecture underlies consistently successful critical-system development.

Critical-safety functions should be designed, tested, encapsulated, and executed independently of control operations code insofar as possible. Design these elements first, begin with a carefully thought-out RA/HA matrix that includes hardware, environmental, software, and user-associated risks. Verify the critical backbone subsystems independently before focusing on integration with operational code. To the extent possible, meet specific needs with specific, traceable code. Think of it as defensive programming.

Where possible, always design systems to verify reality: avoid open-loop designs and mirroring hardware with state variables when the physical state is directly measurable. Consider the user and the user interface in terms of how they affect the mission risks and establish explicit use cases; make sure that information conveyed is necessary, clear, and actionable and that feature creep does not overrun function.

Even non-critical embedded systems can benefit from some of these design patterns. All embedded software has a mission; predictably, achieving that mission is what makes a good-quality product. (My kitchen phone still insists I have a new call.)

Failsafe software design means acknowledging a broad spectrum of downside threats and possibilities and carefully bounding the risks. No human engineered system is inherently infallible and we should not design systems on the expectation that we can make them so. Instead we can engineer for anticipated failure modes insofar as possible and for unanticipated failure modes through built-in redundancy and safety factors. By taking considered steps and exercising good practices, we can create critical systems that are robust to physical component failures and unplanned situations, highly user tolerant, and even forgiving of the system's own potential hidden flaws.

CHAPTER 3

Compliance Concerns for Medical Equipment

Brian Biersach and Jeremi Peck

1. Introduction

Medical equipment is highly regulated by all countries, and held to a higher level of safety than nearly all other equipment on the market. The main reasons for this are that medical equipment may be used on patients who are not able to respond to hazardous conditions or pain; an actual electrical connection between the equipment and patient may exist; patient treatment may be based on the output of the medical device; and certain types of medical equipment function as life support, the failure of which could result in the death of the patient. Understanding these requirements early in the design process will result in lower product development cost, faster certification turnaround, and increased product safety.

2. National and International Requirements

2.1. U.S. Requirements

In the United States, the Center for Devices and Radiological Health (CDRH), a branch of the Food and Drug Administration (FDA), separates medical devices into three categories (Class I, II, or III). This separation depends on the degree of regulation necessary to provide a reasonable assurance of safety and effectiveness. The FDA Federal Food, Drug, and Cosmetic Act requires that all medical devices be "safe and effective," and recognizes safety standards as a means to support a declaration of conformity. You can find a list of the FDA-recognized consensus standards at www.accessdata.fda.gov/scripts/cdrh/cfdocs/cfStandards/search.cfm.

Class I devices are subject to the following requirements: premarket notification, registration and listing, prohibitions against adulteration and misbranding, and rules for good manufacturing practices. Class II devices additionally need to show that they meet recognized consensus standards. Class III devices have the same requirements of Class II devices, but require premarket approval from the FDA.

The collection of documents that are submitted to the FDA for Class I or Class II devices is called an FDA 510(k). This shows that the above requirements are met and that the medical device is substantially equivalent to a medical device that was either in commercial

DOI: 10.1016/B978-0-7506-8567-2.00003-2

distribution before May 28, 1976 or has been reclassified as Class I or II. The FDA 510(k) submission is examined by the FDA or an accredited third-party reviewer to determine whether the device is substantially equivalent to the specified predicate device and whether it meets the specified recognized consensus standards. If the device is found to be substantially equivalent, the FDA accepts the claim of substantial equivalency; it can then be legally marketed and sold in the United States.

If the device is not substantially equivalent to a predicate device, due to new technology or differences in intended use, the medical device is considered to be Class III. For a Class III device, the company must present detailed information, such as clinical trial data, statistical data, and compliance evaluation and testing results to the FDA to show that the device is safe and effective. Only the FDA may review Class III submissions; if the FDA finds the information and data adequate, it will grant premarket approval for the device, allowing it to be legally marketed and sold in the United States.

In addition to the legal requirements of the FDA, many "authorities having jurisdiction" (AHJ) and purchasers of medical electrical equipment in the United States require a safety certification mark on the equipment. A safety certification mark is issued by a nationally recognized testing laboratory (NRTL), such as Underwriters Laboratories Inc. (UL), Canadian Standards Administration (CSA), Technical Inspection Association (in English) (TUV), or European Test Laboratory (ETL). The NRTL program is run by the Occupational Safety and Health Administration (OSHA); it recognizes private sector organizations that meet the necessary qualifications specified in the regulations for the program. The NRTL determines that specific products meet consensus-based standards of safety to provide the assurance, required by OSHA, that these products are safe for use in the U.S. workplace.

Underwriters Laboratories is the major product-safety certification organization in North America. Manufacturers of medical equipment submit product samples and information to UL for evaluation to the applicable safety standard(s). Products that meet these requirements are authorized to apply the UL Classification Mark. The applicable standards are UL 60601-1 and International Electrotechnical Commission (IEC) 60601-1 collateral and particular standards, which are discussed later in this chapter.

2.2. European Requirements

All but low-risk, non-measuring, non-sterile medical devices placed on the market in Europe must bear the CE Mark with a notified body's identification number. This mark specifies that the medical device complies with the requirements of the Medical Device Directive (93/42/EEC). A notified body is a third party designated by European authorities to assess products to directives. The Medical Device Directive is essentially European "law" for the requirements of medical devices. The notified body reviews the evidence of compliance to the Medical Device Directive requirements for safety, performance, suitability for intended use, and risk to patients and operators. Manufacturers can choose from several conformity assessment routes, most involving a quality assurance assessment of the manufacturer's facilities. Low-risk, non-measuring, non-sterile medical devices also require a CE Mark, but are allowed to "self-declare" compliance to the Medical Device Directive without the involvement of a notified body.

2.3. Other Countries

Health Canada reviews medical devices to assess their safety, effectiveness, and quality before being authorized for sale in Canada. The system is similar to the European requirements, but the classification system is different.

Other countries have similar requirements to the United States and Europe, and have adopted the same family of medical consensus standards, with the addition of their national deviations. Since the detailed requirements specific to each country have been changing in recent years, it is not practical to address all of them in this chapter. The key is demonstration of compliance to the family of recognized standards.

3. Medical Device Certification

Product certification agencies, such as UL, CSA, TUV, and ETL use recognized consensus standards to evaluate various types of products. These safety standards are documents that define the minimum construction and performance requirements. The base standard that covers medical devices, IEC 60601-1, is accepted by many countries.

The current (second) edition of IEC 60601-1 has two amendments. These amendments contain additions and corrections to the base standard. The standard also has collateral (horizontal) standards, numbered IEC 60601-1-xx, and particular (vertical) standards, numbered IEC 60601-2-xx. The collateral standards include requirements for specific technologies or hazards; they apply to all applicable equipment. Examples are medical systems (IEC 60601-1-1), EMC or electromagnetic compatibility (IEC 60601-1-2), radiation protection in diagnostic x-ray equipment (IEC 60601-1-3), and software (IEC 60601-1-4). The particular standards apply to specific equipment types, such as high-frequency surgical equipment (IEC 60601-2-2), infusion pumps (IEC 60601-2-24), and hospital beds (IEC 60601-2-38).

The UL 60601-1safety standard, published in April 2003, contains the full text of IEC 60601-1 with amendments, and adds U.S. deviations. The U.S. deviations contain national requirements, such as those for the mains circuits, component requirements, lower leakage current limits, enclosure flame ratings, and production line testing. Since these deviations do not conflict with the base standard, equipment that complies with UL 60601-1 also complies

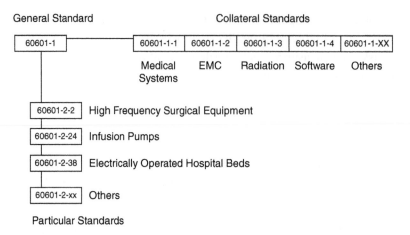

Figure 3.1: Examples of standards for compliance of medical devices.

with IEC 60601-1. The UL 60601-1 standard is an editorial change to the older UL 2601-1 standard; it does not change any requirements.

Figure 3.1 illustrates the organization of the collateral and particular IEC 60601 standards. The base standard, national deviations, and applicable collateral and particular standards are used together to evaluate the medical electrical equipment. All the applicable standards use the same clause numbering, which allows cross-referencing of the requirements between these standards.

4. Philosophy of the Standards

The underlying philosophy of the IEC 60601-1 harmonized standards is that equipment must be safe in normal condition (NC) and single fault condition (SFC). To understand the electrical safety requirements, we need to first define a few terms:

- An *applied part* is any piece(s) of the equipment that can intentionally or unintentionally be brought into contact with the patient.

- *Creepage* is spacing along a surface (as an ant crawls).

- *Clearance* is spacing through the air (as a bug flies).

- *LOP* is a level of protection (level 2 is required by standard).

- *Basic insulation* (BI) is a spacing or a physical insulation barrier providing 1 LOP.

- *Supplemental insulation* (SI) is spacing or a physical insulation barrier providing 1 LOP.

- *Double insulation* (DI) is BI plus SI and provides 2 LOPs.

- *Reinforced insulation* (RI) is a single spacing or physical barrier that provides 2 LOPs.

- *Protective impedance* is a component, such as a resistor, that provides 1 LOP.

- *Protective earth* (PE) is a well-grounded part that provides 1 LOP.

- *Class I equipment* is defined as using PE as 1 LOP.

- *Class II equipment* (known as double insulated) is defined as not using PE, 1 LOP.

For electrical safety, the standard requires 2 LOPs against excessive unintentional current, defined as leakage current, passing through the patient or operator. An insulation diagram (also known as an isolation diagram) graphically depicts the 2 LOPs in a device at each part of the circuit.

Figure 3.2 shows the 2 LOPs between the live part (mains) and the patient (1A and 2A), and between the live part and the enclosure (1B and 2B). In the cases of 1A and 2A, the levels of protection are BI and SI, and for 1B and 2B, they are BI and PE.

The minimum spacing requirements and dielectric (hipot) requirements for these barriers are specified in the base standard. Table 3.1 shows these requirements from the base standard.

If the insulation does not meet both the dielectric and the spacing requirements, it cannot be considered as a level of protection and can be shorted as a normal condition. Note that BI and SI spacing requirements are the same; however, the SI dielectric values are greater than the BI values above 50 V. To be considered protectively earthed, the grounding path of the equipment must pass 15 amps or 1.5 x-rated current for 5 seconds from the protectively earthed part to the earth connection, 0.1 ohms resistance for equipment with a detachable power supply cord or 0.2 ohms for equipment with a non-detachable power-supply cord.

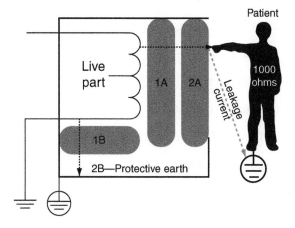

Figure 3.2: Example of protective layers against leakage current through human patients.

Table 3.1: Minimum spacing requirements and dielectric (hipot) requirements for these barriers of the IEC 60601-1 harmonized standards

Creepage and Clearance Requirements (millimeters)						
Voltage	DC	≤15	≤36	≤75	≤150	≤300
Voltage	AC	≤12	≤30	≤60	≤125	≤250
BOP	Creepage	0.8	1.0	1.3	2.0	3.0
	Clearance	0.4	0.5	0.7	1.0	1.6
BI/SI	Creepage	1.7	2.0	2.3	3.0	4.0
	Clearance	0.8	1.0	1.2	1.6	2.5
DI/RI	Creepage	3.4	4.0	4.6	6.0	8.0
	Clearance	1.6	2.0	2.4	3.2	5.0

Dielectric Withstand Voltages (volts)			
Reference Voltage	$0 < V \leq 50$	$50 < V \leq 150$	$150 < V \leq 250$
BI	500	1000	1500
SI	500	2000	2500
DI/RI	500	3000	4000

To demonstrate that medical equipment is safe in normal and single-fault conditions, the following must be addressed when designing and evaluating the medical equipment. These conditions are specified throughout the standard.

Normal conditions (likely to occur)

- Reverse polarity of supply mains
- Failure of insulation less than basic (operational)

Single-fault conditions (could occur)

- Interruption of protective earth
- Interruption of one supply conductor
- Mains voltage on floating (F-type) applied part(s)
- Mains voltage on communication ports
- Failure of electrical components, one at a time
- Failure of mechanical parts, one at a time
- Failure of temperature limiting devices, one at a time

- Shorting of basic or supplemental insulation

- Overload of mains supply transformers

- Interruption and short circuit of motor capacitors

- Locking of moving parts

- Impairment of cooling (fans, vents blocked)

Not evaluated (unlikely to occur)

- Total breakdown of double or reinforced insulation

- Loss of protective earth on permanently installed equipment

- More than one single-fault condition at a time

- Failure of a UL-recognized optocoupler barrier

- Failure of a UL-recognized Y1 capacitor, acting as a barrier

5. Evaluation Process

5.1. Preliminary Evaluation

The process of evaluating medical equipment for compliance to the requirements of standards includes not only the equipment itself, but the user's manual, markings, software (if it mitigates a hazard), biocompatibility of applied parts, and electromagnetic compatibility (EMC). Before submitting equipment for evaluation, the following information should be developed. This is typically done as a preliminary evaluation.

1. Ensure that the equipment fits the scope of the standard.

2. Review whether the equipment fits the scope of any collateral or particular standards.

3. List of all equipment functions and accessories that can be used with it.

4. Check if the equipment can connect to any other equipment (computer, printer, etc.).

    - Any non-medical equipment connected must have IEC certification (evaluated to the applicable IEC standard) or be part of the medical equipment evaluation and meet the medical standard.
    - Computers and other IT equipment are considered to have ground-referenced 50 V in normal condition, mains in single-fault condition on their data ports.

5. Create an insulation diagram (graphic illustration of the required electrical barriers).

6. Determine the equipment classifications from the standard.

7. Document all components that cross electrical barriers, per the insulation diagram.

8. Verify spacings (creepage and clearance), per the insulation diagram.

9. Examine enclosure openings.

   - IEC test finger (access to live parts).

   - IEC test pin (access to live parts).

   - Requirement for a tool to access any live parts.

10. Determine potential mechanical hazards, pinch points, and sharp edges.

11. Determine potential hazards under normal use and foreseeable misuse.

12. Document components that must meet nationally recognized standards.

    - Primary circuit components, up to mains transformer(s).

    - Lithium batteries (also requires reverse charge protection circuitry).

    - CRTs >5 inches.

    - Printed wiring boards with >15 W available.

    - Wiring/tubing with >15 W available.

    - Optical isolators acting as barrier per insulation diagram.

    - Conductive coating process.

13. Verify that component ratings meet the equipment's ratings.

14. List enclosure materials.

    - UL 94 flame-rating requirements for polymeric enclosures if there is >15 W available power in the enclosure.

    - V-2 minimum for mobile, portable equipment.

    - V-0 minimum for fixed or stationary equipment.

15. Verify mains fuse requirements (in the equipment or external power supply).

    - Class I: line and neutral required.

    - Class II with functional earth: line and neutral required.

    - Class II: line only required.

    - Permanently installed equipment: line only required.

16. Verify that protective earth conductors are green with yellow stripe.

17. Verify that wires are secured from hazardous movements.

18. Verify equipment-marking requirements (labels on the equipment).

19. Verify requirements for the user manual (accompanying documents).

20. Take photos of the medical equipment, outside and inside of enclosure.

5.2. Testing

Once you have completed the preliminary evaluation, the safety evaluation of the equipment can continue. One or more samples are required, depending on the equipment type and time requirements. Multiple samples of device components may be needed to perform destructive tests, such as transformers, relays, plastic enclosures, and motors. Once the test lab receives the samples, it can perform the required testing, including electrical, mechanical, temperature, and abnormal condition, as well as essential performance requirements (typically specified in the particular standards section). If software is required for mitigating fire, shock, or mechanical hazards, or if it is specified as required in an applicable particular standard, the IEC 60601-1-4 standard is used to evaluate the software design process and implementation. Electromagnetic compatibility (EMC) testing per collateral standard (IEC 60601-1-2), and the review of biocompatibility documentation are conducted. After the evaluation and testing, the critical component table is developed. Any components that may affect compliance with the requirements of the applicable standard(s) or could have an effect on the testing results are considered critical components.

5.3. Compliance Reports

The documentation developed as a result of a safety evaluation depends on the manufacturer's requirements. The common types of documentation are a letter report, an informative test report, and a certified body report.

A *letter report* is a list of applicable standards used, and a summary of the evaluation, stating whether or not the device complies with the specified standards. An *informative test report* completely documents all the requirements in a standard (N/A, pass, or fail), the test record, an insulation diagram, illustrations, equipment markings, and other applicable information. It is the preferred document for technical files; some hospitals and clinics request it before equipment purchases.

A UL report is an informative test report, with the addition of authorizing the manufacturer to apply the UL/C-UL (United States/Canada) Mark to products covered in the report.

UL conducts quarterly audits using this report to verify that the equipment bearing the UL/C-UL Mark is the same as the equipment that was tested.

A *certified body* (CB) *report* is an informative test report, but contains a certificate from the issuer, who is required to be a member of the IECEE CB scheme. A CB report is used to obtain third-party certification marks (UL, CSA, TUV, VDE, SEMKO (a Swedish testing and certification institute), etc.) in various countries without repeating all equipment testing. The informative test report or CB report is very important to have for the equipment's technical files, as they serve as an international "passport" for the device.

5.4. Common Noncompliances

When medical equipment is submitted for evaluation, designers and manufacturers are expected to be aware that they would need to meet the applicable standards. While this is not always the case, most medical devices require limited changes to meet the standards.

The first and most critical type of noncompliance typically seen is the choice of a power supply. There are a great number of medical power supplies available that are UL recognized to the UL 60601-1 medical standard. Despite this, many devices are submitted with ITE (information technology equipment) power supplies. These power supplies are evaluated to the IEC 60950 standard for ITE equipment, which is different than the medical standard, and will not provide adequate isolation from the mains (wall outlet voltage).

The next type of noncompliance typically seen in markings and accompanying documents, is less critical, but more common. All of the IEC 60601 standards have very specific requirements for markings and inclusions in the accompanying documents. Since most companies have separate departments that create these documents, they are often not aware of the requirements.

Another common noncompliance is inadequate spacing. When designing medical equipment, it is important to be aware that there are minimum spacing requirements for electrical barriers. Inadequate spacings on circuit boards are another typical noncompliance. An example of this is the required barrier between circuitry and accessible parts of the device. This means that there are required spacings between any circuit board, motor housing, connector pins, and so on, and any accessible part of the device. These required spacings are based on the insulation diagram generated in the preliminary evaluation.

Another common noncompliance is misapplied flammability. For equipment with plastic enclosures, there will also be flammability requirements for any plastic that is acting as part of the enclosure (creating a fire enclosure). The plastic chosen for the enclosure must be at least a UL-recognized V-0 flame rating for fixed equipment, or at least a UL-recognized V-2 flame rating for all other types of equipment.

The last common mistake relates to indicator lights. Red indicator lights can only be used for a warning and yellow for caution. Keep this in mind when selecting LEDs for indicator lights.

These common noncompliances can be easily avoided with knowledge of the applicable standards, and they are the major reasons that preliminary investigations of medical equipment are best done in the early design phase.

6. Conclusion

Medical equipment is highly regulated and is required to show compliance to applicable safety standards. Understanding these requirements, as well as the certification and regulatory process in the design phase of the equipment will result in cost reductions in equipment development, faster certification turnaround, and increased product safety.

Bibliography

European Committee for Electrotechnical Standardization (CENELEC). Medical electrical equipment—Part 1, general requirements for safety, EN 60601-1. Brussels: European Committee for Electrotechnical Standardization (CENELEC); 1991.

International Electrotechnical Commission. IEC standards database. Available at: www.iec.ch/searchpub/cur_fut.htm.

International Electrotechnical Commission. Medical electrical equipment—Part 1, general requirements for safety, IEC 60601-1. Geneva: IEC; 1988.

Medical Equipment Compliance Associates. Evaluation package for IEC/UL/CSA/EN 60601-1, Oak Creek, WI: Medical Equipment Compliance Associates, LLC; 2007. Available at:60601-1.com/documents.htm.

Medical Equipment Compliance Associates. Medical equipment compliance: Design and evaluation to the '601' standards. Seminar. Oak Creek, WI: Medical Equipment Compliance Associates, LLC; 2008.

UK Department of Health. Medical device directive: Council Directive 93/42/EEC, of June 1993. London: Department of Health; 1993.

Underwriters Laboratories Inc. Medical electrical equipment—Part 1, General requirements for safety, UL 60601-1. Northbrook, IL: Underwriters Laboratories Inc.; 2006.

Underwriters Laboratories Inc. UL StandardsInfoNet, catalog of UL standards for safety. Available at: ulstandardsinfonet.ul.com/.

U.S. Department of Health and Human Services. Implementation of third party programs under the FDA Modernization Act of 1997. Final guidance for staff, industry, and third parties. Washington, DC: U.S. Department of Health and Human Services; 2001.

CHAPTER 4

Software for Medical Systems

Jeff Geisler

1. Introduction

Safety has long been a prominent concern with food, drugs, and devices. Medical devices are regulated by government in every important economy in the world. With the invention of the computer and the ability to embed computers and software in devices, it was only natural that software would become important in medicine. Sometimes the software *is* a medical device, such as image processing software making a breast cancer diagnosis from a mammogram would be. As software has gained more functionality and touched safety, such as in software systems that keep patients alive, regulatory bodies worldwide have gradually come to realize the importance of regulating software and its construction as well as the more traditional scope of regulating manufacturing processes.

The Food and Drug Administration (FDA), part of the Department of Health and Human Services, is the U.S. government agency with authority over the safety of food, drugs, medical devices, and radiation-producing equipment. To market any medical device or drug in the United States, you must have the approval of the FDA. The rest of the world has similar policies restricting their markets.

The eventual measurement of quality is success in the marketplace. Complex factors are at play in any economy, but by and large if a product meets customer needs they will buy it, and it will win out over its competitors. But this is a lengthy process—the product could spend years in the market and injure many people before it becomes evident that the rewards are not worth the risk. In the 1960s, thalidomide was thought safe and prescribed as an anti-nausea drug. Unfortunately, it was realized too late that if taken at a certain stage of pregnancy, it would cause birth defects. The objective of the regulatory environment is to prevent such disasters before they happen.

The governing law for current good manufacturing practices (cGMP) is found in Section 520 of the Food, Drug and Cosmetic Act. The FDA regulations are a nonprescriptive quality process. The actual regulation derived from the law is not very long. This results in the virtue of comprehensiveness but the vice of generality. Many paths can meet the same goal.

Because the eventual quality of a product is so hard to measure, takes too long, and is too risky, many quality programs do the next best thing, which is to audit compliance to the written policies. What the FDA expects is something similar. To the FDA, you are the experts in your device and your quality programs. Writing down the procedures is necessary—it is assumed that you know best what the procedures should be—but it is essential that you comply with your written procedures. At the same time, the methods cannot be too far off the beaten path, because then you will have to explain how they result in the same safety and effectiveness as standard practices in the rest of the safety-critical industry.

In the wake of political waves seeking to reduce regulation in the U.S. economy, the FDA promises to consider the "least burdensome approach in all areas of medical device regulation" [1]. Companies are free to adopt any policies they wish, but it is still contingent on them to establish with the FDA that the approaches they take will result in a product with adequate safeguards and quality. In practice, few companies are willing to accept the business risk of having their product launch locked up in a regulatory approval cycle.

The actual process for gaining regulatory approval in the United States is known as a *premarket submission*. Companies prepare documentation about the device, its software, the verification and validation activities, and the labeling that goes with the device. One aspect significant about the labeling is the claims for the device, that is, the conditions under which the device should be used, the types of patients and their illnesses that it should be used with, and the outcomes that should be expected. The claims are central to the regulatory process and the FDA thought process. These formulate customer needs and intended use. The standard is high; any company wishing to sell a device that is supposed to make people healthier must be able to prove it through scientific means. This notion of scientific proof underlies what the regulatory bodies are requesting and why they ask firms to do certain things.

Premarket submission must necessarily be late in the product development process, since the company will be documenting what the product can do. Part of this is to submit the product verification and validation documentation and also to perform the final design review, which occurs when the product design moves from development to manufacturing. The last thing a business could want is to have a product ready to manufacture and sell but held up by the approval process. The effect of this is to make companies conservative in their approach to quality programs so as to avoid a lengthy dialog with the FDA.

Rather like getting a legal opinion, in which the lawyer will only speak to the likelihood of an outcome from a court and cannot predict the final result, the FDA will not tell you beforehand which contents of a submission would be acceptable. FDA spokespersons want to avoid being in the position of having suggested that a method would be compliant before actually seeing the results.

This presents a challenge to the engineer developing software for a medical device, because the specifics of the methods for software construction are not spelled out. I will have more to say about this later, but the rigor of the methods and the amount of documentation required also varies with the "level of concern" for the safety of the software. Many of the methods described elsewhere in this volume are also appropriate for medical devices, and will be found acceptable to the medical regulatory bodies. This chapter hopes to inform the reader of methods the author has found that result in high-quality software and establish that quality to the regulatory bodies.

After all, it is not as if we are trying to get away with something. We all want software that works, software that—in the words of the FDA—is "fit for its intended purpose and meets customer needs." We are obliged to build safe devices. In part, this is an ethical issue—as good engineers, we don't want our devices to hurt anybody. But safety is also important to the business because of potential product liability. Practices that will result in the highest-quality software will also be acceptable to the regulatory bodies. For software development, aside from the level of documentation, there are not many practices required by the regulatory bodies over and above what engineers should be doing anyway.

Approaching regulatory bodies in an adversarial relationship is not helpful. They are a fact of life in modern economies; whether we like it or not, they serve a valuable function. Without regulations, markets will engage in a race to the bottom. Even if a company wants to be ethical, they will not remain competitive if their competitors are allowed to cut corners. Regulation establishes a floor for behavior beneath which companies cannot go.

1.1. Verification and Validation

Software verification and validation (V&V) forms a large part of the scope of software development for medical devices. Loosely, V&V is the activity of establishing that the software does what is intended. Often in informal usage, the words "verification and validation" are interchangeable or even redundant. Over the years a type of clarity has emerged, but one could still dispute where the line is drawn between the two.

The FDA definition of software validation is "confirmation . . . that software specifications conform to user needs and intended uses." The phrases "user needs" and "intended uses" are commonplace in the FDA guidance documents. The FDA definition is related to the principle of showing effectiveness: not only did you accomplish what you set out to do, but your actions were the right thing to do. In other words, what you built met user needs.

Verification has a narrower definition. It is the act of demonstrating that design outputs match design inputs. The FDA sees software development (or any engineering development for that matter—see the discussion about design control in Section 3) as a step-wise refinement of project artifacts until development achieves the final deliverable, executable code. Each step starts with design input. The design is further refined, and the result is the design output. So, for example, one phase would be software coding. Coding starts with the software design as the design input. (The software design itself was the design output of the design phase.) Engineers then write the code that satisfies the design. Verification activities would be anything you did to verify that the code fulfills the intent of the design. This might be demonstration, such as dynamic testing. It might be a code review or inspection, in which the team examines the code and the design and ensures that the code implements the design, or that the design is changed to reflect the realities of the actual implementation. Activities performed in the normal course of software development such as requirements reviews, design reviews, unit testing, static tests, and build audits are all examples of verification.

Once you have passed through all the phases of development, the sum total of verification activities conducted at each phase sustain the conclusion that the software is validated. Verification comprises the individual steps, whereas validation is the sum of the whole.

Hence it is a bit mistaken to think that software validation is something that is done—or can be done—at the end of a project. To meet the standards expected by the regulatory bodies, the software will have had to be built with thorough verification activities right from the start. For this reason, it is difficult to subcontract software, because the subcontractor would have to satisfy regulatory requirements just as if you were implementing the development in house. So the problem becomes one of subcontractor management to regulatory standards. (This is part of the subject of the use of off-the-shelf or third-party software, about which I have more to say later on.)

Example 4.1 Precisely Wrong

While not a software failure, the development of the Hubble space telescope is a stellar example of the difference between validation and verification. The instrument had a troubled history—it was seriously behind schedule, and then was further delayed by holdups in the Space Shuttle program. When NASA finally flew the Hubble, they discovered that the mirror was wrong and could not focus [2].

Now this mirror is one of the most precise objects ever created. It was polished to within 10 nanometers—that is to say, 100 atom diameters—of its specification. This is 4 parts per billion in a 2.4-m mechanical structure. Yet when it made it to space, it didn't work.

It turned out that a mistake had been made during its manufacture. A reference tool to establish the aspherical form of the mirror had been mismeasured and put in the wrong place. So when the technicians polished the mirror to within 100 atom diameters of precision, they polished it wrong. In other words, the mirror matched its verification—the output matched the input to an amazing degree. But the input was wrong, and the mirror failed at its primary purpose—to take clear pictures of the universe. This is a failure in validation.

(The NASA report [2] is interesting reading for the commonplace failures in the quality system that allowed the mistake to occur. Among those that also apply to software are ignoring the evidence when the system did not go together right, relying on a single measurement, failing to follow the quality plan, having QA people report to the project manger instead of being independent, and failing to specify the expected result, so that inexperienced personnel were unable to tell that something was amiss. I am happy to report that in later missions to the Hubble, astronauts were able to retrofit a focus adjustment device that allowed the telescope to take some of the most spectacular pictures of all time.)

To sum up:

- Verification is showing that you did what you intended to do.

- Validation is showing that what you did was the right thing to do.

What I have described is the meaning of verification and validation at the system level. Confusion on the part of some readers about the definition of verification and validation results from the FDA itself, in its description of software validation in the guidance document. I believe the FDA has overloaded the definition. There is validation, by which the FDA writers mean system validation as I have just described. There is also software validation, which is the sum total of verification activities—the construction of software according to software development policies. Technically, the definition for software validation is the same as any other kind of validation—the software conforms to "user needs and intended uses." It is just that, in this case, the "user" is the rest of the system. If the software complies with its requirements, it has satisfied its "user's needs." Hence one could call, with complete legitimacy, the document with the tests in which you demonstrate that the software meets its requirements the *software validation protocol*. In fact, at a recent conference devoted to the development of medical device software, I informally polled the audience as to what was called the test protocol which established that the software meets its requirements. Half called it software validation, and the other half called it software verification.

1.2. Life Cycle Model

While the FDA recognizes that "[t]here are a variety of life cycle models, such as ... waterfall, spiral, evolutionary, incremental" [3], their methods for documenting the software development process assume a waterfall model. This makes sense; the waterfall model may not be accurate, but it is a useful method for describing the deliverables and the activities that occur during software development. (As they say, "All models are wrong; some models are useful.") I will follow the same practice, partly for reasons of modeling my presentation after the FDA and ANSI methods, but also because the waterfall is a useful way to discuss process.

To refresh the reader, the waterfall model is the classical software development life cycle. (See Fig. 4.1.) Software development begins with vague customer needs and wants. These are formalized into requirements during the requirements analysis phase. This is input to the architectural design phase, where the structure of the software is

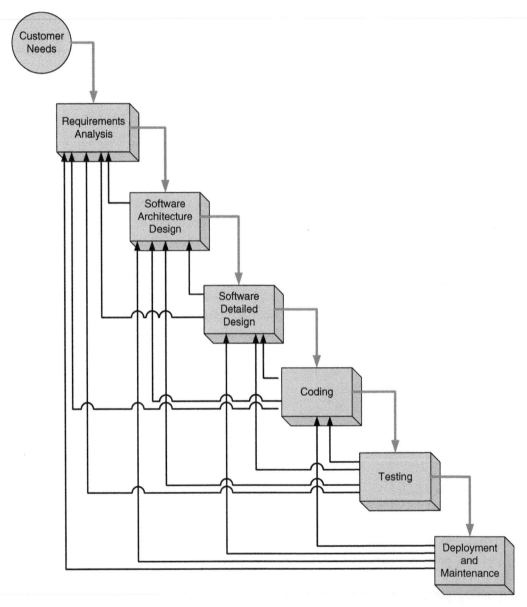

Figure 4.1: Waterfall model of the software life cycle.

created. With a software architecture in hand, engineers next develop the detailed design. Once all the designs are known, class definitions complete, methods named and prototyped in pseudo code, and a data dictionary written, the engineers can begin the coding phase. After the code is complete, the software enters the test phase, and once the unit, integration, and software system testing is concluded, the software is deployed for use by the customer. After that, it is in the maintenance phase. This development sequence is represented by the light-colored arrows leading to the right in the diagram.

The trouble with the waterfall model is that it does not match the reality of the way that systems in general and software in particular is developed. It is rarely possible to define requirements in sufficient detail up front, leave them alone while the team develops the design, and then put all that away and actually start writing code.

After all, requirements originate in the vague wishes of the customers; many things they can imagine or hope for cannot be done. We are constrained by time, resources, technology, competitive intellectual property, and sometimes even by the laws of physics. Moreover, who can imagine that technology, our knowledge, or the marketplace will stay still while we spend months in requirements analysis?

In my experience, we often do not know what we can build until we try to build it. Construction is the real test of requirements. For a software life cycle to have any hope of reflecting reality, it must acknowledge the iterative nature of development. Indeed, more recent descriptions of the waterfall model show backward-pointing arrows between the phases, as you can see in the diagram. The drawing looks complex because you can return to any of the previous phases at any time.

Moreover, the phases are not always that distinct. Requirements sometimes shade into design, and vice versa. For example, if I am writing a custom interface between an embedded controller and a Graphical User Interface running on a laptop, I have quite a bit of design freedom in what the data exchange looks like. But once decided upon, if one or the other components doesn't precisely comply, the interface won't work. So is this design or requirements? It depends on your point of view and what you need it for at the moment.

The regulatory bodies allow that you are not restricted to a waterfall model; you are free to choose a different life cycle model. "Medical device software may be produced using any of these or other models, as long as adequate risk management activities and feedback processes are incorporated into the model selected" [3]. So, the risk management, feedback, and quality assurance activities are the keys to being able to assert that the software is valid, not the technical details of the order in which all of this was accomplished.

More modern agile approaches that recognize and take advantage of the naturally iterative unfolding of software development are the ones more likely to succeed. Develop requirements in layers of gradually increasing detail to get an idea of the scope of the project. This enables negotiating the trade-off of features, time, and resources before investing in a huge amount of detailed analysis. Perform an architectural design to understand the responsibilities of subsystems and the way that project task partitioning will allocate across the team. Then implement in stages.

The FDA has been tasked by Congress to consider the "least burdensome approach in all areas of medical device regulation" [1]. They will consider alternatives, such as a process like extreme programming (XP), which de-emphasizes formal requirements analysis and design in favor of user stories. It becomes a function of the level of concern of the software; such methods might be appropriate if there is no risk of harm from the software system. However, when the level of concern is moderate or major, there will be the expectation that methods of sufficient rigor were used to ensure that risk management activities were carried out.

2. The Medical Regulatory Environment

The FDA is the regulatory body in the United States with oversight of medical devices, among other things. Given that the U.S. medical market is a large part of the world's medical market, medical device manufacturers seeking maximum markets will be interacting with the FDA.

The worldwide standardization body is International Organization for Standardization, or ISO [4]. The name "ISO" is not actually an acronym for anything. It is instead derived from *isos*, the Greek word for "equal," as in "isosceles triangle," and reflects the organization's aim to equalize standards between countries. More than 150 nations are members, including all of the developed countries, so it is truly international.

Closely related to the ISO is the International Electrotechnical Commission (IEC); in fact, the two have published many standards together. As indicated by the name, IEC is specific to electrical and electronic devices. ISO is more concerned with quality systems for a broad range of products, whereas the IEC has standards for particular devices. (The IEC has a standard for "rumble measurement on vinyl disc turntables," for example.)

Example 4.2 Guidance Documents

The FDA has a very good website, www.fda.gov, with many useful documents concerning medical device development in general and software for medical systems in particular. The most useful of these are the guidance documents, which serve to further explain the meaning of the regulations.

General principles of software validation. Final guidance for industry and FDA staff. January 11, 2002. Available at: www.fda.gov/MedicalDevices/ DeviceRegulationandGuidance/ GuidanceDocuments/default.htm	This is the most important for software development. If you read none of the other guidance documents, you should still read this one.
Design control guidance for medical device manufacturers. March 11, 1997. Available at: www.fda.gov/MedicalDevices/ DeviceRegulationandGuidance/ GuidanceDocuments/default.htm	This document provides detail about the design control process itself, which is of larger scope than just software development. It is a description of good engineering practices that the FDA expects to be followed in the development of any medical device. Software development has to follow the same guidelines, as well as further specialized activities.
Guidance for the content of premarket submissions for software contained in medical devices. May 11, 2005. Available at: www.fda.gov/ MedicalDevices/DeviceRegulationandGuidance/ GuidanceDocuments/default.htm	This document contains the criteria for establishing the level of concern of the software under development. It also describes the documentation necessary for the premarket submission, not just the list of documents but the type of content that they should contain. This will be important if you are responsible for putting together the software documentation for the premarket submission.
Guidance for industry, FDA reviewers and compliance on off-the-shelf software use in medical devices. September 9, 1999. Available at: www.fda.gov/ MedicalDevices/DeviceRegulationandGuidance/ GuidanceDocuments/default.htm	This document tells you what the FDA expects you to do if you use third-party software in your medical device. It also has the most thorough discussion of software hazard analysis among the guidance documents.
ANSI/AAMI/IEC 62304:2006. Medical device software—Software life-cycle processes. Available from the webstore at www.iso.org/iso/store.htm.	This is the official international standard. It restates much of the information in the FDA guidance documents, but is crisper and less subject to interpretation. This is a valuable addition to the FDA documents, but does require purchase.

IEC Medical Standards

The standards important in the medical world begin with IEC 60601:

- IEC 60601-1 Medical electrical equipment—Part 1: General requirements for safety.
- IEC 60601-1-1 Medical electrical equipment—Part 1-1: General requirements for safety—collateral standard: Safety requirements for medical electrical system.
- IEC 60601-1-2 Medical electrical equipment—Part 1-2: General requirements for safety—Section 2: Collateral standard: electromagnetic compatibility—requirements and tests.
- IEC 60601-1-4 Medical electrical equipment—Part 1-4: General requirements for safety—collateral standard: programmable electrical medical systems.
- IEC 60601-1-8 Medical electrical equipment—Part 1-8: General requirements for safety—collateral standard: general requirements, tests, and guidance for alarm systems in medical electrical equipment and medical electrical systems. First edition in August 2003.

(Continues)

Example 4.2.   Cont'd

- IEC 60601-1-9 Medical electrical equipment—Part 1-9: General requirements for basic safety and essential performance—Collateral standard: requirements for environmentally conscious design. First edition in July 2007.

These are mostly hardware standards that your device would need to comply with in general. IEC 60601-1-4 is the relevant standard for devices containing microprocessors. IEC 60601-1-8 may also be relevant if your software or device has alarms. This document spells out international standards for physical features of alarms, such as the loudness and frequency characteristics of an audible alarm, its cadence for different alarm priorities, and the colors to use for LEDs.

Other IEC standards would come into play if your product used components from the list of IEC standards. For example, if your device used a lithium battery, you would need to comply with IEC 60086-4 Primary batteries—Part 4: Safety of lithium batteries.

Safety Laboratory Markings

Safety markings, such as the UL, CSA, and CE marks, are much more relevant to overall electrical system safety than to software concerns. Nevertheless, if you are producing an electrical product, some form of a safety mark is required by almost all countries. Underwriters Laboratory (UL) is preferred in the United States, CSA is the Canadian standard, and the CE mark is required for EU countries.

UL is largely a testing standard. The CE mark, on the other hand, can be obtained through documenting a quality process. It is not necessary to have ISO 9000 certification to get a CE mark, but if you do, you will have completed 80% of the effort for getting the mark [5].

One difference between the CE mark and the FDA is that it is sufficient to show that a product is safe for the CE mark. The FDA creates a greater burden of proof by requiring that the product also be shown to be effective; in other words, that using the product results in some benefit to the user. This sometimes results in products being released in Europe before the United States, because effectiveness is much harder to demonstrate.

Approval for the CE is similar to ISO certification. You find (and pay) a notified body (such as TUV or BSI) to represent you to the competent authority—a representative arm of the EU member-state that monitors the activities of the notified bodies. This body will assess your product and company in much the same way that the FDA would assess it, with greater scrutiny accruing to higher-risk devices. The same kinds of testing and design documentation would be suitable for both.

Worldwide regulatory bodies embarked upon a major effort in the 1990s to harmonize medical device regulatory requirements. This has taken the form of the Global Harmonization Task Force (GHTF); it includes "representatives of the Canadian Ministry of Health and Welfare; the Japanese Ministry of Health and Welfare; FDA; industry members from the European Union, Australia, Canada, Japan, and the United States, and a few delegates from observing countries" [6]. There are four subgroups; the task force working on harmonizing quality-system requirements has had the most success.

Many countries defer to ISO standards for quality systems and IEC standards for safety. The regulatory bodies have established cross-approval processes, so generally what has been approved by one agency is acceptable to the other, with perhaps some tweaking. For example, the FDA requires a risk analysis but does not specify the method; the European agencies expect the risk analysis to be conducted per ISO 14971. Following the precepts of ISO 14971 is completely acceptable to the FDA. Generally then, since you know you will need to follow ISO 14971 for Europe, you might as well use the same analysis and structure to satisfy the FDA. "ISO 9001 is the most comprehensive because it covers design, production, servicing, and corrective/preventive activities. The FDA GMP requirements are slightly more extensive because they include extensive coverage of labeling, [sic] and complaint handling" [7].

There are a couple of significant reasons the FDA regulations are still separate from the ISO standards. In the first place, the FDA regulations and guidance documents are free. The ISO/ANSI/IEC documents are protected by copyright and require purchase, typically around $100. Charging for the standards is even more inconvenient than it seems. A full set is not cheap, and, although they are available as electronic documents, you cannot create more than one copy. A company would prefer, of course, to share them to lower the cost, but they can't just print them whenever someone needs to consult a copy.

The second issue is that "FDA does not believe that ISO 9000:1994 alone is sufficient to adequately protect the public health.... Through the many years of experience enforcing and evaluating compliance with the original CGMP regulation, FDA has found that it is necessary to clearly spell out its expectations" [8]. This is because the FDA sees itself as more of an enforcement agency than ISO seems to be. Thus, the FDA wants to define the practices that its staff expects to see, not just recommend good practice.

In the next section, I provide an overview of quality systems from the perspective of both FDA regulations and ISO simultaneously because there is much overlap. At the same time, this provides an opportunity to point out the ways that they differ. Details of the quality system and the submission process are really the bailiwick of a company's regulatory compliance personnel. However, there are occasions when the quality system interfaces to software development, so it is useful to know the regulatory context in which a medical manufacturer must operate.

2.1. Worldwide Quality System Requirements

The actual regulations for medical manufacturers doing business in the United States are codified in U.S. 21 Code of Federal Regulations (CFR) 820, known as the QSR (for quality systems regulations). The ISO general quality system requirements are described in ISO 9001:2001. There is a supplement to the general standard in ISO 13485:2003, which comprise additional requirements for medical devices. Together, these two international standards are equivalent to the QSR. There are no extra requirements in ISO 13485 that have relevance to design engineers, or to software engineers in particular, that have not already been covered by the FDA's QSR.

The regulations are very high-level requirements that medical manufacturers must implement in order to have satisfactory quality systems. Many of these requirements are specific to manufacturing processes, such as device configuration management, documentation control, process control, and material handling. Developers of medical device software will be only indirectly concerned with many of these policies, although the software development process has to fit into the general scheme. Of greatest interest to software engineers is §820.30, Design Control (Subpart C).

Most of this chapter is focused on design control and its application to medical software development. But first I wish to briefly cover general principles of the regulations for medical manufacturers and their interface to software development. These are the sorts of practices that any medical manufacturer will have to follow.

The general policies are known as good manufacturing practices, or GMP. (Since this has been revised in 1996, the current policies are known as cGMP for current good manufacturing practices.) In other words, the QSR codifies cGMP. These are principles of mature product development and manufacturing firms; while medical manufacturers must comply, most manufacturers would benefit from following these practices, for they by and large make sense for manufacturing quality products.

The QSR and the ISO standard cover much the same ground, but the organization of the standards differs slightly. I will use the organization of the QSR, summarizing the policies of both organizations in the following sections.

2.2. Subpart A: General Provisions

The first section in both documents is an introduction that covers scope and definitions, plus some legal information. The applicability of the QSR is "the design, manufacture, packaging, labeling, storage, installation, and servicing of all finished [medical] devices intended for human use" [8]. "Finished" means that it does not apply to companies which manufacture parts that go into a finished device, but they are nevertheless encouraged to follow the same guidelines, where appropriate.

2.3. Subpart B: Quality System Requirements

This part of the QSR and the ISO standard cover organizational principles. Any company that wishes to manufacture medical devices must establish a quality system. The term "establish" has special meaning to the FDA—it means define, document, and implement.

The documentation step is important. Companies must document and follow their processes.

The quality system must state a quality policy, and the company has to have the organization to support the quality system. This includes trained personnel, with sufficient time to do the work. There needs to be enough time in the production schedule to allow for inspection, testing, and verification. The quality organization includes an executive to oversee the quality system. One of the executive's responsibilities is management review. The FDA is more specific and has more details about management review of the quality system than ISO. The FDA encourages an internal review to evaluate compliance, but staff will not ask to see the results under normal circumstances, in order not to discourage forthright reports. One part of review is internal audits of the quality system. Normally, these would not be shown to the FDA. ISO-notified bodies such as the TUV are less adversarial, and can ask to see internal audits, since they are seeking to work with the manufacturer to create compliance.

Both QSR and ISO require the establishment of a quality plan. The guidance from ISO is much more specific; the FDA in this case accepts the ISO standards. This is a case where ISO goes further than FDA and the FDA finds that completely acceptable. In another case, personnel, the FDA goes further than ISO. The FDA specifically instructs that personnel who build or test the device be made aware of defects that may occur if they don't do their jobs right, or the types of error they may see so as to be on the lookout for quality problems.

2.4. Subpart C—Design Controls

The next part of the QSR, Subpart C, concerns design controls. This is the most pertinent part of the regulation for software engineers. In fact, the point of this chapter is to explain how a software development process can be designed to comply with Subpart C. It deserves a section of its own, so discussion of design controls appears in a subsequent expanded section, and in the meantime I press on with an overview of cGMP.

2.5. Subpart D—Document Controls

Document control is an important function in any manufacturing organization: it is the principal method for controlling change. Controlling change is necessary because the reason for any change in a medical device must be understood. In some cases, the change has to be validated, that is to say, it is not sufficient to claim "new and improved"; you must actually be able to show "new and improved" with actual clinical data.

The purpose of document control is to collect all descriptive documents that say how to make a product. These are the drawings for how to make pieces of it, the vendor list of whom to buy components from, the bill of materials (BOM) that tells you how many of what to build the product, and the standard operating procedures (SOPs) that tell you how to manufacture it. The SOPs extend to describing the process for controlling the process, that is, the procedures to change the documents in document control, or other procedures having to do with the quality system itself.

Note: Software has its own set of SOPs that describe the software development process. These are an extension of the quality system SOPs.

Changes to the document set are often known as engineering change orders (ECOs). Various elaborate schemes for managing the ECO process exist, but they are beyond the scope of this chapter. There are electronic systems to manage documents as data elements in a database, complete with electronic signatures. They are expensive and complex, but convenient in many ways. If you are looking for one (or building your own), keep in mind the fact that, since they are part of the quality system, electronic document control systems must themselves be validated. See Section 6.1, Software of Unknown Provenance (SOUP), for guidelines of how conduct this validation. They must also comply with 21 CFR 11, Electronic Records; Electronic Signatures, which are policies to ensure the validity of electronic records.

The ultimate deliverable for document control is the device master record (DMR). This is the sum total of documentation that tells how to build the product. The goal would be to have enough detail (including capturing cultural knowledge, which would then no longer be cultural) that someone else could build the product from the documentation alone.

Document control is especially significant to the FDA since the documents support the ability to trace all of the materials that went into manufacturing a product. In the event of a product recall, the FDA would want to be able to trace all the concerned lot numbers so that deviating product could be sequestered or pulled from the market. You as the manufacturer would want to have enough detail retained about configurations to limit the extent of the recall.

2.5.1. The Interface of Software to Document Control

The interface among software development, software configuration management (SCM), and document control will vary with the business and the applications. Software has its own need for configuration management (see Section 3.11, Software Configuration Management Methods) that needs to accommodate a huge number of changes with a limited amount of overhead. Most of the time the ECO process is enormously more cumbersome than software development could use and still stay productive.

The way I've seen it work is for the interface between SCM and document control to be at very specific boundaries. For example, documents that are used outside the software group are good candidates for release to document control. The alphabet soup of development documents should be in document control—they define what the product does as much as the executable. Most documentation that will be included in the premarket submission will be in document control.

When software is a component in the medical device, the released software is treated as a virtual part under document control. It is given a part number and a place in the BOM. Revising the software requires an ECO.

If software is the medical device and your firm is not otherwise a manufacturer, other systems may work for you. It will, as always, be a function of the level of concern for the software. What is required is compliance with control methods. You have to be able to ensure that you know exactly what software configuration went into the device, and that a change cannot have occurred without proper approvals and validations, as necessary.

2.6. Subpart E—Purchasing Controls

Purchasing controls are procedures to make sure that purchased products conform to their specifications. The FDA is not regulating the component suppliers, so they explicitly require the device manufacturers to exercise control themselves. This mostly has to do with raw materials and components, and so on, but where it is important to software is that it also covers contractors and consultants, who may be providing software. It would also include off-the-shelf (OTS) software.

Firms who use contractors must document the selection of a contractor and why the contractor is able to perform the job. They must also define the control mechanisms they will use to ensure quality, and review the work of the contractor at intervals to confirm that requirements are being met. Finally, the contracting company must document and maintain quality records for acceptable contractors.

One form of the purchasing controls is in the data that the company uses to buy the specific items that will go into the medical device and the methods to ensure that what was delivered was what was agreed to. Purchasing controls like this would prevent something like the Chinese toy recall of 2007. For software, for instance, the purchase agreement would need to state which version of an OTS package was being used in the device.

The FDA regulations spell out that a supplier should notify the manufacturer of changes in the product they are providing so that the manufacturer has an opportunity to evaluate how the change may affect safety and effectiveness. They warn that suppliers who don't notify the manufacturer may be unacceptable. ISO is slightly less strict in this regard, but ISO indicates that it is a good practice.

The degree to which you were to specify an exact version is a function of the safety or effectiveness of the finished device. For example, it may be possible to simply specify a generic version of Windows, such as Windows 2000 or XP. But because Windows comes in so many variants and installations, making testing each combination difficult or impossible, a generic specification is probably not suitable for the moderate or major level of concern, and may be unwise for minor-level-of-concern software.

2.7. Subpart F—Identification and Traceability

The purpose of identification is to prevent the use of the wrong component or material in manufacturing. The method to provide for this is to make each batch or unit traceable with a control number. For software, this implies version labels, although firmware could be identified with a part number on the program storage medium or board.

While it is not part of 21 CFR 820, the FDA also has regulations for tracking which devices go into which patients [9]. This is necessary so that if a problem is found in production, the manufacturer can find the products and warn the recipients or recall if needed. For example, Medtronic recently determined that some defibrillator leads were prone to fracture and issued a voluntary recall. By identifying the patients who had the leads, their doctors can plan to adjust the defibrillator settings at the next visit [10].

2.8. Subpart G—Production and Process Controls

Production and process controls are really a manufacturing subject. They comprise the methods that a manufacturer uses to make sure that the products they are making consistently meet specifications. The subpart covers manufacturing and inspection procedures, environmental control, cleanliness and personal practices of personnel, contamination control, building, and equipment.

Insofar as a manufacturer uses automated data processing systems for production or in the quality system, "it is necessary that software be validated to the extent possible to adequately ensure performance" [11]. This does not mean that you have to conduct software V&V to the extent that you would for product software. Often, source code and design documentation are unavailable. What you must do is confirm that the software meets your needs and is fit for the purpose you have in mind. This is a matter of determining your requirements for the OTS software and conducting black-box tests to verify "that it will perform as intended in its chosen application" [11]. You would not need to test every feature, but you would want to convince yourselves that the features important to correctly manufacturing or inspecting your device are working properly, or that you are using the device properly. For more about this, see Section 6.1.2, Third Party Validation.

2.9. Subpart H—Acceptance Activities, and Subpart I—Nonconforming Product

These subparts discuss receiving, in-process, and finished device acceptance testing and inspections. In an interesting difference between ISO and FDA, ISO allows release of product prior to full verification under an "urgent use provision," provided that the manufacturer keeps track of it in case of a recall. The FDA does not permit urgent use; devices must have a completed set of final acceptance activities, including signed records, before release.

Presumably "urgent use" could include the release of beta software to a customer site to fix a serious problem, before the full software V&V had been completed. This would be allowed by ISO but is forbidden by FDA. To them, the advantage is not worth the risks.

If the inspection finds that a subcomponent, material, or product does not meet its specifications, the nonconforming product subpart deals with "the identification, documentation, evaluation, segregation, and disposition" [12] of such items, so that there is no danger of mixing it with released products. Generally, the nonconforming product must be specially tagged and kept in a protected area away from the acceptable product. The company must have a documented procedure for deciding what to do with the deviating items: reworked, accepted by concession, used in alternative applications, or scrapped. A nonconforming product accepted by concession must have a justification based on scientific evidence, signed by the person authorizing its use.

2.10. Subpart J—Corrective and Preventive Action

Methods to provide feedback when there have been problems are a part of any quality system. For the FDA, this is the corrective and preventive action process, also known as CAPA. The purpose is to identify root causes for quality problems and provide correctives so that they don't happen again. This is risk based—problems deserve investigation equal to the significance and potential risk.

CAPA is a very serious matter to the FDA and other regulatory bodies; for one thing, it is subject to audit and almost always looked at by the auditor. It is a prime role belonging to the QA/RA (quality assurance/regulatory affairs) departments of a medical manufacturer. Any death associated with a device is relevant, even if caused by user error. This is because human factors should have been considered in the design of the device, and misuse is a failure of the human factors design. Manufacturers would be expected to evaluate whether a redesign or more cautions and warnings in the manuals are called for.

You could extend a software problem reporting process to cover all the issues that could occur in manufacturing a device and in handling customer complaints. Usually you would not want to; you don't want to confound and overwhelm the quality data from the rest of the organization with minor software defects. At the same time, you don't want to make the process of closing software defects so cumbersome that you discourage their reporting. The rigor for follow-through on a CAPA issue is often quite high, and CAPAs are supposed to be reviewed in the quality-system management review meeting. You may not wish to find yourself explaining to the CEO what a race condition is and how you verified that you had fixed it. (A race condition occurs when operations can occur in any order but must be done in a certain sequence to be successful.)

A tiered system is likely best, where only software defects discovered after release of the software go into the CAPA system. These defects would be duplicated into the CAPA system if they were determined to be of threshold severity; customer complaints or other quality issues needing a software change would go into the software problem reporting system to inform the software team of actions that need to be taken.

2.11. Subpart K—Labeling and Packaging Control

The one thing to note for software in this section is the special meaning of "labeling" and its importance to the FDA. This is not just the manufacturer's label with the model and serial number of the device. Labeling is any textual or graphical material that is shipped with or references the device. It includes the user's manual, physician's guide, instructions for use, container labels, and advertising brochures. The text on a screen or in a help system in a software-controlled device is labeling.

Labeling is also part of the premarket submission process. (The premarket process is further discussed later in this chapter.) You want to be careful that you don't make claims in the use of the product that have not been validated through clinical use. While doctors, based on their own judgment, can use devices as they see fit, the manufacturer cannot make claims for therapy that are not backed up by science.

Labeling is also an area where the FDA differs slightly from ISO. The FDA requires an examination for accuracy of the labeling, and a record that includes the signature of the inspector, documented in the device history record. The FDA does this because its data show that, even with the rules, there have been numerous recalls because of labeling errors. ISO is not so strict; while you can retain label inspection records, you are not required to do so.

2.12. Subpart L—Handling, Storage, Distribution, and Installation

Now we are into a part of the regulation that is really more about issues related to manufacturing and distributing devices. This doesn't have much to do with designing the products or the software that goes into them. Policies about handling, storage, and distribution are designed to make sure that devices are properly taken care of before they arrive in customer's hands, and that only a qualified released product is shipped.

As for installation, the manufacturer must have directions for correct installation and must ship them with the device or make them readily available. These should include test procedures where appropriate. The installer is also required to maintain records, subject to audit, that show the installation was correct and the test procedures were followed. In another difference from the FDA, ISO does not address installation as a separate subject.

If the medical device requires an elaborate software installation, these requirements could be an issue. If the manufacturer performs the installation, it would have to keep records. Third parties can do the installation, but then are considered to be manufacturers, and are subject to their own record requirements. You would want to provide comprehensive installation instructions—it would not be sufficient to assume that the user would know to double-click *setup.exe*. You could also provide a signature form with instructions to document the installation and retain the record. You might need to provide a small validation to confirm that the software installed correctly, and have the users sign, date, and keep the form in their records. Hospital bioengineering departments are usually comfortable with this.

2.13. Subpart M—Records

This is an area where the FDA has much more to say than the ISO. The FDA requirements would fit under ISO—there is nothing inconsistent—but ISO is much less specific. So if you intend to market in the United States, you would need to comply with the detail the FDA describes. For this reason, I will summarize the FDA record-keeping policies.

There are five classifications for record files:

- Design/device history file (DHF)

- Device master record (DMR)

- Device history record (DHR)

- Quality system record (QSR)

- Complaint file

The DHF is not discussed in this section, but is nevertheless an important record, and second only in importance to the DMR for design engineers. I have more to say about this in Section 3, on design controls, since the DHF is really the documentation of the design history. These are the records of design reviews, technical reviews, verification activities, analysis, and testing that collect the history of the design. ISO has no specific requirement for a DHF.

Creating the DMR is the whole point of the design phase. It is the collection of drawings, specifications, production processes, quality assurance methods and acceptance criteria, labeling, and installation and servicing procedures. In other words, it is the set of documentation that tells a manufacturer how to build a device. These specifications must all be controlled documents under the document control procedures. Software specifications are explicitly included as a component of the DMR.

The DHR comprises the data for each particular device. This must identify each device or lot, the labeling, dates of manufacture, and the acceptance tests used to show that the device was manufactured according to the DMR.

The QSR is the set of SOPs, compliant with FDA regulations that are not specific to a particular device. These describe the quality system itself, and the practices, policies, and procedures used to ensure quality in the device and its manufacture. It is important to occasionally analyze the procedures to determine whether they are inadequate, incorrect, or excessive [7].

The complaint file is a much bigger deal for the FDA than for the ISO. In the ISO guidance document, it is merely referenced as an example of a type of system record. ISO 13485 extends this to a near-equivalent of FDA requirements.

The FDA specifies that each manufacturer will maintain a complaint file and establish "procedures for receiving, reviewing, and evaluating complaints by a *formally* designated unit" [13]. "*Any* complaint involving the possible failure of a device, labeling, or packaging to meet its specifications shall be reviewed, evaluated, and investigated. . . ." [13] (emphasis added). Even if you decide an investigation is not warranted—because it has been reported already, for instance—you must document why no investigation was made and who made that decision.

Medical care providers are required to report any death, serious injury, or gross malfunction to the FDA and the manufacturer. The manufacturer must determine whether the device failed to meet its specifications and how the device might have contributed to the adverse event. Beyond this, there is also a specific list of the data that the manufacturer must keep with respect to complaints.

A software defect could show up as a customer complaint (also sometimes known as a customer experience report, or CER). Normally you would want the complaint handling to be separate from your software problem reporting system, because of the formality required to deal with customer complaints. But expect the complaint file to be the origin of some of the defects in the software problem reporting system.

All of these records need to be carefully maintained, backed up where electronic, and made readily available to the FDA for inspection. You should keep records for the expected lifetime of the device, but not less than 2 years from date of release.

2.14. Subpart N—Servicing and Subpart O Statistical Techniques

Where servicing is required by the product, both FDA and ISO require documented service procedures and records. The FDA is more specific than ISO about exactly what data go into the service report.

Servicing is an opportunity to collect metrics on the process capability and product characteristics. Collecting statistics is a recommended good practice, but it is up to the manufacturer to decide its appropriateness.

2.15. Post-Market Activities

2.15.1. Audits

A key practice of quality systems in general is the conduct of audits. The purpose of an audit is to have an independent reviewer assess how well the organization is following its procedures and meeting its quality goals. Both FDA and ISO recommend internal audits, at least annually. For ISO, it is necessary to have an independent audit by the notified body in order to retain certification. These are audits that you pay for, so they tend to be more congenial than an FDA audit. They are reviewing your compliance and suggesting ways to help you better comply with the letter and spirit of the quality systems.

The FDA, on the other hand, is fulfilling its regulatory duty. FDA staff take a more adversarial, even suspicious, approach to protect public safety. They are less likely to give you the benefit of the doubt. Plus they have had many years of experience with product safety and failure, and have learned what kinds of practices may lead to injury. They are an enforcement agency and are not to be disregarded. They have badges.

They can look at the DHF and the change records. "The holder of a 510(k) must have design control documentation available for FDA review during a site inspection. In addition, any changes to the device specifications or manufacturing processes must be made in accordance with the Quality System regulation" [14].

As a software engineer, it is unlikely that you would have direct interaction with an auditor in all but the smallest of companies. It is usually the responsibility of the QA/RA part of the organization. Nevertheless, if it does come up, there are a few things to remember about handling an audit.

It is best to think of it as a kind of legal interaction. First of all, don't offer information. Answer the questions they ask, but don't volunteer more than they are asking for. It can provide an opportunity for them to dig further or suggest ways in which your organization is weak.

Second, just answer the questions factually. Don't speculate or offer an opinion about how effective a process is or how well something works. Don't guess. It is okay to say, "I don't know" (although that may cause them to write you up for poor employee training).

Finally, and in a related vein, don't get caught up in hypothetical questions. If they ask you what you would do in a certain situation that has not occurred, the best answer is, "We have not dealt with that." If they press, the answer is, "We would follow the procedure."

You don't need to stonewall or be a jerk, but you also don't need to express your concerns. That's their job.

3. Design Control Explained

3.1. Purpose of Design Control

The quality system regulation describes at a high level the process for developing customer needs into a marketable medical device. This process is found in Subpart C of 21 CFR 820 [15] for the United States. It is known as design controls to the FDA but goes by the singular design control in ISO Q9001, Section 4.4. It encompasses designing a new product and transferring that design to manufacturing. The objective of design control is to "increase the likelihood that the design transferred to production will translate into a device that is appropriate for its intended use" [16].

Design control is an outline of a new product development process and is general to all aspects of product development, mechanical and electrical, not just software development. The regulation itself is not very long—about 700 words—and not very specific. The software development process needs to fit into the overall process, but software will have extra phases and detail.

Almost all Class I medical devices are exempt from design control, unless they contain software. (For a discussion of device risk classes, see Section 5.7, Device Risk Classes.) Class II and Class III devices are always subject to design control. Since the purpose of this chapter is to discuss software for medical devices, any process we propose will be subject to the design control regulation because any software-controlled medical device is subject to design control.

The design control process is like the waterfall model for software development but even simpler (Fig. 4.2). There are four phases: design input, where the customer needs are determined and described in engineering language; device development, where the design is

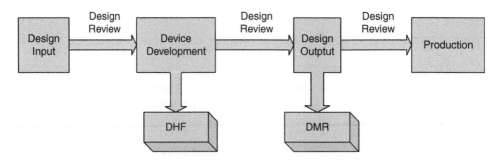

Figure 4.2: Design control process.

fleshed out, and the design process captured in the DHF; design output, where the final design is captured in the DMR; and production, where the device is repeatedly and reliably manufactured for sale. The transition between phases is always qualified by a design review.

It is possible for a simple medical product with a handful of requirements to go through the design control process in a single phase of design input, device development, and design output. From the get-go, software has more phases, so it is only meaningful to discuss design input, device development, and design output as the boundaries between the phases in the waterfall model. One set of design outputs becomes design inputs to the next phase. Thus in a very high-level sense, software development follows the design control model, but sometimes people are confused because there is so much more detail in the software development process. The design control process was invented to encompass all kinds of projects. A simple project like the development of an extension to the application of a catheter (a new claim for what the catheter could do, with some minor changes) is something that might have a single design input/design output phase.

For more complex devices, this really becomes a stepwise refinement of specifications to more and more detail until the final device is fully specified. Once the team has reviewed the design input requirements and found them acceptable, for example, you can begin the iterative process of translating those requirements into a software architecture. Once the architecture is verified as a correct response to the high-level requirements, this output then becomes the input to the next phase of software design. "Each design input is converted into a new design output; each output is verified as conforming to its input; and it then becomes the design input for another step in the design process. In this manner, the design input requirements are translated into a device design conforming to those requirements." [16]

I will first discuss design control as described in the regulation in the general sense in which it applies to product development as a whole. Further on, in Section 5, Software V&V in the Context of Design Control, I describe the way software development fits in to the design control process. (This is my own stepwise refinement of QSR → design control → software development process.)

In terms of the regulations, those who develop software to control a medical device "shall establish and maintain procedures to control the design of the device in order to make certain that specified design requirements are met" [17]. Notwithstanding language about the "least burdensome approach" and some freedom in selecting software-development life cycles, compliance to the design control regulations will probably look like a traditional, rigorous approach to the software development process, with a focus on documenting the process steps and holding reviews at each phase of development.

It is especially important that the design of software be carefully controlled. The FDA recognizes that the quality of software is not dependent on its manufacture, which is contrary to the case with physical devices. There is no process variability to duplicating software—its manufacture is usually trivial. What is critical is the quality of the construction in the first place.

Because it is so easy to change, some fall into the fallacy of thinking that software is easy to correct and thus does not require controls as stringent as the ones for hardware. The FDA holds the more sophisticated view that, "[i]n fact, the opposite is true. Because of its complexity, the development process for software should be even more tightly controlled than for hardware" [1]. Insofar as the goal of design control is to ensure that the design output meets customer needs and serves its intended purpose, controls are even more important. Software can more easily adapt to new knowledge about customer needs, which is a great strength, but the process must ensure that real needs continue to be met.

A feasibility study does not need to meet the requirements of design control, but when you have decided to produce a design, you will have to create a plan so that the device will meet the quality requirements. It is tempting to just develop the prototype, and then reverse engineer the design control into the project. This is not the best method. The FDA has found that "[u]nsafe and ineffective devices are often the result of informal development that does not ensure the proper establishment and assessment of design requirements that are necessary to develop a medical device that is safe and effective for the intended use of the device and that meets the needs of the user" [16, 18].

One final note: the FDA is prohibited by law from determining the safety or effectiveness of a device by looking at the design control procedures that a manufacturer uses—safety and effectiveness determination is the purpose of the premarket submission. However, they do audit the SOPs for an adequate implementation of design control. So the device could be safe and effective, but inadequate procedures could still land you in trouble.

3.2. Project Planning

The first step in design control is project planning. Project planning is necessary for the organization to determine where it is going, and even more important to have an idea what it might look like to arrive there. It is crucial to have the conception of what the project goals are, and what it means to meet customer needs and be fit for intended use.

It is not specified in the design control guidance, but I've always found it helpful if the project plan starts with a mission statement or vision of what accomplishing the project means. A mission statement is a description in 50 words or less of the goal of the project. GE HealthCare has a concept of CTQs—the list of features that are critical to quality. These are the features—no more than five or six—that the product must have to fulfill customer needs and achieve the product goals for the company.

The purpose of the project vision is to focus the development team on what needs to happen, and more important, on what does *not* need to happen. The essence of good design is deciding what to leave *out*. A strong vision statement can focus decisions when a room full of bright people start brainstorming all the things a product *could* do. Usually, doing everything in the world is not a very wise approach to medical product development. It takes longer, and delays the day when our products reach the market and start helping people. It adds to complexity, and higher complexity reduces reliability in devices that must be safe. And it drives engineers nuts.

It is not easy to write a good vision statement, and many are so meaningless that it may lead you to be cynical about the whole idea. They can end up as a mom-and-apple-pie statement that is hard to argue with. For example, a company could start a project to "build the world's best word processor." Unfortunately, that does not leave any basis for deciding what to leave out. Any idea that anybody comes up with must be included—we are trying to build the "best," after all, and how can it be best if it doesn't have every feature imaginable? But projects like this may never end, and even if they do, they are often late and cluttered with features that get in the way of the essential function.

Note: One way to screen out happy talk is to apply a test for information content. To do this, negate the statement and see if it is obviously false. So negate the original example, "We will *not* build the world's best word processor." Clearly, no one would set such a lousy goal.

A better vision would be to "build the word processor that allows writers to get words on paper in the easiest, fastest way." Or "build the word processor that allows casual users to format the look of the document in the easiest way." These have enough content that they communicate the decision about what the product is going to be and what it is not going to be.

A second purpose of planning is to reduce the number of false starts and distracted pathways. This is, of course, advantageous to the company; it is certainly not in the interest of the development organization to pay for development that will never be used. But it also enhances quality. Even though by its nature development will often be iterative, we don't need to indulge more iterations than we have to. Change management is difficult and important to control, but the easiest way to control change is not to have any. Attention lags the tenth time you've done something, and errors creep into upstream documents when it is necessary to revise downstream documents in response to new knowledge.

Of course, there will always be new knowledge, and the last thing we want to do is ship the wrong product come hell or high water. No plan survives contact with reality, but that does not mean that planning is not useful.

Finally, the purpose of planning is to reduce the pressure to compromise quality when projects are behind schedule. The FDA recognizes that deadlines have contributed to defects introduced when designs were not carefully considered due to lack of time, and these defects have resulted in injury [16]. Project plans should emphasize accomplishing quality goals over calendar goals. Good plans let managers make supportable decisions when it is necessary to compromise the project to meet deadlines.

The amount of detail in the project plans is going to vary with organizational needs, the size and complexity of the project, the work habits and personalities of the team, and, as always, the level of concern of the device or software. It is not necessary to plan every last detail. In the FDA's own words, "Each design control plan should be broad and complete rather than detailed and complete. The plan should include all major activities and assignments such as responsibility for developing and verifying the power supplies rather than detailing responsibility for selecting the power cords, fuseholders [sic] and transformers" [1].

It is important for the plan to describe the interfaces between the contributors and the stakeholders that have input to the design process. To be effective, the plan should establish the roles of the groups involved in the design and the information they share. A typical plan would include:

- Proposed quality practices

- Methods to assess quality

- Policies for record-keeping and documentation

- Sequence of events related to design.

3.3. Design Input

Once preliminary plans are formulated, the device design begins with the design input phase. The starting design input is the as-yet unformed user needs or amorphous marketing concepts. A vision statement will have gone a long way toward deciding what the product will be. The project vision statement, if you use one, will be the first formal description beginning to refine the vague wishes of customers into a product concept. With this, the task of the engineer is to turn the user needs into design input requirements with enough detail and formality that they can serve as input to subsequent phases in the development process.

A distinction that has caused me confusion is naming the phases. Design input is not a thing or deliverable, but a phase of the development process. It has inputs and deliverables (outputs), but they should not be confused with design input—what is really meant is the phase.

In the traditional software waterfall model, the design input phase maps to the requirements analysis phase. So the input is the customer needs and user requirements, and the output is the software requirements specification (SRS). From the project perspective, where design control is describing whole product development, not just software, the input to the design input phase is equivalent—customer needs. The output of the design input phase will vary with the complexity and nature of the project.

Projects of moderate complexity that involve both software and hardware have more elaborate needs for what has to happen in the design input phase. It is still requirements analysis, but now the requirements are for a more complex system containing elements of hardware and software that have to interact to satisfy the product requirements.

The output of the design input phase is design description documents. These documents define the product's:

- Functional and performance characteristics

- Physical characteristics

- Safety and reliability requirements

- Environmental limits for safe use

- Applicable standards and regulatory requirements

- Labeling and packaging requirements

Human factors analysis and testing should be conducted and used as input to define the function and performance of both hardware and software, including testing the instructions for usability [19]. The specifications should be quantified whenever practical. Where that is difficult, the design document should identify the parts of the design that require further analysis and testing. The document should record any incomplete, ambiguous, or conflicting requirements and provide a mechanism for resolving them.

Developing the system design requirements document is often a function of a systems engineering group. Many organizations may not have the wherewithal to sustain an entire engineering department devoted to systems engineering, yet the systems engineering must occur. It is where the electrical engineers, mechanical engineers, and software engineers get together and allocate requirements to subsystems.

Often you have the design freedom to decide where best to implement a requirement. For example, if you had a requirement to make an alarm sound at multiple frequencies so that the aurally impaired stood a greater chance of hearing it, you could try to find a sound transducer that would produce the sound at the desired frequencies and loudness, or you could create sound waveforms in software in your main processor, or you could devote a special processor to it. Each has tradeoffs in product cost, development time, power usage, and so on. But if the process does not happen, it would be easy for a software development team to fail to schedule time to develop waveforms to output to a driver circuit that the hardware engineers had designed.

System engineering becomes especially important in the development of requirements around hazards. I'll have more to say about the subject in Section 4 on risk management.

It can be difficult to collect all necessary and correct requirements during the design input phase. As Nancy Leveson has observed, "On many projects, requirements are not complete before software development begins. . . . To avoid costly redesign and recoding, the requirements specification and analysis should be as complete as possible as early as possible. Realistically, however, some of the analysis may need to be put off or redone as the software and system development proceeds" [20]. For this reason, write the system design requirements document so that it can change as issues are resolved and new knowledge gained. The document (or the design input process) should define what would trigger an update, and who is responsible. At the conclusion of the design input phase, the system design document is subject to design review and requires "the date and signatures of the individual(s) approving the requirements" [19].

The FDA realizes that "[a]s the physical design evolves, the specifications usually become more specific and more detailed" [21]. Nevertheless, if possible, the team should strive to get the upstream requirements analysis as correct as possible. As is commonly known, the later in the development process a defect is found, the more expensive it is to correct. (This is especially true for medical devices, which are subject to postmarket surveillance and complaint handling regulations, as explained in Section 2, The Medical Regulatory Environment.) Furthermore, correcting a defect later may have ramifications for the project as a whole. It is much easier to get it all right when thinking about it, especially for hazards, than to go back and reconstruct all the thinking that went on during the risk analysis. Often risks interact in complex ways; a superficially insignificant change in one subsystem may have influences on the rest of the system that are poorly understood.

At the same time, it is possible to spend almost unlimited time in analysis, contemplating behaviors and risk probabilities that cannot be known without some level of prototyping. Perhaps prototyping is needed before the design input phase is complete—even then, what you know about is a prototype and not necessarily the thing that will be built.

Better to not expect to resolve every to-be-determined (TBD) or think that you will capture every requirement. The first drafts of the requirements are documents useful for refining the project planning documents. It is beyond the scope of this chapter to discuss scheduling methods and project planning in depth. But generally what happens is a negotiation of the set of requirements—the features the product has—versus the time and resources available to develop them. During this negotiation it is vital to have a sense of what the requirements are and that is what I mean by the utility of the draft requirements as input to the project planning.

Stepwise refinement of the requirements will occur throughout the development of the project, but this will require effort to maintain consistency between the downstream documents and the upstream documents. Moreover, one of the deliverables at the end of the project is a traceability matrix (see below) in which you establish that all requirements have a corresponding test (and vice versa). So if you had discovered new requirements during development, you would want to be sure that they make their way back into the upstream documents.

One thing to make things easier is to resist tightening the bolts too early. A lot of this is just the use of standard good coding practices. Keep things decoupled. Use accessors and mutators to get at data so that the data itself can change. Keep the data abstraction at a high level. Use identifiers for constants; don't use magic numbers. You can put off deciding a lot of things like thresholds because they will be easy to change in a solid software design. But you can't put them off forever—sooner or later you will have to establish why something has one value and not another. While it can be easy to change in software, it can be difficult to validate that a change is justified. Changing the number may be trivial, but the validation of the value of the number is crucial. (Remember the Hubble example!) "The device specification will undergo changes and reviews as the device evolves. However, one goal of market research and initial design reviews is to establish complete device requirements and specifications that will minimize subsequent changes" [7].

There are several more things that I want to say here about requirements management.

You can use word processor–based documents and it will probably work for small projects. As project complexity increases, you would want to look at a requirements management database. This would let you view the requirements from different perspectives. For example, a system hazard analysis may want to view requirements from the perspective of which requirements are essential to the function of the device. But normally that is not the way to organize requirements as input to the design process—you would want related requirements close to each other. A requirements database would let you run reports to view the requirements in a narrative that made sense for the application at hand, without changing or repeating the requirements themselves. You can organize the requirements in different ways for different audiences.

Furthermore, establishing traceability from a database is much easier compared to tracing in word-processed documents. About the last thing most projects need when they are close to delivery is a bunch of time spent laboriously tracing requirements from narrative documents to test documents.

Next, normalize requirements. Try to describe a requirement only once, so that if it changes you have fewer places to have to change it. Use pointers to reference the associated requirements or requirements that are presented in greater detail. For example, you could define the values of constants in a source file and use that as the requirements specification. This will make you take care to supply good comments!

This is another spot where a database will help by letting you define a requirement only once, even though it may exist exactly as written in multiple downstream documents.

Normalizing requirements is complicated by the premarket submission process. You cannot assume that a reviewer will have access to all the documents, or the ability and time to understand your system. It is most helpful to reviewers if the documents in the review package can stand by themselves. This makes the documents repetitive: they become more comprehensible to your reviewers but more defect-prone to you. Again, a database would help to contain the boilerplate that can then be added to each document as needed.

A requirements database has enormous advantages, but don't forget that you have to perform some level of SOUP validation. But overall it makes the system more resilient to change; change is necessary and inevitable, so installing a system that makes change management easier, less costly, and less likely to result in error is sure to be a good thing.

Requirements management is often so complex that there is a great temptation to wait until the system is built and then document the requirements that it fulfills. A retrospective method like this is not against the rules; indeed, with legacy projects it might be the only way to create the upstream documents at all. Nevertheless, it is not the best design approach and should be avoided if possible. What really needs to happen is that the product should be designed according to the principles of good engineering practices from the beginning.

3.4. Design Output

The design output is the response to the design input. The deliverables that constitute the design output are part of the project planning process, the nature of the project, and the standard operating practices of your company. "The total finished design output consists of the device, its labeling and packaging, and the DMR" [22]. The DMR, or device master record, the complete set of documentation that describes how to build the product. Hence it would consist of things like schematics, assembly drawings, BOM, work instructions, source control drawings that describe how to buy parts, and test specifications for manufacturing and inspecting the device and its subcomponents. The DMR is subject to design review, and the date and signature of any individuals who approve the output must be documented.

The general rule is that design output is the deliverable of a design task in the development planning document. (This is another reason why the planning is important.) For software, the equivalent type of design output is typically source code, the executable, operator's manuals, and documentation of verification activities. One principal verification activity is to review the design output against the design input to confirm that the output matches the input.

In other words, the software implements the requirements. This is usually captured in an acceptance test, software validation protocol, or software system test protocol to demonstrate that the requirements are met.

3.5. Design Review

The design review is a key practice of the design control process. Design reviews should occur at major milestones, such as passage between phases of the design control process. To the FDA, "Design review means a documented, comprehensive, systematic examination of a design to evaluate the adequacy of the design requirements, to evaluate the capability of the design to meet these requirements, and to identify problems" [16]. The purpose of design review is to:

- Assess the quality of the design

- Provide feedback to the designers on possible problems

- Assess progress toward completion of the project to determine if passage to the next stage of development is warranted

When a device is available from initial manufacturing production, the development team will perform a final design review to confirm that the device meets the design specifications as described in the DMR. This is the most important of the design reviews, and it is essential that the review be against a product from pilot production, because the purpose of the final design review is to establish that the manufacturing arm of the organization can in fact routinely manufacture the product that the engineers designed.

Design review will be a formal meeting. The authors of the work product or subsystem designers should be present, as well as the designers of interfacing subsystems when applicable. It is important that at least one reviewer be disinterested, that is to say, should not have direct responsibility for the work product under review, to avoid conflict of interest.

The output of the review should be a list of action items. While it is possible that such a fine job was done that there are no issues to resolve, the FDA is wise enough to realize that the more likely explanation for such an outcome is that the review process is inadequate. FDA staff expect to see issues arise as a sign of due diligence to the design, especially in upstream design reviews. What is important, though, is that those issues be recorded and resolved and tracked to closure. The design review should have minutes in the DHF documenting that the review took place and the issues raised.

The minutes should include:

- Moderator and attendees

- Date and design phase or stage

- Agenda

- Work product under discussion

- Problems and issues identified

- Follow-up report(s) of solutions

- *Or* specifies that the next review covers the solutions and remaining issues

Closure needs more discussion, because this turns out to be the hard part. It is usually not wise to hold off publishing the minutes until the items are resolved—they may involve significant rework or not be solved until subsequent phases. Yet it is vital not to lose sight of issues, or to let them build up into a "project bow-wave" of unhandled problems that must be addressed when the project is under pressure to ship.

One means of following issues to closure is to use the *software problem resolution* (SPR) process described later in this chapter. Issues could be entered as defects and treated with the same workflow as bugs. They would be resolved when new revisions of the documents containing corrections are released under normal document control procedures. Using the SPR process means that all issues—mechanical, electrical, manufacturing, and process, not just software—would be part of the same anomaly resolution process. In this case, it would be the *system* problem resolution process.

What I have described here is the design review process requirements as spelled out in the QSR and the design control guidance. The label *design review* has special meaning to the FDA. It is a system-level review that occurs between the phases of a design control project and especially the manufacturing readiness review. It usually takes several days and involves many personnel, including management.

Reviews closer to subsystems that occur throughout the device development that are confirmations that design output meets design input are really better called *design verification*. This is because they are not meant to be definitive, comprehensive, or multidisciplinary, as design reviews would be. It is helpful to refer to these as *technical reviews*. It is useful to make the distinction between design review and technical review, because some of the policies for good technical reviews are inconsistent with design review under the QSR. For more on conducting technical reviews, see Section 5.1, Verification Methods.

3.6. Design Verification and Validation

Designs are subject to design V&V. Design review is one type of design verification. Others include calculating using alternative methods, comparing the new design to a proven design, demonstrating the new design, and reviewing the design deliverables before release.

The manufacturer should conduct laboratory, animal, and in vitro tests, and carefully analyze the results before starting clinical testing or commercial distribution. "The manufacturer should be assured that the design is safe and effective to the extent that can be determined by various scientific tests and analysis before clinical testing on humans or use by humans. For example, the electrical, thermal, mechanical, chemical, radiation, etc., safety of devices usually can be determined by laboratory tests" [7]. These are usually tests specified by IEC standards.

Software system testing (sometimes known as software validation) is also conducted in the design verification phase. I have more to say about this in Section 5.2, Software System Testing. Two things are important to remember at the regulatory level. First, the software system tests need to be conducted on validated hardware, that is, hardware from pilot production that has been tested to show that it meets its specifications. Second, any tools or instruments used for the software system testing must be calibrated under the policies in Subpart G, Production and Process Controls.

And in keeping with the FDA emphasis on labeling, during verification all labeling and output must be generated and reviewed. Instructions or other displayed prompts must be checked against the manufacturer's and the FDA's standards and vis-à-vis the operator's manual. Testers should follow the instructions exactly to show that they result in correct operation of the device. Warning messages and instructions should be aimed at the user and not written in engineer's language. Any printouts should be reviewed and assessed as to how well they convey information. Patient data transmitted to a remote location should be checked for accuracy, completeness, and identification.

The FDA makes no specific mention of it in the regulations, although in its guidance document mentions that "[d]esign verification should ideally involve personnel other than those responsible for the design work under review" [23]. This is a requirement for ISO. Certainly it is a good practice to have an independent tester or reviewer. It would help to have someone not intimately familiar with the device, and hence less likely to infer information that isn't there when reviewing the labeling, for example.

Design validation is also necessary for the device to show that it "conform[s] to defined user needs and intended uses" [24]. Design validation follows design verification and must be conducted on actual devices from manufacturing production using approved manufacturing procedures and equipment. This is because part of what you are validating is that the complete design transfer took place and manufacturing can build the devices repeatably.

Not all devices require clinical trials, but they all must have some sort of clinical evaluation that tests them in a simulated or, preferably, actual medical use environment by the real customers, users, and patients that the device is intended to help.

3.7. Design Changes

The regulatory bodies recognize that evolution is an inherent part of product development. At the same time, it is important that changes made after the initial design inputs have been decided on must be reviewed, validated, and approved. This is something that makes getting the design inputs as right as you can so important. You don't have to keep track of changes very early in the project (although it may help to do so—see Section 3.11, Software Configuration Management Methods). But once the design is approved in the input phase of the design review, you need to document any subsequent changes to the design.

There are multiple reasons that a product may change after the design phase or during production:

- Discovery of errors or omissions.

- Manufacturing or installation difficulties.

- Product feature enhancements.

- Safety or regulatory requirements have changed.

- Obsolescence.

- Design verification makes change necessary.

- Corrective action makes change necessary.

If the change causes revisions to the system design input document, you would re-release it in the document control system. Any change should be reviewed for the impact that it may have on the rest of the system, especially its impact on completed design verification. You should also have methods to ensure that changes are communicated to stakeholders. Finally, for products already on the market, you must consider whether regulatory approvals such as a 510(k) are required. See Section 5.7.3, When Is a 510(k) Required?

An easy trap to get into is to allow R&D personnel to change a released device without following the change control process. This circumvents necessary evaluation and review and has been known to result in ineffective or hazardous devices. It is a sign that production is not operating in a state of control. All changes to production devices must be made according to approved change control procedures [7].

3.8. Design History File

The DHF is the repository for the work products developed during the design control process. This is a specific requirement for the FDA but only implied for ISO. The DHF captures the evolution of the device specifications that form the DMR. In addition, it contains evidence that the design process conformed to the design plan and design control procedures. It contains the records of verification activities showing that the final design meets the device specifications.

An intent of the DHF is to capture the history of the design process, rather than just the output of the design process. It is valuable to record not just what decisions were made but why they were made. This is useful in that when circumstances change, you don't have to repeat the analysis work that led to making a particular decision.

It sometimes happens that an issue comes up, and we remember that we decided the issue one way, but don't remember why we rejected alternatives. If the alternatives are captured in the DHF, we are able to revisit the decision after circumstances change. For example, we might not have used an algorithm because the processor resources were inadequate, but now we have changed processors and the algorithm would be advantageous. If the original analysis is available in the DHF, we can use that instead of duplicating the work to determine resource requirements of alternative algorithms. "This information may be very valuable in helping to solve a problem; pointing to the correct direction to solve a problem; or, most important, preventing the manufacturer from repeating an already tried and found-to-be-useless design" [7].

For the non-software disciplines, the DHF is found in the company document control system (for a history of important deliverables that have multiple revisions) and a parallel archive for one-off documents, such as meeting minutes. Software artifacts could be stored in the document control system or R&D archives like everybody else's document. This is a function of the SOPs for your organization. If software were but a small part of a product, the company document control system might be sufficient. Generally, however, software needs to be archived in a version control system (VCS). The overhead of writing and approving ECOs and so on for software is usually overkill. It is easy to imagine a modest team of developers making hundreds of changes a day—which would overwhelm any document control system. The beauty of a VCS is that not only can we make the changes easily, but because our artifacts are often text, we can identify every change down to the semicolon or space character at the end of a line.

Our company policies define the VCS as the software DHF. Any electronic document that goes in the DHF will have a copy in the VCS. As far as the FDA is concerned, "Diverse records need not be consolidated in a single location. The intent is simply that manufacturers have access to the information when it is needed" [16]. So there is no reason to create policies that cause the process to get in the way of saving the information needed to show the evolution of the design and its verification activities.

The DHF is a legal record; as such, it means that if something isn't written, it is more difficult to assert that it happened. Hence, any meeting in which the team decided something important should have minutes written, circulated, and copied to the DHF. A handy way to do this without a lot of overhead is to set up a special e-mail address in the company address book, and then send minutes to the e-mail DHF and carbon copy the meeting attendees.

Note: "Something important" is an evaluation made by the team. One of the frustrations of trying to determine how far to take documentation is that the FDA will say the design should go far enough that the implementer does not have to make ad hoc decisions [1, 3]. Well, anyone in software development knows that you make hundreds of decisions a day—what to name variables, where white space is appropriate, what needs a comment, what data structure to use, and so on, ad infinitum. What exactly is ad hoc, and what is at the designer's discretion? This is up to your company policies and the evolved culture. Just remember to capture the policies in SOPs, and have methods to ensure compliance.

One way to capture design notes and test results is by using an engineering notebook. The practices you use are the same ones you would use to fill out a notebook as evidence that you were the first inventor on a patent. The notebook should have bound, numbered pages— usually companies use official notebooks dispensed from document control and assigned to each engineer.

Write in dark ink, and title each page. Do not skip pages. Line out unused portions of a page. Do not use a shorthand unless you also include a glossary. If you make a mistake, line out the incorrect entry, initial and date the line out, and add a correction. Sign and date each page as you fill it in. It is a good practice to have someone review the entries, sign, and date as well.

You can tape printouts or photos into the notebook. To do so, initial and date across the hard copy, the tape, and the underlying paper. (The intent is to make it hard to falsify the hard copy by replacing one printout with another.) If the data can't be conveniently included in the notebook, be sure to provide a reference for traceability. Test output could be archived in your VCS and the reference to the particular revision written in the lab notebook.

3.9. Change Control

Device development is an evolutionary process. While we try to do our best to do it right the first time, we don't always succeed. A hallmark of an effective quality process is that it does in fact discover defects. This is true of design review as well—it should result in the need to make changes. (Otherwise, the review process itself needs a look in case it is not having the quality results that it should.) The issues raised need to be documented and their solution tracked to completion. This is the scope of change control.

The FDA's design control guidance says, "For a small development project, an adequate process for managing change involves little more than documenting the design change, performing appropriate verification and validation, and keeping records of reviews" [16]. The objectives are to:

- Track corrective actions to completion.

- Ensure that changes did not create new problems.

- Update the design documentation to match the revised design.

One of the assumptions underlying the expectations of the regulatory bodies about change control is that, as the product evolves, the rigor you apply to evaluating a change increases. This only makes sense. You will have accrued a lot of testing, verification, and analysis that a change made willy-nilly could invalidate. While it should be captured in the system hazard analysis, it may be possible to forget that a particular approach was not taken because of its risks to safety. If you have conducted the design input phase properly and held design reviews to accept the design output, you should trust your designs and make sure that any changes are for the better.

Most companies use the document control system already in place to control changes in the engineering disciplines other than software. Especially during development, use of the document control system for software would have excessive overhead and at the same time insufficient precision. It is generally a better practice to use a VCS (the tool) to perform software configuration management (SCM—the process).

Software problem reporting—the origin of the changes that occur to software, and SCM are significant issues in software development. We now move from the design control in general to the particulars that matter to software.

3.10. Software Change Control in the Medical Environment

True to form, the FDA does not require the use of a VCS, in case your project is simple enough not to need it. Nevertheless, in order to readily create the deliverables expected for a design control project, a VCS would be very helpful.

A VCS tracks the history of changes to software and stores it in archives. Archives contain the current copy of the file and enough information to retrieve any revision of the file, back to the original. Along with the changes, archives contain descriptions of the changes, information about who made the changes, and the dates and times when the changes were made. You can show the difference between any two revisions, which is very useful for analysis about how much regression testing a change may require.

The scope of artifacts to version control might be larger than you think. Of course, source code and project documentation (SRS, SDD, etc.) would go into version control. But it is also important to keep versions of test results, design worksheets, integrated development environment workspaces and projects, and really anything that you might want to reference in the future.

In addition, it is important to archive the tools that you use to create your software. At the very least, keep track of the installation disks and documentation, even if you don't put it in version control. You want to be able to re-create from the source any version that you release.

Consider archiving support documentation like datasheets for electronic components at the revision level that you used to write the software. Generally the datasheets are available on the Internet, but they could in future be hard to find, or hard to find an obsolete revision that someone might need for maintenance. Also, it then becomes a simple matter to find the exact reference in your own VCS. These are items that easily forgotten, yet crucial for answering questions years in the future.

Generally, derived files would not be archived, except for an official release. However, it is handy to have the executables (and the corresponding map file) for intermediate releases of a project, such as the out file for firmware. Then rebuilding to get an older version for distribution is not necessary.

You can be expected to re-create any older version of software. The FDA reports the story of a medical device whose software was suspect. The manufacturer was unable to produce the source code for the device because a contractor had developed it and only delivered a master EPROM. After a contract dispute, the contractor withdrew with the rights to the source. The device was subsequently recalled and all known units destroyed [16].

This story is an illustration of the usefulness of VCS or some kind of source code archive to the manufacturer, but is also an object lesson in the issues with subcontractor management and software of unknown provenance (see Section 6.1, Software of Unknown Provenance).

3.11. Software Configuration Management Methods

The essential features of a VCS follow:

Control access. Limit access to archives with assigned privileges. Only individuals with the correct authorization can make certain changes—very important to the regulatory bodies.

Track activity. Create an audit trail for each file, telling who changed a file and why it changed.

Control revision storage. Provide a central, backed-up repository where current copies of all revisions are stored.

Manage concurrent access. Any development that requires more than a single individual will need some method to prevent one engineer overwriting the changes of another. So-called *pessimistic* version control uses locks to restrict access to a file to one user at a time. *Optimistic* VCSs use a branching model along with merge tools to allow concurrent development.

Promotion model. Promote revisions to different baselines on successful passing of a project milestone. In combination with control access, this enables increased control of changes as the product matures.

Difference reports. Produce a revision history of the software and show the differences between released versions to support evaluation of limited regression testing.

Part of development planning is to create an SCM plan. (This is one of the documents that would go into a premarket approval [PMA]. See more about this in Section 5.9, Software Documentation Requirements for Premarket Submissions.) The plan will be more or less elaborate depending on the complexity of the project and the number of people working on it. At the very least, it should explain the structure of the software archives and the policies that determine what files go where.

For a one-person project, it could be as simple as stating the version control method used. For especially simple projects, this could be use of the company document control system (although I believe even the simplest of projects benefits from a VCS).

Larger teams are going to require policies to manage concurrent development. One method is to lock files. Then only the person modifying the file can change it; others have to wait their turn. This prevents any possibility of losing changes or introducing inconsistencies when the same code is changed by two different people. This method works fine in small teams working on highly decoupled software.

Projects that require greater collaboration benefit from a VCS that supports a branching model. Branching allows you to make changes and test them in a single baseline, and propagate the change to other baselines. You could, for example, fix a bug in a released version of software, and then transfer that bug fix to development branches without having to copy and paste every change by hand.

Systems that support branching come with powerful merge tools that can integrate changes made by two different people to the same baseline file. They can usually do this automatically; if the code was changed in two different ways in the same location, you may have to merge by hand to select the correct version.

If you use a branching model, your SCM plan should explain what the branches are and what conditions would cause a new branch or work to commence on a dedicated branch. For example, you might have three branches as follows:

Main—Each project will have a main branch. For simple projects, development can be done on the main. The main should be reserved for working code only. Releases will be marked with a label.

Development—Projects of greater complexity or involving more than one engineer may have one or more development branches to control delivery of changes to the main branch. These are for major work that might "break the main" for a period of a few weeks or months. This supports the idea of keeping a working version of software available on the main for interim verification purposes.

Released—Projects may have but do not require a release branch. Release branches are for significant work on a released baseline that will not be integrated back into the main line of development.

This scheme could be expanded on for larger and larger projects. It may be part of the project plan to assign major branches to teams for particular features. These branches could have sub-branches for smaller teams, and branches on those that are owned by individual engineers. There might be multiple release branches for different releases of the product. Discussion of the release branch strategies and policies is something that should go in the software maintenance plan, another control document required as part of a PMA.

There are a handful of policies that are worth enforcing. First of all, no changes get made outside the VCS. People have to check out and check in code to deliver their work—no exceptions. Don't "jump the main." That is, don't make a change in a development baseline that you merge to a release baseline without merging to the main first. Integrate the changes often; aim for at least once a month. If development is going off on a branch for longer than a few weeks, it probably means that you should re-plan the effort to take on less at once. Having parallel development go on for a long time makes the merge, when if finally comes, more complex and much more difficult to debug because so many files will have changed. Require changes to be tested on the development branch before being merged to the main. Keep the main functional. It should be deeply embarrassing to anyone who breaks the main.

Require a comment with any change so that no changes are made without a reason (the FDA hates that). The manager should occasionally audit the comments to make sure they actually contain information. It is useless to say "changed threshold to 180" when that is apparent from the difference report. Instead the comments should say *why* the change was made. (This may be provided by the software problem reporting process; see the next section.) View the file differences before checking them in. This could go as far as not allowing the author to check in changed files, especially to a released baseline, requiring the approval of a reviewer first, who is the one who actually submits the change.

Organizations and individuals might differ in their attitudes on how often to check in code. On the one hand, when checked in, you have a version in the backed-up repository. On the other hand, it is nice to rely on checked-in code to at least compile. The number of times that a file is revised can be a metric for a file that may be a source of trouble. There may be perfectly good reasons. Or it may be that it has been difficult to get it right, or that it is coupled across the system so many people have had to touch it. On the other hand, having a file checked out for 2 months is not getting the best use out of the VCS or iterative development—there should have been intermediates or steps along the way that showed progress (and testing) toward completion. Revision numbers are free. Make it clear to your team that you expect to see a stepwise refinement of the software until delivery (but many fewer changes after!).

Archiving of directory versions by the VCS is advantageous. Without this feature, you may not be able to move files to more sensible locations. This is because even though you can get all of the correct revisions with a version label, if the files are not in the correct directories you will be unable to perform a build without modifying the build files. Another feature to look for is the ability to mark files *obsolete* so that they do not appear in the source code tree any longer, but are still available if they are ever needed.

Version control is the way that we manage changes. Changes arise from the issues, anomalies, and errors that we have observed in the product. Regulatory bodies and auditors are very interested in seeing that these are tracked to conclusion. Hence there needs to be a software problem resolution (SPR) process that is the origin of the changes we archive in the VCS. This is the subject of the next section.

3.12. Software Problem Resolution

The SPR process is the method by which software issues, observations, change requests, and defects are recorded and followed to their conclusion. Recording defects and tracking them to their resolution is a crucial part of any quality system and this is especially true when meeting design control requirements. However, the FDA has almost nothing to say about SPR, except for the general QSR requirements. ANSI 62304 does have a section describing the elements of a successful process [25].

The ANSI standard calls for seven steps:

1. Prepare reports, including an evaluation of the problem type, scope, and safety criticality.

2. Investigate the problem and initiate a change request if needed.

3. Advise stakeholders.

4. Use the configuration management process.

5. Maintain records including verification.

6. Analyze the problems for trends.

7. Verify the software problem resolution.

Two things to note: first, the process does not need to apply until the software is in the software system test phase. You don't have to use an extensive SPR process during development, when you expect to find bugs. Once the software is thought to be complete, however, you need to start tracking and measuring the defects.

Second, you don't have to implement a fix unless the problem is related to safety. You still have to provide a hazard analysis about the safety impact and a rationale about how the problem does not affect safety.

The following is a narrative description of a software problem resolution process that would satisfy ANSI 62304, including requirements for change control. You could, of course, design a more elaborate process to fit a larger team or other needs of your organization. The diagram in Fig. 4.3 is a graphical representation of the workflow. Please refer to it for the states and transitions described in subsequent sections. Note that I assume the use of some kind of defect management database. You do not have to use a database, but it makes metric collection much easier and is more robust and secure than other methods might be. As with any off-the-shelf software, you should validate the defect tracking database to show that it is fit for its intended purpose.

3.13. Problem Evaluation

3.13.1. Creating the Issue

The software problem resolution process begins with software observation reports (SORs). SORs arise as observations or problems discovered during software development. You don't have to abide by all of the policies of ANSI 62304 until software is close to release, but it may be valuable to use the process anyway, especially for issues that you are not going to address right away. It is also a valuable way to store requests for future modifications and enhancements to existing software.

3.13.2. Evaluation Phase

Once an issue is created, the first step is evaluation. From here, it can be dropped, deferred, or accepted into the software corrective action process for a fix. Members of the software team and interested stakeholders form a body sometimes known as a change control board (CCB). These would include product managers representing marketing, someone representing the financial side of the business, representatives from other disciplines as needed, a regulatory representative, and a quality representative, and someone representing the software team, usually the team leader. In other words, this is usually a management team. They meet as often as necessary to resolve requests for changes to a project. They would examine the anomaly reports and make dispositions about what to do about them.

The CCB takes a global view of the project and software and decides the importance and impact of a change. This is a negotiation. If defects were simply a matter of noncompliance to requirements, it would be easy, and such an overweight process would not be needed. But

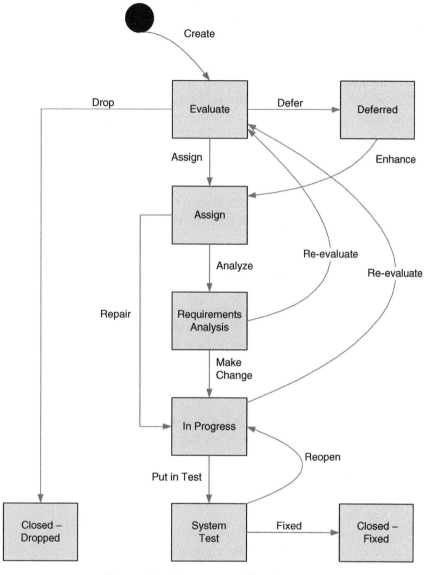

Figure 4.3: Process workflow for SPR.

somebody has to decide what to do when a requirements defect is discovered or there is a request for change to requirements. Remember, many things may have changed while the project was in development, not least of which is the discovery of new hazards that must be addressed. The CCB is the mechanism to decide on necessary changes—whether to change requirements, create a workaround, delay the project to fix the perceived problems, conduct more testing or research, hire more staff—all the things necessary to get the right solution into the customer's hands.

The board should hold regular meetings to evaluate whether the observation is a genuine software issue and assign attributes, such as priority, as well as consider the relevance to safety that the problem may pose, using the risk management process. *Regular* means an interval comfortable for your business and development projects. You would want to meet often enough to avoid going through a huge list. It would not be unusual for the pace to increase as the software nears release. Software under intensive testing may need weekly or even daily evaluation meetings.

The evaluation reviews the description of the problem to determine whether it requires a software change. Meeting participants also review the fields with data for accuracy, fill in missing information if possible, and assign the issue type, priority, and risk analysis. The outcome of the meeting is to move the issue through the workflow. Using the CCB satisfies the ANSI 62304 requirement that no changes to a configuration item are made without approval [25].

3.14. Outcomes of the Evaluation Phase

The following sections describe the transitions out of the evaluation phase. The issue will either not result in a software change, or it will enter the corrective action process. If the decision is that no software change will be made at this time, the individual reporting the problem should be informed of the decision.

3.14.1. Defer

Many issues and suggestions are good ideas but cannot be immediately accommodated by the project without additional planning. Such issues can be deferred as a future enhancement to the product. During project planning phases, these issues will be reviewed and returned to the evaluation phase or left parked in the deferred state. You could also use an attribute for future enhancement if you preferred.

3.14.2. Drop

Sometimes it is clear during evaluation that a problem does not deserve any more attention because:

- The issue is a duplicate of something already in the SPR system.

- The problem is mistakenly assigned to software when it is not a software issue.

- The software is working as specified but the observer misunderstood this.

- The suggestion is a bad idea that would compromise safety.

Such problems can be put in the closed–dropped state. A reason should be given for why no action is needed, including an evaluation indicating that safety is not compromised by taking no action. If the problem is a duplicate report, it is nice to provide a link to the duplicated issue.

The team should take care to give each issue a fair hearing. Problems should stay open in the system, perhaps in a deferred state, unless it is quite clear that they do not contain additional information that is not available elsewhere.

3.14.3. Repair

A nonconformance issue occurs when the software does not meet its stated requirements (also known as a discrepancy, or bug). These enter the corrective action process directly for repair.

3.14.4. Analyze

When an issue is determined to be a change request—that is, the software is compliant with the requirements but the requirements themselves need a change—the software will enter the corrective action process at the requirements analysis phase.

3.15. Corrective Action Process

The software corrective action process is summarized in this section. This process is engaged for defect repair to released software. Significant changes in requirements or design may require a full release in another iteration of software development.

3.15.1. Assign Phase

During the assign phase the software team decides how best to go about addressing the issue. The issue is assigned to an engineer or team for further work. The priority of the problem will govern the urgency with which it is addressed. Plans are made for configuration management, such as assigning a branch for the change. This phase could include effort estimates, start and stop dates, and other information that might provide interesting metrics to use in improving the corrective action process, or that is important from a project planning standpoint.

3.15.2. Requirements Analysis Phase

During requirements analysis, the person assigned the issue will evaluate the impact of the change, update the requirements documents, and advance the issue to the in-progress phase with the make-change transition. Appropriate SHA and SRS reviews should be conducted. You may want to "batch" these at periodic review intervals for software control documents.

Once research into the requirements is complete, it is also possible to find that it is of much larger scope than the CCB initially thought, or to have unintended consequences. The results of the research should be added to the description, and the issue sent back to the evaluation phase with the new data for reconsideration by the CCB.

3.15.3. In-Progress Phase

This is where the bulk of software change takes place: design, coding, and unit testing. If your process warranted it, there could be more detailed states inside the general process. The individual assigned the issue modifies the software design documentation, if required, to reflect the design change. The team conducts appropriate technical design and unit test reviews. The software files are checked out of the designated branch, modified, unit tested and checked back in again to the proper branch.

Once the engineer is satisfied that the problem has been corrected, it is put into system testing. As with requirements analysis, it may happen that the engineer discovers the issue has ramifications beyond what was apparent during the evaluation. An issue can be sent back to the CCB for re-evaluation, if necessary.

3.15.4. System Test Phase

The system test state is where the change is tested in the full system. The fix is delivered to a baseline, the software is built, and appropriate integration and system tests are conducted. The engineer and a quality assurance representative should work together to write the test cases for the specific defect. This is because black-box tests, which are usually what the software system test protocol consists of, may not be appropriate for verifying all repairs. The test methods with pass/ fail criteria are entered into the database, executed, and the results noted. It is worth considering whether the new test cases should be added to or modified in the software system test protocol.

Depending on the size and resources of your organization, you may have a policy that the defects found during the development phase of a software project can be tested and closed by the engineer. If the software is in system test or released, however, an independent person not responsible for delivery of the project must verify the fix. This confirms that all changes were verified, and that only approved changes were made, as required by both the FDA and ANSI 62304.

3.16. Outcomes of the System Test Phase

3.16.1. Reopen

If the software does not pass the software system test successfully, the defect is returned to in-progress status. The tester should enter the test conditions and why the software failed the test. Once returned to in-progress status, the defect could be returned to the evaluation phase for further analysis about the fix and rescheduling if needed.

3.16.2. Fixed

Once the change *has* been successfully tested, the defect is marked closed–fixed. Open SORs relating to changes in document deliverables can be closed after a new version of the document is checked into the VCS or released. Problems should not be deleted from the database once they are closed. As I have defined the process, should the issue arise again, you would have to create a new issue. You could use various policies. This ends the corrective action process.

3.17. Reports

If you use a defect-tracking database, there are any number of reports to run and metrics to collect. ANSI 62304 requires you to analyze defects to detect trends and take corrective action in your development process [25], so metrics are not optional.

Probably most important is the found/closed rate for defects. You want the trend for open defects to be downward over time, especially as you get close to release. Also, at some regular interval, the software team should meet to review all non-closed issues, to make sure that issues are not stuck in some intermediate state, such as in progress.

3.18. Software Observation Reporting and Version Control

The SPR process and VCS are closely related. Especially for released software, no changes should be made that have not been approved by the team, because it is so important to make sure that a validation has taken place and the change has not compromised safety. So the VCS is the repository of the change, and an SOR is the origin of the change.

Integrating the two is worth considering because it provides complete traceability of every change in the VCS to an approved change. Some project management techniques suggest that you enter software requirements as defects in the SOR database. They are defects because the requirements are not met yet. Then you can use the reporting features of the SOR database to show you the level of project completion.

You do want your system to allow for minor improvements and refactoring, so that the software quality improves as it matures. But the regulatory bodies disapprove of changes for "improvement" that are not validated by scientific evidence that they are in fact better. And a "minor improvement" that disengaged a hazard mitigation would be a defect indeed.

4. Risk Management

One of the key activities during the design input phase is risk analysis. Risk analysis looks at *hazards*, the kinds of things that can go wrong, and evaluates the *harm* they may pose if the hazard occurs. The hazard and the harm are the *risk*. Furthermore, the analysis attempts to quantify the probability of the risk occurring, so as to focus effort on reducing the likelihood or severity of the worst outcomes. "Because … what can go wrong probably will, all foreseeable uses and modes of the system over its entire lifetime need to be examined" [20].

The output of the risk analysis process will be system and software requirements, serving as input to the design output phase wherein solutions to the risks posed by the device are designed (risk control measures).

Risk management is really a life-cycle activity, however. During upstream development activities, a team will perform a preliminary risk analysis. The purpose of this is to identify the major risks and guide the design. It is a top-down approach attempting to discover general risks. The team determines how the system could detect each failure and what can be done about it.

The diagram in Fig. 4.4 shows the risk management process. There has been some confusion about the names; the flowchart makes it clearer. The process starts by brainstorming the hazards that the system may represent and their possible severity. Each hazard is examined for possible causes. This part of the activity is the hazard analysis. Next, the probability of occurrence is estimated for each hazard. The risk analysis part comprises the addition of the probability to the hazards. For hardware and system risks, the team uses various techniques, such as researching the probabilities or using historical data, to estimate the likelihood that a particular hazard will occur. For software, because the failures are systematic, the probability is usually assumed to be 100%, that is, worst case [3]. This means that software hazards must be mitigated with control measures; you cannot rely on low likelihood to make the system adequately safe [26]. (The FDA's Center for Devices and Radiological Health [CDRH] uses the term "hazard analysis" to emphasize that software risks should be managed based on severity.) In other words, software hazard analysis does not include risk analysis, so guidance language simply uses software hazard analysis (SHA).

After identifying and quantifying the risks, the team develops risk control measures to reduce the hazards, and then evaluates the results to determine whether the device is sufficiently safe. This is an iterative process, working through all identified hazards and their causes and developing mitigations. It is also important to review the mitigations to verify that they have not introduced other hazards. For example, you might require a button to be held down for a period of time to reduce the risk that someone inadvertently presses it while moving the device. But by requiring the delay, you may have introduced a problem with untrained or forgetful users who cannot command a desired action because they don't know that they must keep the button pressed.

As each subsystem is designed, the risk analysis becomes more detailed. As a subsystem, software is subject to the same analysis procedures. It should be evaluated for failure modes and the effects that those failure modes could have on the rest of the software and the system as a whole.

One useful starting place for the risk analysis is a checklist. The regulatory guidelines for device design often take the form of a checklist. One limitation to the use of checklists is that

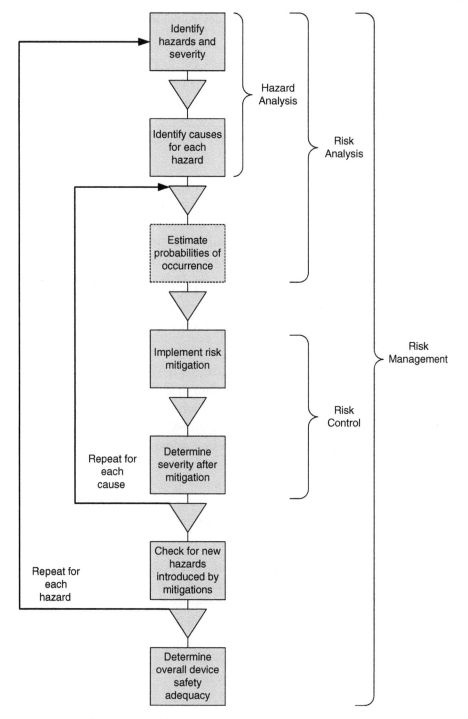

Figure 4.4: Risk management process.

they do not address unforeseen ways in which a device could fail, so used alone, they do not constitute a complete risk analysis.

What is needed in addition to the checklist is a process that analyzes the causes and consequences of a *critical event*. A critical event is something that could go wrong. The analysis should examine the possible causes of a critical event for insight into other types of critical events, and to prevent hazards as early in the design as possible. The process should also look forward to the possible hazards that may originate from the event, in order to provide a risk mitigation if warranted.

During the design output phase, when the response to the design input is being generated, the top-down approach should be supplemented with a bottom-up approach using more formal methods. The FDA will expect to see a formal method of risk analysis as part of the premarket submission. Not surprisingly, the extent of the risk analysis and the care with which it is conducted is a function of the risk class of the device.

The FDA does not specify the method for risk analysis, but international bodies require ISO 14971:2007, "Medical devices—Application of risk management to medical devices." The FDA will accept ISO 14971, so following its procedures will simplify the approval process in foreign markets as well as satisfy the FDA.

Some of the formal methods include fault tree analysis (FTA) or failure mode and effects analysis (FMEA), or the more comprehensive failure modes and effects criticality analysis (FMECA). It is possible to estimate the quantitative probability of a risk, and these methods might include this risk analysis. When performing an FMEA at the level of individual components in an electronic assembly, the mean time between failures might be known and can be used for the estimate. As a rule, however, the probability of generalized risks is often unknowable beforehand, dependent as they are on the system design and problem domain. Extensive effort devoted to detailed quantification may be questionable, especially if it detracts from identifying as yet unknown risks or diminishes the design effort for mitigating the known risks.

Risk analysis is the process of identifying the risks and probabilities; once identified, risk control is the process of reducing the harm of a hazard as much as is practicable. Ultimately, the regulatory body with jurisdictional authority over the device evaluates the threshold that the device must meet. For example, failure of a pacemaker to generate the electric charge to pace the heart would be critical because it would necessitate surgical intervention to repair. Failure of the programming device that lets the physician set the heart rate would be of lesser concern, since presumably a backup would be available.

The risk analysis must consider the system as a whole. Few hazards are presented entirely by software; instead, it is the interaction of software and transducers or other hardware that could create hazardous conditions. The divide-and-conquer principle is applied; the subsystems are analyzed for hazards particular to their nature. However, software has its own types of hazards. Software can:

- Fail to do something it was supposed to do. It failed to perform an operation it should have, or failed to perform it in a timely manner, or failed to recognize or prevent a hazard that was the software's responsibility.

- Do something that it was not supposed to do. It performed an unintended operation, or at the wrong time.

- Give the wrong answer. It shows data for the wrong patient or stale data, performs the wrong response to a hazard, or produces a false positive or false negative.

Risk control is the endeavor to reduce the risk to an acceptable level. Methods of risk control, in order of preference, follow:

- Design the risk out.

- Implement protective measures.

- Provide adequate information, such as warnings [3].

Eliminating the risk by design is the most preferred approach. For example, a left ventricular assist device that augments the patient's natural heart by pumping additional blood could overpressure the ventricle and injure it. Overpressure could be eliminated by design by using a spring for the pumping action whose physical characteristics are such that it cannot create more pressure than a ventricle could bear.

To provide a software example: memory allocation routines are complex and may result in fragmented memory, so that a request for memory could fail. A method to mitigate by design is to allocate all memory at compile time and not use an allocation routine at all. (This is easier in embedded systems, where you are usually aware of all the resources that the system will require.) Hence you would prefer designs that avoid things like malloc() or recursion, or that otherwise have nondeterministic behavior.

Using protective measures to reduce the risk means designing in interlocks, alarms, and defensive code to protect against a risk or notify the user if it occurs so that she could intervene. For example, an x-ray system has a safety interlock so that software alone cannot command x-rays; instead, a hardware switch must also be depressed at the same time. As another example, software may poll a bit in an analog to digital converter to signal the completion of an operation. The code should include a loop counter and a timeout, so that the software can exit the loop if something goes wrong and the bit is never true.

The final line of defense is to provide warnings in the labeling that caution against improper or hazardous use. This may be the only option for some things: knives have sharp edges essential to their function, for instance. This really should be the last resort. Proper human factors analysis will show that risk reduction by design is a much more inherently safe approach.

Once the risk management is completed, you need to document the results, preferably in a tabular form. This should include descriptions of the hazards, the severity level, the software cause, the risk control means, the verification methods, and finally the severity level after the risk control method mitigates the risk. All software-related hazards should be reduced to the level where they present no risk of injury to the patient or user—in other words, reduced to the minor level of concern.

One wrinkle that can influence the result of risk analysis is institutional self-interest. Regulators can make two types of errors. They could be too accepting, and allow a dangerous device onto the market, whose problems are only revealed after it has done injury. Or, they could delay launch by insisting on extensive risk analysis and onerous risk management procedures, thereby denying patients the use of devices that could benefit them. If the regulator allows a device to "slip through the approval process that later proves harmful to some people some of the time, a hue and cry is sure to follow. Look no further than the recent public backlash against the FDA after several deaths were linked to Vioxx...." [27]. The regulators are on the chopping block for not having foreseen an unforeseeable event. The second risk, not treating patients who could use the device rarely gets the regulators in trouble. Moreover, big manufacturers may acquiesce because it provides a barrier to entry to firms without the wherewithal to deal effectively with the regulatory bodies.

It is vital to think about hazards and mitigations early in the development process. Risk control measures are a major source of requirements; you want to avoid project and schedule surprises because hazards were discovered late. It is also more difficult to take a systemic view late in the project. Pasting the mitigations on at the end runs the risk of unintended consequences, and hence unsafe systems.

Example 4.3 Hazard Analysis Example

Hazard analysis is crucial but can also be difficult to grasp, so I will provide a fault tree analysis example based on a closed-loop insulin pump. While hazards are specific to devices, many times medical devices share a subset of the same types of software hazards, and many of the risk management techniques are the same.

Usually, a hazard analysis will be recorded in a table. Each row is a specific software failure that could lead to a system hazard. Multiple software failures could lead to a single system hazard, so the system hazard occurs many times. The columns capture the analysis that went into assessing the hazard, determining mitigations, and tracing through requirements and test plans to establish that the mitigations were met. The columns follow:

- Software hazard ID—It is useful to uniquely identify each software hazard.
- System hazard—The actual hazard that the software failure will lead to.
- Software failure—The specific failure of software that causes the system hazard.
- System effect of software failure—The effects of the software failure at the system level.
- System-level risk control—Risk controls at the system level, such as interlocks.
- Software-level risk control—Specific risk controls implemented in the software, such as watchdogs or self-tests.
- Severity—Level of severity of the hazard.
- Probability (optional)—Chance of the hazard occurring after mitigations. This is assumed to be 1 for software.
- Residual risk acceptability—The acceptability of the risk after mitigation.
- System safety requirements—Requirements imposed by the risk mitigation at the system level.
- Software safety requirements—Software requirements imposed by the risk mitigation.
- System requirements verification—Trace to specific tests that show that the system safety requirements have been met.
- Software requirements verification—Trace to software system tests that show that the safety requirements have been met.

Let's examine the hazards presented by the example insulin pump.

Insulin pumps are currently available for the treatment of diabetes. They are worn under clothing and have a catheter shunt that goes under the skin. The patient uses a glucosometer to measure her blood sugar, and combined with what she has eaten and expects to eat, can calculate how many units of insulin to deliver from the pump. The next logical development would be to close the loop with a detector that assesses the need for insulin and automatically provides an insulin dose.

Two immediately obvious hazards are delivering too much insulin or delivering too little. If our device also had a readout that displayed the blood glucose determination, we have a third—an incorrect or stale readout that causes a patient or clinician to deliver inappropriate therapy. In a software-controlled device, these *system hazards* can all be caused by software failure, as well as other types of failures, such as the shunt coming out or the display failing.

(The third hazard—displaying wrong data causing wrong therapy—is common in medical device software, and in many cases is the most serious hazard they present. There are often alarms associated with the data, which presents the possibility of failing to alarm or alarming when it should not. Failing to alarm is not in and of itself a hazard, but is instead failure of a risk control in place to prevent a hazard. As such, it will not have system-level risk controls or system safety requirements, but it should be analyzed in the software hazard analysis because it leads to software requirements.)

Severity is an assessment of the extent of the injury if the hazard were to occur. These will differ by device. For a blood pump, for example, the most severe outcome might be output less the 2.0 liters per minute with no opportunity to intervene. For an x-ray machine, it might be excessive dose resulting in tissue necrosis and long-term tissue damage.

Severities are usually classified in levels and used as a shorthand. For diabetics, getting too much insulin is worse than getting too little. The patient can usually tell when she has too little

(Continues)

Example 4.3. Cont'd

and will have a backup insulin injection to manage the failure of the device. Too much insulin, on the other hand, can cause unconsciousness and the need for hospitalization. For the insulin pump, the following severities are possible (these are for illustration only and are not exhaustive):

S4—Excessive pump output without alarm
S3—Excessive pump output with alarm
S2—Insufficient pump output without alarm
S1—Insufficient pump output with alarm

Left Half of Software Hazard Analysis Table

ID	System Hazard	Software Failure	System Effect of Software Failure	System-Level Risk Control	Software-Level Risk Control
1A	Insufficient pump output	Failure of user/patient to control pump speed	User/patient commands low pump speed resulting in hazard	1. Low-speed fault alarm 2. Hardware not under software control that ensures minimum speed	1. User/patient watchdog 2. User/patient self-tests 3. Alarm logic separate from pump speed control
1M	Insufficient pump output	User/patient self-test software repeatedly generates false test failures resulting in repeated resets	Loss of pumping due to constant reset by user/patient	Audible alarm from constant resets	User/patient should eventually stop trying and shut down.
1P	Insufficient pump output	User/patient RTOS software failure results in faulty operation of application software	User/patient commands low motor speed due to RTOS fault resulting in hazard	1. Pump operation alarm 2. HW not under software control that ensures minimum speed	1. User/patient watchdog 2. User/patient self-tests 3. Alarm logic separate from pump speed control

(Continues)

Left Half of Software Hazard Analysis Table: Cont'd

ID	System Hazard	Software Failure	System Effect of Software Failure	System-Level Risk Control	Software-Level Risk Control
2D	Failure to alarm for a hazardous failure	User/patient calculates misleading flow value	Alarm not issued for high/low-flow condition	Not applicable; alarm does not cause a hazard	1. User/patient watchdog 2. User/patient self-tests 3. Alarm logic separate from pump flow calculations
4C	Administration of inappropriate therapy	Software failure results in corruption of critical pump parametric data	Uncommanded or incorrect/corrupt pump configuration setting	Pump flow alarm	1. User/patient watchdog 2. User/patient self-tests 3. Alarm logic separate from pump speed setting 4. Redundant storage of configuration data 5. Integrity check on configuration data

Right Half of Software Hazard Analysis Table

Severity	Residual Risk Acceptability	System Safety Requirements	Software Safety Requirements	System Requirement Trace to Verification	Software Requirement Trace to Verification
S1	Acceptable	1. The pump shall provide a low-speed alarm	1. The user/patient shall provide a periodic strobe to the watchdog		

(Continues)

Example 4.3.   Cont'd

Right Half of Software Hazard Analysis Table: Cont'd

Severity	Residual Risk Acceptability	System Safety Requirements	Software Safety Requirements	System Requirement Trace to Verification	Software Requirement Trace to Verification
		2. The pump shall provide a low-flow alarm	2. The user/ patient shall perform start-up and periodic self-tests, including periodic memory test 3. The low-flow and low-speed alarm software shall be independent of the speed-control software as much as possible (separate tasks, separate memory regions)		
S1	Acceptable	An alarm independent of the user/ patient shall be generated for multiple resets of the user/ patient	The watchdog shall maintain a reset counter and determine when it should stop trying to recover and shut down (alarm generated by separate hardware in response to shut-down)		

(Continues)

Right Half of Software Hazard Analysis Table: Cont'd

Severity	Residual Risk Acceptability	System Safety Requirements	Software Safety Requirements	System Requirement Trace to Verification	Software Requirement Trace to Verification
S1	Acceptable	1. The system shall have a pump operation alarm 2. The system shall have a HW controller that ensures minimum pumping in the event of software failure	1. The user/patient shall provide a periodic strobe to the watchdog 2. The user/patient shall perform start-up and periodic self-tests, including periodic memory test. 3. The low-flow and low-speed alarm software shall be independent of the speed-control software as much as possible (separate tasks, separate memory regions)		
n/a	n/a	n/a	1. The user/patient shall provide a periodic strobe to the watchdog		

(Continues)

Example 4.3. **Cont'd**

Right Half of Software Hazard Analysis Table: Cont'd

Severity	Residual Risk Acceptability	System Safety Requirements	Software Safety Requirements	System Requirement Trace to Verification	Software Requirement Trace to Verification
			2. The user/ patient shall perform start-up and periodic self-tests, including periodic memory test 3. The low-flow and low-speed alarm software shall be independent of the speed control software as much as possible (separate tasks, separate memory regions)		
S1	Acceptable	1. The pump shall provide a low-speed alarm 2. The pump shall provide a low-flow alarm	1. The user/ patient shall provide a periodic strobe to the watchdog 2. The user/ patient shall perform start-up and periodic self-tests,		

(Continues)

Right Half of Software Hazard Analysis Table: Cont'd

Severity	Residual Risk Acceptability	System Safety Requirements	Software Safety Requirements	System Requirement Trace to Verification	Software Requirement Trace to Verification
			including periodic memory test		
			3. The low-flow and low-speed alarm software shall be independent of the speed control software as much as possible (separate tasks, separate memory regions)		
			4. The user/patient shall store configuration data in NVRAM and RAM, and use NVRAM values in the event of an integrity check		
			5. The user/patient shall check the integrity of configuration data and use NVRAM values in the event of a mismatch		

(Continues)

Example 4.3.   Cont'd

The fault tree analysis (FTA) carries on in this same vein, looking at a hazard and identifying the failures that could cause it to occur. This may result in some repetition, especially of the mitigations, but each subcomponent should be analyzed so as to ensure completeness. Excessive pump output could have the same software failure causes, but would have a different severity rating. A real software hazard analysis would have far more than five items, depending on the complexity of the device.

I have filled in the requirements columns for readability. You would want to use a trace to the unique requirement IDs to avoid stating requirements in more than one place. It is useful to flag safety requirements somehow so that the team understands their origin and importance—they should not be changed lightly! Also, fill in the traces to the test plans showing that the requirements were met.

FTA is a top-down approach. You may also wish to do a bottom-up approach—something more like a failure modes and effects analysis (FMEA). In this case, you would be looking inside the code for the types of ways that software can fail, such as overrunning a queue, and assessing what hazard could occur as a result.

---

5. Software Verification and Validation in the Context of Design Control

We have reviewed the general requirements for design control at the project level. We have touched upon some possible software life cycle models. How does software fit into the design control process? In particular, what does software verification and validation look like?

Sometimes it is useful when talking about the role of V&V in the software life cycle to refer to the *software V* (Fig. 4.5). In this diagram, software development proceeds down the left leg of the V, and corresponding verification activities proceed up the right leg. So, requirements analysis precedes architectural design, which then leads to detailed design. Once the detailed design is complete, the implementation and coding phase begins. Then, proceeding up the V, the correct operation of the units is confirmed with unit testing. Units are brought together into integration tests, which verify the design. Finally, the system software test confirms that all the requirements are met.

To quote the guidance document, "Software verification provides objective evidence that the design outputs of a particular phase of the software development life cycle meet all of the specified requirements for that phase" [1]. In other words, verification is showing that design output meets design input. Testing—by which we mean a demonstration of operation or dynamic functional tests—is of course a vital component of verification. But verification is broader than functional testing, as it also includes inspections and analyses. This is

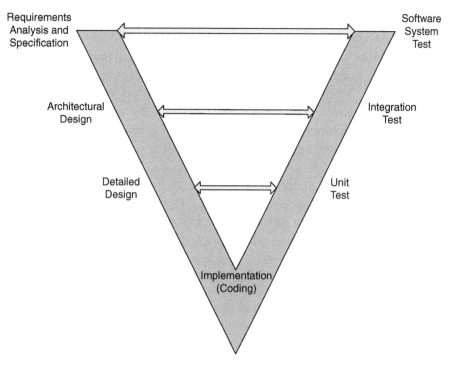

Figure 4.5: The software V.

significant because not all requirements are functional requirements that can be demonstrated by execution. These requirements are no less important, and should not be left out because they are not testable by the conventions of what has been thought of as "testing." Thus "[a]ny approach which establishes conformance with a design input requirement is an acceptable means of verifying the design with respect to that requirement" [16].

Software validation, on the other hand is "confirmation ... that software specifications conform to user needs and intended uses, and that the particular requirements implemented through software can be consistently fulfilled" [1]. This is a full life cycle practice and depends on the verification steps taken throughout the development of the software, rather than one activity done at the end. It includes "evidence that all software requirements have been implemented correctly..." [1].

In terms of the software V, verification is what happens in analysis between neighbors on the V. For instance, the architectural design should trace from and implement the software requirements. Verification is also the corresponding testing across the V. Validation, on the other hand, is the whole V—the sum of all the verification activities, plus measures to suggest that the trend of defects toward zero is sufficient to lead to a level of confidence that the software is validated.

There is some confusion possible between software validation and other kinds of validation, and in particular, design validation. It is unfortunate that the FDA uses the same word. In particular, the final test establishing that all software functional requirements are met is sometimes called the "software validation." This is a legitimate interpretation of the guidance, which says, "A primary goal of software validation is to then demonstrate that all completed software products comply with all documented software and system requirements" [1]. However, I prefer to follow the ANSI 62304 standard in this instance, and call the testing which establishes that software requirements are met "system software testing." This avoids confusion with design validation. (At a conference dedicated to software for medical systems, I asked in an informal poll of the audience what people called the testing in which you demonstrate that the software requirements are met. Half call it verification; the other half validation. I don't think it matters much to the FDA what you call it, provided that your standard operating procedures or other planning documents clearly select a definition and stick to it. But the matter is avoided by calling it the system software testing.)

Software validation is a component of system design validation, but design validation must involve the customer and clinical evaluation to some extent, which would make it difficult if not impossible to confirm the correctness of all software requirements. You could hardly open the box and attach an oscilloscope to establish that the RTOS clock was running at 1 kHz, for example. On the other hand, some companies have made the mistake of thinking that design validation ends at software validation. To sum up, design validation for systems with software consists of:

- Conformance of software to its requirements (verification).

- Trace of software requirements to system requirements (we didn't just meet the requirements, we had the right set).

- Validation that the device meets customer needs and is fit for its intended use. This involves *both* hardware and software.

After release, at the top of the V, the software enters the maintenance phase. Maintenance should be addressed as part of your software development planning. Generally speaking, you would use the same techniques, with a greater emphasis on regression testing. Software development in maintenance needs to be conducted with the same rigor as the original construction, or with even more care since the engineers who did the original work are often not available in the maintenance phase.

The software design input phase is the left half of the V. This is where you develop the plans to accomplish the life cycle activities, write the SRS, create an architectural design and a detailed design, and begin implementation.

Software development, as structured by the design control process, begins with quality planning. This is by and large expressed in the SOPs you have created for software development at your firm. These should describe the tasks for each life cycle activity and the methods and procedures for accomplishing them. For example, you might plan for a subset of the software team to write the SRS, and then have it reviewed by the whole team. You will want to define the inputs and outputs for the activities and the criteria for evaluating that the outputs match the inputs. It is important to describe the resources, roles, and responsibilities for the life cycle tasks.

This planning should include a SCM plan and the software problem resolution process that will be used. It should also include a software quality assurance plan (SQAP) that defines the quality assurance activities (such as are described in the next section) that will be used. For instance, you could define the method of unit testing as using test harnesses where practicable, or some form of automated testing. This is a high-level test strategy plan.

You want also to create a software V&V plan (SVVP). This is a more tactical plan that describes the breakdown into test plans per subsystem or per major feature, and details the test resources needed and so on. This document will have to evolve as the software design evolves, and you gain more knowledge about the types and content of integration testing.

The planning at this stage should be high level. You won't have enough information to break the software development into work packages—not before the architectural design, for instance. But the idea is to plan the life cycle activities in general terms, such as the architectural design phase, without knowing about the internal details, until the design has evolved enough to allow you to plan in more detail.

Once you have general plans and policies in place, you develop the software requirements. These should follow—and trace to—the system design requirements. A major component of this phase is the hazard analysis, the evaluation of the level of concern of the software systems, and whatever software requirements derive there for risk control.

The hazard analysis is a direct input to the design process in another sense. One of the outputs of the hazard analysis is the level of concern of the software. You need this to know the appropriate level of risk-control measures to apply to the software. Since you can't test quality, software must be constructed according to a careful life-cycle process in order to be safe [25].

Because the higher levels of concern require more exhaustive testing and documentation, it is also useful to partition the safety-critical software into its own software system. This will allow you to focus your attention on the most critical parts, and minimize the amount of effort overall.

With the software requirements in hand, you can begin the architectural design. It is probably beneficial to begin this design before the requirements are finalized, since trying to design from them is one of the first verifications that the requirements are feasible and complete.

Neither the FDA guidance documents nor the ANSI 62304 provides complete direction as to the dividing line between what is a software requirement and what is design, merely granting that it is "an area of frequent confusion" [25]. This is hard to know without experience or some examples. ANSI notes that software requirements often include software functional specifications that define in detail how to meet the software requirements, even though different designs might meet the requirements. The FDA guidance more clearly assigns this to design [1].

The architectural design is where you decide the software subsystems and modules (classes), and design the major interfaces and data structures. Further design refinement is carried out in the detailed design, where you specify algorithms and methods. The level of detail that seems to be expected is at the pseudo-code level or similar expressive artifacts, especially for major-level-of-concern software.

The architecture and detailed design is generally verified by evaluation, that is to say, technical review. But once we start coding the next stage of software development, we can apply a broader range of functional tests.

Software testing takes two broad forms: white-box and black-box testing. White-box testing—also known as glass box, open box, clear box, or, to the FDA, structural testing—is, as the name implies, a method of looking inside the code structure for the testing. It assumes that the tester knows what the software is supposed to do. It is a good way to look for errors of *commission*, and is the way that we confirm software is not working incorrectly.

Black-box testing—also known as functional, behavioral, or closed-box testing—tests the expected outputs against the inputs, without assuming any knowledge of what's in the box. It cannot confirm that all parts have been tested, or that the software is not appearing to give the right answer for the wrong reasons. On the other hand, it is testing against the specification, and hence can catch errors of *omission*, where the requirements have not been met.

An aspect of white-box testing that is important to the FDA, and is in fact a powerful testing technique, is coverage. The basic idea is that complete testing would examine the condition of the software in every possible state it could be in. Obviously, this is impossible for all but the most trivial programs, so there are types of coverage that can be separately achieved.

- Statement coverage, where you execute every line of code in the program.

- Branch and condition coverage, where you test every program logic condition possible, and hence execute all branches.

- Loop coverage, where you execute all loops with variable termination conditions zero, once, and multiple times.

- Data flow coverage, where you watch the value of data as it flows through the program execution.

- Path coverage, where you take all the possible paths through the software.

The completeness of the coverage, both in extent and type, is a function of the level of concern of the software. The goal is 100%, but clearly this is difficult to achieve, especially for path coverage. Furthermore, 100% coverage generally cannot be achieved in black-box or functional tests. For one thing, it is difficult to create conditions in a functional test to throw an exception or invoke defensive code. The error handling is there, after all, to protect against software gone awry—without injecting errors under a debugger you can't see it happen, precisely because it is not supposed to happen. And you wouldn't want to leave out error handling or defensive coding because you can't write a functional test to make it happen! So coverage is meant to be performed as a white-box test, primarily at the unit level [1]. We will see later in the section unit testing methods that can accomplish 100% coverage.

While appropriate at the unit level, white-box testing quickly becomes cumbersome, especially in real-time systems where the hardware can't wait for humans to step through every line of code. Black-box approaches, on the other hand, are useful at all levels of testing.

There are several approaches to take to functional testing with increasing level of effort:

- Normal use, where you show that the software works with usual inputs, including forcing the software to generate typical outputs. This is necessary but not sufficient for testing medical software.

- Robustness, where you show that the software responds correctly to incorrect inputs. This is where you think of input equivalence classes, boundary conditions, and other analysis to derive the incorrect inputs.

- Combinations of inputs, where you test the behavior of the software when presented with multiple inputs in different combinations. Like path coverage, complex software can have tens of millions of combinations. Statistical testing methods that generate test data for you are generally necessary for accomplishing this type of testing.

Demonstration is not the only type of test than can occur. Indeed, some vital requirements cannot be safely demonstrated in production code at all, but have to be tested with parameterization or other white-box tests. For example, a watchdog is the final defense against errant software. But you wouldn't want to leave test code in that tests the watchdog, on the off chance that a software failure happens to engage the watchdog test, causing the watchdog to trip when it should not. Nor is it acceptable not to test the watchdog, since it is an important part of the hazard mitigation. You could put a jumper in the wire to the signal that services the watchdog, and remove it in testing to simulate that the software failed to service the watchdog and verify that the watchdog resets the software. But a jumper could fall off in the field, causing artificial resets that had nothing to do with software failure. It is better to remove the jumper to enable the watchdog.

The testing will be a function of how elaborate the watchdog design is. In its simplest form, you could create a test harness with an infinite loop that locks up the software and verifies that the watchdog does in fact reset the processor. If your design incorporates a real-time operating system, your watchdog may take the form of a thread monitor which verifies that threads run as often as they should, no more and no less (within limits). I have found that a design like this, implemented early in the design of the project, will be verified repeatedly during subsequent development as engineers make mistakes elsewhere in the system, locking up threads and causing the watchdog to bite. This is an instance of error handling that when implemented early helps to debug the system while validating itself.

White-box and black-box testing techniques, as well as technical reviews, are the methods used to verify correctness through the construction phase. In the next section, I will describe the specific types of testing at various levels up the right side of the software V that are expected for any medical product.

5.1. Software Verification Methods

5.1.1. Example Review Process

As I mentioned in Section 3.5, Design Review, the FDA assigns special meaning to design review that is not the same as the term is often used in software development. We think of "design review" as the review of a software design; it may drill down to the level of a code inspection or walkthrough. These less comprehensive review activities the FDA would usually refer to as design verification. Because design review is something conducted at major milestones or is a project phase review, it is often an extensive undertaking. Of course, the extent is a function of the complexity of the product under review, but for moderate complexity projects, it will probably involve most of the development team and take several days.

The software engineering body of knowledge is ahead of the curve in the review process. It has been applied to software systems since Michael Fagin's seminal work in the late 1960s. The design reviews that the FDA describes shares similarities with the Fagin-style reviews, but are a bit less formal. Shifting gears slightly, I'm going to talk about the design-verification style of review.

Designs benefit from well-conducted inspections and reviews. This is especially true perhaps of software, where the creation is entirely abstract and so must be evaluated analytically.

It has been argued that formal inspections are one of the best methods of finding defects. These could take the form of full-on Fagin formal reviews [28, 29]. These are formal scheduled meetings with trained participants taking on specific roles. First, there is the moderator. Her role is to schedule the meeting, distribute the materials, and follow up on the issues raised to their closure. During the meeting, she will keep the meeting on track and moving forward. Training in facilitating meetings and fulfilling the role of moderator is considered essential. The recorder is tasked with writing down the issues and producing the minutes. One to several reviewers add their technical expertise. Finally, the reader reads each line of the document out loud for comment by the other participants.

The author may or may not attend; this is a subject of some controversy. To be sure, management should not attend. The reason is that having management present will interfere with the give and take necessary to achieve the highest-quality output. A manager's presence could inhibit criticism if team members don't want to make each other look bad. Or in dysfunctional organizations, the manager's presence could cause unnecessary friction between the reviewers and authors if one side wants to make the other look bad.

Most of the literature data are based on Fagin-style reviews, and the claims for its value are outstanding. To some degree, one wonders if its utility is in part a function of the nature of the documents reviewed that formal inspections are usually applied to. That is, if you have a thickly populated hierarchy of software documents, it is easy to imagine a lot of mistakes. For example, if there is system requirements spec, the parent of a subsystem spec, the parent of a software spec, which then repeats the information in great detail in the software design description, and beyond this there is a detailed software design description, then the code expressed as models, followed by the actual implementation of the models in a programming language, and then test plans that work themselves all the way back up to the system level, there are many points for failure. This is one reason for flattening the documentation where possible. So what may be causing the utility of the Fagin review is precisely this duplication of data. If the same requirement language, more or less, has to be traced through five levels of documents, then it is easy to see how it could get out of whack. The Fagin review examines these traces, accomplishing software quality, but not necessarily in the best way. Rather than attempting to inspect quality in to the work product, it would be better if the process emitted the right result by design. Moreover, it is the job of the organization to create products that work, not create reams of documentation that may distract from the achievement of safe, effective systems.

So, you may not need formal inspections, but there are several rules of thumb for reviews.

- Reviews should not last longer than 2 hours. Nor should there be more than one review meeting per day. We are after the best-quality work we can get, and people's attention flags after 2 hours. (Besides, that is generally as long as an author can keep his mouth shut.) If the review has not been completed in 2 hours, reschedule. Sometimes with a substantial piece of work, such as a requirements specification, it is worthwhile to schedule routine meetings until the work is done. You might schedule a 2-hour meeting every day at 10:00 for the next 2 weeks.

- Reviews should focus on finding defects, not solutions. They should not be occasions for designing, brainstorming, blaming, pontificating, or other distractions. It is the designer's job to come up with solutions. If your organization has trouble staying on track, it may be helpful to use a moderator. Decide how long discussion can go on before an item becomes an issue. To make good progress, discussion should normally not be longer than a minute or two. Use an egg timer if you have to. If the item cannot be resolved in 2 minutes of discussion, it is an issue. Write it down and move on.

- If your team cannot resist designing, you may benefit from a concept invented by NASA called the third hour. Reviews invite bright people with lots of ideas and opinions, so it is not surprising that they want to contribute. Plan an hour after the regularly scheduled meeting to consider solutions to the issues raised.

The formalism of a Fagin review may seem like overkill—and often is, when you are participating in one. It is perhaps best reserved for the most safety-critical components of the system. But there is one virtue in trying a few Fagin reviews. If a review process is a new concept in your organization, or you find that it is not very effective, using a formal process with a trained moderator can help the team learn what is expected out of a review and how reviews can be made more effective. Once a reviewing culture has been inculcated, then it is possible to relax the formality somewhat.

For instance, it would not be necessary to have an assigned reader drone through the document. Instead, the reviewers could project a document on a screen, read it for themselves, and suggest improvements that can be made in real time. (Don't be too hasty though. Some changes may have ramifications to downstream documents or other parts of the system and thus need more careful consideration. These sorts of issues should still be captured in the minutes or issue log, and allowed to marinate for further solution, not in the heat of the moment.)

In smaller companies or projects, there may not be enough people to fulfill all the roles prescribed in a formal process. For reviews to be conducted at all, a less formal structure will be needed.

Review styles less formal than Fagin, but still face-to-face, are usually referred to as walkthroughs. This is a process wherein the authors guide the reviewers through the work product, and the review team, including the author, attempt to discover and resolve defects.

Because much of communication is non-verbal, a face-to-face meeting is usually the best method for controversial material, such as requirements analysis in the early phases of a project, and reviews of high-level designs. Depending on the sophistication of your team both at reviewing and understanding each other's work, you could use a pass-around review instead of a meeting [16]. Routine code reviews could be conducted as a pass-around, for example, so that the reviewers can review at a convenient time and their own pace.

To conduct a pass-around review, the author checks the work products into the VCS (they are not distributed as hard copy). Next, the author or SQA schedules a meeting in the calendar software for the completion date for the review. This is a virtual meeting, not an actual meeting, just to remind people about the deadline. The reviewers respond with an e-mail with their issues. The author responds to each issue in turn with an explanation, or a fix, writing the response in the return e-mail. The response e-mail is attached to a review cover sheet, signed, and archived.

Such a review method works well with small teams. If there are a dozen people providing input, the collating effort required of the author can be excessive. In this case, a walkthrough meeting is more productive, since the same issue only has to be brought up once.

If you wanted more formality than the e-mail record, you could enter defects discovered in review into the defect tracking system. This is not needed in my organization; the author is trusted to fix the issues in response to the review, and then check in the reviewed work product. Only defects that were perhaps not in the scope of the work product under review, or that were not something the author could address by herself, such as changes to requirements, would be entered into the defect tracking system so as to reduce the process overhead of managing the response to review. What you don't want to see happen is to have so much overhead or trouble caused by reviews that they cease to be effective, that is, people avoid pointing out errors because it is too disruptive to the team or too much trouble to fix it. This is a quality malaise that works against the goal of making the best products possible in a timely way.

How antagonistic reviews are is going to be a function of the culture at your company and the personalities of the people making up the team. There should be disagreement—otherwise, one would question the value of the review. But this can be reduced and time saved by holding informal review meetings long before the more formal methods, both so that the reviewers understand the work product when they get it, and so that it is not a shock when the first formal meeting is held. Also, plan time for issues that arise [16].

The value of informality is recognized by the FDA. "[T]he manufacturer should expect, plan for, and encourage appropriate ad hoc meetings as well as the major design review meetings" [7]. Decisions from these ad hoc meeting don't necessarily require formal documentation, unless an important issue was resolved [16]. You can write some minutes for these meetings, summarizing the subject under review and the decisions. That way, you capture that cultural knowledge and get credit. These belong in the DHF. Routine activities that are part of your day-to-day job do not need to be documented [7].

At the same time, too much formality can reduce the delivered quality of a system. "Persons who are making presentations should prepare and distribute information to help clarify review issues and help expedite the review. However, the intent of the quality system is not that presentations be so formal and elaborate that designers are spending excessive time on presentations rather than on designing a safe and effective device" [20].

5.1.2. Unit Testing

The standard and guidance documents do not go into much detail about what constitutes a unit test, or even what qualifies as a unit, for that matter. For minor and moderate level-of-concern software, unit tests will not need to be supplied with the premarket submission. You are still expected to perform unit testing according to the methods you have outlined for the project in the SVVP. This gives you the freedom to define the unit test as whatever your process says it is.

A unit test could be as simple as a code review and whatever ad hoc testing the author thought suitable. This is satisfactory for software of minor level of concern. Another method that has become popular recently is automated testing. This idea evolved as one of the elements of extreme programming in which writing test functions confirm that the answer returned from a class method is the answer that the programmer expected [30]. (And may be the secret of its success. Some of the notions of extreme programming are, well, extreme, and not applicable to embedded or safety-critical systems, but the virtue of thorough unit testing is one of its great strengths.) Each routine has a test written for it, and these are collected in a test suite. Developers can run an automated test after each build to ensure that any recent changes have not broken any tests. This makes it a good way to do regression testing, especially because the cost is low. Test tools are evolving that even write the test for you, but an engineer still must be involved to evaluate their adequacy and what to do when they complain of a mistake. These can generate extensive reports that can be used to supply the requirement to provide test report summaries.

However, automated testing is almost solipsistic in the way it looks only inside the box. It does not test the fundamental assumptions of the requirements, and cannot check for errors of omission. If the engineer misinterpreted the requirements, it is likely he will misinterpret the tests. So it is by no means sufficient testing by itself. At the very least it requires a review to verify that the test methods meet the requirements. It may help to have someone other than the author write the tests, but nevertheless there is a danger of passing the test and being quite wrong, especially in terms of meeting the customer needs and being fit for the intended purpose. Integration testing and software system testing are still required.

Automated testing can also have only limited application. Usually embedded systems are interacting with physical input/output. Without a method of supplying to the test drivers the ability to observe and record the results of the operation, it is not very meaningful to write automated tests. Usually this feedback is provided by having a human observer. The tester presses a key and sees that an LED lights. It is possible to simulate the key press, but without a way to sense the corresponding action, the simulation is not very useful.

There are methods to address this but they involve the construction of special hardware. For example, you build a simulator that provides input to the software under test and evaluates the output. The simulator then runs a test script, providing the input and looking for a certain output. You might create your own drivers for a touchscreen, for example, that you can program to provide touches at random location and irregular time intervals. You could let this run for a while and at least verify that the GUI did not hang up.

One team built a simulator with photo detectors at physical locations corresponding to LEDs in the device under test, which is a clever way to observe the test output. You might also be able to use LabView, but keep in mind that LabView implementations would have to satisfy third-party software validation processes.

5.1.3. Unit Testing with Test Harnesses

Another process that has been successful in the embedded space is to write test harness for each unit. These test harnesses provide a framework in which to do whatever type of testing is called for. The test harness can be written to automate the tests, providing simulated input and verifying the correct output (especially useful for regression testing). The test harness can be used for performance testing of the module's subroutines, for example, by strapping clocks around the call to measure how long they take to execute. It is nice to do this in a test harness so that the results are not skewed by other things that might be going on, like servicing interrupts or allocating some CPU cycles to the execution of another thread.

You can write stress tests into the test harness, testing boundaries, memory limits, resource pools, or executing as fast as possible by throwing messages at a routine far faster than the physical system would generate them. And a test harness is the ultimate white-box test, allowing all source lines to be stepped through with a debugger; all logic, including error handling, to be executed; and all outputs to be observed to verify that the code is not working incorrectly. Hence you can exhaustively test the permutations and achieve 100% statement, branch, and condition coverage, if not path coverage, at least for the unit under test.

The principle of this approach to unit testing is to only implement a handful of requirements at a time. Only these few requirements are met, but they are concretely met and completely verified. The implementation can be used in a working system to start integration testing or even to get a subset of functionality into a user's hands to begin the validation of the requirements, that is, that task of showing that the requirements meet customer needs.

In this method, the unit is the fundamental organizing component of a project during the implementation phase. It is the basic work product and its completion is the measure of project progress. The definition of *unit* is "the minimal testable set of functionality." (This also fits with the definition of a unit as described in ANSI 62304 §B.5.4.) A unit may map to a file or module, if the definition of a module is source code and an associated header file. But it does not have to and sometimes cannot. Multiple files may be required for interrupt service, for example. It would not be prudent to violate other rules of good software design and decoupling just to fit a program structure into some kind of unit bag. The goal is nevertheless to organize the code and its implementation based on its testability.

Every module gets a test harness, if possible. Test harnesses are archived work products in the VCS that are retained throughout the software-product life cycle. It can actually be included in the source code, activated with the use of conditional compilation. (In C this could take the form of #ifdef TEST_HARNESS. Then when building the test harness, define TEST_HARNESS to the preprocessor.) It could be a separate artifact with a different extension. (So a unit might consist of test.h, test.c, and test.th.) Either method ensures that the code that is integrated is identical to the code that is tested, both verifying the unit testing and making integration easier.

The following code listing is an example of a simple routine to calculate the median of a set of data and its associated test harness. The main routine of interest is TakeMedian(). It happens to need a sorted array in order to do its work, so the SelectionSort() method is also supplied. (This could alternatively be in its own library.)

Two test regimes are needed: for a data set with an even number of elements, where the two middle values are averaged, or an odd sized data set, where the middle element is selected as the median. In addition, two flavors of data are tested—one with widely dispersed data, and one with narrowly dispersed data—to reveal problems with duplicated values in the data set, or mixing negative and positive, for example.

The `printfs` show the expected result. This could as easily be an automated test, where the computer checks the result and reports pass or fail. Also, the test harness could be extended to test different data set sizes, especially the null data set, or one or two elements only. This is left as an exercise for the reader (Fig. 4.6).

Each unit is completed when it passes a code and test review. The code review establishes that the software implements the requirements that it was supposed to, and the test review is an opportunity to verify that the tests are comprehensive and correct. Since it is in the VCS, the test code is available whenever the unit requires retesting—in the event of refactoring or porting to new hardware, for example, or any other needed regression testing. It also provides an example of the way to use the routines implemented in the unit and the boundaries of their use, since boundary conditions should be captured in the test harness.

One issue is whether to maintain test harnesses and what to do about changes that occur after the unit test is completed. You only have to comply with your own policies, and you may want to make the degree of review and regression testing a function of the risk class of the module. In my own practice, I do not require test harnesses to be maintained in the general course of development. Since there is in my projects at least as much test software as application software, such a policy could easily double development time—and not for a particularly beneficial result. The test harness code is available in any case as a starting place

```
/*********************************************************/
/**
 *    Name:    SelectionSort
 *    Descr:   Sort array of ints using simple selection sort
 *             algorithm. This is adapted from Algorithms in C
 *             by Robert Sedgewick, pp. 96-98.
 *    Inputs: a - array of data to sort.
 *            n - length of array.
 *    Output: Array sorted in place.
 *********************************************************/

void
SelectionSort( int a[], const int n )
{
    int i, j, min, t;

    for (i = 0; i < n - 1; i++)
    {
        min = i;
        for (j = i + 1; j < n; j++ )
        {
            if (a[j] < a[min])
                min = j;
        }
        // swap
        t = a[min];
        a[min] = a[i];
        a[i] = t;
    }
}


/*********************************************************/
/**
 *    Name:    TakeMedian
 *    Descr:   Derive the median of a data set of ints.
 *    Inputs: a - array of data of interest.
 *            n - length of array.
 *    Output: Median, rounded down.
 *            Note that array will be sorted after processing
 *            with this function.
 *********************************************************/

int
TakeMedian( int a[], const int n )
{
    int median;
    int middleIdx;

    SelectionSort( a, n );
    middleIdx = n / 2;
```

(Continues)

```
        if (n % 2)          // n odd
        {
            median = a[middleIdx];
        }
        else                // n even - average two middle elements
        {
            median = (a[middleIdx] + a[middleIdx - 1]) / 2;
        }
        return median;
}


/*****************************************************/
#ifdef TEST_HARNESS
// This is where the test harness begins.
#define SAMPLE_SIZE 10

int Test1[SAMPLE_SIZE] =
{
    -31,     1811,    12123,   18044,   2800,
    4835,    298,     932,     -522,    263
};

int Test2[SAMPLE_SIZE] =
{
    3821,    3820,    3819,    3818,    3817,
    3820,    3817,    3821,    3818,    3820
};


int main()
{
    int median;
    int i;
    int test[SAMPLE_SIZE];

    // odd n, widely dispersed data set
    for (i = 0; i < SAMPLE_SIZE; i++)
        test[i] = Test1[i];

    median = TakeMedian( test, SAMPLE_SIZE - 1 );
    printf( "median = %d (should be 1811)\n", median );

    // even n, widely dispersed data set
    for (i = 0; i < SAMPLE_SIZE; i++)
        test[i] = Test1[i];
    median = TakeMedian( test, SAMPLE_SIZE );
  printf( "median = %d (should be 1371)    \n", median );
```

```
            median = TakeMedian( Test2, SAMPLE_SIZE );
            printf( "median = %d (should be 3819)\n", median );

            return 0;
    }


    #endif // #ifdef TEST_HARNESS
```

Figure 4.6: Unit test code example.

if something changed sufficiently to need retesting, although it may have to be refactored to adapt to changes made in the main line code in the course of integration.

Note: Just because it is test code does not mean that the code quality can be any less than required for the level of concern of the software! [1]

You can allow minor changes to be tested in integration tests—it should not be necessary to retest an entire module because a data type changed from signed to unsigned. The same is true of reviews. If code reviews are going to have the most value, they need to be conducted early in the life of the unit, when design errors are still fixable. What is necessary is to have a policy for evaluating whether changes have been sufficient to warrant a follow-on review.

It is key that the testable code be *minimal*. The whole idea is to limit how much influence external code can have on the unit under test. Write the test harness to stand alone. Use stubs in the test harness code to reduce the connections to called functions and/or to simulate their functionality. If you follow the principle of minimalism, and make an effort to reduce the linkages, the smallest amount of data that the unit needs will become obvious. You can then make the data local in scope and provided by the test harness. At the least you will understand what external data the module is coupled to. If the data are complex, you may find it necessary to use test scripts or files (also archived in your VCS) to provide simulated data.

As for what guides the organization into units, it will depend on your requirements, architectural design, implementation language, inclination, and experience. If you are developing a multitasking system, it is often smart to implement a thread per unit. The test harness can simulate input messages to the thread, and stub out the output thread. The input messages can be controlled with a script and hence different circumstances can be simulated that might occur rarely in actual use. The output can be verified against the input—you can use the computer to automate this testing if you choose to.

A unit usually implements only a small subset of requirements. Depending on the SRS level of detail, this might be on the order of 2 to 8. The cohesion of a limited set means that the project leader can assign the development engineer a unit of functionality to implement. It simplifies the problem to the level of providing one cohesive, distinct service at a time. This is also usually in a small enough chunk to enable project tracking and oversight.

Another virtue of being focused on units is that you can use a phased approach to implementation. For example, say that the unit of functionality you have decided to work on is the implementation of the analog-to-digital (ADC) conversion routines. The first phase may see the ADC implementation using a polled mode and minimal configuration of the hardware. This lets you verify that what may be a new hardware design is in fact working. The next phase is to initialize the device for interrupt operation. Since interrupts can be tricky to get running, there is an advantage in already knowing what to expect from the device. There is less to debug all at once.

Once you have established that the ADC interrupts are working and giving you expected data, you can add channel calibration. This is also the time when you might consider characterizing the way that the hardware and software work together. You can measure the linearity, signal to noise, response to step input, and other parameters important to validating the functionality of the system. You can focus on the components under study, reducing the interaction with the rest of the system that might confound the results and interfere with debugging.

There are some things to watch out for with a phased approach. You don't want to get stuck in doing things differently than the final application so that you are misled into thinking you know something you don't. You also don't want to be doing the same thing, just in a different way. These worries should diminish with experience.

The requirements that a unit implements should be well bounded. So should the amount of effort. It should be possible to implement a unit on the order of a few days to a week. It might be only a few hours if that constitutes as sensible unit. This is a short time frame, so there is only so much code a person could write. This means that units tend to be a few hundred lines in length, including the test harness. Shorter modules are far more manageable than 5000 to 10,000-line modules. It is also much easier to get someone to review a short module. It is going to be difficult to do an effective review of 5000 lines of code, and usually too late to do anything about it if there are issues. It is much cheaper and easier to throw away 50 or even 500 lines than 5000 lines. Human nature (and project exigencies) makes it hard to discard the latter, regardless of how unsuitable it may be.

The small size and the short amount of time spent on development also encourage good cohesion, and especially loose coupling. Coupling is obvious in the list of files that the test harness links. (This could actually be used as a metric.) Sometimes the list is quite large and it is inconvenient to specify each file that has to be linked to create the test harness. For once, an inconvenience is a virtue: it encourages the engineer to loosen the coupling. It is an inducement to decouple through refactoring interfaces or hiding data, or even developing a well-structured library. Faced with the alternative of linking everything every time, software engineers think a bit more about global data, linkages, and coupling.

As more and more functionality is implemented, the units are bolted together into more comprehensive integration tests. These have test drivers of their own, and ought to have formal test plans. The purpose of the integration test is to implement and verify a subsystem. The purpose of the test plan is twofold: it is a reviewed document that establishes that the verification is adequate and the testing is complete, and it is a necessary element in the premarket submission for higher level-of-concern software.

To extend the ADC example from above, the final package might have five units. The first unit tests the initialization of the ADC hardware and the setup for interrupt service in the application. It outputs raw analog values from the ADC channels. The next unit tests the calibration of the analog data. It did not use raw data in its unit test; instead, test data were provided to it to demonstrate the properties of the calibration transfer function. For example, it was verified that the calibration could handle the complete range of data that could be presented to it, including error values. It was also verified that the calibration transfer function would transform the data correctly, with complete understanding of rounding errors and the effect of bit errors in the ADC on the final result.

Another unit is required to interface to the transducers connected to the ADC. General-purpose output is needed to control the transducers for wraparound tests, or it is necessary to control a multiplexer to read multiple channels.

A fourth unit accepts the output from the ADC and filters it. The unit test of this consists of providing known input to the filter and examining the output. This can be done analytically by writing a console application to read an input file, process with the filter, and then write it out again. Or you could use a desktop application, such as an Excel spreadsheet—certainly Excel would be a good way to see the results of the filter. (Just remember to verify that the output of the third-party tool is correct—see Section 6.1, Software of Unknown Provenance.)

Finally a unit takes data from the filter and displays it. The embedded application does not need to display the data, but debugging and verifying the system greatly benefits from returning to an analog form of displaying the data so that human brains can easily process what has happened in the "black box." The displayed trace can then be compared to the raw data seen in an oscilloscope trace and the processing verified and characterized for things like response to noise, ringing, response to a step function, and phase shift through the filter. This is an example of writing code that is necessary for developing the product but not for the final product. Sometimes in R&D you'll write a lot of code that you are not going to keep—it is necessary to write the code to verify the system. This display driver application could be the integration test harness for the ADC subsystem. The test plan would consist of this characterization of the whole thing, from signal input to the output in the display unit.

Note that each of these units was separately demonstrated with test data before integration into the whole. The engineer already had experience with the end result if something was haywire—if the ADC hardware was faulty, for example. You can trace the source of the problem, because you have already determined what outputs will occur from what inputs. Instead of seeing some value that did not seem to change, the experience allows the engineer to know that he is seeing the ADC rail, or is excessively noisy, or is otherwise not getting the input expected. The system is testable. You can apply a stimulus before or after the transducer and check its output. If something unexpected is going on, the test harness is available for a suspect unit. You can rebuild the test harness, poke it with bad data, and explore what is going on.

As more and more of the system is implemented, the same activity occurs over and over again. More and more subsystems are built, and these are integrated into larger and larger components with corresponding integration test applications. The integration tests become the deliverables for integration milestones wherein you demonstrate that parts of the system or particular features are working. (I am assuming some iterative development life cycle here.) Eventually, the delivered system is the integration of all the subsystems. They have been fully tested all along the way, down to the unit level. You are able to provide the unit, integration, and system test documentation for the premarket submission, if needed. If problems show up in integration or if hardware breaks, you have the test harness available to investigate suspect code. If there are changes to requirements, or if defects arise, you have the test harnesses to explore the defects or retest the changed requirements. The system is decoupled so that you can make reasonable arguments about how much regression testing to apply, and you have the integration test drivers and test plans available so that you are not starting from scratch when regression testing is needed.

5.1.4. Static Tests

Static tests are analyses performed without executing the software. Probably the most used static test is a code review. This could take the form of a highly formal inspection or a more informal code review; the rigor you use should be related to the level of concern of the software. Per ANSI 62304, code review is not optional on moderate or major level-of-concern software, since it is the basis of the acceptance criteria applied to a unit before it is acceptable in integration [25].

Code review should take place against a coding standard in order to realize aimed-for code characteristics. The coding standard is one of your software process control documents. It should spell out the desirable code characteristics, such as understandability, flexibility, and maintainability. It should describe layout conventions, such as spaces or tabs and the tab indent, as well as the bracing style, so that code developed by several team members will be consistent. Naming conventions are also important, so that the same thing doesn't get several names, which can lead to confusion. You should capture policies on complexity and the length of subroutines, and forbid language practices that are error-prone.

There is some difference of opinion on when to perform code reviews. Should they be carried out before doing any testing, or do you review the tested result? There is even the idea that code reviews are better than nothing, but not by much.

The answer, as in so many things, is that it depends. Both the culture of the organization and maturity of the team with the coding standard will be influential. If you have a new employee, or someone who is a bit of a cowboy, you might want an experienced engineer to review the code before a lot of time is invested in testing it. If there are big changes, refactoring to shorter routines, for example, you would have to conduct the testing over again.

On the other hand, in my company we tend to review after the code has been written and tested, so that the code review functions as a test review also. This works because we have technical design reviews before the implementation. Without the design reviews, a code review is sometimes the first time anybody sees a design—and by then, so much has been invested in time, testing, and emotional energy—that the team may tend to commit to bad designs. In any case, the policies that describe your review process should spell this out; it is then a matter of following your own guidelines.

For what it's worth, the guidance document suggests that code review be done before "integration and test" [1]. This is ambiguous with respect to whether it is done before unit testing. What is important is that the results of the review methods and results be captured in the DHF. Review checklists are a handy way to do this. See many of the good texts on code construction for sample checklists.

As for keeping track of whether the code review took place, there are several methods. The most rigorous would be to restrict the ability of the author to submit changes to the VCS. Someone else would have to do it for him, only after checking that the code review had been done, or doing the code review as part of the check in. Normally this would only be needed for released software. A similar method would be one that restricts who may promote a code change to a branch. Then software quality assurance would verify that the code had been reviewed before integrating changes from development branches and rebuilding the software.

Some VCSs have a promotion model attribute that is applied to archives. Then code modules would be promoted to reviewed status; if the file is subsequently changed it drops out of reviewed status. It is then necessary to run a report before release to verify that all of the software is in the proper promotion group. You could do the same thing with a label. You apply a label such as "Reviewed" to any code at the revision that the team reviewed. If the code changes and needs another review, you move the label. As in the case for promotion groups, some kind of audit is required to verify that the released code has all been reviewed.

A last method would be to wait until the software is ready for release, and then conduct all the code reviews at once on the final code. This might work for small projects. However, it is generally unwise. Projects are usually under some time pressure close to release. About the last thing you want is a lot of pressure to conduct code reviews. It is unlikely that the reviews will serve the purpose of finding defects—no one will want to change anything and start the process over, no matter how flawed the software might be. To get the maximum benefit out of reviewing code, the earlier in the process the better.

Some tools have emerged to verify compliance to a coding standard. It is quite common for modern C compilers to have an option to verify compliance to the Motor Industry Software Reliability Association (MISRA) C language guidance. MISRA C is a subset of C that disallows use of its more dangerous features.

A program called *lint* has been the classic static testing tool for C and C++ code. It performs a very detailed syntactic analysis and can add strong data typing to the rather fast and loose C-type of environment. Unfortunately, it can be about as much trouble to get running as any bugs it finds—it is not necessarily superior to code review. For deeply embedded systems in which typing is problematic, it can be quite inconvenient.

Recently, other tools have emerged that purport to evaluate the software for semantic errors; run-time errors, such as divide by zero; out-of-bounds array access; and memory leaks, among other things. The most important automated static test is to put the compiler on a high error level and make sure that you have a clean compile. Modern compilers usually have quite good error checking that is more forgiving than *lint*. Any errors in the build would need to be documented in the software build and release documentation. For example, when I build I get a bogus warning about an overlapped memory segment, a result of having located constants in memory so that I can check their validity during runtime. My build documentation explains why this is bogus and safe to ignore.

5.1.5. Integration Tests

Integration testing is overloaded with two meanings. It refers both to integrating software with hardware and the integration of multiple modules into software subsystems.

To accomplish integration testing in the first sense of the word, you demonstrate the hardware and software working together to produce correct output. This is often done by providing a known input to the hardware and observing the result when it is processed by the software. For instance, you could use a function generator to simulate a square wave at the signal input to the hardware at a certain frequency and level. After signal conditioning in the hardware and whatever conversion, calculation, or other processing occurs in the software, the signal should be comparable or representative of what the original input looked like.

Software integration testing in the more usual sense is the testing of the aggregation of software units into larger subsystems. I have already described how it is a natural incremental progression from a unit focused approach to development. Of course, the project does not have to develop that way. In any case, integration testing should concentrate on data and control transfer across both internal and external interfaces.

In integration testing, you need to verify that the software performs as intended and that the testing itself is correct. You must document the results, especially for moderate and major level-of-concern software. This should include expected results, actual results, the pass/fail criteria, and any discrepancies. The tester must be identified, and you should retain the test plan and records of which version, which hardware, and so on, so that the test could be repeated. If the testing is going to be part of a regulatory submission, you need to identify any test equipment and its calibration, per the QSR.

It is acceptable to combine integration testing and system software testing, although you would normally only do this for simpler, less risky projects.

5.1.6. Regression Testing

Regression testing is testing engaged after a change is made to the software. Its purpose is to demonstrate that the software still meets its requirements and that no new defects have been introduced. This is a subject of great interest to the FDA staff because they have observed that, of the product recalls attributable to software defects, 79% of those defects were introduced as a result of changes made to the software after deployment [1].

Where the level of concern of a component is high, regression testing is particularly important to verify that safety requirements are still being met. The requirements assigned to software for detection or mitigation of hazards must still be met after a change.

The minimum requirement for regression testing is a test protocol that demonstrates the major functionality of the system and the safety features, where practicable. This should be conducted after each build. Some organizations use the regression protocol as a "smoke test" to assess the quality of daily builds as a means to discover problems as soon as they are integrated.

The regression testing needed as software naturally evolves is part of the utility of developing software in units with test harnesses. Loosely coupled systems are easier to change, and easier to test. It is easier to create a justification for limiting the testing to the modules that changed, rather than repeating every test for the project. This is also a function of the nature of the change and the criticality of the software.

5.1.7. Auditable Build

By definition, the DMR is supposed to provide the information to trace everything that went into the manufacture of a device to its point of origin. This extends to source code. In the event of a customer complaint or recall of your device, you may be required to re-create from source the software that went into the device. Being able to do so might save you from rewriting the whole thing from scratch if there is a minor error to fix.

To support this, one consideration to have in mind when releasing software is to audit the build so that you can be sure you have all the elements needed to re-create all executables and other deliverables from their sources (usually in the VCS). You want to be able to produce the records of any released configuration item [25].

The ultimate method would be to install everything—operating system, code generation tools, VCS, and source code—on a new or clean computer. As a rule, it is probably sufficient to build in a clean directory. This assumes that you were careful about documenting the special setup steps that the tools might have needed. For example, it is easy to forget years after you installed a compiler that the command line needs environment variables set to specific values. If you were to do a full install, you would be assured that you understood all of the setup steps needed. A document that describes the setup steps to install the tools needed for software development is a useful document to have for this reason; it is also handy when you have a new employee or get a new computer.

Depending on the criticality of your device, in addition to the auditable build, you may want to consider locking away in a closet the build computer, or at least its hard disk. Since most of us are performing development in commercial operating systems and computers, even with the full source code and tool chain, it is possible to have trouble running tools on newer operating systems. The pace of change is so fast in the commercial systems that sustaining identical functionality for years is difficult if not impossible.

5.2. Software System Testing

The pinnacle of software testing is the software system test. This is where you show that all of the software requirements have been met [25]. You do this by providing input ("Apply calibrated reference heat source to the probe end"), expected output ("View the reading on the display"), and the pass/fail criteria. (Verify that the display reads between 39.8 and 40.2 degrees C—for this simple example of a digital thermometer.)

This is also known in some companies as the software validation. Because software validation is a larger subject, namely, the construction of software using a defined, rigorous process, I prefer the term software system test, as defined by ANSI 62304.

Software system testing usually consists of a set of test protocols executed when the software is complete. These can be organized in different ways—they could be organized by the software system specification, in other words, map to the SRS, or they could be organized by features that cross SRS boundaries. You are free to organize them as you like; test management software would probably be helpful. You do need to make sure that, whatever organization you use, you can show that all requirements have a corresponding test somewhere.

This brings us to our first difficulty. There are other kinds of requirements besides functional ones. For example, assume that we want to specify the initial settings of registers in the microcontroller, so that unused general purpose I/O pins are set to inputs to use the lowest power. They are unused and hence by definition have no function. Or we might want to specify the development language or the hardware that the software is designed to run on. How do we test these requirements?

Not all requirements are functional requirements. We don't want to leave them out of our requirements specification (a solution I have seen more than once) because the system design would be wrong if we did not include them. But they cannot be verified with a black-box test.

These types of requirements will have to be verified by other means, usually an inspection. When the traceability analysis verifies that a requirement has been met, it can point to a code review or unit test that provides the evidence. And they can be done early. Not all requirements have to wait for the final software system test to be verified. (You would have to be sure however, through a label or promotion model in the VCS, that the software had not changed since its inspection.)

I have already mentioned that there is no one defined way to organize your software system test protocol. However, there is an interesting formalism that has been the case at all the medical software firms I am familiar with. I am not exactly sure why this formalism evolved—it is not specified in any of the literature. I believe that it originated in the practice of having tests describe expected results before the testing including pass/fail criteria, plus FDA requirements for approval signatures. It makes more sense for statistical tests applied to hardware components, such as in testing the burst pressure of a balloon catheter. It has been adapted to software testing, where the outcome is much more deterministic, so this style of testing doesn't make quite as much sense. Nevertheless, this is what it looks like.

The software system testing consists of three phases. In the first phase, you write the test protocol, have it reviewed, and get it approved. Of course, since the best review is to actually execute it, you usually have a good idea of what the test results are going to be. Once approved, it is executed and the results recorded. This must be by someone who is not the author and is not responsible for delivery of the product, to avoid blind spots and conflicts of interest. Then the test results are summarized and a final report written. (I have known some firms to type up the test results to avoid questions from the premarket submission reviewers about reading the handwriting of the testers.)

The test results must include the following:

- The date of the test.

- Identification and signature of tester.

- Identification of the software configuration under test.

- Identification of the test articles—the serial number and configuration of the medical device used for the tests, for example.

- Identification of test tools and equipment used. This equipment must be calibrated.

The objective is to supply enough information so that the test could be repeated some time in the future to get the same results.

In spite of our best efforts, it may not happen that the system passes all the tests. Sometimes the test plan itself was misinterpreted or in error. You don't have to keep iterating the test plan until it passes. What you do instead is to write an exception or deviation to the test. For each test that failed, you write an explanation for why it failed and provide a risk assessment. In the case of a test plan failure, this may be a simple matter of redefining the test procedure and rerunning the test.

If you plan to release the software with the defect in it, the risk assessment must provide a rationale for how the software is safe and effective in spite of the presence of the defect. You will have to provide a list of the anomalies in the software along with any premarket submission.

You can also use the SPR process to record the test plan failures, and manage the test plan failures with the normal procedures. One good way to deal with defects in the test plan is to explain in the exception report that the test was modified, rerun to show that the software itself is correct, and has had a defect entered for the test procedure that will be corrected at next revision of the test protocol. This method closes the loop on the defect.

Once you have completed the software system testing, there is a step left that is peculiar to medical software—certification of the software. Like the formalism in the test protocols, it is not exactly clear where the need for this originated. I believe it relates to requirements in the QSR for signatures affirming that the quality policy has been followed. (It may be a method to criminalize noncompliance to the QSR. Were you to sign such a document and the

FDA could prove that you had not followed the process, you might be susceptible to a charge of perjury.)

The certification is a written statement that the software complies with its specified requirements and is acceptable for operational use. In it, you also certify that the software development process of your organization was followed. You should also include a checklist of the control documents created during the software development. These deliverables are related to the level of concern of the software and will have been spelled out in the software development plan. The checklist is to ensure that all are complete in the release package.

The certification itself should read something like the following. The actual titles and signatures depend on your company's quality policy and organization.

> *My signature affirms that the software was developed according to procedures in SOP-XYZ and test results demonstrate that the product meets its specification.*
>
> *I specifically certify that:*
>
> *I have verified that all required deliverables are on file.*
>
> *The documents in the DHF reflect that fact that the software followed the software development life cycle.*
>
> *All tests in the software system test protocol were performed and meet the specifications for safe and effective software.*
>
> *Approved by: _____ Software Engineering Manager, date*
>
> *Certified by: _____ QA Manager, date*

With the certification in hand, the software deliverables archived in the DHF and the software release itself in the DMR, we are ready for the system design validation that will demonstrate that the device itself is safe and effective.

FDA cautionary note: "Testing of all program functionality does not mean all of the program has been tested. Testing of all of a program's code does not mean all necessary functionality is present in the program. Testing of all program functionality and all program code does not mean the program is 100% correct! Software testing that finds no errors should not be interpreted to mean that errors do not exist in the software product; it may mean the testing was superficial" [1].

We can only be diligent. But don't assume because all the tests were passed that everything is perfect. We must take care that the tests test what we mean them to test. Getting the tests right is about as hard as constructing the software in the first place.

5.3. System Validation (Acceptance Tests)

Insofar as software validation means establishing that the software meets its requirements, software validation can be done in the laboratory. But where validation without the software qualifier means design validation—that is, showing that the product meets customer needs and is fit for its intended use—is something that must be done by the customers themselves in actual use. For medical devices, this means clinical trials or clinical evaluation.

The design of clinical trials is outside the scope of this paper, and beyond my expertise as well. Medical professionals and statisticians need to be involved. A big issue in design validation is the labeling, or claims for what the device is good for. If the marketing literature makes a claim for an application of the device, the claim has to be substantiated with clinical data. Medical manufacturers cannot get away with the type of intellectual sloppiness we are used to in other types of advertising. It is insufficient to claim "new and improved"; you will have to demonstrate it. So some caution about claims is warranted. You don't have to work outside your specified range. And you also cannot control how the doctor uses the device. At least in the United States, the doctor can use devices and drugs "off-label," that is to say, contrary to the indications in the labeling that accompanies the device. But to do so is to put himself and his practice at risk, and, as a practical matter, attempting a legal defense that the warnings should have prevented misuse is usually not sufficient to exonerate a firm from blame.

5.4. Traceability

Traceability is an analytical activity in which you establish the link between requirements, design elements, implementation, and especially to tests. Remember, the software system test is a demonstration that all of the software requirements have been fulfilled, so it is directly traced from requirements. And it is not sufficient to just generate a matrix that shows each requirement number has a corresponding number in the test document. There needs to be a review of the tests against the specific requirements to verify that the test tests the requirement correctly.

Requirements should give rise to design elements. If there are design elements without corresponding formal requirements, it is a mistake one way or another. It could be a sign of tacit requirements that need to be stated explicitly so that they are not forgotten upon further development, and are written down to ensure that they are tested. Or it could be a sign of gold-plating at the design level. Gold-plating is bad for several reasons. First, the system should not have unnecessary elements that may cause failure—the system is more elaborate than it needs to be and complexity works against reliability [16]. Second, from the point of view of the project, implementation that was not part of the requirements is something that the company did not sign up to pay for.

Traceability is particularly important to risk management. If the output of the risk analysis process is hazards that require mitigation, it is crucially important to verify that the software does indeed mitigate the hazards. It is the worst kind of mistake to identify a hazard, attempt to mitigate it, and fail to properly design a safe system because of lack of oversight during testing that allowed an identified hazard to occur. It is hard enough already to get these systems right and ensure that they are safe. To fail to verify that we do what we promise in the requirements is the worst kind of mistake.

5.5. Metrics

As for any quality system, metrics are valuable to the FDA. Software is highly complex and unlikely to be defect free. It is impractical to test forever, and difficult to know whether the software is of sufficient quality to be safe and effective. This is where metrics are important. The number of defects, their discovery rate, and their severity should all be on a downward trend before release. In addition, the types of defects are of interest to detect problems and trends, and manufacturers are required to analyze defects for trends [25].

As usual, the FDA does not specify which metrics are important or how to collect them, but there are implicit requirements for metrics to establish the other types of evidence that the FDA is looking for, such as the list of anomalies and their effect on the system. One thing the FDA realizes is Deming's insight that it is more expensive to test quality into a product rather than to design it in. For software, it is impossible to test quality in for all but the simplest of projects. Software is enormously complex; the system could exist in trillions of states, cannot be shown to be correct past any level of complexity greater than the trivial, and is unlikely to be 100% bug-free. The metrics will show that the software is on a path toward a minimum number of defects.

This is not a discussion of software metrics, and it is not the main focus of the FDA or the design control process. The purpose of metrics is to establish that the trends are in the right direction and that the process is in fact controlled. You can of course create as elaborate a set of software metrics as you like. There are whole books on the subject. However, a little is worth a lot, and a lot is not necessarily worth a lot more. In fact, a lot can be worth far less, if the time the team spends collecting and analyzing metrics subtracts from the time available to get the software itself as correct as it can be.

Nor are all defects created equal. Some obviously are more serious than others. It is essential to focus the team's energies on the most serious matters. This is just a commonplace of criticality partitioning. Your policies should establish what defect severity means, usually based on the harm that could arise if the defect were to occur. Broad categories are better— you don't want to have to spend a lot of mental energy deciding what the classification is. And you should give severity benefit of the doubt. It is also useful to separate the priority of a defect from its severity. Priority is a business matter; severity it an objective matter.

Sometimes it is important to the business to fix a nuisance bug that is irritating an important customer, whereas the policy will usually say that a nuisance bug would not prohibit shipping the product, provided that the nuisance does not create a hazard. From the point of view of the regulatory bodies, they don't care much about nuisance bugs provided that the product is safe. The exception is difficulty of use, because human factors analysis would tell us that something that is not easy to use is prone to user error, and the errors themselves may create hazards.

Major defects would be something like a system reset or hang. These will be evident from the hazard analysis. Moderate defects are less critical to delivery of the essential function of the device. Minor defects are things like nuisances, misspellings, cosmetic errors, some label in the wrong color, and so forth. (Insofar as labels for some parameters may be required to be a certain color by international standard or convention, it may be a moderate defect if one label were the wrong color, whereas if another label were the wrong color it would be only minor.)

The defect analysis does not usually belong only to the software team. In the first place, the defect may not even be software related; unfortunately, software is mostly what users see, so they begin by blaming the software. But in many cases it could be a hardware defect, a faulty system, or a misunderstanding on the part of the user. In the second place, the software team is probably too self-interested to make correct judgments about priority.

If you are extremely lucky, the requirements are well understood and beautifully written, so that the only kinds of defects you have are noncompliance of the implementation with the requirements. But usually this is not the case; the toughest bugs are bugs in requirements. This is where the CCB comes into play.

The purpose of the defect metric is to establish a trend toward zero. For the highest level of concern, there should be no defects of major or catastrophic severity. You may also have the policy that there are no defects of moderate severity. If you plot the number of open defects versus time, you should see a negative slope. The line should converge on zero defects— some use this as the metric to determine when testing is complete and they can ship product. The line will have brief jumps when more effort is focused on testing—this could be normalized away by using a metric like the number of defects found per hour of testing. More time should pass between each defect found as bugs are being fixed.

What you don't want to see is a positive slope. It is entirely possible in an unstable system to see more defects introduced as other defects are fixed. This is a sign of trouble, and usually also a sign of poor design. It may be necessary to go back to an earlier phase in the software life cycle in order to fix design errors that are causing too many defects.

You don't want to see the number of changes increase as the software gets close to release (Fig. 4.7 A, B). The size of the triangles represents the size of the change. The dark triangle

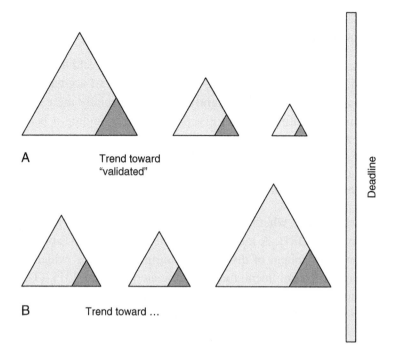

Figure 4.7 A, B: The change trend.

in the corner is the number of re-injection errors. Re-injection errors are mistakes you made while making other changes. The industry standard is about 1 in 8; the FDA experience is 80% [1]. (The error re-injection rate for your organization is another metric that would be good to have. It probably varies with the phase of development or the work product. Knowing it would help you estimate schedules and determine the amount of regression testing that is called for.)

You want the rate of change to the software to be trending down, as in Figure 4.7 (A), so that the re-injection errors get smaller as well. A mistake that I have often seen is the urge to ram in changes at the last minute, as in Figure 4.7 (B). Each change has a risk of introducing a defect elsewhere in the system. Last-minute changes have the same re-injection error rate as any other change—in fact, they are probably worse, because people will be taking shortcuts with testing under schedule pressure. So while it appears that a lot of work gets done at the last minute, it really just extends deadlines, because the defects are genuine—and it is far worse to have your customers discover your defects rather that discover them yourself and have a chance to fix them. The stabilization phase represented by 4.7 (A) is still going to happen, but it will happen after release, when it is the most expensive to fix.

The CCB can play a role here in limiting willy-nilly change. (Sometimes this board is the source of the problem, but that is another matter.) It is important for the board to remind themselves of the definition of a defect: does it *so* interfere with the form, fit, or function of the device that it *cannot* be used for its intended purpose?

Early in the project, it is okay to be proactive in fixing defects, and accepting a low bar for what constitutes a defect. Late in the project, after the accrual of a great deal of verification and testing, changes should not be made lightly. This is especially the case because under time pressure, the risk of creating a defect elsewhere and not finding it because of the limitations of regression testing is so high.

5.6. FDA Regulatory Approval Process

The first thing to determine is whether your product is a medical device, and will thus fall under the purview of the CDRH. A medical device is an instrument, apparatus, or implant used in the diagnosis or treatment of disease, including accessories, which is not a drug. It includes in vitro diagnostic equipment and the associated reagents [31]. (The technical definition is, of course, more elaborately legal than what I have stated here.) If a product meets the definition of a medical device, "it will be regulated by the Food and Drug Administration (FDA) as a medical device and is subject to premarketing and postmarketing regulatory controls" [32]. Other medical products are also regulated by the FDA, but they are regulated by other centers within the umbrella agency, instead of CDRH.

Since devices may contain software, the software in it is regulated by the FDA. Furthermore, software that is itself the medical device is regulated. Software for use in blood bank management must meet FDA regulations, as does software used in the quality system or for manufacturing medical products. This last is important—it means that software used to control the process to manufacture the device also should meet the V&V requirements we've been discussing.

5.7. Device Risk Classes

The FDA classifies devices into three risk classes related to the risk presented by use of the device, but the classification is really a method to define the regulatory requirements. The risk class also governs what type of approval process the device will go through. "Most Class I devices are exempt from Premarket Notification 510(k); most Class II devices require Premarket Notification 510(k); and most Class III devices require Premarket Approval [PMA]" [33]. There are various exemptions and limitations on exemptions which move the regulatory requirements boundaries around. Thus some Class II devices are not required to have a 510(k), whereas others require a PMA. In any case, all medical devices are subject to a baseline set of general controls, as defined in the QSR. The general controls mean that "[a]ll medical devices must be manufactured under a quality assurance program, be suitable for the intended use, be adequately packaged and properly labeled, and have establishment registration and device listing forms on file with the FDA" [3].

Class I is the lowest-risk class. These present minimal potential harm in their use. They are also often simple devices. Examples include exam gloves, bandages, hand-held surgical instruments, and preformed gold denture teeth. Most external prosthetics fall in this category, such as hearing aids, artificial limbs, and eyeglasses. Artificial hands are an example of a device that is even exempt of GMP requirements except for records and complaints. Medical library software or other nondiagnostic software would be considered Class I.

The FDA considers general controls sufficient for Class I devices. However, Class I devices with software are subject to design control. Devices of greater complexity, risk, or uncertainty may not have sufficient assurance of safety and effectiveness solely through general controls. These will be Class II or Class III devices.

Devices where we know how to establish safety and effectiveness—because we have done so for similar products—will generally be Class II with a path to market using 510(k). If we don't have enough information to ensure safety and effectiveness, the device will be Class III and will usually require PMA before it can be sold. Class III devices are riskier: "Class III devices are usually those that support or sustain human life, are of substantial importance in preventing impairment of human health, or which present a potential, unreasonable risk of illness or injury" [35].

It is possible for a malfunction in a Class II device to cause serious injury or death, but its primary function is not life-support, which is what distinguishes it from Class III. Generally, diagnostic equipment is Class II. The reason for this is that a malfunction in such a device could result in incorrect data. The clinician could mistakenly use this incorrect data and deliver wrong therapy. Ultimately, however, the clinician is still responsible for using judgment when relying on the device and its information.

Pulse oximeters, blood pressure machines, and x-ray machines are all Class II devices. X-ray machines are interesting because not only do they provide images for diagnosis, but they also deliver harmful radiation. Hence they have an additional hazard in that a malfunction could result in a harmful dose that injures the patient. Medical devices that emit radiation—such as x-ray, ultrasound, or laser—are also regulated by CDRH and have special standards they must meet in addition to the usual QSR standards.

The highest-risk devices and the ones requiring the most scrutiny are Class III. Pacemakers are Class III because they pace the heart and might cause injury or death if they fail, whereas a device that tests pacemaker leads is Class II devices because it provides diagnostic information. Apparently because of their troubled history, silicone breast implants are Class III.

Higher-risk classes, by their nature, require more risk analysis and risk management. In addition, to establish that the risk analysis process has been conducted with due diligence, the higher risk classes require more stringent design control, more thorough documentation of the development process, and more extensive verification. Moreover, much of this additional documentation will have to be supplied in the premarket application [3]. I will have more to say about the ramifications of risk class on the software process later.

5.7.1. Determining Your Device's Risk Class

While the amount of risk presented by a device is a general guideline to what classification it might be, you can't rely on guesswork. The classification is based at least in part on how comfortable the agency is with the technology of a device and its history, that is, how long it has been on the market. For instance, you might think a gas machine that ventilates a patient during surgery and is used to deliver anesthetic agents would be a life-support apparatus and hence Class III. So it was 15 years ago; it is now Class II.

The best place to go for more information about devices, risk classes, and the premarket application process is the CDRH's Device Advice website at www.fda.gov/cdrh/devadvice/. Here you will find the Product Classification database (under the CDRH Databases tab) [36]. You can search in this database for the keywords that describe your device. The keywords are not always obvious, so you may have to make several attempts. For example, the patient ventilator is known as "gas-machine, anesthesia." The "review panel" field is the medical specialty applicable to the device. For example, I selected "anesthesiology" and browsed the 174 entries to find "gas-machine, anesthesia."

The search will describe the risk class for the device, the submission type, and whether it is exempt from GMP requirements or has special requirements. There is usually a link to a regulation number, which is the actual Code of Federal Regulations Title 21 definition of the device and its classification.

Novel devices that are not in the database will require PMA. The next section will address the premarket application process necessary to get the device on the market in the United States.

5.7.2. Premarket Submissions

The process for getting approval for the sale of a medical device in the United States can take one of two pathways. A simplified application process known as a premarket notification or 510(k), after the section of the federal law that allows it, is available for many products already on the market. If a medical device is "equivalent" to a device that was available in 1976, companies can pursue a 510(k) approval path. The purpose of the 510(k) submission is to demonstrate that your device is as safe and effective as a device already being marketed.

In other words, you show that your device is "substantially equivalent" to another device. This other device is commonly known as the "predicate device."

A device is substantially equivalent if it:

- Has the same intended use as the predicate; *and* has the same technological characteristics as the predicate

  or

- Has the same intended use as the predicate; *and* has different technological characteristics and the information submitted to FDA. Does not raise new questions of safety and effectiveness; *and* demonstrates that the device is at least as safe and effective as the legally marketed device [14]

Once FDA personnel are convinced of the substantial equivalence, they will send a letter allowing the device to be marketed. If the FDA determines the device is *not* substantially equivalent, the manufacturer may submit another 510(k) with additional data, petition for reclassification, or submit a PMA.

There is a fourth pathway for novel but lower-risk-class devices. Prior to 1997, all new devices were Class III by default and required a PMA. The risk class could only be lowered by reclassifying the device. Nowadays, applicants can use the de novo process. This is a written request within 30 days of the judgment of not substantially equivalent that argues for a summary classification into Class I or II. If successful, a 510(k) is sufficient, and the device can now serve as the predicate for other devices as well.

5.7.3. When Is a 510(k) Required?

Obviously, the first time that a device is offered on the market, if it is not exempt or a PMA, it requires a 510(k). This needs to happen at least 90 days before the device goes on sale, to give the FDA sufficient time to respond.

It is not the intent of the FDA that every change must have a 510(k). However, products evolve. There are two significant changes that would trigger a 510(k). First, if the intended use of the device changes. If you make a new claim in the labeling for an application, you would need a 510(k) supporting that claim. So, for instance, if you had a device that screened for liver cancer, and you discovered it could also screen for stomach cancer, you would need another 510(k) to substantiate that the device is indeed capable of screening for stomach cancer.

For this reason, one needs to be careful about claims of "new and improved." Unlike the people selling laundry soap or public policy, the FDA will require that you offer scientific evidence that the device is in fact improved.

The second type of change that needs a 510(k) would be something that "could significantly affect its safety or effectiveness" [14]. It is up to you and your team as experts in the problem domain and the device to determine whether a change affects safety and effectiveness, but the decision should be documented in the change control records and is subject to audit [7]. The addition of a new alarm might be a reason for a 510(k). Certainly the deletion of a previously detected alarm would need a 510(k) to justify why the alarm is no longer needed.

For software, minor bug fixes would not normally require a 510(k). Changes to a GUI, such as rearranging the controls or refining the text of messages would probably not need one. Changes to requirements would have to be assessed on a case-by-case basis. New requirements that supported a new therapy or indication for use would require a 510(k).

Devices going through the investigation device exemption (IDE) or PMA require approval for changes in advance from the FDA. You do this by making a supplemental IDE or PMA submission.

By the way, companies do not get to decide whether their device is substantially equivalent. Sometimes it is obvious and the FDA is comfortable with the new device and its predicate. Sometimes it is less so. One company that the writer is familiar with sought a 510(k) on an innovative ultrasound technique for breast imaging and cancer screening using transmission technology, claiming other ultrasound imaging devices as the predicate. Unfortunately for this company, the FDA did not accept its predicate since all other ultrasound machines work with reflective technology. The company has since backed off the claim and is using reflective technology. They could also have switched to a PMA, but they would have been required to supply substantially more clinical data. A regulatory approval strategy unacceptable to the FDA can seriously disrupt corporate financial plans and the bottom line.

5.7.4. Premarket Approval

Remember that the FDA's objective is to evaluate both the safety and effectiveness of a device. This is one way to look at the dividing line between the two approval pathways. If a similar device is already on the market, effectiveness has been established by the market itself—the need for further clinical trials is lessened. The 510(k) is available for devices like this. The judgment of whether a 510(k) is sufficient is complicated by the risk of harm presented by the device. Even if a new product is similar to a device already on the market, if it "support[s] or sustain[s] human life, [is] of substantial importance in preventing impairment of human health, or which present[s] a potential, unreasonable risk of illness or injury" [37], the FDA will require a PMA application to ensure that the device is safe and effective.

The PMA is really about science. While it is necessary to have approval before marketing the device, the PMA is the phase where the manufacturer presents laboratory and clinical data to establish through statistically valid scientific means that the device's benefits exceed the risks and that it is effective at achieving what it promises.

Brand-new devices will first have to gain an IDE (not interactive development environment) to get approval for use on humans in the United States. Effectiveness of the device is as yet unknown at the start of the clinical trials; the IDE is a method to "try out" new devices to "collect safety and effectiveness data required to support a Premarket Approval (PMA) application or a Premarket Notification [510(k)] submission to FDA" [38]. (An IDE would be rare for a 510(k), unless the device were being investigated for a new intended use.)

A major component of the IDE is a clinical plan that defines the types of patients the device can be used in and the disease indications that it is hoped the device will alleviate. The IDE must be approved by an institutional review board—a clinical committee that evaluates the scientific basis of the study—or by the FDA if the device involves significant risk. In addition, patients must give their informed consent, the device must be labeled for investigational use only, and there are monitoring and reporting requirements. With an approved IDE, the manufacturer can ship the device for investigational use without complying with the quality system regulations, with one important exception: the device must still comply with the requirements for design control.

It is still necessary to provide in the IDE evidence that the device is safe. Any hardware standards for electrostatic discharge (ESD), electromagnetic interference (EMI), susceptibility to other electrical devices in the use environment, and leakage current must be met. A system and software risk analysis is required for the IDE, and the software V&V rigor must be sufficient for the level of concern for the software.

5.7.5. Software and Premarket Applications

The preceding is an overview of the regulatory process, which as a rule not the direct responsibility of software engineers but rather the firm's regulatory organization. It is still useful to know what you may face if you have an idea for a new device. If you have regulatory responsibility, you would want to look further at the CDRH Device Advice website.

Developing the documentation for the premarket application—whether 510(k) or PMA—is an essential part of the development process for medical devices. The objective is to develop the documentation supporting the assertion that the software is validated as one piece of the larger picture of the device as a safe and effective medical product.

It is helpful to make the submission as reviewer friendly as possible. Keep jargon to a minimum and use an organization that makes sense. For complex products, the submission will probably be evaluated by several reviewers. A test document ought to be understandable by itself, although for software, it is going to be hard not to point to the upstream documents.

In tests, use a minimum of handwriting and don't do things that would prompt questions, triggering another review cycle. Provide enough information that the reviewers will not feel compelled to ask for more.

The documentation requirements are based on the level of concern for the software, not whether the approval process is 510(k) or PMA. And the documentation and process requirements are no more than are required to build safe, quality software that works, anyway.

5.8. Software Level of Concern

There is an equivalent way to the risk class of the device in thinking about risk as applied to software, known as the *software level of concern*. This is an evaluation of the extent of the harm that could result to which the software could contribute through failure or latent design flaws. The lowest level of concern is *minor*—failures are unlikely to cause injury to the operator or patient. *Moderate* level of concern occurs when there is a risk of minor injury. When the failure could cause serious injury or death, the software is of *major* level of concern. Failure includes providing incorrect diagnostic data that result in wrong therapy. Serious injury in this context means permanent bodily damage or the need for surgical intervention to prevent permanent damage [3]. (ANSI/AAMI/IEC 62304:2006 has a similar classification for the risk of software, but uses Class A for minor level of concern, Class B for moderate, and Class C for the highest level of concern.)

The level of concern is distinct from the risk class of the device, although it is related to it. A device itself could be Class II, not requiring PMA or clinical trials, but have major level-of-concern software in it. A higher level-of-concern device could have lower level-of-concern software in it, but the software will always take on the level of concern of the device. However, once software is a major level of concern, it is always a major level of concern, even if risk control measures reduce it to a minor level. This is because you must always consider the way that risk control measures could fail and hence not remain a minor level of concern.

An anesthesia gas-machine, which provides life support during surgery, would be one example. Any premarket submission would require documentation for major level-of-concern software. On the other hand, even minor level of concern software requires documentation for major level of concern if it is a component of a Class III device. Remember, the risk class of the device I, II, or III governs the extent of clinical data needed and the postmarket surveillance. The level of concern of the software governs the rigor of the development process and the extent of the documentation provided for the premarket submission.

To figure out the level of concern for the software, go to the CDRH document, "Guidance for the Content of Premarket Submissions," [3] and answer the questions in Tables 1 and 2.

You will want to write down your rationale for choosing the level of concern. It is the first component of the premarket submission. I have usually seen this done as a repetition of and answer to each of the questions posed.

Not all the software in the device or system has to have the same level of concern. The architecture could be designed such that the safety-critical software is isolated from less critical software, so that a failure in the lower-risk software could not cause harm. "Isolated" means that a malfunction in one piece of software could not possibly affect another—in other words, there has to be some measure of memory protection or better yet, the applications should run on different processors.

This notion of safety partitioning allows the team to concentrate resources on the safety-critical software, rather than doing a superficial job across all of the software because the standards are too high. It also lets you change the less critical software more easily or at least with less extensive regression testing. So, for example, the system design might have a processor devoted to safety-critical functions, and a separate GUI running on a different processor. The two communicate over a serial link. As long as the interface does not change, and the safety-critical software has been shown to meet its safety requirements regardless of failure of the communications link, the criticality of the GUI has been reduced. This is an advantage for the overall safety of the device, since the user interface can be more readily adapted to correct flaws in the human factors design. You will, of course, have to justify the risk classification in the level of concern statement of the different components if you use this strategy.

For hardware, risk analysis includes an estimate of the probability of a hazard occurring. Because software does not fail in statistical ways, software risk analysis is conducted assuming that the probability is 100% [3, 25]. That is, if the failure *can* occur, it is assumed *to* occur. Hence software cannot rely on a low probability of occurrence to reduce the need for risk control. Since risk controls for software include a careful, rigorous process, software of higher-risk classes must always be constructed with the process appropriate for that level of risk [25].

Assuming that a software failure will always occur makes sense at one level—if there is a logic flaw in the software, it will *always* fail when presented with the failure-inducing inputs. On the other hand, there are failures that software can do nothing about—the failure of a CPU register or stack memory during runtime. These are hardware failures and treated with hardware probabilities of occurrence. The risk of errant software is also mitigated with hardware risk-control measures, such as safety interlocks and watchdogs.

ANSI/AAMI/ISO 62304 describes the process requirements for the different risk levels of software. This will generate control, testing, and verification documentation. The FDA's "Guidance for the Content of Premarket Submissions for Software Contained in Medical Devices" describes the documents that you must provide as part of the premarket submission.

These documents emerge from the software development process for the risk class of software that is part of the regulatory submission. If you are actually putting together the premarket submission, I highly recommend the guidance. However, to provide an overview of what the content looks like, I discuss each level of concern in the following sections.

5.9. Software Documentation Requirements for Premarket Submissions

5.9.1. Premarket Documentation, Minor Level of Concern

The bare minimum documentation required for the premarket submission for all software-controlled devices is the same required for minor level-of-concern software-controlled devices. For higher levels of concern, some of the chapters would have more detail.

- A statement of the level of concern for the software, and the rationale supporting why that level of concern was chosen.

- An overview of the features and environment in which the software operates.

- The device risk analysis for both hardware and software, and the control measures where appropriate.

- Summary of the software functional requirements. This does not need to be the full SRS for minor level-of-concern software. You still should write a full SRS, of course; here you summarize it for the sake of the reviewers.

- Traceability from requirements, especially hazard mitigations, to V&V testing.

- The system- or device-level testing, including the pass/fail criteria and the test results.

- A software revision history. This should summarize major changes to the software and internal releases made during development. It includes a statement of the software release version label and date.

5.9.2. Premarket Documentation, Moderate Level of Concern

Moderate level-of-concern software requires more documentation and more detail in the submission.

- Statement of the level of concern.

- An overview of the features and environment in which the software operates.

- The device risk analysis, as above.

- The complete SRS.

- An architecture design chart. A diagram or similar depiction that shows the relationship of major design entities in the software to each other and to hardware and data flows.

- The software design description. This is the document that describes the *how* in relation to the *what* in the SRS. This should have enough detail to make it clear that the design implements the requirements.

- Traceability analysis, as above.

- A summary of the software development environment that explains the software life cycle and the processes used to manage the life cycle activities, including a summary of SCM.

- The complete software system test protocol including the pass/fail criteria and the test results. You will also need to include a summary of the activities conducted for unit-, integration-, and system-level V&V.

- A software revision history. This should summarize major changes to the software and internal releases made during development. It includes a statement of the software release version label and date.

- A report of the software defects remaining in the device along with a statement about their impact on safety or effectiveness, including what effect they may present to operator usage and the human factors environment.

5.9.3. Premarket Documentation, Major Level of Concern

Major level-of-concern software includes everything in the moderate category, but with more detail. The required testing documentation is much more extensive. In particular, you must include the test protocols for unit and integration testing in addition to the software system test protocol. The documents for major level-of-concern software follow:

- A statement of the level of concern.

- An overview of the features and environment in which the software operates.

- The device risk analysis.

- The complete SRS.

- An architecture design chart.

- The software design description.

- Traceability analysis, as above.

- A summary of the software life cycle development plan and the processes used to manage the life cycle activities. Include an annotated list of the control documents generated during development, such as release letters, and the detailed SCM plan and software maintenance plan. These documents could just be copies of your SOPs.

- A description of the activities conducted for unit-, integration-, and system-level V&V. Include the actual unit-, integration-, software system–test protocols with the pass/fail criteria, test results, a test report, and summary.

- A software revision history. This should summarize major changes to the software and internal releases made during development. It includes a statement of the software release version label and date.

- A report of the software defects remaining, as above.

5.9.4. Premarket Documentation, Off-the-Shelf Software

As the software industry matures and software grows in complexity, the argument to use off-the-shelf (OTS) software in medical devices rather than building your own becomes ever more compelling. It has advantages for adding features, time to market, and even reliability. But it does put a burden on the manufacturer to provide evidence that off-the-shelf (or software of unknown provenance) used in the device is of quality commensurate with the level of concern of the software. There is more about V&V of OTS and SOUP software in Section 6.1, of the same name. In this section, I describe the minimum requirements for the documentation to include in the premarket submission.

Fully identify the OTS software, including versions, and what product documentation the end user will get. Describe the hardware specifications that the software is tested with. Explain why the OTS software is appropriate for the device and outline any design limitations.

Describe the configuration and the installation, and plans for controlling the configuration and changing it if an upgrade is needed. For general-purpose computers and operating systems, describe the measures in place to prevent the use in the device of software that the medical application has not been tested with. It may be necessary to disable storage devices to prevent the downloading of nonspecified software.

Provide a description of what the OTS software does—an equivalent to the SRS, but for the OTS software. Include a discussion of the links to outside software, such as networks.

Describe the V&V activities and the test results that demonstrate the OTS software is appropriate for the device hazards associated with it.

Finally, you have to include a hazard analysis of the OTS software as part of your system hazard analysis. This is conducted according to the procedures outlined previously in Section 4, Risk Management. This will result in a list of hazards presented by the OTS software, the steps you are taking to mitigate those risks, and the residual risk.

If the residual risk ranks as a major level of concern, you will have to include more documentation than that described so far. You are expected to:

- Show that the OTS software developer used product development methods that are adequate for a safety-critical device. This is best done with an audit of their construction policies, which are expected to meet FDA requirements for major level of concern software.

- Demonstrate that the V&V is adequate for a safety-critical device, including the V&V that you engaged to qualify the OTS software.

- Provide assurances that the software can still be supported if the original developer disappears. This is best done by owning the source code.

Note: If an audit of the OTS developer's design processes is not possible and the OTS software remains at major level of concern after mitigation, it may not be appropriate in a medical device application [26].

5.9.5. The Special 510(k)

The FDA has gained enough confidence in design control that, for certain types of changes, a full documentation package for the 510(k) is not required. Another advantage is that the FDA will respond in 30 days [40]. The device must have been designed and approved for market originally with a 510(k), and the manufacturer must provide a statement to that effect. The changes will also have to be made according to design control and kept in the DHF and DMR, subject to audit. But when you submit the 510(k), you only have to include the documentation for the change. You should provide the regression testing which validates that the change did not have unintended consequences, but you only have to provide the "test plans, pass/fail criteria, and summary results rather than test data" [3].

This is a good option for a bug-fixing release. You cannot use a special 510(k) for either of two reasons. First, there cannot be a change in the intended use of the device. In fact, you should include in the submission the new labeling with the changes highlighted so the FDA can check on this. The second reason that would result in traditional 510(k) is a change in the fundamental scientific technology. This would be such things as changing a manual device to automatic, or incorporating feedback into the function of the device. These would require a full 510(k).

5.10. The Review Process and What to Expect from the FDA

Once you have assembled your documentation package, you can submit it to the FDA, along with a fee. (The fee is reduced for small businesses.) You submit two copies, one of which can be electronic. Don't bind them, because the FDA will just rip them up and rebind them for distribution to the reviewers.

Within 2 weeks you will receive an acknowledgement of receipt and a K number, which is a tracking number for submission. Within 30 days, the FDA will tell you if your submission is not administratively complete, that is, if they think you have left something out. For a 510(k), after 90 days you can start asking the FDA for a status report on your submission. They may have questions; this starts a review cycle. Normally you have 30 days to respond to the questions, although you can request an extension. There is no statutory end to the cycle; if FDA staff remain unconvinced, you will have to answer their questions or withdraw the application and resubmit.

The PMA is similar but the time frame is 180 days. For more information, see the Device Advice web pages at www.fda.gov/MedicalDevices/DeviceRegulationandGuidance/ HowtoMarketYourDevice/PremarketSubmissions/PremarketApprovalPMA/default.htm

6. Special Topics

6.1. Software of Unknown Provenance

Software of unknown provenance (FDA uses the term "Software of Unknown *Pedigree*") is, according to the ANSI 62304 definition, a "SOFTWARE ITEM that is already developed and generally available and that has not been developed for the purpose of being incorporated into the MEDICAL DEVICE (also known as 'off-the-shelf software') or software previously developed for which adequate records of the development PROCESSES are not available[.]" In other words, SOUP comes in two flavors: third-party software (OTS, off-the-shelf), *or* internal software that has not been developed using rigorous methods.

Compliance to the process standards that we have been discussing may lead one to think that it is necessary to build every part of the software, not just the application, in order to ensure adequate quality. Companies have done just that—they built not only the application, but the chip set, the compiler for the chip set, the operating system, and the rest of the peripherals.

The industry has moved away from self-development to OTS software because it makes a lot of sense to do so. It is less costly to buy than to build; you can get products to market faster if you don't have to wait to build it yourself; there is an improvement in the quality because you are buying product (an RTOS, say) from developers who are expert in that problem domain instead of doing it yourself. Moreover, OTS software will have the benefit of many thousands of hours of concurrent testing in many different environments to establish its reliability [41]. It is a basic economic principle for businesses to focus on what they are good at and outsource what they can.

Using OTS software puts a burden on the medical device manufacturer to validate that the software thus used is fit for its intended purpose and used correctly. There is guidance from the FDA about what practices need to be applied [26]. This is discussed in more detail in Section 6.1.2, Third-Party Validation. The documentation for the premarket submission was described in Section 5.9, Software Documentation Requirements for Premarket Submission.

The other form of SOUP is internal software, whether legacy or prototype software. It may be software that has been in use for some time. This "legacy code" was perhaps developed before the revisions to the standards and guidance documents that tightened up development processes or was software used in a lower-risk-class device that is migrating to a higher level of concern.

Prototype software is used to explore indeterminate requirements or demonstrate device feasibility. It is a valuable method for reducing project risk. It can be developed using lighter-weight, agile methods with reduced paperwork and much less error handling or hazard analysis. Or, it could use languages or development environments unsuitable to the final product. The idea is to generate code more quickly so as to determine the scope and feasibility of the project.

However, because prototypes are used to refine and validate requirements, they may have been developed without any formal requirements analysis. Nor are they often developed according to any formal design control. Of greater concern, they are usually done well before the hazard analysis—they can be used as a way to explore the hazards—and hence prototype software is not, by definition, safe.

6.1.1. Prototypes—Throwaway or Evolutionary?

As mentioned, prototyping is a way to reduce project risk. But it can introduce risks to the project as well, if the prototype creates expectations of performance and completeness that are not warranted, or if the company attempts to ship what amounts to a prototype.

Throwaway Prototypes—There are a couple of approaches to take to prototyping. The first is the development of a throwaway prototype. The prototyping is done with methods or languages faster than the final methods, but that are not appropriate for the final delivery. It is particularly valuable for exploring user interfaces. You could, for instance, mock up a user interface in Visual Basic running under Windows that mimics what you will later build for a small screen on an embedded device with much less memory and CPU power. There are no design control issues because it is not going into the final product. Instead, it can be part of the design input or design verification in that it demonstrates the methods and requirements that are going to work.

There are risks with throwaway prototyping. Some might object that it costs too much to do the work in the prototype, throw it away, and then do the work for real. But the point is that you are making a prototype because "it is cheaper to develop a throwaway prototype, learn lessons the cheap way, and then implement the real code with fewer mistakes" [42] than to make the mistakes in your expensive system, and then throw it away anyway or, worse yet, accept the bad solution because it is the only one you had time for.

It is also possible that the real work gets delayed while engineers are playing with the prototype. For this and the cost reason above, it is important to abandon the prototype as soon as it has answered the questions about feasibility or correctness of design it was asked to answer.

Prototyping can be used as an excuse to not follow the rules. But best practices for any software development apply to prototypes as well. For example, use of the VCS on a rapidly changing prototype allows us to back up when a new idea is in the wrong direction. Documenting the code helps us understand it a few weeks later when we want to improve it.

Prototyping is also sometimes used as a way of pressing on with development without talking to customers—their opinions are messy, after all, and they may want something different from what we want to build for them. A huge amount of formality is not required, but it almost never makes sense for engineers in the back room to be cooking something up without having done a preliminary assessment of what it is the customer wants. Numerous products fail because they provide solutions to problems people don't have. (My food was fine before it was genetically modified, thank you very much.)

The prototype may create unrealistic expectations for how fast development can proceed, or what the final performance might be (because the prototype doesn't have to do any real work). Consider "crippling" the prototype in some obvious way so that non-technical people don't take it in their heads that the software is more complete than it is.

But the principal danger is in keeping the throwaway prototype. It sometimes happens that businesses think the prototype is good enough and it would take too long or cost too much to "redo" it. This is even a temptation to the developers. Did the prototype really get thrown away or did the code creep back in? You must be sure to get agreement among the stakeholders that you are building disposable software that will be thrown away. It is *not* the final product.

The FDA allows that it is reasonable to use a prototype to explore feasibility of an approach before developing the design input requirements. Design control does not apply during the feasibility phase. But they warn against "the trap of equating the prototype design with a finished product design" [16]. Reviewers have learned enough to know what to look for.

Evolutionary Prototyping—If there is an insurmountable likelihood that someone in authority is going to want to ship the prototype, it may be advisable to use this second prototyping method. Evolutionary prototyping is a type of life-cycle model. It is especially valuable when you don't know exactly what it is you need to build. In it, you implement the riskiest or most visible parts of the system first, and then iteratively refine those subsystems while evolving the rest of the system. You don't discard code—you evolve it into the delivered system. Unlike throwaway prototyping, an evolutionary prototype's "code quality needs to be good enough to support extensive modification, which means that it needs to be at least as good as the code of a traditionally developed system" [42].

One of the ramifications of evolutionary prototyping is that you will have to revisit the requirements throughout the development process. As you develop the system and provide answers to what the requirements should really be, you will have to align the upstream documents with the implementation. This may lead you to think that it is best to wait until the end to capture the requirements retrospectively. Not so! Even if the requirements are uncertain, it is still worth analysis to the extent that the team's knowledge will support it. Requirements become more of a planning tool: this is what we plan to build next, not what we think will be in the final product. The team needs to understand which requirements to build first, and which are likely to change.

Unit test harnesses fit in nicely with evolutionary prototyping. Each unit can be constructed and tested to the highest standard in isolation. You may not know when to sound an alarm, but you can build and test the software to drive a digital to analog converter to make a multi-frequency alarm sound. If the principles of evolutionary prototyping are followed, the code quality will be sufficient to provide a good input into the retrospective V&V process described below.

There are a few things to look out for. As with throwaway prototyping, it is important to manage expectations. Often, quick progress is made on the visible parts of the system. But in safety-critical systems, substantial and difficult work is not visible, yet must be done for the final product. You must make the stakeholders aware of this.

As with any evolutionary development, there is a risk of poor design. You can only change the requirements so much before the design starts to creak. Evolutionary development is still the best method to deal with changing requirements, but you may have to plan some time for refactoring a design when it becomes unwieldy under the weight of changed requirements and expectations. For these reasons, it is important to use experienced personnel for evolutionary prototyping projects, who can better anticipate the ways that requirements may change and build more maintainable, robust designs.

Finally, evolutionary prototyping runs a risk of poor maintainability. Sometimes, if it is developed extremely rapidly, it is also developed sloppily, without due consideration for requirements scrubbing and proper code maintenance requirements. Again, the unit test approach helps here. But don't let evolutionary prototyping become a mask for code-and-fix development. You still have the tools of the retrospective V&V, if that is required, but you want code quality sufficient at the entry to that process so that you don't end up throwing it away.

In any case, software not developed according to the software development process of the firm must nonetheless have an equivalent process to validate it for use in the final product. We will discuss what these processes look like next.

6.1.2. Third-Party Validation

If the firm uses software for the manufacture of medical devices, their design, or in the quality system itself, that software also requires validation. This could seem like an overwhelming task, but it need not be. Remember that the purpose of validation is to show that software is fit for its intended purpose. Most of the time, this means that the third-party validation called for is a matter of demonstrating that the software fulfills the needs of your organization and provides the correct, reliable answer, not that it is necessary to validate every implemented functionality.

Computers and desktop applications are universal in modern business practice; it is hard to imagine being able to do our work at all, let alone remain competitive, if we don't take advantage of the tools that are available. Businesses cannot afford to build all their own applications, nor does it make much sense to do so. A medical equipment company is much better off concentrating on its core competency—medical devices—than attempting to reinvent the wheel (in spite of what your engineers may sometime argue!). Businesses use Microsoft Word to create their process instructions, or may use Access to keep a database of defects. Excel is popular for modeling systems and analyzing data. Do we have to validate the feature clutter of Word? Bioengineers may use the Internet to look at frequently asked questions about our devices—do we need to validate the Internet?

The short answer: No. Windows, Linux, and real-time operating systems are being used in medical devices without the entire operating system having been validated. It is, as usual, a function of the level of concern of the software. Windows 95 is a bad choice for a Class III medical device expected to keep a patient alive. However, it is an adequate choice for a GUI front end to configure a device, provided that the configuration is established to work within the specified limits.

All the features of the operating system will not have been validated. But the device itself will still have to be validated—that is, shown to work for its intended purpose—and the device's interaction with the underlying operating system needs to be validated, so that we convince ourselves that we have correctly used the OS services.

I think of third-party validation as showing that the way we have set up and use third-party software is correct. It is probably not necessary to choose every menu item and go through every dialog box and press every button to validate an application. We can argue that the vendor has done a satisfactory amount of functional testing. (If this is not the case, the product is so buggy we wouldn't use it anyway, and the market will winnow out the defective products.) What we need to show is that the way we use the application, and the way it interacts with our process, and the process itself provides the answers we expect.

As mentioned, a VCS of some kind is required for software development. Because it is so essential to the software development process, and it is used as the repository of the code that will go into the device, it needs to be validated as well.

I would argue that I don't need to exercise every feature or push every toolbar button and show that it does what the manual says. In fact, I don't care that much whether the GUI works entirely correctly. What really matters is that the tool keeps track of my software and that I am using it correctly.

In the above, I wrote about the methods for managing the software configuration. What is necessary in the validation is to show that the pieces of the VCS that you choose to use do what you intend. So, if you are going to use branches, you would want to show that branching the code gets the version you expect in the location you expect to find it. This is your requirement. Create a test that establishes that this is what happens. Clone the reference files to the branch location and use a difference tool to show that they are the same. Open one or two in an editor and verify that the version is what you expect to see. Change a file on the branch, check it back in, and verify that the information about the check in is as expected. Verify that the code in the main branch is still untouched. Change a file in the same line in both the branch and the main. Integrate the branch change from the main and show that the merge detects a conflicting change; then when you fix it (using the automated tools or whatever process the VCS supports) examine the results and establish that the conflict was solved in a satisfactory way.

These don't need to be elaborate tests—remember, the vendor did some testing, or they would likely not still be in business. But you want to convince yourself and others that the software is installed correctly and that you are using it correctly.

What is important is that you follow the validation principles and document what you have done. So, you want to write down the requirements—that is to say, the functions you expect the tool to do for you. Then you want to create tests that will demonstrate that the function works. Execute the tests and archive the results in the DHF.

If you use a product for data analysis or modeling, you could of course validate the whole system—but this is going to be difficult because you cannot do the type of white-box testing and construction analysis called for in the guidance documents. But that is not really necessary. Depending on how critical the application is to the safety of your device, you can make four arguments about validation.

- You can validate it all yourself with formal tests. These will of necessity be black-box tests unless you can get access to the source.

- You can seek a certification from the vendor that the software has been validated in a manner similar to the design control policies. Some vendors are stepping up to this, but expect to pay for it.

- You can argue wide distribution and validation through use. That is, if you have been using a product successfully for a long time, and many other people do as well, it can be assumed to work satisfactorily.

- You can argue irrelevance. If you use a word processor to create process instructions, it is merely a tool to create the written document. The written document is the thing that matters, not the typist or the printer it was printed on.

What is important is to validate that your models are correct. Excel is an excellent product; I have no doubt it provides the right answer 99.999999% of the time, provided that the equations are correct. This is where the difficulty lies—in showing that the equations are correct, the cell references are the right ones, the constants have sufficient decimal places and so on. Validating Excel is like validating arithmetic—I think we can trust the Greeks for that one. What we have to do is show that *we* used it correctly. Many times I have seen people pull out their models without examining the assumptions or even the correctness of the implementation. So rather than worry extensively about what Excel is up to, if you establish that the model gives expected results, you can conclude that Excel did not introduce errors.

What you do want to avoid is the story (okay, I have it second hand) about a large medical company that paid someone to type data from an Excel spreadsheet into a program they used for statistical analysis, because no one had validated CTRL-C/CTRL-V in Windows. Aside from the fact that this sounds so absurd that I wonder how it can be true, why did not someone (1) take, oh, 20 minutes and perform a validation if they were so worried, or (2) realize that having a human retype the data is many times more error-prone than relying on the software could possibly be?

To sum up: off-the-shelf software requires some level of validation. Write down the requirements that you expect the software to fulfill for you. Create some tests to show that the software meets these requirements, and document them. What you are doing is showing why it is that you think your tools are giving you the right answer.

I want next to discuss a couple of issues that arise when the process does not work as smoothly as we could wish. These are retrospective validation and the validation of software of unknown provenance.

6.1.3. Retrospective V&V

Retrospective verification and validation is the reality of systems with legacy code, and sometimes occurs in new product development, when for business reasons the product was not developed strictly according to the regulatory guidelines or my recommendations in this paper. It is allowed, but realize that it is slightly off-color and against the spirit of proper software development. Software should be constructed from the start using careful methods in order to ensure the highest quality.

Sometimes there is a temptation to think that all we have to do is clean up a prototype, perhaps software provided under contract, wave a magic quality wand, and test in quality. This seems like a way to get a product to market early and can create the illusion of sufficient quality to be a safe product. However, this does not lend itself to an evaluation of what the product *should do*—meet customer needs. Instead, it results in what the product *can do* being accepted as the requirements. Worse, the testing may demonstrate what the prototype product *does do* instead of what it *ought to do*.

Nevertheless, there are times when a retrospective V&V is warranted, or at least safe, especially if it is applied to evolutionary prototyping. However, sometimes due to circumstances beyond our control, we are now faced with legacy code or prototype software that we are expected to incorporate into our device, what do we do?

The difference between legacy and prototype is that we're stuck with legacy—we have a product and we need to provide a retrospective V&V because it was not done during the construction. With prototype software, we have the opportunity from a project point of view to take what the prototype has taught us about the requirements and the design, throw the prototype away, and start over doing the formal work required for a safety-critical device.

Software of *unknown* provenance implies that what is desired is software of *known* provenance. Provenance in this context means not documentation about what the product does (reverse engineering or design recovery) but the documentation of the process used to develop the software. In particular, the evaluation of provenance needs to establish:

- That the software is safe and its ability to cause harm was analyzed and reduced

- That practices were used to ensure adequate quality and maintainability

- That functionality is correct by tracing the implemented result back to the requirements.

We cannot change the design processes used if we accept the product as is. All we can do is enhance them with a post hoc analysis. Just because the provenance is unknown or the software was developed without formal rules, does not mean that it is unusable or lacks quality. The objective is to establish that what we have is good enough to be used in a medical device—that it is safe and effective.

The first step is to proceed with a plan. The problem with the SOUP is probably that it was developed without a plan in the first place. The retrospective validation is an opportunity to rectify the oversight. As with any design control process, start with a documented plan to evaluate the adequacy of the SOUP. There are four stages to the plan:

- Determine what the SOUP does.

- Perform a risk analysis.

- Determine the validity of the requirements the SOUP implements.

- Evaluate the quality attributes of the SOUP.

First, determine what the SOUP claims to do. Find out what you can about the requirements—from the implementation, if necessary—to gain an understanding about the problem the SOUP intends to solve. Figure out how the SOUP works—what the underlying design is. Look into the code and see the way it was constructed. Evaluate any test data or test results—is there something that provides evidence that the SOUP actually works? This information may not exist separately from the code. It may be necessary to reverse engineer the higher level concepts from the implementation.

Perform a risk analysis as you would on any subsystem. Start by evaluating the level of concern of the SOUP. Remember, the higher the level of concern, the more critical the risk evaluation and the care for construction. If the SOUP is of low level of concern, the problem of accepting it becomes less.

The risk analysis should evaluate both the risks in the SOUP itself and any indirect risks it may impose through its own failure. Determine whether the SOUP has an unacceptable risk of a hazard, or whether it can interfere with other parts of the system designed to mitigate hazards. Consider whether the impact of a failure in the SOUP can be reduced through safety partitioning. If SOUP can be segregated to a low level of concern, the risk it presents because of its lack of provenance becomes less. If the software is of moderate or major level of concern and its risk is unacceptable, *throw it away*. Learn from it, yes, but build it from scratch with the correct risk mitigations.

Once you have determined what the requirements that the SOUP actually implements are, compare these to the requirements needed by the medical device. Are the requirements correct for the intended purpose? Can they be made close enough? If not, *don't use it*. Also, if the SOUP came from a non-medical environment, evaluate how it will function in the medical device's environment. For example, an application might freely share its information with anyone, whereas in a medical environment, some of that information might be electronic protected health information and require controls on whether it can be seen or not.

Finally, consider the quality of the SOUP. Establish which quality attributes matter and the acceptance criteria for adequacy. For example, your coding standard could use Pascal-style variable names, but the SOUP is written in lower case with underbars. This would not affect the quality, really, so you would make an exception. There are other quality attributes, of course, that may have varying levels of importance to you.

Providing provenance to SOUP is a method to incorporate prototypes, external custom code, and legacy code into medical products that were developed without formal controls. This can reduce project risk and speed time to market, but requires a documented process in its own right, and focused hazard analysis, for the product risk to be acceptable.

6.2. Security and Privacy—HIPAA

The Health Insurance Portability and Accountability Act (HIPAA) is U.S. law primarily concerned with portability of health insurance coverage when people change jobs. It also establishes standards for healthcare transactions. Where it is of interest from the point of view of software development is the intent of the HIPAA to protect the privacy of patients and the integrity and privacy of their medical records.

6.2.1. Who Must Comply

Protection of privacy is mostly the responsibility of the healthcare provider [43]; unless you are in the business of providing software that directly handles patient records for reporting or billing, compliance to the provisions of the HIPAA is usually indirect. The healthcare provider will be doing the heavy lifting, but the security provisions may impose requirements on the software that you are creating for their use. (Or it may provide market opportunities for devices useful for protecting medical data or authenticating users.)

The security aspects of the HIPAA are known as the *security rule*. The Department of Health and Human Services (HHS) under the U.S. government has published a series of introductory papers discussing the security rule on the website, www.cms.hhs.gov/SecurityStandard/. Quoting from the web page, "[the] rule specifies a series of administrative, technical, and physical security procedures for covered entities to use to ensure the confidentiality of electronic protected health information."

The "covered entities" that the rule applies to are "any provider of medical or other health care services or supplies who transmits any health information in electronic form in connection with a transaction for which HHS has adopted a standard" [44]. The "transactions for which HHS has adopted a standard" is a reference to the Electronic Data Interchange (EDI) definitions having to do with health care that HHS has enumerated.

In fact, there is some ambiguity about to whom the security rule applies. There is an exemption for researchers, for example, provided they are not actually part of the covered entity's workforce. Insofar as a researcher *is* a covered entity and deals with Electronic Protected Health Information (EPHI), they would have to comply. Hence, companies researching whether their products are safe and effective in clinical trials would also have to comply if they access EPHI.

This also applies to vendors who have access to EPHI during "testing, development, and repair" [45]. In this circumstance, the vendor is operating as "business associate," and must implement appropriate security protections. The methods for doing so are flexible, however, so it ought to be possible for the covered entity and the business associate to come up with reasonable methods.

One simple method to achieve compliance with the security rule for vendors or researchers is to "de-identify" the data. "If electronic protected health information [EPHI] is de-identified (as truly anonymous information would be), it is not covered by this rule because it is no longer electronic protected health information" [45]. By making the data anonymous, it is no longer technically electronic protected health information, and thus not subject to the regulations.

Not everything is EPHI anyway. If the data are not in electronic form, they are not covered by the security rule, which does, after all, only apply to *electronic* protected health information. "Electronic" in this sense are data stored in a computer which itself can be programmed. The issue is the accessibility of the computer, not so much the physical format of the data. Therefore, personal phone calls or faxes are exempt; whereas a system that returned a fax in response to a phone menu system would be EPHI and subject to the rule [45].

Patients themselves are not covered entities and thus are not subject to the rule [45]. It is nice to know that you are allowed to see your own health data, and discuss it with your doctor.

So even though your data may not be subject to the security rule, you would nevertheless want to make reasonable efforts to protect its data against loss, damage, or unauthorized access, if only to prevent competitors from seeing it. But you would not be required to maintain a complete security process including security risk assessment and a security management plan.

The provisions of the security rule may not be directly applicable to a medical device manufacturer. Nevertheless, they will be important to your customers. It may be necessary to provide the technical security solutions so that your customer can implement the required administrative policies. On the other hand, if the purpose of your software is to provide EPHI data handling, you will find that your customer is required to obtain satisfactory written assurances from your business that you will safeguard EPHI. You will need to follow the full set of regulations in the security rule including security risk assessment and a security management plan. If your hardware or software has access to EPHI, the healthcare provider will have to assess whether you also need to comply [46].

6.2.2. Recommended Security Practices

We have established some guidelines for determining the extent to which the security rule may impact your business. We next turn to a discussion of the type of issues that might be important.

Malicious Software. One aspect that may affect anyone providing software into the medical environment is the requirement for the "covered entity [to] implement: *'Procedures for guarding against, detecting, and reporting malicious software.'* Malicious software can be thought of as any program that harms information systems, such as viruses, Trojan horses or worms" [46]. The reasoning is that malicious software could damage, destroy, or reveal EPHI data. This means that your customers will require of you assurances that your software is not an open door to malicious code that could harm the provider computer network or other devices. You may be required by the customer to provide assurances that your installation software is protected from viruses.

If your device is connected to the Internet, it may be necessary to provide anti-virus software along with regular updates to prevent just such an occurrence. It is probably insufficient to trust the healthcare provider employees to always engage in appropriate safe computing— you might want to consider using an input device special to your device or somehow protected from general use lest it acquire a virus and infect your system. For example, rather than using a standard USB thumb drive, you could use a device that does the same thing but with a custom connector, so that it could not be plugged into an unknown computer that may be infected with a virus.

Malicious software is a more significant issue for software written to run on general-purpose computers. It is less an issue for many embedded systems whose programs execute from read-only memory and hence are difficult or impossible to infect.

Administrative Support. While monitoring log-ins and manage passwords is generally the responsibility of the healthcare provider, device makers sometimes want to limit the access to functionality in the device (i.e., information relevant to engineering or system diagnostics). If the engineering mode provided access to EPHI, a single password to your device that could not be changed would not be an adequate security safeguard.

The administrative policies of covered entities may also require regular reviews of information system activities for internal audits. To do this, they may need your device or software to provide records of log-ins, file accesses, and security accesses [45].

Physical Security. You must have the ability to back up the data or restore it in the event of a disaster, that is, somehow get the data out of the device and into a secure facility if the data are part of health information. For example, if your device contains "electronic medical records, health maintenance and case management information, digital recordings of diagnostic images, [or] electronic test results," [46] the healthcare provider would need to be able to archive this information. It is also important to provide for obliterating EPHI data from your device at end of use or disposal.

As for physical safeguards, you would want to avoid doing anything that would make it impossible for an organization to impose some standards. For example, you wouldn't want to broadcast EPHI or make it available on a web page or some other method such that restricting it to only the people who need to know it becomes impossible.

This extends to physical media that might be used to store EPHI. The provider has to establish rules about how the media goes into or out of the facility, how it is re-used, and how it is disposed of so that protected data are not revealed to unauthorized personnel. In the case of re-use, "it is important to remove all EPHI previously stored on the media to prevent unauthorized access to the information" [47]. If you are making a storage device, the provider may want to be able to identify each device individually so that they can track them.

Risk Analysis. As is the case with risk analysis for the safety of the software or the device, depending on how close you are to the EPHI data, you may need to carry out a formal risk assessment, wherein you evaluate the potential threats and vulnerabilities to those threats and develop a risk management plan in response [48].

Threat is twofold: unauthorized access or loss of data. Both must be guarded against. CMS has a good discussion and example of risk analysis as applied to security concerns. Interestingly enough, many of the same issues and analytical practices are relevant to device risk analysis. The example has good hints for both. The document HIPAA Security Guidance for Remote Use of and Access to Electronic Protected Health Information, available at www.cms.hhs.gov/SecurityStandard/ is a useful specific discussion of remote vulnerabilities and possible risk management strategies.

The security rule is enforced by the Office for Civil Rights; violation may bring down civil monetary penalties, not to mention possible tort awards. Moreover, there is something of an ethical obligation for healthcare providers and others in the medical industry to exercise due care with private information.

While security is often not a direct concern to the manufacturers of medical devices, as information technology evolves and the desire to share information from individual diagnostic devices increases, it will become increasingly important. In addition, there are best practices for protecting data—such as guarding against viruses, unauthorized access, or data corruption—that are the sorts of things we should be doing anyway. We want our medical devices to be of the highest quality and serve customer needs; some measure of data integrity ought to be a given.

7. Summary

The development of software for medical systems is not significantly different from rigorous methods traditionally applied for other safety-critical systems. It comprises the stepwise refinement of requirements and design artifacts until delivery of the final system. The quality systems are focused on customer needs and intended uses to the point of requiring scientific evidence that customer needs are fulfilled and the device is fit for its intended purpose.

There are extensive regulations, standards, required procedures, and guidance documents with only minimal interaction with software. I presented an overview of quality system regulation and ISO 9001. Thanks to the Global Harmonization Task Force, requirements for the U.S. market and the rest of the world are not very different. You would clearly want to know more about aspects of the regulations if you had regulatory responsibility in your firm, but for the purposes of software development it is really design control and the guidance documents related to software that are of the most interest.

Design control provides a waterfall model of development for purposes of discussing the phases of development and the deliverables development should produce. It consists of design input, where you refine customer needs into engineering specifications; design output, where you build designs that meet the specifications; and design reviews to establish that the design output meets the design input. An important life cycle activity for any safety-critical system is hazard analysis and risk management. This is a major source of requirements during the design input phase, and must be constantly verified for correctness and completeness during the design output phase.

Since software is so easy to change, software change management, including a software problem resolution process as the source of software changes, is critical to show that defects are chased to their conclusion, and no changes are made that are not fully verified and validated. Verification and validation are important activities throughout the development process, and we have seen some of the methods and deliverables associated with this. To close the loop on requirements, we have seen how important it is to trace requirements through implementation and verification to show that what we said we would do, we in fact did.

The same kinds of principles apply to off-the-shelf software. We cannot control the construction of software obtained from third parties. It is preferable if it was constructed using techniques like design control. But if we don't know or can't prove that, we supply a risk assessment to determine whether we can use it or have to build it properly ourselves.

The principal interaction of software development with the regulatory system, aside from audits, is the premarket submission. These take three forms, with the intensity of the documentation based on the level of concern of the software. The level of concern is minor if the software by itself cannot cause harm to the patient or user. If the software could cause a minor injury, the level of concern is moderate. If the software, by itself, could cause serious injury or death, the level of concern is major. The goal of the software risk management process is to reduce the risk after mitigations to a minor level of concern. The purpose of the premarket submission documents is to show that the software will reduce the risk to a minor level of concern, through the application of careful construction and design control procedures to ensure that careful construction.

Preparing the premarket submission should be trivial. Provided that all the V&V activity has been captured in the DHF, the submission is just a matter of collecting the documents from this file. This is the ideal world, but for some reason there is a great temptation to stay in "feasibility"—when design control doesn't apply—until the whole thing is working, and then trying to apply enough retrospective V&V to get the design through the approval process.

But writing it all up and then reverse engineering the requirements and other deliverables out of the completed product is not the most efficient way to operate. This may superficially appear to be the case, because there is overhead with change management and with writing things down in multiple places and then having to trace through and find and fix every occurrence when a change is called for.

But if the change management is so cumbersome, perhaps you need to revisit your change management methods and tools. The process ought not to be so obnoxious that the main effort in our working lives is to circumvent it. The test of process is that it should make our lives easier. We version code from the beginning of development so that we can backtrack if we go down a flawed pathway. We do code reviews close to construction so that we can find defects when the code is fresh in our minds and before we invest a lot of time and effort in testing. We build in units so that we can debug them in a simple, controlled environment, rather than debug the units when we have integrated them with the other subsystems, confounding the unit bugs with the integration bugs.

The irony of software quality is that it serves to shorten schedules. I cannot say it better than Steve McConnell:

The General Principle of Software Quality is that improving quality reduces development costs . . . . The single biggest activity on most projects is debugging and correcting code that doesn't work properly. Debugging and associated refactoring and other rework consume about 50 percent of the time on a traditional, naïve software-development cycle . . . . Reducing debugging by preventing errors improves productivity. Therefore, the most obvious method of shortening a development schedule is to improve the quality of the product and decrease the amount of time spent debugging and reworking the software. [42]

In circumventing or avoiding software quality, you are missing out on the true power of design control. It is not some overbearing process that the FDA came up with to make engineers' lives difficult. It is not even a regulatory hurdle designed to protect established firms as a barrier to entry to competitors (although some would have you believe that is its main purpose). In fact, it is just good engineering practice.

Requirements are important. Why would a development team not want to decide what they are going to build? Wouldn't it benefit everyone to have as much agreement up front as possible? Then the marketing representatives can be careful about over-promising. The engineers can build what their customers want rather than guessing, gold-plating, or just building whatever they like. The business can avoid paying more for development than the plans call for. Managers can make reasonable decisions about which product to pursue based on good estimates of the relative costs and risks.

To develop requirements after the fact is missing their value. I am going to say something extreme here. Requirements are vitally important as a planning and negotiating tool before product development begins. Their utility declines as the product is built. By the time the product is finished, they are useless: if you want to know how the product works, go use it.

This is an exaggeration of course, and biased to the point of view of a design engineer. There are more customers for requirements than the development staff, including the testers and test plan writers. They need to know what the development team intended written in a human language, so that they can compare the result with the assertion of performance.

But my point is that the upstream phases should not be short-changed. They have enormous value for creating the right product, for implementing it correctly, and for keeping to schedules. They comprise the means through which we build products that are safe, effective, and fit to their purpose.

I have done my utmost to ensure that the information in this chapter is accurate. I have proposed processes and methods that I believe will result in validated software that is safe, effective, and appropriate for its intended use. It is my intent that these processes be adequate to establish to regulatory bodies that the software is safe and effective. But the circumstances that you will see with your device and your company vary from mine. No method guarantees approval. Remember to use your own judgment, and be diligent.

8. FAQS

Mine is a very simple device that cannot harm a patient, but does contain software. Do I have to comply with the QSR?

- Yes, any device automated with software must meet the requirements of 21CFR §820.30 design controls, including Class I devices.

My software is an addition to a finished medical device. Do I have to comply with the QSR?

- Yes, a product meant to augment the performance of a medical device is an accessory and as such is subject to the QS regulation.

I am making a modification to existing software in a legacy product that was designed without design control. Do I have to bother?

- Yes. The design control requirements apply to changes to existing designs even if the original design was not subject to the requirements.

I used Microsoft Excel to help me design my product. Do I have to validate Excel?

- No, but you do have to validate that the answers Excel provided were the correct ones through an inspection, sample comparison to hand calculations, or some reasonable method to verify that it was calculating correctly and you were using in correctly.

Am I finished tracing when I have traced requirements to design and tests?

- No. You want to trace from the design back to requirements as well. This is so that you can uncover tacit requirements or gold-plating. [1]

How do I achieve 100% code coverage in a functional test? Some of the conditions are defensive code that should never appear.

- Coverage testing is best done in white-box (structural) testing, stepping through the code on a debugger.

I have requirements that cannot be tested with functional tests, such as a requirement for the initial state of microcontroller registers. Should I leave out the requirements or the tests from the software system test plan?

- Not all requirements are functional requirements. Other requirements are the processor that the software will run on, or the development language. These still need to be verified, but they will be verified by means other than functional tests, such as an analysis or inspection.

What are the differences between the software documentation requirements for a 510(k) submission versus a PMA?

- The software documentation requirements are governed by the level of concern of the software, not the premarket submission. Each risk class of software has specific documentation requirements, and it does not differ with the premarket process.

Okay then, what are the software documentation differences between moderate and major levels of concern software?

- Mostly the provision of unit test data. For a full discussion, see Section 5.9.3, Premarket Documentation, Major Level of Concern.

My software is merely a diagnostic function for the device to help field service troubleshoot it (hence, minor level of concern). But the device itself is Class III. Does my software have to be treated as major level of concern?

- Software associated with a major level of concern device is itself major level of concern. This means that it requires a full hazard analysis. Of course, if it cannot harm anyone, it will already have achieved reduction to minor level of concern.

What are the differences between the FDA and other international regulatory bodies?

- They differ slightly in detail, with the FDA being more specific. (The FDA requires a design history file, for example.) Most of the differences are relative to the quality system for the organization as a whole. There are no practical differences between the methods that you would use to produce quality software for the United States versus the rest of the world.

How long can I expect before I get a response to my premarket submission?

- The FDA is required to provide a response to the premarket submission within 30 days. This can be as simple as an acknowledgment that the package is in order. More likely, it is a list of further questions, asking for further responses from you. PMA normally takes 180 days for approval, although it may take longer; 510k takes 90 days, according to the FDA. In practice, the rounds of requests for more information can take an unpredictable amount of time.

My device has proprietary algorithms protected from competitors as a trade secret. Do I have to show them to the FDA?

- While FDA inspectors may ask for design control procedures, blank forms, and required design review and design verification or validation records, where confidential information appears it may be blacked out. You will have to show the information if the inspection is related to a marketing submission, however. [7]

How can I most easily comply with the security policies of the HIPAA?

- Not everyone falls under the limits of the HIPAA and not all data are protected. But if you have any doubts, the easiest thing to do is to "de-identify" the data. If the data cannot be associated with an individual, it is exempt from the security rule.

Is encryption required for electronic protected health information?

• No, it would not be essential for a dial-up connection. But it should be seriously considered for transmitting medical data, especially over the Internet.

What are the sanctions for violating the HIPAA?

• The Office for Civil Rights is the enforcement arm, and this agency can levy civil financial penalties. In addition, companies may fire workers whom they feel have violated HIPAA policies.

References

[1] Center for Devices and Radiological Health. General principles of software validation. Final guidance for industry and FDA staff, Washington, DC: FDA; 2002. Available at: www.fda.gov/MedicalDevices/DeviceRegulationandGuidance/GuidanceDocuments/ucm085281.htm.

[2] Allen L. The Hubble Space Telescope optical systems failure. NASA Technical Report, NASA-TM-103443, Washington, DC: National Aeronautics and Space Administration; 1990. Available at: ntrs.nasa.gov/archive/nasa/casi.ntrs.nasa.gov/19910003124_1991003124.pdf.

[3] Center for Devices and Radiological Health. Guidance for the content of premarket submissions for software contained in medical devices, Washington, DC: FDA; 2005. Available at: www.fda.gov/MedicalDevices/DeviceRegulationandGuidance/GuidanceDocuments/ucm089543.htm.

[4] International Organization for Standardization. Home page. Available at: www.iso.org/iso/home.htm.

[5] Seeley RS. Exporting: getting small device companies through the CE marking maze. MD&DI (October). Available at: www.devicelink.com/mddi/archive/95/10/014.html. 1995.

[6] Trautman KA. The FDA and worldwide quality systems requirements guidebook for medical devices. Milwaukee (WI): ASQ Quality Press; 1997.

[7] Lowery A, et al. Medical device quality systems manual: a small entity compliance guide, Washington, DC: Center for Devices and Radiological Health; Available at: www.fda.gov/MedicalDevices/DeviceRegulationandGuidance/PostmarketRequirements/QualitySystemsRegulations/MedicalDeviceQualitySystemsManual/default.htm. 1996.

[8] U.S. 21 CFR §820.1 Scope. Guidance document.

[9] U.S. 21 CFR §821. Medical Device Tracking Requirements.

[10] Food and Drug Administration. Statement on Medtronic's Voluntary Market Suspension of Their Sprint Fidelis Defibrillator Leads, Oct 15, 2007. Available at: www.fda.gov/NewsEvents/Newsroom/PressAnnouncements/default.htm.

[11] U.S. 21 CFR §820.70(i) Automated processes. Guidance document.

[12] U.S. 21 CFR §820.90(a) Control of nonconforming product.

[13] U.S. 21 CFR §820.198 Complaint Files.

[14] Available at: www.fda.gov/MedicalDevices/DeviceRegulationandGuidance/HowtoMarketYourDevice/PremarketSubmissions/PremarketNotification510k/default.htm.

[15] Available at: www.gpoaccess.gov/CFR.

[16] Center for Devices and Radiological Health. Design control guidance for medical device manufacturers, Washington, DC: FDA; 1997. Available at: www.fda.gov/MedicalDevices/DeviceRegulationandGuidance/GuidanceDocuments/ucm070627.htm.

[17] U.S. 21 CFR §820.30(a) General.

[18] U.S. 21 CFR §820.30 Design controls. Guidance document.

[19] U.S. 21 CFR §820.30(c) Design input, Guidance document.

[20] Leveson NG. Safeware: system safety and computers. Upper Saddle River, NJ: Addison-Wesley; 1995.

[21] U.S. 21 CFR §820.30(d) Design output.

[22] Global Harmonization Task Force. GHTF Guidance, 4.4.7 Design verification.

[23] U.S. 21 CFR §820.30(g) Design validation.

[24] ANSI/AAMI/IEC 62304:2006. Medical device software—Software life cycle processes.

[25] Center for Devices and Radiological Health. Guidance for industry, FDA reviewers and compliance on off-the-shelf software use in medical devices, Washington, DC: FDA; 1999. September 9, Available at: www.fda.gov/MedicalDevices/DeviceRegulationandGuidance/GuidanceDocuments/ucm073778.htm.

[26] Economics focus: an unhealthy burden. The Economist. June 30, 2007, p. 88.

[27] Freedman DP, Weinberg G. Handbook of walkthroughs, inspections, and technical reviews. 3rd ed. New York: Dorset House; 1990.

[28] Wiegers K. Peer reviews in software: a practical guide. Boston: Addison-Wesley; 2002.

[29] Beck K. Test-driven development: by example. Boston: Addison-Wesley; 2003.

[30] Food and Drug Administration. Available at: www.fda.gov/MedicalDevices/DeviceRegulationandGuidance/Overview/ClassifyYourDevice/ucm051512.htm.

[31] Food and Drug Administration. Available at: www.fda.gov/MedicalDevices/DeviceRegulationandGuidance/Overview/ClassifyYourDevice/ucm051512.htm.

[32] Food and Drug Administration. Available at: www.fda.gov/MedicalDevices/DeviceRegulationandGuidance/Overview/default.htm.

[33] Food and Drug Administration. Available at: www.fda.gov/MedicalDevices/DeviceRegulationandGuidance/Overview/ClassifyYourDevice/ucm051549.htm.

[34] Food and Drug Administration. Available at: www.fda.gov/MedicalDevices/DeviceRegulationandGuidance/Overview/GeneralandSpecialControls/default.htm.

[35] Food and Drug Administration. Product classification. Search classification database. Available at: www.accessdata.fda.gov/scripts/cdrh/cfdocs/cfPCD/classification.cfm.

[36] Food and Drug Administration. Available at: www.fda.gov/MedicalDevices/DeviceRegulationandGuidance/HowtoMarketYourDevice/PremarketSubmissions/PremarketNotification510k/ucm070201.htm.

[37] Dunn WR. Practical design of safety-critical computer systems. Solvang, CA: Reliability Press; 2002.

[38] McConnell S. Rapid development: taming wild software schedules. Redmond, WA: Microsoft Press; 1996.

[39] U.S. 45 CFR 160.103 General Adminstrative Requirements. Available at: www.gpoaccess.gov/CFR/retrieve.html.

[40] www.cms.hhs.gov/EducationMaterials/Downloads/Security101forCoveredEntities.pdf, vol. 2, Paper 1, p. 2.

[41] Federal Register, vol. 68, no. 34. Thursday, February 20, 2003. Rules and regulations 8361.

[42] U.S. Department of Health and Human Services. Security standards: administrative safeguards. HIPAA Security Series, vol. 2. Paper 2. Washington, DC: DHHS; May 2005, rev. March 2007. Available at: www.cms.hhs.gov/EducationMaterials/Downloads/SecurityStandardsAdministrativeSafeguards.pdf.

[43] U.S. Department of Health and Human Services. Security standards: physical safeguards. HIPAA Security Series, vol. 2. Paper 3. Washington, DC: DHHS; February 2005, rev. March 2007. Available at: www.cms.hhs.gov/EducationMaterials/Downloads/SecurityStandardsPhysicalSafeguards.pdf.

[44] U.S. Department of Health and Human Services. Basics of risk analysis and risk management. HIPAA Security Series, vol. 2. Paper 6. Washington, DC: DHHS; June 2005, rev. March 2007. Available at: www.cms.hhs.gov/EducationMaterials/Downloads/BasicsofRiskAnalysisandRiskManagement.pdf.

Bibliography

Beck K. Extreme programming explained. Boston: Addison-Wesley; 2000.

Boehm B, Turner R. Balancing agility and discipline: a guide for the perplexed. Boston: Addison-Wesley; 2004.

Burnstein I. Practical software testing. New York: Springer; 2003.

Food and Drug Administration. CFR—Code of Federal Regulations Title 21. Quality system regulation 21 CFR Part 820. Available at: www.gpoaccess.gov/CFR/retrieve.html.

Food and Drug Administration. Device advice: device regulation and guidance. Home page. Available at: www.fda.gov/cdrh/devadvice.

International Organization for Standardization. ISO 14971:2007. Medical devices—Application of risk management to medical devices.

McConnell S. Code complete 2nd ed. Redmond, WA: Microsoft Press; 2004.

Plum T. Reliable data structures in C. Cardiff, NJ: Plum Hall; 1985.

CHAPTER 5

Best Practices in Spacecraft Development

Chris Hersman and Kim Fowler

This chapter covers the practices and processes for developing unmanned robotic spacecraft beginning with a section on regulations and standards. This chapter *does not* address the design of man-rated spacecraft. These standards represent proven practices from government and industry. Following the regulations and standards are two sections covering examples of company processes and documentation that are typical of a successful spacecraft mission. In general these practices are implemented to meet the relevant requirements outlined in the regulations and standards, while enforcing consistency across projects for different customers. Examples of documentation of standard practices as well as project-level documentation are identified. Finally, this chapter wraps up with a case study of a successful NASA mission and the obstacles encountered during the project. The focus of this chapter will be on the aspects that contribute to reliable and successful missions.

1. Regulations and Standard Practices

Resources for many government regulations and standard practices that relate to the development and launching of space missions can be found online. Listed below are relevant documents with online sources cited. In some cases additional related information may be found on the referenced website. For each regulation or standard practice listed, a summary description is given, along with an explanation of how it relates to the development of a space mission.

Government standards and regulations tend to hold sway in developing spacecraft. Commercial standards are less developed but generally tend to follow the government standards.

1.1. Government Regulations

The U.S. government agency that issues most spacecraft-related regulations is the National Aeronautics and Space Administration (NASA). NASA regulations come in the form of NASA Procedural Requirements (NPR), NASA Policy Directives (NPD), and NASA Technical Standards.

Doi: 10.1016/B978-0-7506-8567-2.00005-6

1.1.1. *Project Management*

1. NPR 7120.5, *NASA Space Flight Program and Project Management Requirements* (http://nodis3.gsfc.nasa.gov/). The purpose of this document is to establish "the requirements by which NASA will formulate and implement space flight programs and projects." Topics covered include program and project definitions, program and project life cycles, reviews, roles and responsibilities, lines of authority, project and program phases, and templates for program and project plans. (See Figs. 5.1 and 5.2 for a graphical depiction of the elements and timeline used in systems engineering to define requirements, roles and responsibilities, and project phases.)

2. NPR 8000.4, *Agency Risk Management Procedural Requirements* (http://nodis3.gsfc. nasa.gov/npg_img/N_PR_8000_004A_/N_PR_8000_004A_.doc). The purpose of this NASA Procedural Requirements (NPR) is to provide the minimum requirements for the planning and acquisition of NASA facility projects. Risk management includes two complementary processes: risk-informed decision making (RIDM) and continuous risk management (CRM). This NPR establishes requirements applicable to all levels of the

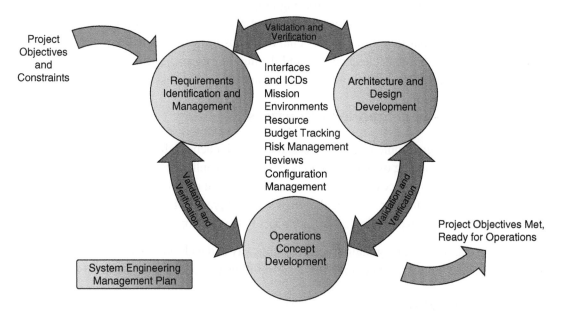

Accomplishing mission objectives requires a consistent set of requirements, design and an operations concept. The operations concept uses the design to meet the requirements. Producing the design and then operating it to meet the requirements must be done within the cost and schedule constraints. Validation, Performance Predictions, Analysis, and Trade Studies are used to develop and optimize the total system.

Figure 5.1: Systems engineering functions, with interrelationship of major system engineering functions (while this figure derives from the military world, it has elements that fit developments for spacecraft and space instruments). Goddard Procedures and Guidelines, "Systems Engineering," DIRECTIVE NO. GPG 7120.5, p. 6. Available at: http://spacecraft.ssl.umd.edu/design_lib/GPG7120.5.pdf

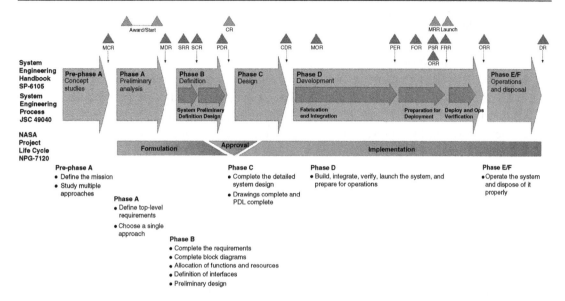

Figure 5.2: Systems engineering life cycle relationship with project life cycle, with major goal of each phase.

Agency. It provides a framework that integrates the RIDM and CRM processes at all levels. This NPR also establishes the roles, responsibilities, and authority to execute the defined requirements.

3. NPD 8610.24B, *Launch Services Program (LSP) Pre-Launch Readiness Reviews* (http://nodis3.gsfc.nasa.gov/). NASA is accountable for program mission success, which includes launch success. NASA assesses and certifies the readiness of the launch vehicle, payload support hardware and software, and preparation of the launch site infrastructure through a structured prelaunch review process. NASA conducts a Launch Services Program (LSP) prelaunch review; it entails the following:

- Launch vehicle readiness review (LVRR)—Certify readiness for integration of the spacecraft and launch vehicle; typically conducted before the mission readiness review (MRR).

- Flight readiness review (FRR)—Update the mission status, close out actions from both the LVRR and the MRR, and certify readiness to initiate the launch countdown. The FRR is held about 3 days before launch.

- Launch readiness review (LRR)—Update the mission status, close out actions from the previously held FRR, authorize approval to proceed into launch countdown, and sign the certification of flight readiness (COFR). The LRR is held 1 day before launch.

- Final commit-to-launch poll—Confirm readiness to launch approximately 5 minutes before launch. A "go" statement is required from all parties polled to enter into the terminal count. Mandatory launch constraints cannot be waived after start of the terminal launch countdown.

The mission spacecraft usually has a parallel set of prelaunch reviews for the spacecraft and ground system elements as well as status of the launch service. The Spacecraft MRR is typically held after the LVRR. The project may hold other reviews deemed appropriate and necessary to prepare for launch; examples include System Requirements Reviews, Critical Design Reviews, Design Certification Reviews, Preship Reviews, Ground Operations Reviews, Project Manager's Reviews, and safety reviews.

4. NASA *Cost Estimating Handbook* 2002 (http://cost.jsc.nasa.gov/NCEH/index.htm). The NASA Cost Estimating Handbook (CEH) provides a balance between documenting processes and providing basic resources for cost estimators without setting a tone of strict guidance. It is a top-level overview of cost estimating as a discipline, not an in-depth examination of each and every aspect of cost estimating. It recognizes the nature of NASA systems and the NASA environment. This handbook claims that cost estimation is part science, part art and that it is a starting point for accurate, defensible, well-documented estimates that are consistently presented and can be easily understood.

1.1.2. Systems Engineering

1. NPR 7123.1, *NASA Systems Engineering Processes and Requirements* (http://nodis3.gsfc. nasa.gov/). This document is a NASA Procedural Requirement (NPR) and it provides requirements to perform, support, and evaluate systems engineering. It defines systems engineering as a "logical systems approach performed by multidisciplinary teams to engineer and integrate NASA's systems to ensure NASA products meet customers' needs." It claims that applying this approach to all elements of a system and all hierarchical levels of a system over the complete project life cycle will help ensure safety and mission success, increased performance, and reduced cost.

2. NPR 7120.6, *Lessons Learned Process* (http://nodis3.gsfc.nasa.gov/). This NPR establishes the requirements for the collection, validation, assessment, and codification of lessons learned submitted by individuals, NASA directorates, programs and projects, and any supporting organizations and personnel.

3. *NASA Lessons Learned Database* (http://llis.nasa.gov/). The NASA Engineering Network is a knowledge network that promotes learning and sharing among NASA's engineers. It gives public access to search the NASA Lessons Learned database system, which is the official, reviewed learned lessons from NASA program and projects. The information in the database for each "lesson learned" is a summary of the original driving event, as well as recommendations. NASA uses these recommendations for continual improvement through training, best practices, policies and procedures.

4. SP-6105, *NASA Systems Engineering Handbook* (http://education.ksc.nasa.gov/esmdspacegrant/Documents/NASA%20SP-2007-6105%20Rev%201%20Final%2031Dec2007.pdf). "The objective of systems engineering is to see to it that the system is designed, built, and operated so that it accomplishes its purpose in the most cost-effective way possible, considering performance, cost, schedule and risk." This handbook attempts to communicate principles of good practice and alternative approaches rather than specify a particular way to accomplish a task. It provides a top-level implementation approach to the practice of systems engineering, which unique to NASA. It has six core chapters: (1) systems engineering fundamentals, (2) the NASA program/project life cycles, (3) systems engineering processes to get from a concept to a design, (4) systems engineering processes to get from a design to a final product, (5) crosscutting management processes in systems engineering, and (6) special topics relative to systems engineering. Appendices supplement the core chapters and provide outlines, examples, and further information to illustrate topics in the core chapters.

5. Defense Acquisition Guidebook, DODD 5000.1, DODI 5000.2 (https://akss.dau.mil/dag/GuideBook/PDFs/GBNov2006.pdf). The Department of Defense has three principal decision-making support systems, all of which were significantly revised in 2003. These three systems, illustrated in Fig. 5.3, provide an integrated approach to strategic planning, identification of needs for military capabilities, systems acquisition, and program and budget development follow:

   • Planning, Programming, Budgeting and Execution (PPBE) Process—Strategic planning, program development, and resource determination process. The PPBE

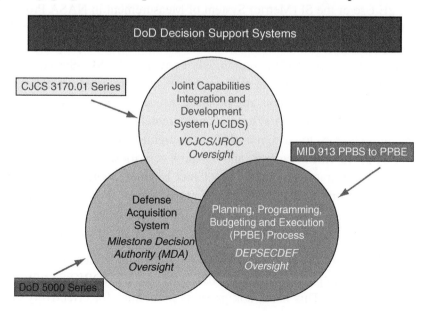

Figure 5.3: DoD decision support system.

process aids the crafting of plans and programs that satisfy the demands of the National Security Strategy within resource constraints.

- Joint Capabilities Integration and Development System—The systematic method established by the Joint Chiefs of Staff for assessing gaps in military joint war-fighting capabilities and recommending solutions to resolve these gaps.

- Defense Acquisition System—The management process to acquire weapon systems and automated information systems. Although the system is based on centralized policies and principles, it allows for decentralized and streamlined execution of acquisition activities. This approach provides flexibility and encourages innovation, while maintaining strict emphasis on discipline and accountability.

6. MIL-HDBK-1547 Technical Requirements for Parts, Materials, and Processes for Space and Launch Vehicles (http://store.mil-standards.com/index.asp?PageAction=VIEWPROD&ProdID=99). The purpose of this handbook is to establish and maintain consistent and uniform methods for development of technical requirements for electronic parts, materials, and processes used in the design, development, and fabrication of space and launch vehicles. It provides a common basis for estimating of Application Information, Design and Construction Considerations, and Quality Assurance Provisions for the proposed Design Application. It also establishes a common basis for comparing and evaluating of industry practices for related or competitive designs. This handbook is intended to be used as a tool to increase the performance and reliability of the system under design.

7. NPD 8010.2E Use of the SI (Metric) System of Measurement in NASA Programs (http://nodis3.gsfc.nasa.gov/displayDir.cfm?t=NPD&c=8010&s=2E). This document presents NASA policy for systems of measurement to be used on NASA programs/projects. The International System of Units (commonly known as the SI—Systeme Internationale—or metric system of measurement) is the preferred system of weights and measurement for NASA programs and projects. All new programs and projects covered by NPR 7120.5 shall use the SI system of measurement for design, development, and operations, in preference to customary U.S. measurement units, for all internal activities, related NASA procurements, grants, and business activities. Exceptions to this requirement may be granted by the NASA Chief Engineer, where use of SI units is demonstrated to be impractical, adds unacceptable risk, or is likely to cause significant inefficiencies or loss of markets to U.S. firms. Special emphasis shall be placed on maximum use of SI units in cooperative programs with international partners.

8. MIL-HDBK-881 *Work Breakdown Structures for Defense Materiel Items* (http://www.acq.osd.mil/pm/currentpolicy/wbs/MIL_HDBK-881A/MILHDBK881A/WebHelp3/MILHDBK881A.htm). This handbook presents guidelines for effectively preparing,

understanding, and presenting a Work Breakdown Structure (WBS). Its primary objective is consistent application of the WBS for all programmatic needs (including Performance, Cost, Schedule, Risk, Budget, and Contractual). It is intended to provide the framework for Department of Defense (DoD) Program Managers to define their program's WBS and guidance to defense contractors in their application and extension of the contract's WBS. Section 1 defines and describes the WBS. Section 2 provides instructions on how to develop a Program WBS in the pre-award timeframe. Section 3 offers guidance for developing and implementing a Contract WBS and Section 4 examines the role of the WBS in the post-award time frame.

1.1.3. Design

NASA Reliability Preferred Practices for Design & Test See http://www.klabs.org/DEI/ References/design_guidelines/nasa_reliability_preferred_practices.htm. This website by the Office of Logic Design provides short (four- to eight-page) summaries of reliability design and test practices which have contributed to the success of NASA spaceflight missions. It provides more than 100 preferred practice summaries in PDF format within the categories of natural space environment, reliability design, reliability analysis, and hardware test. Brief descriptions of some of these summaries are listed in Table 5.1.

1.1.4. Mission Assurance and Safety

Mission assurance and safety represent two aspects of reducing risk within the project. Mission assurance activities represent the measures taken to improve the probability that the mission will achieve its objectives. Safety processes are implemented to assure that the risk of potential hazards to personnel and external environments are minimized. The two aspects are grouped together as a single topic because they are often interrelated. For example, a spacecraft with a propulsion system must prevent the inadvertent release of propellant, because it is hazardous to personnel on the ground. Similarly, avoiding inadvertent release of propellant in flight is important because loss of propellant could result in failure of the mission to achieve its objectives. This section covers some of the government regulations relating to mission assurance and safety.

1. NPD 8700.1C, *NASA Policy for Safety and Mission Success* establishes NASA requirements for safety and mission success, including who or what is protected, who is responsible for protecting it, how risks are managed, and how information is communicated.

2. NPD 8700.3A *Safety and Mission Assurance (SMA) Policy for NASA Spacecraft, Instruments, and Launch Services* establishes safety and mission assurance requirements for these flight elements.

Table 5.1: Summary of Best Practices from NASA Office of Logic Design

Title and File Name	Comments on Practice and Benefits
	Design
1201: EEE Parts Derating 1201.pdf	**Practice:** Derate applied stress levels for electrical, electronic, and electromechanical (EEE) part characteristics and parameters with respect to the maximum stress level ratings of the part. The allowed stress levels are established as the maximum levels in circuit applications. **Benefits:** Derating lowers the probability of failures occurring during assembly, test, and flight. Decreasing mechanical, thermal, and electrical stresses lowers the possibility of degradation or catastrophic failure.
1202: High-Voltage Power Supply Design and Manufacturing Practices 1202.pdf	**Practice:** Thoroughly test high-voltage power supply packaging on flight configured engineering models, in a simulated spaceflight environment, to evaluate corona effects. **Benefits:** Process controls on design, manufacturing, and testing operations reduce component failure rates and improve reliability. The goal is production of power supplies that will operate in space for the mission duration.
1203: Class S Parts in High-Reliability Applications Practice No. PD-ED-1203	**Practice:** Use Class S and Grade 1 or equivalent parts in all applications requiring high-reliability or long life to yield the lowest possible failure rates. **Benefits:** Low parts failure rates in typical circuit applications result in significant system reliability enhancement. For space systems involving serviceability, the mean-time-between-failure (MTBF) is greatly extended, which significantly reduces maintenance requirements and crew time demands.
1204: Part Junction Temperature Practice No. PD-ED-1204	**Practice:** Maintain part junction temperatures during flight below 60°C. (Short-term mission excursions associated with transient mission events are permissible.) **Benefit:** Reliability is greatly increased because the failure rate is directly related to the long-term flight temperature.
1206: Power Line Filters 1206.pdf	**Practice:** Power line filters are designed into power lines (power buses) at the inputs to payloads, instruments, subsystems, and components. **Benefits:** Power line filters minimize the flow of conducted noise currents on power buses emanating from hardware that could interfere with the proper operation of other hardware also operating on the same power buses. Additionally, power-line filters minimize the flow of noise currents on power buses into hardware that could interfere with the proper operation of that hardware.

(Continues)

Table 5.1:   Cont'd

Title and File Name	Comments on Practice and Benefits
1207: Magnetic Design Control for Science Instruments Practice No. PD-ED-1207	**Practice:** Design flight subsystems with low residual dipole magnetic fields to maintain the spacecraft's total static and dynamic magnetic fields within science requirements. **Benefit:** Provides for a magnetically clean spacecraft, which increases the quality and accuracy of interplanetary and planetary magnetic field data gathered during the mission.
1210: Assessment and Control of Electrical Charges 1210.pdf	**Practice:** Provide protection against electrostatic charges, discharges, and lightning strikes by shielding and bonding space systems, structures, and their components in accordance with Standard Payload Assurance Requirements (SPAR-3) for GSFC Orbital Projects. This reliability practice does not cover Electrostatic Discharge (ESD) control due to an energetic, space plasma environment. **Benefit:** The Earth's space environment (geospace) is uniquely comprised of dynamic and complex regions of interacting plasmas, ionized particles, magnetic fields, and electrical currents. Proper grounding/bonding of the space vehicle's shell and its electronic equipment can provide protection against lightning strikes in geospace, and also can eliminate or control most of its internal electrical and electrostatic hazards. This results in lower failure rates and significant reliability and safety enhancement of space systems and space vehicles.
1211: Combination Methods for Deriving Structural Design Loads Practice No. PD-ED-1211	**Practice:** Design primary and secondary structural components to accommodate loads which include steady-state, transient dynamic, and vibroacoustic contributions at liftoff. **Benefit:** The probability of structural failure during launch and landing is significantly reduced.
1212: Design & Analysis of Circuits for Worst Case Environments and Part Variations Practice No. PD-ED-1212	**Practice:** Design all circuits to perform within defined tolerance limits over a given mission lifetime while experiencing the worst possible variations of electronic piece parts and environments. **Benefit:** The probability of mission success is maximized by assuring that all assemblies meet their mission electrical performance requirements at all times.
1213: Electrical Shielding of Power, Signal, and Control Cables 1213.pdf	**Practice:** All wiring harnesses, cables, and wires on payloads, instruments, subsystems, and components are well shielded, including the use of connector types that provide tight EMI back shells or other means for attaching shields. This practice assumes that all efforts have been made to develop a design which requires minimum shielding.

(Continues)

Table 5.1:   Cont'd

Title and File Name	Comments on Practice and Benefits
	Benefit: High-performance shielding on wiring harnesses, cables and wires minimizes radiated emissions from hardware that could be picked up by itself or other hardware and interfere with proper operation. Shielding also minimizes the sensitivity of hardware to radiated emissions, from itself or other hardware, that could interfere with proper operation.
1214: Electrical Grounding Practices for Aerospace Hardware 1214msfc.pdf	**Practice:** Electrical grounding procedures must adhere to a proven set of requirements and design approaches to produce safe and trouble-free electrical and electronic circuits. Proper grounding is fundamental for reliable electronic circuits. **Benefits:** Grounding procedures used in the design and assembly of electrical and electronic systems will protect personnel and circuits from hazardous currents and damaging fault conditions. Benefits are prevention of potential damage to delicate spaceflight systems, subsystems and components, and protection of development, operations, and maintenance personnel.
1215-1: Preliminary Design Review Practice No. PD-ED-1215.1	**Practice:** Conduct a formal preliminary design review (PDR) at the system and subsystem levels prior to the start of subsystem detail design, to assure that the proposed design and associated implementation approach will satisfy the system and subsystem functional requirements. **Benefits:** The PDR will provide for increased assurance that the proposed design approach, and the manufacturing and test implementation plans, will result in an acceptable product, with minimal project risk.
1215-2: Hardware Review/Certification Requirement Practice No. PD-ED-1215.2	**Practice:** A Hardware Review/Certification Requirement (HR/CR) Review is conducted prior to the delivery of flight hardware and associated software to evaluate and certify that the hardware is ready for delivery and that it is acceptable for integration with the spacecraft. **Benefit:** The HR/CR provides a structured review process for assessing the status of flight hardware and screening for unresolved defects prior to delivery for integration.
1215-3: Critical Design Review for Unmanned Missions Practice No. PD-ED-1215-3	**Practice:** Conduct a formal Critical Design Review (CDR) of hardware, software, and firmware at the subsystem and system levels. Schedule the review prior to the start of subsystem fabrication and assembly to assure that the design solutions satisfy the performance requirements established in the development specifications. Establish this review as a standard reliability engineering practice for flight hardware. **Benefits:** The CDR provides increased assurance that the proposed design, and the planned manufacturing and test methods and procedures, will result in an acceptable product, with minimal project risk.

(Continues)

Table 5.1: Cont'd

Title and File Name	Comments on Practice and Benefits
1215-4: Common Review Methods Practice No. PD-ED-1215-4	**Practice:** Conduct technical reviews to validate engineering designs using a common, consistent approach which has been proven to lead to reliable and quality products. A technical review is an evaluation of the engineering status of products and processes by an independent group of knowledgeable people. Although major technical reviews for a project differ in their content and timing, there are practices common to most reviews which may be defined to assure review success. These practices provide a common framework for planning, conducting, documenting, and evaluating the review process. **Benefits:** Standards established for common review methods are presently supporting reliability assurance by emphasizing early detection and correction of deficiencies through the increased use of working level, peer reviews (detailed technical reviews) in preparation for major design reviews. The standards also assure that reviews are scaled in accordance with criticality, complexity, and risk, and that the review process is optimized to produce results of value to the mission.
1215-5: Pre-Ship Review Practice No. PD-ED-1215-5	**Practice:** Prior to shipment of hardware or software, conduct a pre-ship review at the completion of the fabrication or build and testing of the item to be shipped. This review is scheduled as part of the overall technical review program as defined in a project review plan. Preship review is held at the supplier or NASA facility where the item was made and tested. **Benefits:** Preship review ensures the completeness and readiness of each item of hardware and, if applicable, any associated software or firmware, prior to release for shipment to another facility. By imposing this requirement, any discrepancies or unresolved problems may be identified and corrected while the item remains under supplier purview. This review is beneficial because it provides an independent assessment of product readiness by knowledgeable people not directly involved in the fabrication and test activity.
1216: Active Redundancy np1216.pdf	**Practice:** Use active redundancy as a design option when development testing and reliability analysis show that a single component is not reliable enough to accomplish the function. Although active redundancy can be applied to various types of mechanical and electrical components and systems, the application detailed in this practice illustrates an approach using a traveling wave tube amplifier in a spaceflight application. **Benefits:** Provides multiple ways of accomplishing a function to improve mission reliability.

(Continues)

Table 5.1: Cont'd

Title and File Name	Comments on Practice and Benefits
1217: Structural Laminate Composites for Space Applications 1217msfc.pdf	**Practice:** The creation of reliable structural laminate composites for space applications requires precision design and manufacturing using an integrated, concurrent engineering approach. Since the final material characteristics are established at the same time the part or subassembly is fabricated, part design, fabrication development, and material characterization must proceed concurrently. Because composite materials are custom-tailored to meet structural requirements of the assembly, stringent in-process controls are required to arrive at a configuration with optimum physical and material properties. **Benefits:** Conscientious adherence to proven procedures in the design, manufacture, and test of aerospace structural composites will result in low rejection rates and high product integrity. In specific applications, successful composite design provides design flexibility, increased strength to weight ratio, dimensional stability under thermal loading, light weight, ease of fabrication and installation, corrosion resistance, impact resistance, high fatigue strength (compared to metal structures with the same dimensions), and product simplicity when compared to conventional fabricated metal structures.
1219: Vehicle Integration/ Tolerance Build-up Practices 1219msfc.pdf	**Practice:** Use master gauges, tooling, jigs, and fixtures to transfer precise dimensions to ensure accurate mating of interfacing aerospace hardware. Calculate overall worst-case tolerances using the root sum square method of element tolerances when integrating multiple elements of aerospace hardware. **Benefits:** Using prudent and carefully planned methods for specifying tolerances and for designing, manufacturing and mating major elements of aerospace hardware, will result in a cost-effective program with minimal rejects and waivers, and will avoid costly schedule delays due to potential mismatching or misfitting of major components and assemblies.
1220: Demagnetization of Ferromagnetic Parts Practice No. PD-ED-1220	**Practice:** In those cases where spacecraft science requirements or attitude control systems impose constraints on the magnetic characteristics of components and the use of ferromagnetic material cannot be avoided, perform a complete demagnetization of the ferromagnetic parts, individually, prior to assembly. **Benefit:** In an unassembled state, ferromagnetic parts can be exposed to stronger AC demagnetizing fields, as high as 60 mT (600 Gauss), thus assuring a lower level of remanent magnetization than can be achieved after the parts are mounted on assemblies. Attaining a low level of remanent magnetization minimizes the adverse effects of

(Continues)

Table 5.1: Cont'd

Title and File Name	Comments on Practice and Benefits
	unwanted fields. In those cases where magnetic compensation may be required, the ability to apply high-level fields to an unmounted part enables the utilization of techniques to stabilize the magnetic moment of the part.
1221: Battery Selection Practice for Aerospace Power Systems 1221msfc.pdf	**Practice:** When selecting batteries for space flight applications, the following requirements should be considered: ampere-hour capacity, rechargeability, depth of discharge (DOD), lifetime, temperature environments, ruggedness, and weight. Many batteries have been qualified and used for space flight, enhancing the ease of selecting the right battery. **Benefits:** Selection of the optimum battery for space flight applications results in a safe, effective, efficient, and economical power storage capability. The optimum battery also enhances launch operations, minimizes impacts to resources, supports contingency operations, and meets demand loads.
1222: Magnetic Field Restraints for Spacecraft Systems and Subsystems 1222.pdf	**Practice:** Control magnetic field disturbance of spacecraft systems by avoiding the use of components and subassemblies with significant magnetic dipole moments. **Benefit:** Limits magnetic field interference at flight sensor positions and minimizes magnetic dipole moments that can increase magnetic torquing effects that place additional loads on attitude control systems.
1225: Conducted and Radiated Emissions Design Requirements Practice No. PD-ED-1225	**Practice:** Initially, the design requirements for each subsystem are established so that all nonfunctional emissions will be at least 9 dB below the emission specification limit. **Benefits:** By initially selecting a 9-dB margin, the probability of complying with the electromagnetic compatibility (EMC) specification during system test is high.
1226: Thermal Design Practices for Electronic Assemblies Practice No. PD-ED-1226	**Practice:** Ensure that thermal design practices for electronic assemblies will meet the requirements of the combined ground and flight environmental conditions defined by the spacecraft mission. Special emphasis should be placed on limiting the junction temperature of all active components. Proper thermal design practices take into consideration the need for ease of operation and repairability to enhance overall system reliability. The environmental conditions that the spacecraft encounters, both on the ground and in flight, are designed to include adequate margin. The use of proper thermal design practices ensures that the assemblies will survive the expected environmental conditions.

(Continues)

Table 5.1: Cont'd

Title and File Name	Comments on Practice and Benefits
	Benefit: Constraining the electronic component junction temperature through proper design practices will ensure that the assemblies can withstand the mission's environmental conditions.
1227: Controlling Stress Corrosion Cracking in Aerospace Applications 1227msfc.pdf	**Practice:** This practice presents considerations that should be evaluated and applied concerning stress corrosion and subsequent crack propagation in mechanical devices, structural devices, and related components used in aerospace applications. Material selection, heat treat methods, fabrication methodology, testing regimes, and loading path assessments are presented as methods to reduce the potential for stress corrosion cracking (SCC) in a material's operational environment. **Benefits:** Selection of materials, heat-treating methods, fabrication methodologies, testing regimes, and loading paths that are not susceptible to stress corrosion cracking will promote fewer failures due to SCC and will eliminate downtime due to the change-out of components.
1228: Independent Verification and Validation of Embedded Software 1228msfc.pdf	**Practice:** To produce high-quality, reliable software, use independent verification and validation (IV&V) in an independent, systematic evaluation process throughout the software life cycle. Using the IV&V process; locate, identify, and correct software problems and errors early in the development cycle. **Benefit:** The use of IV&V processes ensures that computer software is developed in accordance with original specifications, that the software performs the functions satisfactorily in the operational mission environment for which it was designed, and that it does not perform unintended functions. Identification and correction of errors early in the development cycle are less costly than identification and correction of errors in later phases, and the quality and reliability of software are significantly improved.
1229: Selection of Electric Motors for Aerospace Applications 1229msfc.pdf	**Practice:** Careful attention is given to the specific application of electric motors for aerospace applications when selecting motor type. The following factors are considered in electric motor design: application, environment, thermal, efficiency, weight, volume, life, complexity, torque, speed, torque ripple, power source, envelope, duty cycle, and controllability. Brushless direct current motors have been proven to be best all-around type of motors for aerospace applications because of their long life, high torque, high efficiency, and low heat dissipation.

(Continues)

Table 5.1: Cont'd

Title and File Name	Comments on Practice and Benefits
	Benefit: Selection of the optimum electric motor for spaceflight operations results in a safe, reliable, effective, efficient, and economical electric motor power source for spaceflight. Brushless direct current motors provide the lightest-weight alternative for most applications.
1230: System Design Analysis Applied to Launch Vehicle Configuration 1230msfc.pdf	**Practice:** Use design management improvements such as matrix methods, quality techniques, and life cycle cost analyses in a systematic approach to systems analysis. **Benefit:** The use of advanced design management methods in each program phase of major launch vehicle developments will maximize reliability and minimize cost overruns. Significant improvements in user satisfaction, error-free performance, and operational effectiveness can be achieved through the use of these methods.
1231: Design Considerations for Lightning Strike Survivability 1231.pdf	**Practice:** Implement lightning survivability in the design of launch vehicles to avoid lightning-induced failures. **Benefits:** Experience learned from the Atlas/Centaur and Space Shuttle flights serve to emphasize the importance of the implementation of the proper protection/design enhancements to avoid and survive natural or triggered lightning for all launches.
1233: Contamination Control Program 1233.pdf	**Practice:** Apply a contamination control program to those spacecraft projects involving scientific instruments that have stringent cleanliness level requirements. **Benefits:** This practice enables spacecraft to meet these stringent cleanliness level requirements of state-of-the-art scientific instruments. It also serves to maintain the inherent efficiency and reliability of the instrument by minimizing degradation of critical surfaces and sensors due to undesired condensation of molecular and accumulation of particulate contamination layers.
1236: EEE Parts Selection Guidelines for Flight Systems pded1236.pdf	**Practice:** Use highest-reliability EEE parts available, consistent with functional requirements, program cost, and schedule constraints, for spaceflight systems. **Benefit:** One of the most important considerations in designing reliable flight hardware is selection and use of the highest-quality possible components. Proper selection, application, and testing of EEE components will generally contribute to mission success and provide long-term program cost savings. An effective EEE parts program has helped many projects in achieving optimum safety, reliability, maintainability, on-time delivery, and performance of program hardware. The resulting reduction in parts and part-related failures saves program resources through decreased failure investigation and maintenance costs.

(Continues)

Table 5.1:   Cont'd

Title and File Name	Comments on Practice and Benefits
1238: Spacecraft Electrical Harness Design Practice 1238.pdf	**Practice:** Design and fabricate spaceflight electrical harnesses to meet the minimum requirements of the GSFC Design and Manufacturing Standard for Electrical Harnesses. **Benefit:** Designing and testing flight harnesses in accordance with the requirements of the GSFC Design and Manufacturing Standard for Electrical Harnesses enhances the probability of mission success (reliability) by ensuring that harnesses meet high standards of quality as well as the electrical and environmental requirements of spaceflight missions. The occurrence of early failures is minimized.
1239: Spacecraft Thermal Control Coatings Design and Application Procedures 1239.pdf	**Practice:** Select and apply thermal coatings for control of spacecraft and scientific instrument temperatures within required ranges and for control of spacecraft charging and RF emissions. **Benefit:** This practice enhances the probability of mission success by controlling temperatures of flight hardware as well as spacecraft charging and RF emissions over the life of the mission.
1240: Identification, Control, and Management of Critical Items 1240msfc.pdf	**Practice:** Initiate the preparation of critical items lists (CILs) early in programs to identify and potentially eliminate critical items before the design is frozen and as an input to hardware and software design, testing, and inspection planning activities. Utilize CILs during the operational portion of the life cycle to manage failures and ensure mission success. **Benefits:** Early identification, tracking, and control of critical items through the preparation, implementation, and maintenance of CILs will provide valuable inputs to a design, development, and production program. From the CIL activity, critical design features, tests, inspection points, and procedures can be identified and implemented that will minimize the probability of failure of a mission or loss of life.
1241: Contamination Budgeting for Space Optical Systems 1241msfc.pdf	**Practice:** Use preplanned contamination budgeting for each manufacturing/assembly, testing, shipping, launch, and flight operation and meticulously test optical systems using witness samples throughout the process to track actual contamination against total and incremental allocations. **Benefit:** Budgeting of a specific amount of the established allowable contamination to the major elements and operations during fabrication, assembly, testing, transportation launch support, and launch, and on-orbit operations of space optical systems will preclude jeopardizing the scientific objectives of the

(Continues)

Table 5.1: Cont'd

Title and File Name	Comments on Practice and Benefits
	mission. Budgeting of contamination to major elements will ensure that the cleanliness of the optics and instruments will remain within designated optical requirements for operations in space. Reliability of the scientific objectives is increased by limiting the contamination allowed to the optical systems during each operation, which ensures that contamination during orbital operations is within specification.
1243: Fault Protection Practice No. PD-ED-1243	**Practice:** Fault protection is the use of cooperative design of flight and ground elements (including hardware, software, procedures, etc.) to detect and respond to perceived spacecraft faults. Its purpose is to eliminate single point failures or their effects and to ensure spacecraft system integrity under anomalous conditions. **Benefits:** Fault-protection design maximizes the probability of spacecraft mission success by avoiding possible single failure points through the use of autonomous, short-term compensation for failed hardware.
1244: Design Practice to Control Interference from Electrostatic Discharge (ESD) Practice No. PD-ED-1244	**Practice:** Minimize the adverse effects of electrostatic discharge (ESD) on spacecraft by implementing the following three design practices: 1. Make all external surfaces of the spacecraft electrically conductive and grounded to the main structure. 2. Provide all internal metallic elements and other conductive elements with an "ESD conductive" path to the main structure. 3. Enclose all sensitive circuitry in an electrically conductive enclosure—a "Faraday cage." **Benefit:** The first two practices should dissipate most electric charges before a difference in potential can become high enough to cause an ESD. If a discharge occurs, the third practice lowers the coupling to sensitive circuits, reducing the probability or severity of the interference.
1245: Magnetic Dipole Allocation Practice No. PD-ED-1245	**Practice:** Magnetic dipole allocation is an empirical method for initiating control of spacecraft magnetic contamination. The practice is necessary for missions which incorporate instruments to measure low-level magnetic fields. **Benefit:** Control of the net magnetic dipole of the spacecraft will assure the integrity of magnetic field measurements made during the mission. Measurement of the individual contributions from various assemblies, subassemblies, and components allows the identification of the major dipole sources. The major contributors can then be evaluated for corrective action,

(Continues)

Table 5.1: Cont'd

Title and File Name	Comments on Practice and Benefits
	and they can be monitored individually to ensure that they are at the lowest level of magnetization at the time of installation on the spacecraft.
1246: Fault Tolerant Design Practice No. PD-ED-1246	**Practice:** Incorporate hardware and software features in the design of spacecraft equipment which tolerate the effects of minor failures and minimize switching from the primary to the secondary string. This increases the potential availability and reliability of the primary string. **Benefits:** Fault tolerant design provides a means to achieve a balanced project risk where the cost of failure protection is commensurate with the program resources and the mission criticality of the equipment. By providing compensation for potential hardware failures, a fault-tolerant design approach may achieve reliability objectives without recourse to nonoptimized redundancy or overdesign.
1247: Spacecraft Lessons Learned Reporting System Practice No. PD-ED-1247	**Practice:** Develop a spacecraft lessons learned file (LLF), a quick, but formal record of significant occurrences during design, implementation, and operation of spacecraft and support equipment. Provide fast and convenient traceability for knowledge capture of significant events to guide future spacecraft managers and engineers in recognizing and avoiding critical design problems. Maintain the system as a living problem-avoidance database for all flight project activities. **Benefits:** The spacecraft LLF is a quick-reference document that preserves the NASA knowledge base, providing engineers and scientists with brief summaries of meaningful events that offer valuable lessons. Within the LLF, lessons of interest can be accessed through a keyword search, with more detailed information accessible from the referenced problem/failure report or alert documentation. The LLF serves as a repository of valuable information, including lessons that were learned at great expense, which would otherwise be lost following personnel turnover. The JPL LLF activity is performed in coordination with the NASA headquarters LLF program.
1248: Spacecraft Data Systems (SDS) Hardware Design Practices 1248.pdf	**Practice:** Use a standard SDS in spacecraft where possible that utilizes a standard data bus and spaceflight-qualified versions of widely used hardware and operating software systems. **Benefit:** This practice enhances reliability of the SDS and the probability of mission success by simplifying the design and operation of the SDS system and providing capability to work around spacecraft and instrument problems.

(Continues)

Table 5.1: Cont'd

Title and File Name	Comments on Practice and Benefits
1249: Electrostatic Discharge (ESD) Control in Flight Hardware 1249.pdf	**Practice:** Apply an electrostatic discharge (ESD) control program to all spaceflight projects to ensure that ESD-susceptible hardware is protected from damage due to ESD. **Benefit:** This ESD control practice significantly enhances mission reliability by protecting susceptible flight and critical flight-support electronic parts and related hardware from damage or degradation caused by ESD and induction polarization charge (IPC) during the prelaunch phases of the mission.
1250: Pre-Flight Problem/Failure Reporting Procedures 1250.pdf	**Practice:** A formal procedure is followed in the reporting and documentation of problems/failures occurring during test, prelaunch operations, and launch operations for both hardware and software. A separate system, the "spacecraft orbital anomaly report (SOAR)," is used for the reporting, evaluation, and correction of problems occurring on-orbit (see Practice No. PD-ED-1232). **Benefit:** This practice significantly enhances the probability of mission success by ensuring that problems/failures occurring during ground test are properly identified, documented, assessed, tracked, and corrected in a controlled and approved manner. Another benefit of the PFR procedure is to provide data on problem/failure trends. Trend data may then be analyzed so that errors are not repeated on future hardware and software.
1251: Instrumentation System Design and Installation for Launch Vehicles 1251msfc.pdf	**Practice:** Instrumentation systems and related sensors (transducers), particularly those designed for use in reusable and refurbishable launch systems and subsystems, are analyzed, designed, fabricated and tested with meticulous care in order to ensure system and subsystem reliability. **Benefits:** The benefits of implementing these reliability practices for instrumentation system and related sensors are (1) consistent performance and measurement results, (2) minimum need for continuous or periodic calibration, (3) avoidance of and resistance to contamination, and (4) reduced necessity for repair or replacement in repeated usage.
1255: Problem Reporting and Corrective Action System 1255ksc.pdf	**Practice:** A closed-loop problem (or failure) reporting and corrective action system (PRACAS or FRACAS) is implemented to obtain feedback about the operation of ground support equipment used for the manned spaceflight program. **Benefits:** The information provided by PRACAS allows areas in possible need of improvement to be highlighted to engineering for development of a corrective action, if deemed necessary. With this system in place in the early phases of a program, means are

(Continues)

Table 5.1:   Cont'd

Title and File Name	Comments on Practice and Benefits
	provided for early elimination of the causes of failures. This contributes to reliability growth and customer satisfaction. The system also allows trending data to be collected for systems that are in place. Trend analysis may show areas in need of design or operational changes.
1258: Space Radiation Effects on Electronic Components in Low-Earth Orbit 1258jsc.pdf	**Practice:** During system design, choose electronic components/devices that will provide maximum failure tolerance from space radiation effects. The information below provides guidance in selection of radiation hardened (rad-hard) solid state devices and microcircuits for use in space vehicles which operate in low-Earth orbits. **Benefit:** This practice provides enhanced reliability and availability as well as improved chances for mission success. Failure rates due to space radiation effects will be significantly lower, and thus system downtime will be much lower, saving program cost and resources.
1259: Acoustic Noise Requirements Practice No. PD-ED-1259	**Practice:** Impose an acoustic noise requirement on spacecraft hardware design to ensure the structural integrity of the vehicle and its components in the vibroacoustic launch environment. Acoustic noise results from the propagation of sound pressure waves through air or other media. During the launch of a rocket, such noise is generated by the release of high-velocity engine exhaust gases, by the resonant motion of internal engine components, and by the aerodynamic flow field associated with high-speed vehicle movement through the atmosphere. This environment places severe stress on flight hardware and has been shown to severely impact subsystem reliability. **Benefit:** The fluctuating pressures associated with acoustic energy during launch can cause vibration of structural components over a broad frequency band, ranging from about 20 Hz to 10,000 Hz and above. Such high-frequency vibration can lead to rapid structural fatigue. The acoustic noise requirement assures that flight hardware—particularly structures with a high ratio of surface area to mass—is designed with sufficient margin to withstand the launch environment. Definition of an aggressive acoustic noise specification is intended to mitigate the effects of the launch environment on spacecraft reliability. It would not apply to the space station nor to the normal operational environment of a spacecraft.
1260: Radiation Design Margin Requirement Practice No. PD-ED-1260	**Practice:** Design spacecraft hardware assemblies with the required radiation design margin (RDM) to ensure that they can withstand ionization effects and displacement damage resulting from the flight radiation environment. The term "margin"

(Continues)

Table 5.1: Cont'd

Title and File Name	Comments on Practice and Benefits
	does not imply a known factor of safety, but rather accommodates the uncertainty in the radiation susceptibility predictions. The reliability requirement to survive for a period of time in the anticipated mission radiation environment is a spacecraft design driver. **Benefits:** The RDM requirement is imposed on assemblies or subsystems to ensure reliable operation and to minimize the risk, especially in mission-critical applications. The general use of an RDM connotes action to overcome the inevitable uncertainties in environmental calculations and part radiation hardness determinations.
1261: Characterization of RF Subsystem Susceptibility to Spurious Signals Practice No. PD-ED-1261	**Practice:** Reliable design of spacecraft radios requires the analysis and test of hardware responses to spurious emissions which may degrade communications performance. Prior to hardware integration on the spacecraft, receivers and transmitters are tested to verify their compatibility with respect to emissions of conducted radio frequency (RF) signals and susceptibility to these signals. This reliability practice is applied to receivers and transmitters located in the same subsystem and to those installed in different subsystems on the same spacecraft. This early test to identify and resolve radio compatibility problems reduces the risk of uplink/downlink degradation which might threaten mission objectives. **Benefits:** This practice validates the compatibility of spacecraft receivers and transmitters. If electromagnetic compatibility problems are identified early in radio design, solutions can be developed, implemented, and verified prior to the integration of the hardware on the spacecraft.
1262: Subsystem Inheritance Review Practice No. PD-ED-1262	**Practice:** Conduct a formal design inheritance review at the system, subsystem, or assembly level prior to, or in conjunction with, the corresponding subsystem preliminary design review (PDR). The purpose of the inheritance review is to identify those actions which will be required to establish the compatibility of the proposed inherited design, and any inherited hardware or software, with the subsystem functional and design, requirements. **Benefit:** Use of inherited flight hardware or software may reduce cost and allow a spacecraft designer to avoid the risk of launching unproven equipment. However, the designer often lacks full information on the many design decisions made during

(Continues)

Table 5.1: Cont'd

Title and File Name	Comments on Practice and Benefits
	development, including some which may cause incompatibility with current spacecraft requirements. Subsystem inheritance review (SIR) probes inheritance issues to help ensure that the proposed inherited item will result in an acceptable and reliable product with minimal mission risk.
1263: Contamination Control of Space Optical Systems 1263msfc.pdf	**Practice:** Contamination of space optical systems is controlled through the use of proper design techniques, selection of proper materials, hardware/component precleaning, and maintenance of cleanliness during assembly, testing, checkout, transportation, storage, launch, and on-orbit operations. These practices will improve reliability through avoidance of the primary sources of space optical systems particulate and molecular contamination. **Benefit:** Controlling contamination of space optical systems limits the amount of particulate and molecular contamination which could cause performance degradation. Contamination causes diminished optical throughput, creates off-axis radiation scattering due to particle clouds, and increases mirror scattering. Controlling molecular contaminates minimizes performance degradation caused by the deposition of molecular contaminants on mirrors, optical sensors and critical surfaces; improves cost-effectiveness of mission results; and improves reliability.
1272: Manned Space Vehicle Battery Safety 1272.pdf	**Practice:** This practice is for use by designers of battery-operated equipment flown on space vehicles. It provides such people with information on the design of battery-operated equipment to result in a design which is safe. Safe, in this practice, means safe for ground personnel and crew to handle and use; safe for use in the enclosed environment of a manned space vehicle and safe to be mounted in adjacent unpressurized spaces. **Benefit:** There have been many requests by the space shuttle payload customers for a practice which describes all the hazards associated with the use of batteries in and on manned spaceflight vehicles. This practice is prepared for designers of battery-operated equipment so that designs can accommodate these hazard controls. This practice describes the process that a design engineer should consider in order to verify control of hazards to personnel and the equipment. Hazards to ground personnel who must handle battery-operated equipment are considered, as well as hazards to space crew and vehicles.
1273: Quantitative Reliability Requirements Used as	**Practice:** Develop performance-based reliability requirements by considering elements of system performance in terms of

(Continues)

Table 5.1: Cont'd

Title and File Name	Comments on Practice and Benefits
Performance-Based Require-ments for Space Systems 1273.pdf	specific missions and events and by determining the requisite system reliability needed to achieve those missions and events. Specify the requisite reliability in the system specifications in quantitative terms, along with recommended approaches to verify the requirements are met. Require the system provider to demonstrate adherence to the reliability requirements via analysis and test. **Benefits:** Quantitative reliability requirements provide specific design goals and criteria for assuring that the system will meet the intended durability and life. Early in the design process, the system developer will be required to consider how the design will provide the requisite reliability characteristics and must provide analyses to verify that the delivered hardware will meet the requirements. Assessment of the early design's ability to meet quantitative reliability requirements will support design trades, component selection, and maintainability design, and help ensure that appropriate material strengths are used as well as the appropriate levels and types of redundancy.
	Analysis
1301: Surface Charging/ESD Analysis Practice No. PD-AP-1301	**Practice:** Considering the natural environment, perform spacecraft charging analyses to determine that the energy that can be stored by each nonconductive surface is less than 3 mJ. Determine the feasibility of occurrence of electrostatic discharges (ESD). ESD should not be allowed to occur on surfaces near receivers/antenna operating at less than 8 GHz or on surfaces near sensitive circuits. For this practice to be effective, a test program to demonstrate the spacecraft's immunity to a 3 mJ ESD is required. **Benefit:** Surfaces that are conceivable ESD sources can be identified early in the program. Design changes such as application of a conductive coating and use of alternate materials can be implemented to eliminate or reduce the ESD risk. Preventive measures such as the installation of RC filters on sensitive circuits also can be implemented to control the adverse ESD effects.
1302: Independent Review of Reliability Analyses Practice No. PD-AP-1302	**Practice:** Establish a mandatory closed-loop system for detailed, independent, and timely technical reviews of all analyses performed in support of the reliability/design process. **Benefit:** This process of peer review serves to validate both the accuracy and the thoroughness of analyses. If performed in a timely fashion, it can correct design errors with minimal program impact.

(Continues)

Table 5.1:   Cont'd

Title and File Name	Comments on Practice and Benefits
1303: Part Electrical Stress Analyses Practice No. PD-AP-1303	**Practice:** Every part in an electrical design is subjected to a worst-case part stress analysis performed at the anticipated part temperature experienced during the assembly qualification test (typically 75°C). Every part must meet the project stress derating requirements or be accepted by a formal project waiver. **Benefit:** Part failure rates are proportional to their applied electrical and thermal stresses. By predicting the stress through analysis, and applying conservative stresses, the probability of mission success can be greatly enhanced.
1304: Problem/Failure Report Independent Review and Approval Practice No. PD-AP-1304	**Practice:** Problem/failure (P/F) reports are reviewed independently and approved by reliability engineering specialists to ensure objectivity and integrity in the closure process. This practice assures that the analysis realistically bounds the extent of the P/F, and the corrective action and its verification are successfully accomplished. The key elements are: 1. Analysis must address the problem. 2. Corrective action must address the analysis and the problem. 3. Analysis must address the effect on other items. 4. Corrective action must have been implemented. 5. Item must have passed the gate that caused the P/F—the hardware/software must be successfully retested. **Benefit:** Any independent review process increases the level of compliance of the subject process. It also broadens the scope and depth of experience available for each individual issue without the need for a large supporting staff at each supplier organization. Also, an in-place independent review structure improves the rate of data flow for a given level of effort.
1305: Risk Rating of Problem/Failure Reports Practice No. PD-AP-1305	**Practice:** Problem/failure (P/F) reports are assigned a two-factor set of ratings: a failure effect rating and a failure cause/corrective action rating. The composite rating is used to assess the hardware/software residual launch and mission risk. The high-risk P/F reports are labeled "red flag." **Benefit:** Risk rating enables management to focus on the issues with the highest probability of impacting mission success. Project management is provided with visibility to a concise subset ($<5\%$) of a large information base focusing on the key problematic areas in a timely fashion.
1306: Thermal Analysis of Electronic Assemblies to the Piece Part Level Practice No. PD-AP-1306	**Practice:** Perform a piece part thermal analysis that includes all piece parts in support of the part stress analysis. Also include fatigue sensitive elements of the assembly such as interconnects (solder joints, bondlines, wirebonds, etc.).

(Continues)

Table 5.1: Cont'd

Title and File Name	Comments on Practice and Benefits
	Benefit: Allows the thermally overstressed parts to be identified and assessed for risk (instead of just the electrically overstressed parts). Allows the design life requirements of the thermal fatigue sensitive elements (solder joints, bondlines, wirebonds, etc.) to be quantified.
1307: Failure Modes, Effects and Criticality Analysis (FMECA) Practice No. PD-AP-1307	**Practice:** Analyze all systems to identify potential failure modes by using a systematic study starting at the piece part or circuit functional block level and working up through assemblies and subsystems. Require formal project acceptance of any residual system risk identified by this process. **Benefit:** The FMECA process identifies mission critical failure modes and thereby precipitates formal acknowledgment of the risk to the project and provides an impetus for design alteration.
1308: Electromagnetic Interference Analysis of Circuit Transients Practice No. PD-AP-1308	**Practice:** Network circuit analysis programs are valuable tools in the analysis of switching circuit transients which are capable of generating conducted and radiated electromagnetic interference (EMI). The analysis is performed to insure that disruptions or degradations due to EMI do not occur. EMI is capable of disrupting the normal operating environment of an electronic circuit or degrading the performance of such a circuit. **Benefits:** Circuit analysis for the purpose of evaluating the conducted and radiated EMI from a switching circuit has resulted in the proper design of switching circuit electronics. The devices connected to electronic switching circuits will not be adversely affected by transient currents and associated radiated fields generated by such currents.
1309: Analysis of Radiated EMI From ESD Events Caused by Space Charging Practice No. PD-AP-1309	**Practice:** Modeling is utilized for the analysis of conducted and radiated electromagnetic interference (EMI) caused by an electrostatic discharge (ESD) event. The modeling requires the combined use of a SPICE, or other circuit analysis code and a wire antenna code based on the method of moments, and is primarily applicable to wires, cables, and connectors. **Benefit:** The use of a combined SPICE circuit analysis code and a method of moments code for the study of possible conducted and radiated EMI resulting from an ESD event, allow the assessment of EMI noise coupling onto electronic circuit interfaces.
1310: Spurious Radiated Interference Awareness Practice No. PD-AP-1310	**Practice:** Unexpected interference in receivers can be avoided in a complex system of transmitters and receivers by performing an intermodulation analysis to identify and solve potential problems. Various emitters may be encountered during system test, launch, boost, separation and flight. There are a large

(Continues)

Table 5.1:   Cont'd

Title and File Name	Comments on Practice and Benefits
	number of these harmonics and intermodulation products from which potential sources of spurious radiated interferences are identified by a computer aided analysis and corrective measures evaluated. **Benefit:** Spurious radiated interference can be identified and evaluated during the design phase of the project. Solutions can be proposed and implemented in the design phase with far less impact on cost and schedule than when changes are required later.
1312: The Team Approach to Fault-Tree Analysis 1312msfc.pdf	**Practice:** Use a multidisciplinary approach to investigations using fault-tree analysis for complex systems to derive maximum benefit from fault-tree methodology. Adhere to proven principles in the scheduling, generation, and recording of fault-tree analysis results. **Benefits:** The use of the team approach to fault-tree analysis permits a rapid, intensive, and thorough investigation of space hardware and software anomalies. This approach is specifically applicable when the solution of engineering problems is urgent and when they must be resolved expeditiously to prevent further delays in program schedules. The systematic, focused, highly participative methodology permits quick and accurate identification, recording, and solution of problems. The resulting benefits of the use of this methodology are reduction of analysis time, and precision in identifying and correcting deficiencies. The ultimate result is improved overall system reliability and safety.
1313: System Reliability Assessment Using Block Diagramming Methods 1313.pdf	**Practice:** Use reliability predictions derived from block diagram analyses during the design phase of the hardware development life cycle to analyze design reliability; perform sensitivity analyses; investigate design tradeoffs; verify compliance with system-level requirements; and make design and operations decisions based on reliability analysis outputs, ground rules, and assumptions. **Benefit:** Reliability block diagram (RBD) analyses enable design and product assurance engineers to (1) quantify the reliability of a system or function, (2) assess the level of failure tolerance achieved, (3) identify intersystem disconnects as well as areas of incomplete design definition, and (4) perform tradeoff studies to optimize reliability and cost within a program. Commercially available software tools can be used to automate the RBD assessment process, especially for reliability sensitivity analyses, thus allowing analyses to be performed more effectively and timely. These assessment methods can also pinpoint areas of concern within a system that might not be obvious otherwise and can aid the design activity in improving overall system performance.

(Continues)

Table 5.1: Cont'd

Title and File Name	Comments on Practice and Benefits
1314: Sneak Circuit Analysis Guideline for Electromechanical Systems 1314msfc.pdf	**Practice:** Sneak circuit analysis is used in safety-critical systems to identify latent paths which cause the occurrence of unwanted functions or inhibit desired functions, assuming that all components are functioning properly. It is based upon the analysis of engineering and manufacturing documentation. Because of the high cost of a sneak circuit analysis, it should be conducted only in areas where there is a high potential for a hazard. **Benefit:** Identification of sneak circuits in the design phase of a project prior to manufacture can improve reliability; eliminate costly redesign and schedule delays; and eliminate problems in test, launch, on-orbit, and protracted space operations. Sneak circuit analysis can also be beneficial in identifying drawing errors and design concerns.
1316: Thick Dielectric Charging/ Internal Electrostatic Discharge (IESD) Practice No. PD-AP-1316	**Practice:** Dielectric compositions used in such spacecraft materials as circuit boards, cable insulation and thermal blankets will build up an imbedded charge when exposed to a natural space environment featuring energetic electrons. If the electric field resulting from the imbedded charge exceeds the breakdown threshold for the dielectric, an arc will occur, damaging the dielectric and producing an electromagnetic pulse that can couple into subsystem electronics. Enhance hardware reliability in an energetic electron environment by conducting a materials inventory, resistivity analysis, and shielding assessment. Ascertain material susceptibility to deep dielectric charging and explosive discharge when the material: 1. Is exposed to an energetic electron flux exceeding 2×10^5 electrons/(cm^2-s), and 2. Achieves an imbedded charge density greater than a threshold of 10^{11} electrons/cm^2. **Benefit:** Materials and design structures that represent possible internal electrostatic discharge (IESD) sources can be identified early in the program. Risk to hardware may be reduced through design changes which substitute materials having sufficient conductivity to permit charge bleed-off. Sensitive cable runs may be rerouted or shielded to reduce exposure to energetic electrons. Grounding schemes may be changed to ensure that otherwise isolated conductors are grounded and that grounds are designed to maximize the opportunity to bleed-off the charge from dielectric materials.

(Continues)

Table 5.1:   Cont'd

Title and File Name	Comments on Practice and Benefits
	Test
1401: EEE Parts Screening test_series/new/1401.pdf	**Practice:** Implement a 100% nondestructive screening test on EEE parts prior to assembly, which would prevent early-life failures (generally referred to as infant mortality). **Benefits:** A lower rework cost during manufacturing and lower incident of component failures during flight.
1402: Thermal Cycling np1402.pdf	**Practice:** As a minimum, run eight thermal cycles over the approximate temperature range for hardware that cycles in flight over ranges greater than 20°C. The last three thermal cycles should be failure-free. **Benefit:** Demonstrates readiness of the hardware to operate in the intended cyclic environment. Precipitates defects from design or manufacturing processes that could result in flight failures.
1403: Thermographic Mapping of PC Boards np1403x.pdf	**Practice:** Use thermographic mapping methods to locate hot spots on operating PC boards. **Benefit:** Quick find of electronic components operating at or above recommended temperatures. Also, this technique can validate the derating factors and thermal design via low cost testing versus analysis.
1404: Thermal Test Levels & Durations Practice No. PT-TE-1404	**Practice:** Perform thermal dwell test on protoflight hardware over the temperature range of +75°C/-20°C (applied at the thermal control/ mounting surface or shear plate) for 24 hours at the cold end and 144 to 288 hours at the hot end. **Benefit:** This test, coupled with rigorous design practices, provides high confidence that the hardware design is not marginal during its intended long life high reliability mission.
1405: Powered-On Vibration Practice No. PT-TE-1405	**Practice:** Supply power to electronic assemblies during vibration, acoustics, and pyroshock and monitor the electrical functions continuously while the excitation is applied. **Benefit:** Aids in the detection of intermittent or incipient failures in electronic circuitry not otherwise found. This reliability practice benefits even those electronics not powered during launch.
1406: Sinusoidal Vibration Practice No. PT-TE-1406	**Practice:** Subject assemblies and the full-up flight system to swept sinusoidal vibration. **Benefit:** Certain failures are not normally exposed by random vibration. Sinusoidal vibration permits greater displacement excitation of the test item in the lower frequencies.

(Continues)

Table 5.1:   Cont'd

Title and File Name	Comments on Practice and Benefits
1407: Assembly Acoustic Tests Practice No. PT-TE-1407	**Practice:** Subject selected (large surface area, low mass) assemblies, in addition to the full-up flight system, to acoustic noise. It is imperative on missions with fixed launch windows that acoustic problems on assemblies not be deferred to system level tests. **Benefit:** Acoustic noise tests subject potentially susceptible hardware to a significant launch environment, revealing design and workmanship inadequacies which might cause problems in flight.
1408A: Pyrotechnic Shock Testing (revised to reflect "powered" test mode) Practice No. PT-TE-1408A	**Practice:** Subject potentially sensitive flight assemblies that contain electronic equipment or mechanical devices, as well as entire flight systems, to pyrotechnic shock (pyroshock) as part of a development, acceptance, protoflight, or qualification test program. Perform visual inspection and functional verification testing before and after each pyroshock exposure. Where feasible, perform assembly-level and system-level pyroshock tests with the test article powered and operational to better detect intermittent failures. **Benefit:** Early assembly-level pyroshock testing can often reduce the impacts of design and manufacturing/assembly deficiencies upon program cost and schedule prior to system-level test. Such testing can provide a test margin over flight pyroshock conditions which cannot be achieved in system testing. Conversely, system-level shock testing can be used to verify system performance under pyroshock exposure, thus providing increased confidence in mission success and verifying the adequacy of the assembly-level tests.
1409: Thermal-Vacuum Versus Thermal-Atmospheric Tests of Electronic Assemblies Practice No. PT-TE-1409	**Practice:** Perform all thermal environmental tests on electronic spaceflight hardware in a flight-like thermal vacuum environment (i.e., do not substitute an atmospheric pressure thermal test for the thermal/vacuum test). Moreover, if a compromise is thought to be necessary for nontechnical reasons, then an analysis is required to quantify the reduction in test demonstrated reliability. **Benefit:** Assembly-level thermal vacuum testing is the most perceptive test for uncovering design deficiencies and workmanship flaws in spaceflight hardware. The margin beyond flight conditions is demonstrated, as is reliability. However, substituting an atmospheric pressure thermal test for the thermal/vacuum test can effectively reduce electronic piece part temperatures by 20°C or

(Continues)

Table 5.1: Cont'd

Title and File Name	Comments on Practice and Benefits
	more, even for low-power density designs. The net result of this is that the effective test temperatures may be reduced to the point where there is zero or negative margin over the flight thermal environment.
1410: Selection of Spacecraft Materials and Supporting Vacuum Outgassing Data 1410.pdf	**Practice:** Each flight project provides requirements for defining and implementing a contamination control program applicable to the hardware for the program. The program consists first in defining the specific cleanliness requirements and setting forth the approaches to meeting them in a contamination control plan. One significant part of the contamination control plan is a comprehensive materials and process program beginning at the design stage of the hardware. This program helps ensure the safety and success of the mission by the appropriate selection, processing, inspection, and testing of the materials employed to meet the operational requirements for the application. The following potential problem areas are considered when selecting materials: radiation effects, thermal cycling, stress corrosion cracking, galvanic corrosion, hydrogen embrittlement, lubrication, contamination of cooled surfaces, composite materials, atomic oxygen, useful life, vacuum outgassing, toxic offgassing, flammability, and fracture toughness. The practice described here for the collection and compilation of vacuum outgassing data is used in conjunction with a number of other processes in the selection of materials. Vacuum outgassing tests are conducted on materials intended for spaceflight use, and a compilation of outgassing data is maintained and constantly updated as new materials are tested. This includes materials used in the manufacture of parts intended for space applications. **Benefit:** These test data provide outgassing information on a wide variety of materials and should be used as a guide by engineers in selecting materials with low outgassing properties.
1411: Heat Sinks for Parts Operated in Vacuum np1411.pdf	**Practice:** Perform a thermal analysis of each electronic assembly to the piece-part level. Provide a heat conduction path for all parts whose junction temperature rise exceeds 35°C above the cold plate. **Benefits:** Controlling the operating temperature of parts in a vacuum flight environment will lower the failure rate, improve reliability, and extend the life of the parts.
1412: Environmental Test Sequencing Practice No. PT-TE-1412	**Practice:** Perform dynamic tests prior to performing thermal-vacuum tests on flight hardware. **Benefit:** Experience has shown that until the thermal-vacuum tests are performed, many failures induced during dynamics tests are not detected because of the short duration of the dynamics tests.

(Continues)

Table 5.1: Cont'd

Title and File Name	Comments on Practice and Benefits
	In addition, the thermal-vacuum test on flight hardware at both the assembly level and the system level provides a good screen for intermittent as well as incipient hardware failures.
1413: Random Vibration Testing Practice No. PT-TE-1413	**Practice:** Define an appropriate random vibration test, and subject all assemblies and selected subsystems to the test for design qualification and workmanship flight acceptance. **Benefit:** This practice assists in identifying existing and potential failures in flight hardware so that they can be rectified before launch.
1414: Electrostatic Discharge (ESD) Test Practices Practice No. PT-TE-1414	**Practice:** Test satellites for the ability to survive the effects of electrostatic discharges (ESDs) caused by a space-charging environment. Such environments include Earth equatorial orbits above 8000 km and virtually all orbits above 40 degrees latitude, Jupiter encounters closer than 15 Rj (Jupiter radii), and possibly other planets. **Benefit:** Proper implementation of this practice will assure that satellites will operate in the space charging environment without failure or awkward ground controller operations.
1415: Power System Corona Testing np1415.pdf	**Practice:** Test power system components for corona to ensure that their insulation system will meet the design requirements imposed on the equipment and to verify that the gas discharges are not deteriorating the insulation system. The acceptable corona levels are verified in power system components. **Benefits:** Knowledge of the presence or absence of corona discharge will help in controlling the reliability of high voltage components/ systems. Corona testing can reveal potential and unaccounted-for corona discharges that may shorten the service-life of electrical insulating systems, seriously interfere with high-voltage system operation and communication links, and result in failure and loss of mission objectives.
1416: Radiated Susceptibility System Verification Practice No. PT-TE-1416	**Practice:** Verify that a flight vehicle or system is hardened to the launch, boost, and flight electromagnetic radiation environment by radiating simultaneously, during system checkout, on all major emission frequencies that are known to exist during vehicle operations. Monitor all critical systems for erroneous performance while the spacecraft or system is stepped through all operating modes. **Benefit:** Spurious interferences and responses can be identified during system checkout. After the spurious responses are evaluated, solutions can be proposed, and remedial action taken, if necessary, prior to the actual flight.
1417: Electrical Isolation Verification (DC)	**Practice:** Direct current (DC) electrical isolation verification tests are made as part of the EMC test of hardware prior to final spacecraft

(Continues)

Table 5.1:   Cont'd

Title and File Name	Comments on Practice and Benefits
Practice No. PT-TE-1417	assembly. Flight acceptance isolation retest is required after any hardware rework of subsystems with electrical interfaces that utilize system wiring. **Benefit:** Inadvertent grounds of isolated circuits and ground loops are detected directly by this test. In some cases, such grounds may pass other tests with no apparent degradation. Failure may not occur until the vehicle is subjected to high-level electromagnetic radiation. Since this test requires minimal test equipment and can be performed in a short time, its benefits are achieved at low cost.
1418: Qualification of Non-Standard EEE Parts in Space Flight Applications 1418.pdf	**Practice:** The source for selection of acceptable flight-quality EEE parts for use on Goddard projects is GSFC Preferred Parts List (PPL-20). PPL-20 complements NASA Standard Electrical, Electronic, and Electromechanical (EEE) Parts List (NSPL) (MIL-STD-975) by listing additional part types and part categories not included in MIL-STD-975. Recognizing that it is neither possible nor desirable to include all parts in the GSFC PPL and in the NSPL, the GSFC parts requirements make provision that limited numbers of parts not included in the PPL or the NSPL may be used if it is demonstrated that the parts are acceptable. The acceptability of (5, section 5) nonstandard parts is enhanced by use of the part procurement specifications provided in Appendix E of PPL-20. The acceptability of these nonstandard parts must be demonstrated prior to commitment to design or use. Requests for approval to use nonstandard parts with supporting documentation are forwarded to the appropriate GSFC Project Office for review and approval. The practice described herein is used for demonstrating and documenting the acceptability of nonstandard parts for spaceflight use. **Benefits:** The practice of using approved nonstandard parts that have been appropriately demonstrated to be acceptable for the applications provides for a wider range of parts selection than are available with standard parts. These parts are at a quality level equal to that of Grades 1 or 2 standard parts.
1419: Vibroacoustic Qualification Testing of Payloads, Subsystems, and Components 1419.pdf	**Practice:** Perform acoustic and random vibration testing supplemented with additional sine vibration testing as appropriate to qualify payload hardware to the vibroacoustic environments of the mission, particularly the launch environment and to demonstrate acceptable workmanship. **Benefit:** Adherence to the practice alleviates vibroacoustic-induced failures of structural stress and fatigue, unacceptable workmanship, and performance degradation of sensitive subsystems

(Continues)

Table 5.1: Cont'd

Title and File Name	Comments on Practice and Benefits
	including instruments and components. Implementation of this practice assures that minimal degradation of "design reliability" has occurred during prior fabrication, integration and test activities.
1420: Sine-Burst Load Test 1420.pdf	**Practice:** The sine-burst test is used to apply a quasi-static load to a test item in order to strength qualify the item and its design for flight. **Benefits:** The sine-burst is a simple method to apply a quasi-static load using a vibration shaker and shock testing software. Depending on the complexity of the test item, it often can be used in lieu of and is more economical than, acceleration (centrifuge) or static tests. For components and subsystems, the fixture used for vibration testing often can also be used for sine-burst strength testing. For this reason, strength qualification and random vibration qualification can often be performed during the same test session which saves time and money.
1421: Eddy Current Testing of Aerospace Materials 1421msfc.pdf	**Practice:** Eddy Current Testing (ECT) can be used on electrically conductive material for detecting and characterizing defects such as surface and near surface cracks, gouges, and voids. It can also be used to verify a material's heat treat condition. In addition, wall thickness of thin wall tubing, and thickness of conductive and nonconductive coating on materials can be determined using ECT. **Benefits:** Eddy current testing is a fast, reliable, and cost-effective nondestructive testing (NDT) method for inspecting round, flat, and irregularly shaped conductive materials. Specific processes have been developed to determine the usability and integrity of threaded fasteners. In addition, ECT has the capability of being automated. With proper equipment and skilled test technicians, readout is instantaneous.
1422: Ultrasonic Testing of Aerospace Materials 1422msfc.pdf	**Practice:** Three general methods of ultrasonic testing can be used singly or in combination with each other to identify cracks, debonds, voids, or inclusions in aerospace materials. Each has its own unique application and all require certain precautions or techniques to identify potentially flawed hardware. This practice describes selected principles that are essential in reliable ultrasonic testing. **Benefit:** Careful attention to detail in ultrasonic testing can result in the identification of very small cracks, debonds, voids, or inclusions in aerospace hardware that could be detrimental to mission performance. New ultrasonic technologies are enhancing the accuracy, speed, and cost-effectiveness of this method of nondestructive testing.

(Continues)

Table 5.1: Cont'd

Title and File Name	Comments on Practice and Benefits
1423: Radiographic Testing of Aerospace Materials 1423msfc.pdf	**Practice:** Radiographic testing can be used as a nondestructive method for detecting internal defects in thick and complex shapes in metallic and nonmetallic materials, structures, and assemblies. **Benefit:** Unlike most other nondestructive testing methods, radiographic testing provides a permanent visual record of the defects for possible future use. It can also be used to determine crack growth for use in fracture mechanics to determine critical flaw size in a particular component.
1428: Practice of Reporting Parts, Materials, and Safety Problems (Alerts) 1428msfc.pdf	**Practice:** Ensure that potentially significant problems involving parts, materials, and safety discovered during receiving inspection, manufacturing, postmanufacturing inspection, or testing do not affect the safety or the performance of NASA hardware by reporting all anomalies via ALERT systems. ALERTS and SAFE ALERTS pertaining to these problems are quickly disseminated for impact assessment and, if required, corrective action taken or a rationale developed for "flying as is." **Benefit:** The benefit of the ALERTS system is the reduction or elimination of duplicate expenditures of time and money by exchanging information of general concern regarding parts, materials, and safety problems within MSFC, between MSFC and other NASA centers, between NASA and other government organizations, and between government and industry to assist in preventing similar occurrences. The use of the ALERTS system avoids future failures, rules out fraudulent hardware, helps enhance reliability, and ensures mission success.
1429: Integration & Test Practices to Eliminate Stresses on Electrical and Mechanical Components 1429.pdf	**Practice:** Use proven GSFC practices during the integration and testing of flight hardware to prevent electrical and mechanical overstressing of flight hardware parts and components, thereby, assuring that the "designed in" reliability is not compromised. **Benefits:** These practices prevent the long term degradation and early failure of electrical parts and components due to electrical and mechanical overstressing. Damage due to overstressing may not result in immediate failure and may not be detected by component or assembly level testing but can result in early failures.
1430: Short Circuit Testing for Nickel Hydrogen Battery Cells 1430lerc.pdf	**Practice:** Use short-circuit testing method or response characteristics on nickel/hydrogen (Ni/H_2) battery to characterize the battery impedance. These data are necessary for designing power-processing equipment and electric power fault-protection systems.

(Continues)

Table 5.1: Cont'd

Title and File Name	Comments on Practice and Benefits
	Benefits: Ni/H_2 battery technology is gaining wide acceptance as an energy storage system for use in space applications because of its reliability, weight, and long-cycle expectancy at deep depths-of-discharge (DOD). When a charged Ni/H_2 battery is short-circuited, its short circuit current data can be used to calculate the internal resistance of the cells for the purpose of determining the overall characteristics of the energy storage system. Also, by examining the cell impedance only, a Ni/H_2 battery simulation utilizing low-cost lead-acid cells can be developed.
1431: Voltage/Temperature Margin Testing Practice No. PD-TE-1431	**Practice:** Voltage and temperature margin testing (VTMT) is the practice of exceeding the expected flight limits of voltage, temperature, and frequency to simulate the worst case functional performance, including effects of radiation and operating life-parameter variations on component parts. For programs subject to severe cost or schedule constraints, VTMT has proven an acceptable alternative to conventional techniques such as worst-case analysis (WCA). WCA is the preferred approach to design reliability, but VTMT is a viable alternative for flight projects where tradeoffs of risk versus development time and cost are appropriate. **Benefits:** On spacecraft hardware where risk vs. cost trades permit higher risk (Class C), VTMT is an economical alternative to classical worst case analysis. The major benefits in using VTMT instead of WCA are: 1. Assurance of a systematic method for investigation of potential risks where the parameters are not adequately modeled by worst case analysis. An example is RF circuits which have distributed circuit parameters. 2. Labor savings for units too complex to simulate and which generally require Monte Carlo or root-sum squares analyses. 3. Real-time operation and review of complex circuits, allowing the weighing of alternative design actions. 4. Cost savings from expedited risk assessment. Comparative studies have demonstrated that testing may be completed in less than one-third the time required for analyses.
1432: RF Breakdown Characterization Practice No. PD-TE-1432	**Practice:** Tests are performed to verify that radio frequency (RF) equipment, such as receivers, transmitters, diplexers, isolators, RF cables, and connectors, can operate without damage or degradation.

(Continues)

Table 5.1: Cont'd

Title and File Name	Comments on Practice and Benefits
	Reliability assurance is necessary in both a vacuum environment and at critical pressure with adequate demonstrated margins above the expected operating RF signal levels. **Benefits:** Knowledge of the dielectric breakdown characteristics of RF devices at low pressures or in a near vacuum environment can be used to protect sensitive flight equipment. RF breakdown is a concern because of the low, near-vacuum pressures at which spacecraft are tested and operated. RF breakdown testing is conducted to establish hardware resilience to the application of out-of-spec input signal levels, signal reflections due to mismatches at hardware interfaces, inadvertent evacuation of vacuum chambers during RF input, application of RF signals during the ascent phase of the spacecraft launch vehicle, and so on.
1433: Mechanical Fastener Inspection System 1433.pdf	**Practice:** Applies a formal flight assurance inspection system for mechanical fasteners used in flight hardware and critical applications on ground support equipment (GSE), including all flight hardware/GSE interfaces. **Benefit:** This practice significantly enhances flight reliability by ensuring that mechanical fasteners do not fail during the mission due to inadequate integrity requirements or quality control inspection procedures.
1434: Battery Verification through Long-Term Simulation 1434msfc.pdf	**Practice:** Conduct highly instrumented real-time long term tests and accelerated testing of spaceflight batteries using automated systems that simulate prelaunch, launch, mission, and postmission environments to verify suitability for the mission, to confirm the acceptability of design configurations, to resolve mission anomalies, and to improve reliability. **Benefit:** Since the operational readiness and future performance of spaceflight batteries at any point in a mission are strongly dependent on past power cycles and environments, thoroughly instrumented and analyzed ground testing of spaceflight batteries identical to flight configurations will ensure predictable performance and high reliability of flight batteries.
1435: Verification of RF Hardware Design Performance Practice No. PD-TE-1435	**Practice:** Analyses are performed early in the design of radio frequency (RF) hardware to determine hardware imposed limitations which affect radio performance. These limitations include distortion, bandwidth constraints, transfer function non-linearity, non-zero rise and fall transition time, and signal-to-noise ratio (SNR) degradation. The effects of these hardware performance impediments are measured and recorded.

(Continues)

Table 5.1:   Cont'd

Title and File Name	Comments on Practice and Benefits
	Performance evaluation is a reliability concern because RF hardware performance is sensitive to thermal and other environmental conditions, and reliability testing is constrained by RF temperature limitations. **Benefits:** Identification of hardware-imposed limitations on RF subsystem performance permits designers to evaluate a selected radio technology or architecture against system requirements. In the test phase of the reliability assurance program, it also helps engineers to understand performance characteristics they encounter during testing. RF modeling and verification provides for designed-in reliability in accordance with NASA's project streamlining policy.

3. NPR 8715.7, *Expendable Launch Vehicle Payload Safety Program* (http://nodis3.gsfc.nasa.gov/displayDir.cfm?t=NPR&c=8715&s=7) assists "ELV [Expendable Launch Vehicle] payload projects in achieving safety design objectives and obtaining the necessary safety approvals and to assure that NASA safety policy is satisfied for all ELV payload missions." NASA ELV payloads often incorporate hazards which can pose significant risk to life and property. NASA ELV payload missions require the coordination of efforts among a diverse group of participants who have varying responsibilities and authorities. These missions can present unique challenges to the payload safety assurance process, which often involves numerous organizations internal and external to the Agency.

4. Air Force Space Command Manual 91-710, *Range Safety User Requirements* (http://www.afspc.af.mil/library/launchsafety/index.asp). All range users operating on the AFSPC ranges, including the ER and WR, are subject to the requirements of this volume to ensure safety by design, testing, inspection, and hazard analysis.

5. The *National Environmental Policy Act of 1969* (NEPA) requires U.S. federal agencies to consider the impacts to the environment of proposed projects before taking action. These potential impacts are documented in an Environmental Impact Statement (EIS) and provided to the public for comment as part of the process required by NEPA. This process involves the issuance of a Notice of Intent (NOI) followed by a draft EIS. After public review and comment, a final EIS is published and a record of decision (ROD) is issued.

6. Planetary protection activities at NASA are managed under the Science Mission Directorate at NASA Headquarters. The purpose of planetary protection activities is to twofold: (1) to preserve Solar System bodies from contamination by Earth life and (2) to protect Earth from possible life forms that may be returned from other Solar System bodies. Planetary protection requirements for NASA projects vary depending on the target Solar System body being visited and on the type of mission (for example, planetary flyby, orbiter, lander or rover, sample return). Related documents for planetary protection include NPR 8020.12C, *Planetary Protection Provisions for Robotic Extraterrestrial Missions;* NPR 8020.7F, *Biological Contamination Control for Outbound and Inbound Planetary Spacecraft* (revalidated 10/23/03); and NPD 7100.10E, *Curation of Extraterrestrial Materials.* References on planetary protection may be found at http://planetaryprotection.nasa.gov.

7. NPR 8700.5, *Probabilistic Risk Assessment Procedures Guide for NASA Managers and Practitioners* (http://www.hq.nasa.gov/office/codeq/doctree/praguide.pdf). Probabilistic risk assessment (PRA) serves two purposes:

- To complement the training material taught in the PRA course for practitioners and, together with the *Fault Tree Handbook*, to provide PRA methodology documentation.

- To assist aerospace PRA practitioners in selecting an analysis approach that is best suited for their applications. The material of this procedures guide is organized into four parts:

   A management introduction to PRA is presented in Chapters 1 through 3. It presents an overview of PRA with simple examples after an introduction of the history of PRA at NASA and a discussion of the relation between PRA and risk management.

   Chapters 4 through 14 cover probabilistic methods for PRA, methods for scenario development, uncertainty analysis, data collection and parameter estimation, human reliability analysis, software reliability analysis, dependent failure analysis, and modeling of physical processes for PRA.

   Chapter 15 provides a detailed discussion of the "scenario-based" PRA process using two aerospace examples.

   The only departure of the PRA from the description of experience-based recommended approaches is in the areas of human reliability (Chapter 9) and software risk assessment (Chapter 11). Analytical methods in these two areas are not mature enough, at least in aerospace applications. Therefore, instead of recommended approaches, these chapters describe some popular methods for the sake of completeness.

8. *Fault Tree Handbook with Aerospace Applications*, August 2002 (http://www.hq.nasa.gov/office/codeq/doctree/fthb.pdf). The current *Fault Tree Handbook*, serves two purposes: (1) as a companion document to the training material taught in FTA courses for practicing system analysts; and (2) to assist aerospace FTA practitioners in acquiring and implementing current state-of-the art FTA techniques in their applications. The current version of the handbook contains the following material that was not in the original version:

   - A discussion of the binary decision diagram (BDD) method for solving fault trees that were originally solved only through Boolean reduction and the use of minimal cuts sets

   - An introduction to dynamic fault trees (DFTs) and methods to solve them

   - Illustrations of fault tree analysis in aerospace applications, with detailed description of the models

   - An extended discussion of modeling common cause failures and human errors in FTA

   - Descriptions of modeling feedback loops so as to properly cut such loops in a FT

   - Extended discussion of applications of FTA for decision making, covering applications to operating systems and to systems that are in design

   - Descriptions of absolute and relative importance measures that are obtainable from FTA and that enhance the output and value of an FTA

   - Expanded discussion of success trees, their logical equivalence to fault trees, and their applications

9. Worst-case analysis (http://klabs.org/richcontent/General_Application_Notes/SDE/WCA_Requirements.pdf). The purpose of a worst-case analysis (WCA) is to prove the design will function as expected during its mission. The spirit of analysis is proof: all circuits are considered guilty of design flaws until proven innocent. Here are areas considered by WCA:

   - Part parameters and deratings—Each parameter must be derated from the data book value for the intended environment to compensate for the effects of temperature, age, voltage, and radiation.

   - Timing analysis—Set-up and hold times at all clocked inputs, pulse widths of clocks, and asynchronous set, clear, and load inputs, all clock inputs and asynchronous inputs such as sets, clears, and loads must be shown to be free from both static and dynamic hazards.

- Gate output loading—Show that no gate output drive capacities have been exceeded.

- Interface margins—Show that all of the gates have their input logic level thresholds met.

- State machines—Must be analyzed to assure that they will not exhibit anomalous behavior, such as system lock-up.

- Asynchronous interfaces—Must show either that asynchronous signals are properly synchronized to the appropriate clock or that the circuitry receiving asynchronous signals will function correctly if set-up and hold times are not met.

- Reset conditions and generation—All circuitry must be shown to be placed into a known state during reset.

- Part safety conditions—The analysis must prove that the circuit is designed so as to prevent its parts from being damaged.

- Cross-strap signals between redundant modules—Show that isolation between boxes is actually achieved.

- Circuit interconnections—Show that circuit interconnection requirements are met from the standpoint of signal quality as affected by edge rates, loading, and noise.

- Bypass capacitance analysis—Show that the amount of on-board bulk and bypass capacitance is appropriate for the circuitry.

10. *Failure Modes, Effects and Criticality Analysis (FMECA) Public Lessons Learned Entry: 0795* (http://www.nasa.gov/offices/oce/llis/0795.html). Failure modes, effects, and criticality analysis (FMECA) comprises two separate analyses: failure mode and effects analysis (FMEA) and criticality analysis (CA). FMEA analyzes different failure modes and their effects on the system while CA classifies their level of importance based on failure rate and severity of the effect of failure. The ranking process of CA can use either existing failure data or a subjective ranking conducted by a team of people with an understanding of the system.

NASA originally developed FMECA to improve and verify the reliability of space program hardware. MIL-STD-1629A, which has been canceled, established requirements and procedures for performing a FMECA, to evaluate and document, by failure mode analysis, the potential impact of each functional or hardware failure on mission success, personnel and system safety, and maintainability and system performance. It ranks each potential failure by the severity of its effect so that corrective actions may be taken to eliminate or control design risk. High-risk items are those items whose failure would jeopardize the mission or endanger personnel. The techniques presented in this standard may be applied to any electrical or mechanical equipment or system. Although MIL-STD-1629A has been canceled, its concepts should be applied during the development phases of all critical systems and equipment whether it is military, commercial, or industrial systems/products (see http://www.army.mil/USAPA/eng/DR_pubs/dr_a/pdf/tm5_698_4.pdf).

11. MIL-HDBK-338B *Electronic Reliability Design Handbook* (http://www.relex.com/resources/mil/338b.pdf). Reliability engineering is doing those things which ensure that an item will perform its mission successfully. The discipline of reliability engineering consists of two fundamental aspects: (1) paying attention to detail and (2) handling uncertainties. The traditional, narrow definition of reliability is "the probability that an item can perform its intended function for a specified interval under stated conditions." This narrow definition applies largely to items which have simple missions, such as equipment, simple vehicles, or components of systems. For large complex systems, such as command and control systems, aircraft weapon systems, a squadron of tanks, and naval vessels, it is more appropriate to use more sophisticated concepts such as "system effectiveness" to describe the worth of a system. System effectiveness relates to that property of a system output, carrying out of some intended function, which was the real reason for buying the system in the first place; if the system is effective, it functions well; if it is not effective, it does not function well and attention must be focused on those system attributes that are deficient.

12. Radiation models (http://setas-www.larc.nasa.gov/LDEF/RADIATION/rad_exp_space.html). Designers must address a variety of important radiation effects: dose (which can range from 20 to 30 rad/year), single-event effects (which affect microelectronics), displacement damage, and sensor noise. The analysis can help determine the amount of redundancy, and hence cost, of the spacecraft. (For example, total dose requirements for microelectronics are a common concern; analysis can help determine the amount of shielding necessary to protect the components. In another example, a data collecting processing unit with large memory may not require as much redundancy because the data may indicate if a single-event upset occurs. In contrast, a final example is flight controls, which require more redundancy to keep the spacecraft operating properly.)
In preparing a spacecraft design, engineers may use any number of different radiation environment models, as given in the website:

- International reference ionosphere (IRI)

- International geomagnetic reference field (IGRF)

- AE/AP radiation belt models

- Cosmic ray effects on microelectronics (CREME) model

- Tsyganeko models of the Earth's magnetic field

Lessons learned are also valuable in predicting radiation effects. One useful reference is Poivey C, et al., "Lessons Learned from Radiation Induced Effects on Solid State Recorders (SSR) and Memories," December 2002 (http://radhome.gsfc.nasa.gov/radhome/papers/2002_SSR.pdf).

13. Carosso N, "Contamination Engineering Guidelines" Swales Aerospace (http://400dg.gsfc.nasa.gov/sites/400/docsguidance/All%20Documents/Contam_Eng_Guidelines.doc). This document provides a description of the necessary elements involved in planning, designing, implementing, and verifying an adequate contamination control program for spacecraft and science instrument hardware. The document may be applied to all types of hardware development from individual components to complete subsystem assemblies, to any and all levels of science instrument hardware, up to and including entire integrated spacecraft and launch vehicles.

    Complementary documents are *Contamination Control of Space Optical Systems*, NASA Preferred Reliability Practices, PD-ED-1263 (http://snebulos.mit.edu/projects/reference/NASA-Generic/PD-ED-1263.pdf); and Harkins W, *Selection of Spacecraft Materials and Supporting Vacuum Outgassing Data*, NASA Engineering Network, Public Lessons Learned Entry 0778, February 1, 1999 (http://www.nasa.gov/offices/oce/llis/0778.html).

14. NPR 8621.1 *NASA Procedural Requirements for Mishap and Close Call Reporting, Investigating, and Recordkeeping* (http://nodis3.gsfc.nasa.gov/displayDir.cfm?t=NPR&c=8621&s=1B). The purpose of the NASA mishap investigation process is to determine cause and develop recommendations to prevent recurrence. A notional timeline of the investigation process is as follows:

    • Immediately—24 hours safe site, initiate premishap plans, make notifications, classify mishap

    • Within 48 hours of mishap—Appoint investigating authority

    • Within 75 workdays of mishap—Complete investigation and mishap report

    • Within an additional 30 workdays—Review and endorse mishap report

    • Within an additional 5 workdays—Approve or reject mishap report

    • Within an additional 10 workdays—Authorize report for public release

    • Within an additional 10 workdays—Distribute mishap report

    • Concurrently

        • Within 15 workdays of being tasked—Develop corrective action plan

        • Within 10 workdays of being tasked—Develop lessons learned

15. Root cause analysis (RCA) is a structured evaluation method that identifies the root causes for an undesired outcome and the actions adequate to prevent recurrence (http://klabs.org/DEI/References/design_guidelines/content/nasa_specs/root_cause_analysis_bradley_2003.pdf). Root cause analysis should continue until organizational factors have been identified, or until data are exhausted.

16. NPD 8730.2C *NASA Parts Policy* (http://nodis3.gsfc.nasa.gov/displayDir.cfm?t=NPD&c=8730&s=2C). It is NASA policy to control risk and enhance reliability in NASA spaceflight and critical ground support/test systems, in part, by managing the selection, acquisition, traceability, testing, handling, packaging, storage, and application of the following:

    • Electrical, electronic, and electromechanical (EEE) parts

    • Electronic packaging and interconnect systems

    • Mechanical parts such as fasteners, bearings, studs, pins, rings, shims, piping components, valves, springs, brackets, clamps, and spacers

    • Manufacturing materials affecting the performance/acceptability of parts such as plating, solder, and weld-filler material

17. EEE-INST-002 2003, *Instructions for EEE Parts Selection, Screening, Qualification, and Derating* (http://nepp.nasa.gov/DocUploads/FFB52B88-36AE-4378-A05B2C084B5EE2CC/EEE-INST-002_add1.pdf). Establish baseline criteria for selection, screening, qualification, and derating of EEE parts for use on NASA GSFC spaceflight projects. This document provides a mechanism to assure that appropriate parts are used in the fabrication of space hardware that will meet mission reliability objectives within budget constraints.

    This document provides instructions for meeting three reliability levels of EEE parts requirements based on mission needs:

    • A Grade 1 part is consistent with reliability Level 1. Levels of part reliability confidence decrease by reliability level, with Level 1 being the highest reliability and Level 3 the lowest. A reliability Level 1 part has the highest level of manufacturing control and testing per military specifications.

    • Level 2 parts have reduced manufacturing control and testing.

    • Level 3 parts have no guaranteed reliability controls in the manufacturing process and no standardized testing requirements. The reliability of Level 3 parts can vary significantly with each manufacturer and part type due to unreported and frequent changes in design, construction and materials.

18. Jet Propulsion Laboratory (JPL) standard processes and documents—This is only a partial listing of the documents that JPL uses for developing programs and projects.

Some are not accessible in a public format. If you are a contractor to JPL, you should be able to get the appropriate standards and documents from your primary point of contact.

- *Radiation Effects Group Publications* (http://parts.jpl.nasa.gov/resources.htm).

- D-20348, Rev. A, *JPL Institutional Parts Program Requirements* (http://nepp.nasa.gov/docuploads/8DB633E8-7AA9-4A1C-87DC1135F87B613C/JPL-D-20348.doc). Every electrical, electronic, and electromechanical (EEE) part intended for use in spaceflight shall be reviewed and approved for compatibility with the intended space environment and mission life. This document defines the baseline parts program requirements for all JPL missions, including both spacecraft and instruments.

- D-5703, Rev. 2, *Reliability Analyses for Flight Hardware in Design* (http://dmie.jpl.nasa.gov/cgi/doc-gw.pl?DocRevID=80729&frame=html&mimetype=&dispform=3).

- D-58032 *Flight Project Practices*.

- D-8671 *JPL Standard for Reliability Assurance*.

- D-8091 *JPL Standard for Anomaly Resolution* (http://pbma.nasa.gov/docs/public/pbma/bestpractices/bp_jpl_07.doc). The purpose of this document is to define the guidelines and procedures for an effective problem/failure reporting system. To be effective, the system must ensure that every problem or failure is reported in a timely manner, and that the corrective action will preclude the recurrence of the problem/failure. The system should also ensure that for those special cases in which effective corrective action has not been fully implemented, the residual risk is identified and is acceptable to project/task managers.

- D-560 JPL Standard for Flight Systems Safety.

- D-11119 Alert/Concerns Handbook.

- D-12872, Rev. 1, JPL Process for Tailoring Mission Assurance to Specific Projects, January 1999 (http://trs-new.jpl.nasa.gov/dspace/bitstream/2014/12133/1/01-0005.pdf). This document provides guidance to identify a process for tailoring and integrating mission assurance (MA) activities into JPL flight projects that is consistent with a project's characteristics and resources. Such tailoring process replaces flight hardware classification and any predetermined set of MA prequirements as provided by JPL D-1489 and JPL D-8966.

- Atkins K, Gowler P, "Preparing Project Managers for Faster-Better-Cheaper Robotic Planetary Missions," IEEE Aerospace Conference, November 2002 (http://trs-new.jpl.nasa.gov/dspace/bitstream/2014/10956/1/02-2819.pdf). This paper advocates moving toward "a set of consistent project implementation processes with process owners and process engineering teams focusing on the FBC [faster-better-cheaper] paradigm and ISO objectives." The JPL processes that the paper advocates follow:

  - Define mission/science objectives and data

  - Products

  - Plan the project

  - Plan, manage, and control resources

  - Manage and mitigate risk

  - Secure launch approval

  - Lead and build the team

  - Staff and de-staff projects

  - Plan and execute project acquisitions

  - Provide and manage project information

  - Manage international participation

  - Engage the educational and public community

  - Manage mission assurance

  - Assure product quality

  - Assure product reliability

  - Ensure parts reliability

  - Ensure system safety

  - Manage configuration of project elements

  - Implement project reviews

  - Design project architecture

- Engineer the project

- Engineer mission and navigation systems

- Engineer flight systems

- Engineer mission operations systems

- Design product systems

- Develop hardware products

- Develop software products

- Integrate and test products

- Operate product systems

- Integrate and test mission systems

- Provide operation services

- Infuse and transfer technology

19. PPL-21, Goddard Space Flight Center Preferred Parts List (http://nepp.nasa.gov/ DocUploads/AA0D50FD-18BE-48EF-ABA2E1C4EFF2395F/ppl21notice1.pdf). This document contains a list of preferred parts, additional test requirements for preferred parts, part derating guidelines, screening requirements for nonpreferred parts, details of space radiation effects, and a list of nonpreferred parts that can be procured to GSFC specifications.

20. Aerospace Report No. TOR-2006(8583)-5236, *Technical Requirements for Electronic Parts, Materials, and Processes Used in Space and Launch Vehicles*, November 13, 2006. This document establishes the minimum technical requirements for electronic parts, materials, and processes (electronic PMP) used in the design, development, and fabrication of space and launch vehicles. Application information, design and construction information, and quality assurance provisions are provided.

21. NPR 8735.1, GIDEP Notifications and NASA Advisories (http://www.gidep.org/). The Government-Industry Data Exchange Program (GIDEP) is an information-sharing program to ensure that only reliable, quality parts are used on all government programs and operations. The objective of this policy is to ensure that information about nonconforming or defective items in use at NASA are identified and shared among NASA facilities and with GIDEP as appropriate. This document is intended for use in

acquisition of space and launch vehicles. This document should be cited in the contract statement of work and may be tailored by the acquisition activity for the specific application or program.

The following NASA standards are for certification of technicians to workmanship standards. Typically NASA will levy these requirements on a contractor who is building space-qualified hardware or mission-critical equipment.

22. NASA STD 8739.3, *Soldered Electrical Connections* (http://www.hq.nasa.gov/office/ codeq/doctree/87393.htm). This standard sets forth requirements for hand and wave soldering to obtain reliable electrical connections. The prime consideration is the physical integrity of solder connections. This publication applies to NASA programs involving soldering connections for flight hardware, and mission critical ground support equipment; it does not define the soldering requirements for surface-mount technology (SMT).

    • Prescribes NASA's process and end-item requirements for reliable, soldered electrical connections.
    • Establishes responsibilities for training personnel.

    • Establishes responsibilities for documenting process procedures including supplier innovations, special processes, and changes in technology.

    • For the purpose of this standard, the term "supplier" is defined as in-house NASA, NASA contractors, and subtier contractors.

23. NASA STD 8739.2, *NASA Workmanship Standard for Surface Mount Technology* (http:// www.hq.nasa.gov/office/codeq/doctree/87392.htm). This standard sets forth NASA's requirements, procedures, and documenting requirements for hand and machine soldering of surface-mount electrical connections. It is a complement to NASA STD 8739.3 described above.

24. NASA STD 8739.1, *A Workmanship Standard for Polymeric Application on Electronic Assemblies* (http://www.hq.nasa.gov/office/codeq/doctree/87391.htm). This standard sets forth NASA's technical requirements, procedures, and documentation requirements for polymeric applications for staking, conformal coating, bonding, and encapsulation of components used in electronic hardware.

25. NASA STD 8739.4, *Crimping, Interconnecting Cables, Harnesses, and Wiring* (http:// www.hq.nasa.gov/office/codeq/doctree/87394.htm). This standard provides a baseline for NASA project offices to use when preparing or evaluating process procedures for the manufacture of harnesses and cabling, including crimping of connector pins, for spaceflight hardware or mission-critical ground support equipment.

    • Prescribes NASA's process and end-item requirements for reliable crimped connections, interconnecting cables, harnesses, and wiring.

- Establishes responsibilities for training personnel.

- Establishes responsibilities for documenting process procedures including supplier innovations, special processes, and changes in technology.

- For the purpose of this standard, the term "supplier" is defined as in-house NASA, NASA contractors, and subtier contractors.

1.1.5. Integration and Test

1. MIL-STD-1540B *Military Standard Test Requirements for Space Vehicles* (http://www. everyspec.com/MIL-STD/MIL-STD+(1500+−+1599)/MIL-STD-1540B-_MILITARY_STANDARD_TEST_REQUIREMENTS_FOR_SPACE_VEHICLES_2539/). This standard establishes the environmental and structural ground testing requirements for launch vehicles, upper-stage vehicles, space vehicles, and for their subsystems and units. (Draft E is the latest as of December 2002 but it has ITAR restricted access.)

2. Aerospace Report No. TR-2004(8583)-1, Rev. A, Perl E, ed., *Test Requirements for Launch, Upper-Stage, and Space Vehicles*, September 6, 2006 (http://www.everyspec.com/USAF/TORs/download.php?spec=TR-2004(8583)-1_REV_A.00000936.pdf). This standard establishes the environmental and structural ground testing requirements for launch vehicles, upper-stage vehicles, space vehicles, and their subsystems and units. In addition, a uniform set of definitions of related terms is established.

3. GSFC-STD-7000, *General Environmental Verification Standard (GEVS)* (http://www. goes-r.gov/procurement/flight_documents/GSFC-STD-7000.pdf).

4. *Electromagnetic Effects and Spacecraft Charging* (http://see.msfc.nasa.gov/ee/eepub. htm, NASA/TP-2003-212287). This document is intended as a design guideline for high-voltage, space power systems (>55 volts) that must operate in the plasma environment associated with low earth orbit (LEO). Such power systems, particularly solar arrays, may interact with this environment in a number of ways that are potentially destructive to themselves as well as to the platform or vehicle that has deployed them.

> The first objective is to present an overview of current understanding of the various plasma interactions that may result when a high voltage system is operated in the Earth's ionosphere. A second objective is to reference common design practices that have exacerbated plasma interactions in the past and to recommend standard practices to eliminate or mitigate such reactions.

5. MIL-STD-461E, *Control of Electromagnetic Interference (EMI) Characteristics of Subsystems and Equipment*. This standard covers electromagnetic effects that are both conducted and radiated. Each area addresses specific modes, either emissions or susceptibility, and bandwidths. Chapter 6 in this volume on military development and best practices has more information on MIL-STD-461.

6. *Environmental Compliance/Launch Approval Status System* (http://www.teerm.nasa.gov/ Environmental_EnergyConference2008_files/3Van%20Damme%20Final%20JPL% 20Tools%20for%20NEPA%20Compliance%209-23-08.pdf). Flight project practices (FPPs) establish requirements/processes for satisfying NASA imposed agency-wide requirements (e.g., NPR 8580.1 and NPR 7120.5D). Launch-approval engineering FPP establishes following requirements (i.e., "gate products" associated with standard project milestones) to ensure timely NEPA (and associated) compliance by JPL flight projects:

    *   Environmental compliance and launch-approval status system (ECLASS) form at mission concept review (MCR)

    *   Launch-approval engineering plan at project mission system review (PMSR)

    *   Final NEPA document at preliminary design review (PDR)

7. *Nuclear Safety Launch Approval* (http://pbma.nasa.gov/framework_content_cid_493).

1.1.6. Mission Operations

Mission operations are particular to each spacecraft, launch, orbit, and mission. Several samples of concerns and issues with mission operations follow:

1. NPD 8700.1E *NASA Policy for Safety and Mission Success* (http://nodis3.gsfc.nasa.gov/ displayDir.cfm?t=NPD&c=8700&s=1E).

2. 20060013538 NASA Johnson Space Center, Houston, TX, *Lunar Surface Mission Operations Scenario and Considerations* (http://aero-defense.ihs.com/news/star-06H1/ star-0620-lunar-planetary-science-exploration.htm).

    Planetary surface operations have been studied since the last visit of humans to the moon, including conducting analog missions. Mission operations lessons from these activities are summarized. Characteristics of forecasted surface operations are compared to current human mission operations approaches. Considerations for future designs of mission operations are assessed.

3. *Code S Mission Operations Mission Management Plan*, Rev. 8 (http://www.ssmo_home. hst.nasa.gov/SSMO_Best_Practices_010705/Code%20S%20Mission%20Management%

20Plan%20(Rev.8).doc). This mission management plan (MMP) provides a high-level description of the manner in which the mission operations and mission services (MOMS) contractor will manage the mission operations for those task orders (TOs) that pertain to space sciences missions. This MMP describes our approach to mission management, mission reporting, staffing, training and certification, risk management and best practices, configuration management, IT security, and maintenance of mission and technical records.

4. *Proceedings of the SpaceOps 2008 Conference*, May 15, 2008 (http://www.aiaa.org/agenda.cfm?lumeetingid=1436&formatview=1&dateget=15-May-08).

1.1.7. Summary of NASA-Developed Technical Standards

Table 5.2 contains a summary of NASA technical standards that can be found at the following website: http://standards.nasa.gov/documents/nasa.

1.1.8. Summary of Military Aerospace Standards

Table 5.3 has a summary of military aerospace technical standards; some of these can be found at the following website: http://snebulos.mit.edu/projects/reference/MIL-STD/index.html.

1.2. Industry Standards

Even now industry standards for commercial spacecraft are either proprietary or broadly applicable to many industries. What follows are some of the more prominent standards that may figure into future commercial space programs.

1.2.1. Project Management

Guide to Program Management Body of Knowledge (PMBOK), Project Management Institute (http://www.pmi.org/Resources/Pages/Library-of-PMI-Global-Standards-projects.aspx). A global standard for the industry, which can help project management practitioners prepare for credential examinations, or assist organizations in creating and shaping their project management system. The PMBOK® Guide is not designed to function as a step-by-step, how-to book, but rather to identify that subset of the project management body of knowledge that is generally recognized as good practices. The Fourth Edition continues to reflect the evolving knowledge within the profession of project management. Like previous editions, it represents generally recognized good practice in the profession.

Table 5.2: Summary of NASA Standards

Document Number	Document Title
NASA-GB-8719.13	NASA Software Safety Guidebook
NASA-HDBK-1001	Terrestrial Environment (Climatic) Criteria Handbook for Use in Aerospace Vehicle Development
NASA-HDBK-4001	Electrical Grounding Architecture for Unmanned Spacecraft
NASA-HDBK-4002	Avoiding Problems Caused by Spacecraft On-Orbit Internal Charging Effects
NASA-HDBK-4006	Low Earth Orbit Spacecraft Charging Design Handbook
NASA-HDBK-5010	Fracture Control Implementation Handbook for Payloads, Experiments, and Similar Hardware
NASA-HDBK-5300.4(3J)	NASA Handbook Requirements for Conformal Coating and Staking of Printed Wiring Boards for Electronic Assemblies
NASA-HDBK-6003	Application of Data Matrix Identification Symbols to Aerospace Parts Using Direct Part Marking Methods/Techniques (Supersedes NASA-HDBK-6003b)
NASA-HDBK-6007	Handbook for Recommended Material Removal Processes for Advanced Ceramic Test Specimens and Components
NASA-HDBK-7004	Force Limited Vibration Testing
NASA-HDBK-7005	Dynamic Environmental Criteria
NASA-HDBK-8719.14	Handbook for Limiting Orbital Debris
NASA-HDBK-8739.18	Procedural Handbook for NASA Program and Project Management of Problems, Nonconformances, and Anomalies
NASA-SPEC-5004	Welding of Aerospace Ground Support Equipment and Related Nonconventional Facilities
NASA-STD-(I)-5005	Standard for the Design and Fabrication of Ground Support Equipment
NASA-STD-0005	NASA Configuration Management (Cm) Standard
NASA-STD-2202	Software Formal Inspections Standard
NASA-STD-2202-93	Software Formal Inspections Standard
NASA-STD-2818	Digital Television Standards for NASA
NASA-STD-3000 VOL I	Man-Systems Integration Standards, vol. I
NASA-STD-3000 VOL II	Man-Systems Integration Standards, vol. II
NASA-STD-3000 VOL III	Man-Systems Integration Standards, vol. III
NASA-STD-3001 VOL I	NASA Spaceflight Human System Standard, vol. 1: Crew Health (Superseding NASA-STD-3000, vol. 1, Chapter 7; and JSC 26882, Spaceflight Health Requirements Document)
NASA-STD-4003	Electrical Bonding for NASA Launch Vehicles, Spacecraft, Payloads, and Flight Equipment

(Continues)

Table 5.2:   Cont'd

Document Number	Document Title
NASA-STD-4005	Low Earth Orbit Spacecraft Charging Design Standard (Supersedes NASA-STD-(I)-4005)
NASA-STD-5001	Structural Design and Test Factors of Safety for Spaceflight Hardware
NASA-STD-5002	Load Analyses of Spacecraft and Payloads
NASA-STD-5003	Fracture Control Requirements for Payloads Using the Space Shuttle
NASA-STD-5005	Ground Support Equipment (Superseding NASA-STD-5005a)
NASA-STD-5006	General Fusion Welding Requirements for Aerospace Materials Used in Flight Hardware
NASA-STD-5007	General Fracture Control Requirements for Manned Spaceflight Systems
NASA-STD-5008	Protective Coating of Carbon Steel, Stainless Steel, and Aluminum on Launch Structures, Facilities, and Ground Support Equipment
NASA-STD-5009	Nondestructive Evaluation Requirements for Fracture Critical Metallic Components
NASA-STD-5012	Strength and Life Assessment Requirements for Liquid Fueled Space Propulsion System Engines
NASA-STD-5017	Design and Development Requirements for Mechanisms
NASA-STD-5019	Fracture Control Requirements for Spaceflight Hardware (Superseding NASA-STD-(I)-5019 (interim) and NASA-STD-5007)
NASA-STD-6001	Flammability, Odor, Off-gassing and Compatibility Requirements and Test Procedures for Materials in Environments That Support Combustion
NASA-STD-6002	Applying Data Matrix Identification Symbols on Aerospace Parts (Superseding NASA-STD-6002c)
NASA-STD-6008	NASA Fastener Procurement, Receiving Inspection, and Storage Practices for Spaceflight Hardware
NASA-STD-6016	Standard Materials and Processes Requirements for Spacecraft
NASA-STD-7001	Payload Vibroacoustic Test Criteria
NASA-STD-7002	Payload Test Requirements
NASA-STD-7003	Pyroshock Test Criteria
NASA-STD-7009	Standard for Models and Simulations
NASA-STD-8709.2	NASA Safety and Mission Assurance Roles and Responsibilities for Expendable Launch Vehicle Services; Revalidated/Reaffirmed 08/21/2003
NASA-STD-8719.10	Standard for Underwater Facility and Non–Open Water Operations
NASA-STD-8719.11	Safety Standard for Fire Protection
NASA-STD-8719.13	NASA Software Safety Standard (Rev B W/Ch1 of 7/8/2004)

(Continues)

Table 5.2: Cont'd

Document Number	Document Title
NASA-STD-8719.14	Process for Limiting Orbital Debris (Baseline W/Ch 1 of 9/6/07)
NASA-STD-8719.17	NASA Requirements for Ground-Based Pressure Vessels and Pressurized Systems (Pv/S)
NASA-STD-8719.7	Facility System Safety Guidebook
NASA-STD-8719.9	Standard for Lifting Devices and Equipment; Revalidated/Reaffirmed 10/01/2007
NASA-STD-8729.1	Planning, Developing and Managing an Effective Reliability and Maintainability (R&M) Program
NASA-STD-8739.1	Workmanship Standard for Polymeric Application on Electronic Assemblies
NASA-STD-8739.2	Workmanship Standard for Surface Mount Technology (Baseline with Chapter 1 of 6/6/08); Revalidated/Reaffirmed 06/05/2008
NASA-STD-8739.3	Soldered Electrical Connections (Baseline with Chapter 3 of 6/6/08)
NASA-STD-8739.4	Crimping, Interconnecting Cables, Harnesses, and Wiring (Baseline with Chapter 4 of 7/25/08)
NASA-STD-8739.5	Fiber Optic Terminations, Cable Assemblies, and Installation (Baseline with Chapter 1 of 7/25/08)
NASA-STD-8739.8	Software Assurance Standard (Baseline with Chapter 1 of 5/5/05)
NSS-1740.12	Safety Standard for Explosives, Propellants, and Pyrotechnics
NSS-1740.14	NASA Safety Standard Guidelines and Assessment Procedures for Limiting Orbital Debris

Table 5.3: Summary of Military Standards That Might Apply to Spacecraft

Document Number	Vol.	Rev.	Document Title
MIL-HDBK-5		J	Metallic Materials and Elements for Aerospace Vehicle Structures
MIL-HDBK-17			Composite Materials Handbook
	1	F	Polymer Matrix/Guidelines for Characterization
	2	F	Polymer Matrix Composites, Vol. I: Guidelines for Characterization of Structural Materials
	3	F	Polymer Matrix/Materials Usage, Design and Analysis
	4	A	Metal Matrix Composites
	5		Ceramic Matrix Composites

(Continues)

Table 5.3:   Cont'd

Document Number	Vol.	Rev.	Document Title
MIL-HDBK-217		F	Reliability Prediction of Electronic Equipment
MIL-HDBK-263		B	Electrostatic Discharge Control Handbook for Protection of Electrical and Electronic Parts
MIL-HDBK-340			Application Guidelines for MIL-STD-1540
MIL-HDBK-343			Design, Construction, and Test Requirements for One-of-a-Kind Spacecraft
MIL-HDBK-454		A	Standard General Requirements for Electronic Equipment
MIL-HDBK-814			Ionizing Dose and Neutron Hardness Assurance Guidelines for Microcircuits and Semiconductor Devices
MIL-HDBK-1547		A	Electronic Parts, Materials, and Processes for Space Launch Vehicles
MIL-HDBK-83377			Requirements for Adhesive Bonding for Aerospace and Other Systems
MIL-A-83376		A	Acceptance Criteria for Adhesive Bonded Metal Faced Sandwiched Structures
MIL-A-83577		B	General Specifications for Moving Mechanical Assemblies for Space and Launch Vehicles
MIL-M-38510		J	General Specification for Microcircuits
MIL-P-50884		D	General Specification for Printed Wiring Boards
MIL-PRF-13830		B	Optical Component Inspection
MIL-PRF-19500		M	Performance Specification, Semiconductor Devices
QML-19500		22	Qualified Manufacturers List
MIL-PRF-31032		A	General Specification for Printed Circuit Board/Printed Wiring Board
MIL-PRF-38534		E	Performance Specification, Hybrid Circuits
QML-38534		48	Qualified Manufacturers List
MIL-PRF-38535		F	Performance Specification, Integrated Circuits
QML-38535		17	Qualified Manufacturers List
MIL-STD-202		G	Test Methods for Electronic and Electric Component Parts
MIL-STD-275		E	Printed Wiring for Electronics Equipment
MIL-STD-461		E	Control of Electromagnetic Interference
MIL-STD-462			Measurement of Electromagnetic Interference Characteristics
MIL-STD-750		D, -5	Test Method for Semiconductor Devices

(Continues)

Table 5.3:   Cont'd

Document Number	Vol.	Rev.	Document Title
MIL-STD-810		F	Test Method Standard for Environmental Engineering Tests
MIL-STD-883		F, -5	Test Methods and Procedures for Microelectronics
MIL-STD-889		B	Dissimilar Metals
MIL-STD-975		M	NASA Standard Electrical, Electronic, and Electromechanical Parts List
MIL-STD-1246		C, -4	Cleanliness Levels
MIL-STD-1285		D	Marking of Electrical and Electronic Parts
MIL-STD-1540		E	Test Requirements for Launch and Space Vehicles (restricted access)
MIL-STD-1686		C	Electrostatic Discharge Control Program for Protection of Electrical and Electronic Parts, Assemblies and Equipment
MIL-STD-2000		A	Solder Technology, High Quality, High Reliability

The American National Standards Institute/Electronic Industries Alliance, Standard 748-B, *Earned Value Management Systems* (ANSI/EIA-748). The standard contains guidelines and common terminology for earned value management systems (EVMS). It also contains a discussion on the EVMS process, system documentation, and system evaluation sections that are informative sections providing application and implementation insight. Earned value management (EVM) is a technique for measuring project progress (http://en. wikipedia.org/wiki/Earned_value_management). EVM combines measurements of schedule and cost into integrated metrics that can give early warning of performance problems. EVM promises to improve the tracking of progress of the project and to keep the project team focused on achieving progress.

1.2.2. Systems Engineering

International Council on Systems Engineering (INCOSE) (http://www.incose.org/ ProductsPubs/products/sehandbook.aspx), *Systems Engineering Handbook*, Version 3, of *INCOSE Systems Engineering Handbook*. This handbook represents a shift in paradigm toward global industry application consistent with the systems engineering vision. Developed for the new systems engineer, the engineer in another discipline who needs to perform systems engineering or the experienced systems engineer who needs a convenient reference, the handbook provides an updated description of key process activities performed by systems engineers.

The descriptions in this handbook show what each systems engineering process activity entails, in the context of designing for affordability and performance. On some projects, a given activity may be performed very informally (e.g., on the back of an envelope, or in an engineer's notebook); on other projects, activities are performed very formally, with interim products under formal configuration control. This document is not intended to advocate any level of formality as necessary or appropriate in all situations.

ISO/IEC 15288:2008 (E) and IEEE Std 15288-2008. (See Chapter 6 in this volume on systems engineering in military projects.)

1.2.3. Fault Protection

Chapter 9, Long-Life Systems, pp. 671–690, in Siewiorek DP, Swarz RS, *Reliable Computer Systems: Design and Evaluation,* 3rd edition, contains several case studies that are instructive [1].

Jackson B, A Robust Fault Protection Strategy for a COTS-Based Spacecraft, 2007 IEEE Aerospace Conference. "This paper presents a robust fault protection strategy for a low-cost single-string spacecraft that makes extensive use of COTS components. These components include commercial processors and microcontrollers that would traditionally be considered inappropriate for use in space. By crafting an avionics architecture that employs multiple distributed processors, and coupling this with an appropriate fault protection strategy, even a single-string COTS-based spacecraft can be made reasonably robust. The fault protection strategy is designed to trap faults at the highest possible level while preserving the maximum amount of spacecraft functionality, and can autonomously isolate and correct minor faults without ground intervention. For more serious faults, the vehicle is always placed in a safe configuration until the ground can diagnose the anomaly and recover the spacecraft. This paper will show how a multi-tiered fault protection strategy can be used to mitigate the risk of flying COTS components that were never intended for use in the space environment" [2].

1.2.4. Mission Assurance and Safety

Quality Management Systems (QMSs)—The primary systems are ANSI/ISO/ASQ 9001 and AS9100 *Quality Management System for Aerospace Industry*; CMMI can also be used, particularly if the system has much software development. See the first chapter for more discussion about QMSs.

Company specific elements might include the following:

- Project safety evaluation checklist

- System safety program plan

- Safety assessment report

- Hazard report

- Requirements compliance assessment

1.2.5. Integration and Test

EMI/EMC—MIL-STD-461E is a rigorous standard, but it may not cover extreme frequencies or it may be too difficult to meet for some applications. Commercial standards for EMC are addressed in reference [3]. There are five primary bodies that generate relevant standards: the International Electrotechnical Commission (IEC), CISPR, the European Committee for Electrotechnical Standardization (CENELEC), the European Telecommunications Standards Institute (ETSI), and the Federal Communications Commission (FCC). The international community has worked to harmonize standards around IEC 61000-x set of standards.

1.2.6. Miscellaneous Industrial and Commercial Standards

American Society for Testing and Materials (ASTM)

> ASTM E-595 2005, Standard Test Method for Total Mass Loss and Collected Volatile Condensable Materials from Outgassing in a Vacuum Environment

The Institute for Interconnecting and Packaging Electronic Circuits (IPC):

> IPC-6011 1996, Generic Performance Specification for Printed Boards

> IPC-6012 B 2004, Qualification and Performance Specification for Rigid Printed Boards

> IPC-6013 A 2003, Qualification and Performance Specification for Flexible Printed Boards

> IPC-6018 A 2002, Microwave End Product Board Inspection and Test

J-STD-001 C 2000, Requirements for Soldered Electrical and Electronic Assemblies

J-STD-004 A 2004, Requirements for Soldering Fluxes

J-STD-020 B 2002, Moisture/Reflow Sensitivity Classification for Nonhermetic Solid State Surface Mount Devices

J-STD-033 A 2002, Handling, Packaging, Shipping and Use of Moisture/Reflow Sensitive Surface Mount Devices

1.3. Commercial Off-the-Shelf

As more commercial ventures explore and exploit spaceflight, opportunities grow for the use of commercial off-the-shelf (COTS) modules and subsystems. Commercial components can greatly reduce cost and speed development. COTS provides access to a wide variety of high-performance components; the downside is that COTS almost never is radiation hard. The use of COTS modules and subsystems in spaceflight can be appropriate for short missions and low earth orbits. One example of the use of COTS in spacecraft was the LCROSS mission in 2009; another was the MiTEx mission in 2006.

"LCROSS is a fast-paced, low-cost, mission that leverages select NASA flight-ready systems, commercial-off-the-shelf components ..." [4]. The LCROSS mission will have one portion of the spacecraft to observe another part of the spacecraft impact the moon; the goal is to identify substances, particularly water, on the moon. "The LCROSS science payload consists of two near-infrared spectrometers, a visible light spectrometer, two mid-infrared cameras, two near-infrared cameras, a visible camera, and a visible radiometer [4]".

The Micro-Satellite Technology Experiment, or MiTEx, launched on June 21, 2006 into an elliptical geosynchronous transfer orbit. The Defense Advanced Research Projects Agency, Air Force, and Navy collaborated on this space mission to test technologies that could be incorporated in future military programs. Two major goals for MiTEx were to:

- "Investigate and demonstrate advanced space technologies such as lightweight power and propulsion systems, avionics, and spacecraft structures; commercial-off-the-shelf processors; affordable, responsive fabrication/build-to-launch techniques; and single-string components

- Demonstrate a one-year lifetime for small satellites built using these new technologies and techniques" [5]

The biggest problem with using COTS components is surviving radiation during spaceflight. This is a unique problem because vendors do not perform the reliability and radiation hardness analysis (RHA) on COTS components. As Barnes indicated, you cannot even leverage experience from other high-reliability users like the automotive industry because the total ionizing dose (TID) response depends on the specific fabrication process for the integrated circuits and single-event effects (SEE) depend on circuit design and its dimensions. Furthermore, packaging makes RHA hard to establish: analysis for SEE is difficult on plastic components and multichip modules are difficult to test. Finally, vendors typically change a fabrication process for integrated circuits, which then can reduce radiation tolerance, without informing customers of the changes or of the impacts to their applications [6]. There is no way of predicting radiation response for COTS components without testing them in an ionizing chamber, a nuclear reactor, or an ion beam accelerator.

Barnes suggests the following means to use COTS components for spacecraft:

- Establish RHA with radiation testing

- Disseminate radiation data to designers so they can use it early in project cycle

- Use various shielding techniques

- Use software and hardware mitigation methods

- Use modified commercial designs that are more radiation tolerant [6]

For missions that avoid areas of concentrated radiation, such as the South Atlantic Anomaly, COTS components stand a better chance for survival. If cost and reduced development time drive a mission, then you might tailor the mission to reduce radiation exposure; you can also use a combination of the suggested techniques, such as shielding and mitigation methods to reduce the effects of radiation.

Two other problems with COTS components are the use of prohibited substances (e.g., tin, zinc, cadmium) and counterfeit components. High tin content, particularly in solder, can lead to tin whisker growth and short circuits; you will need special equipment to detect these prohibited materials. Counterfeit electronic components, such as ICs, capacitors, and resistors, with poor quality standards have slipped into the supply chain and have caused major problems when component lots suffer widespread failures; signing up with the Government–Industry Data Exchange Program (GIDEP) can help you identify counterfeit or bad components.

Goodman's summary for a project using COTS subsystems echoes similar experiences found in projects for unmanned spacecraft. "The Space Shuttle Program procured 'off the shelf' GPS and EGI units with the expectation that procurement, development, certification and operational costs would be significantly reduced. However, these projects consumed more budget and schedule than originally anticipated. Numerous and significant firmware changes were required to adapt these units for use on space vehicles. The promise of COTS products is most likely to be fulfilled when the intended application is close to or matches that for which the COTS product was originally designed. Independent verification and validation of receiver software, availability of receiver technical requirements to the Shuttle Program, open and frequent communication with the vendor, design insight and a rigorous process of receiver testing, issue investigation and disposition were keys to resolving technical issues with a complex unit. Modification of an aviation navigation unit for a space application should be treated as a development project, rather than as a 'plug and play' project under a fixed-price contract" [7]. Although this was written in 2002, it is still true, in large part, for current projects. A modified approach to COTS, where products are largely off-the-shelf but have the flexibility for some changes, still has promise to reduce cost and development time.

2. Company Processes

This section covers the processes relating to the regulations and standards listed previously, giving examples of how company processes comply with these requirements. Maintaining a consistent set of processes within an organization while serving the needs of many different customers often requires developing processes that cover the requirements from multiple sources. For example, NASA's NPR 7120.5D calls for a particular review, but a DoD standard calls for another review that may have slightly different objectives or components. By developing an internal process that accommodates the requirements of both sets of these standards, a company can maintain consistency across projects within the organization.

To handle situations where a customer has a specific need or requirement not covered by a company's standard process, a project may have to deviate from a prescribed process. In cases where a project has a legitimate reason for following a different process, it is necessary to obtain a waiver or deviation. A method for obtaining waivers or deviations is an essential part of any formal process management system. An example of such a process is also described in this section.

This section begins with project management, systems engineering, and fault protection processes. Subsequent topics include mission assurance, integration, test, and mission operations.

2.1. Project Management

Good project management forms the foundation of a successful project. Nearly all aspects of the implementation of a space mission fall under the responsibility and authority of the project manager. Therefore, establishing good project management processes is critical to mission success.

Most space mission projects begin with a notice from a sponsor in the form of an announcement of opportunity or a request for proposal (RFP). In some cases the opportunity may involve the production of multiple spacecraft to support an operational space-based capability. In others it may be just a single spacecraft designed to visit unexplored worlds in our solar system. The sponsor may include requirements about the organizational structure or type of development for the implementation of the project. For example, NASA's Explorer, Discovery, Mars Scout, and New Frontiers Programs use an organizational structure in which the principal investigator (PI), a scientist, is ultimately responsible for delivering the proposed scientific results. Thus the PI has ultimate authority over the project. Such PI-led missions have been an effective way to implement successful space missions (e.g., Previous Studies on Lessons Learned from PI-Led Missions) [8]. Regardless of the type of opportunity, the solicitation notice initiates the competition between multiple organizations to propose the best solution and win the opportunity to fulfill the sponsor's need. This proposal process begins with the formation of a project team and the first phase of the project, the concept formulation.

2.1.1. Project Organization: Team Roles and Responsibilities

One of the most important tasks of the project manager is the formation of the project team. As with any competitive endeavor, a successful team starts with a talented mix of players with diverse skill sets that bond together to achieve collective success. Each player has a role to play in reaching the common goal. Forming a team that will be dedicated to the mission from the start provides continuity and knowledge retention throughout the development. With a capable team that takes a vested interest in the project and a talented project manager to coach them, the team can achieve excellence beyond their expectations.

To operate efficiently and effectively, the team requires clearly defined lines of authority. The program or project manager is responsible for establishing those lines of authority and ensuring communication within the team. Documenting and distributing the project structure in the form of an organization chart and written definitions of the roles and responsibilities ensures that the members of the project team know what to do and where to get answers. A sample organization chart is shown in Fig. 5.4 and the responsibilities of individuals in the project organization are listed in Table 5.4.

2.1.2. Communication and Teamwork

In addition to clear lines of authority, communication and teamwork are essential to the successful implementation of a project. Regularly scheduled team status meetings provide an important forum for discussion of technical and programmatic progress. Depending on the phase of the project, the size and frequency of these meetings may vary. During the concept formulation phase, weekly meetings or teleconferences involving the entire project may be

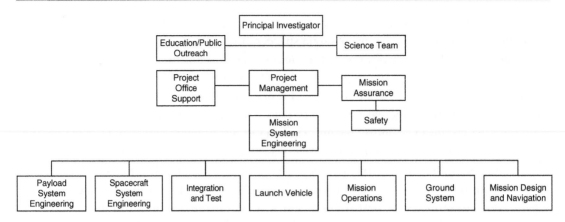

Figure 5.4: Project organization for a principal investigator (PI)–led mission.

Table 5.4: Roles and Responsibilities for a Principal Investigator (PI)–Led Mission

Role	Responsibility
Principal Investigator (PI)	The principal investigator is responsible for the overall success of the mission. He/she serves as the ultimate authority within the project for decisions that could affect the ability to deliver the science results. The PI sets the science goals for the mission and is responsible for developing the plan to meet those goals.
Project Scientist	The project scientist is responsible for implementing the science plan that was developed by the PI. He/she derives the measurement requirements from the science objectives. He/she also leads the science team and reports to the PI.
Education/Public Outreach (E/PO) Lead	The lead for E/PO is responsible for planning, developing, and coordinating programs to educate students and the public about the mission. The E/PO lead works closely with public relations regarding press releases and public events surrounding major mission milestones or accomplishments.
Project Manager (PM)	The project manager is responsible for formulating the project plan and implementing the project according to this plan. The PM reports to the PI. The PM establishes and coordinates the project office for the purpose of directing the project tasks and managing the project cost, schedule, and risk. The PM communicates the project progress/performance to the customer and conducts technical and programmatic reviews of the project.
Deputy Project Manager (DPM)	The DPM assists the PM with the project management responsibilities. In the event the PM is unable to fulfill his/her duties, the DPM may serve as the acting PM in his/her absence.

(Continues)

Table 5.4: Cont'd

Role	Responsibility
Payload Manager	The payload manager is a member of the project office and is responsible for the development of the payload. He/she works closely with the payload system engineer to deliver the required instrument performance on schedule and budget.
Project Office Support	The project office support includes a number of individuals in various areas of expertise. The project office supports the PM with skills necessary to perform scheduling, cost accounting, subcontracting, export control, and administrative tasks.
Mission Assurance Engineer (MAE)	The MAE (sometimes referred to as the performance assurance engineer) is responsible for the development and implementation of the mission assurance plan. The MAE oversees configuration management and enforces the quality standards for the project. MAE approval is required for the release of all project documentation and the closure of all issue reports.
Safety Engineer	The safety engineer reports to the MAE and is responsible for the implementation of system safety and personnel safety plans on the project. These plans include safety training, identification and mitigation of safety hazards, compliance with the range safety requirements, and developing the missile systems pre-launch safety plan (MSPSP).
Mission System Engineer (MSE)	The MSE serves as the lead technical authority on the project and reports to the project manager. He/she is responsible for developing the systems engineering management plan (SEMP) and managing the systems engineering team. The MSE handles the flow-down of the mission requirements to the mission elements. The overall requirements verification plan is also the responsibility of the MSE. The MSE conducts trade studies to evaluate various mission concepts and architecture options. He/she also monitors risks and identifies mitigations and reports recommendations to the PM. The MSE is responsible for managing all system budgets and margins on the mission. These typically include mass, power, RF link, alignment, guidance and control (G&C), data recorder space, downlink volume, etc. In some cases, tracking of these budgets may be delegated to a lead engineer. For example, the mass budget is often updated by the mechanical systems engineer.
Deputy Mission System Engineer (DMSE)	The DMSE assists the MSE in completion of mission system engineering tasks. In the absence of the MSE, the DMSE may serve as acting MSE.
Mission Software System Engineer (MSSE)	The MSSE is part of the mission systems engineering team. He/she is responsible for establishing the process and standards for all software development on the project, including both flight and ground applications.

(Continues)

Table 5.4: Cont'd

Role	Responsibility
Fault Protection Engineer (FPE)	The FPE, also part of the mission systems engineering team, is responsible for the development, implementation, and verification of the fault protections requirements. The FPE also coordinates the development of the onboard fault detection and autonomous responses.
Payload System Engineer (PSE)	The PSE is responsible for the requirements and verification of the instruments. He/she is responsible for coordinating with the MSE regarding the instrument interfaces and reports to the payload manager.
Instrument Lead Engineers	Instrument lead engineers are responsible for the technical development, cost, and schedule of their respective instruments. They report to the PSE.
Spacecraft System Engineer (SSE)	The SSE is responsible for the technical aspects of the spacecraft segment. He/she documents the requirements flow-down from the segment to the subsystems and is responsible for the verification of these requirements.
Subsystem Lead Engineers	Subsystem lead engineers are responsible for technical development, cost, and schedule of the spacecraft subsystem development. The lead engineer assignments are dependent on the spacecraft architecture. For example, some spacecraft may not have a propulsion system and therefore would not require a propulsion lead engineer. Common designations for subsystem lead engineers include structural, mechanical, thermal, power, command and data handling (C&DH), RF, G&C, propulsion, and flight software. Each lead engineer has specific responsibility with regard to her/his respective subsystem.
Launch Vehicle Coordinator (LVC)	The LVC is the project point-of-contact for all launch vehicle–related activities. This person participates in all meetings of the ground operations working group and the payload safety working group. The LVC is also involved in the trajectory review cycle and the coupled loads analysis. He/she is responsible for developing the launch vehicle interface requirements document, which documents the mission-unique requirements for the project.
Integration and Test (I&T) Lead Engineer	The I&T lead engineer is responsible for the integration and testing of the spacecraft and instruments. She/he leads the I&T team and coordinates the schedule of activities involving the spacecraft through launch.
Mission Operations Manager (MOM)	The MOM is responsible for managing the mission operations team and coordinating the in-flight activities on the spacecraft. He/she provides status updates for the project during the operations phase of the mission.
Ground System Engineer	The ground system engineer is responsible for the development of the hardware and software that makes up the ground system. She/he maintains the command and telemetry database and manages the development of many of the ground system software tools used by mission operations to perform the mission.
Mission Design Lead	The mission design lead is responsible for the development of the mission trajectory (or orbit) and associated parameters, such as launch date, C_3 requirements, arrival date, and delta-V budget. He/she works with the LVC and LV provider in the trajectory review cycles.

appropriate. During critical periods of integration and test, daily meetings of the integration and test (I&T) team are necessary. These meetings have peripheral benefits, too. In some cases, the meetings may spawn additional impromptu conversations afterwards leading to the resolution of other issues or potential problems.

When communications between team members can't wait until the next scheduled meeting, knowing who to call and how to reach them is invaluable for solving problems quickly. Since some phases of spacecraft development, such as I&T, require work outside normal business hours, the ability to reach someone after hours could mean the difference between progressing to the next activity and losing a day of schedule reserve. For this reason, every project should maintain an up-to-date contact list, including the individual's role and a work, home, and mobile phone number.

Another essential aspect of project communication is open dialog with the customer. Keeping the customer or sponsor informed of current progress, even when the news is bad, helps build a trusting relationship. Frequent communication can be used to alert the customer of issues which may be beyond the project's control but that the customer could influence. For example, when the New Horizons project experienced an issue with a vendor component, NASA was able to change the delivery sequence of similar components from that vendor to another NASA project to minimize the schedule impact.

Various types of project communication are listed in Table 5.5. Internal communication types include: team meetings, technical interchange meetings, working group meetings, integration and test meetings (daily during I&T), website/network directories, and anomaly reports. Communications with the customer include project reviews and progress/status reports.

2.1.3. Work Breakdown Structure

The work breakdown structure (WBS) is a hierarchical representation of the work required to produce the products and services necessary to complete the project. Dividing the work into smaller deliveries helps facilitate the management of cost, schedule, and risk. NASA's prescribed work breakdown structure as required by NPR 7120.5D is shown in Fig. 5.5.

The top level of the WBS represents the entire spaceflight project. The second level represents divisions at the segment level. The third level (not shown) is typically divided at the subsystem level for the spacecraft. Some segments, such as the spacecraft and payload, may lend themselves to a deeper level of division than other segments. There is flexibility to add elements to the WBS for products or services not covered in the existing elements. The work covered in each element is explicitly described in the WBS dictionary that

Table 5.5: Example of Project Communication

Communication	Recommended Frequency	Description
Progress Reports	Weekly/monthly	Informal weekly status provided by the leads. Monthly progress reported to customer.
Status Meetings	Weekly	
Technical Interchange Meetings	Quarterly	Meetings to exchange information regarding interfaces between instruments and spacecraft. During development these are useful on a quarterly basis.
Working Group Meetings	Twice per year	Payload safety working group and ground operations working group meetings are held for the launch site and the project to communicate safety requirements and launch site support requirements.
Project Reviews	TBD	The timing reviews depends on the complexity of the project and on the sponsor. Typically an important design review occurs at the end of each phase. The sponsor may require intermediate project reviews.
Integration and Test Meetings	Daily/weekly	Again, this depends on the timing. During the integration phase, meetings probably should be much more frequent to cover late-breaking news and events that arise during tests.
Anomaly Reports	As needed	This is project-specific.

accompanies the WBS. Every WBS must have a WBS dictionary that defines in prose the scope of work involved in each element. The WBS dictionary for the elements in Fig. 5.5 can be found in Appendix G of NPR 7120.5D.

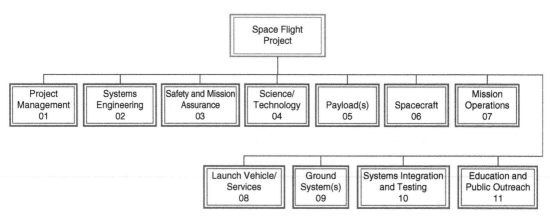

Figure 5.5: Example of a high-level WBS for spaceflight projects.

2.1.4. The Project Schedule

The project schedule is necessary for tracking the progress of the development to identify problems early while there is time to correct them. The schedule tracks the progress of the project using the same breakdown of tasks as the WBS. By tracking the work as it is completed along with the accounting of expenditures, the project can use the schedule to closely monitor the financial status of the project.

An effective method for measuring progress is earned-value management (EVM). The last chapter on systems engineering in military projects begins to address EVM. Good resources exist for EVM, such as the book by Eric Verzuh [9].

Commercial off-the-shelf (COTS) software tools for managing a large development schedule are readily available. Microsoft Project® and Primevera® are two examples. When multiple organizations are involved in a project, it may become necessary to convert file formats from different tools to update the project schedule status. Obstacles such as incompatibilities between scheduling tools could delay the incorporation of updated schedule status and potentially delay the identification of an issue. Enforcing standardization in the scheduling software can help increase the efficiency of schedule updates.

2.1.5. Project Phases

To enable the assessment of technical and programmatic progress of NASA missions, the project life cycle is divided into phases. Although NASA standards clearly define the mission phases, it can be advantageous to divide some of the longer phases into smaller periods to provide better visibility into the progress. Figure 5.6 illustrates a comparison of the project life cycle phases for various project management standards and how they are bounded by the major reviews of the project. The details of each of these phases are described in Table 5.6.

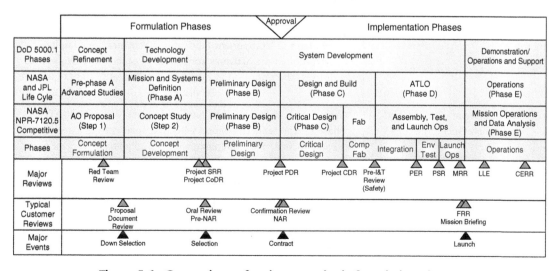

Figure 5.6: Comparison of various standards for mission phases.

Table 5.6: Project Phases for Spacecraft Development

Phase	Description
Pre-Phase A: Concept Formulation	Pre-phase A represents the period when advanced concept are studied. For NASA announcements of opportunity (e.g., for Discovery-class and New Frontiers-class missions), the development of the initial proposal is equivalent to the pre-Phase A period. During this phase, preliminary concepts are explored and trade studies conducted. This phase concludes with the production of a report or proposal describing the science objectives, measurement requirements, and the preliminary mission implementation plans.
Phase A: Concept Development	The activities conducted during Phase A include developing the technical and programmatic requirements and formulating the details of the implementation plan. For NASA announcements of opportunity, the development of concept study report is considered equivalent to Phase A.
Phase B: Preliminary Design	Phase B is the period when the preliminary design is completed and long-lead procurements are begun. Hardware breadboards of new designs are typically built and tested during Phase B. This phase concludes with the Preliminary Design Review (PDR) and the authorization to proceed from the formulation phases (Phases A/B) into the implementation phases (Phases C/D/E/F).
Phase C: Critical Design, Component Fabrication	Also referred to as "final design and fabrication," Phase C marks the beginning of the hardware development and includes the fabrication and test of components and subsystems. Critical design concludes with the critical design review (CDR), after which component fabrication begins. For many projects, Phases C and D are merged as single phase in the project life cycle (Phase C/D), because the transition from Phase C to Phase D is not necessarily marked by a single event in time. For example, the schedule may show some Phase D activities, such as the "system assembly," starting before all of the Phase C "component fabrication" activities are complete. So even though the activities conducted in each of the two phases are distinct, the transition in time between Phases C and D may not be distinct.
Phase D: Integration, Test, and Launch Operations	Also referred to as assembly, test, and launch operations (ATLO), Phase D represents the period when the spacecraft is assembled, tested, launched, and commissioned. Typically Phase D includes the 30 to 60 days of commissioning after launch.
Phase E: Operations and Sustainment	Phase E represents the primary mission operations phase. Often the mission operations phase is further divided into mission phases based on the trajectory and mission-critical events. For example, an interplanetary mission often refers to the long journey to the object of interest as the "cruise" phase. Upon arrival there would be increase

(Continues)

Table 5.6:   Cont'd

Phase	Description
	activity during the "encounter" phase or the "orbit insertion" phase, depending on whether the event was a flyby or an orbital mission. These phase boundaries often represent changes in staffing requirements or changes in the complexity of operations. At the completion of the primary mission, the mission may be closed out or continue for an extended mission.
Phase F: Closeout	Phase F marks the end of the project by implementing the system decommissioning/disposal plan. If additional data analysis is required, it may take place during this phase.

2.1.6. Review Process

Reviews take place at many levels in a project, from detailed circuit board reviews and software code walkthroughs to major mission-level, project-wide reviews. As illustrated in the previous section (see Fig. 5.6), many of the major mission-level reviews form milestones or "gates" between project phases. Customer reviews during the early phases provide input to the selection process during competitive procurements. Later in the project life cycle, technical reviews benefit the project, because they can catch problems or potential problems that may have been overlooked by the development team. Reviews also provide the project with fresh insights from experienced reviewers for potential improvements and lessons learned.

Before going into the purpose and content of individual reviews, some key elements of any effective design review should be mentioned. Every review must have a review board with a chairperson. For small reviews this board may consist of only a few people, but for mission-level reviews the board may have as many as 10 or more members. The chairperson is responsible for selecting the board and conducting the review. To provide an independent viewpoint, the board members should not be involved in the design that is being reviewed. To allow time for a comprehensive review, the design review package should be provided to the review board at least 1 week prior (2 weeks is preferable) to the start of the review. The review package should contain the agenda and a summary of the requirements and the baseline design.

During the review, the chairperson is responsible for making sure that the agenda is completed, the minutes are recorded, and the action items are captured. After the review the chairperson distributes the minutes as a record of the review, including the date, the agenda, attendees, any major decisions, and the list of action items with assignments and due dates. Finally, it is the responsibility of project team to provide timely written response to each action item.

Major milestones illustrated in Fig. 5.6 are explained in Table 5.7.

Table 5.7: Major Project Reviews for Spacecraft Development

Review	Description
Red Team Review	A red team review is an independent review of a proposal draft for the sole purpose of improving the final proposal. It should be held when a complete draft of the proposal can be assembled. The review panel should consist of experienced reviewers from various disciplines, and the panel should evaluate the draft proposal using the same evaluation criteria that the customer will use to evaluate and score the final proposal. Receiving independent feedback that identifies potential weaknesses early helps the project team to strengthen their proposal and increase the probability of an award.
Proposal Document Review	This review represents the customer evaluation of the mission proposal. This review is performed at the conclusion of the pre-Phase A effort. In NASA's announcement of opportunity process, the primary focus is on evaluation of the science, the science implementation strategy, and risk. The results of the evaluation factor into the "down-select" process. If selected, the project is awarded a contract for the next phase of the project (Phase A).
Concept Study Report Evaluation and Oral Site Visit Review	At the conclusion of the Phase A, the concept study report is evaluated according to the published evaluation criteria. The evaluation committee also conducts a site visit, during which members of the project team make oral presentations to the evaluators. The primary focus of this evaluation is to assess the risk associated with mission implementation. Results and recommendations from the evaluation team are used by the selecting official in the selection process.
System Requirements Review (SRR) Program/System Requirements Review (P/SRR)	The purpose of the SRR is to demonstrate that the decomposition of the requirements from the mission objectives is sufficient to proceed with the development of the design concept and performance specifications. This review typically occurs during Phase A. Topics covered in the SRR include the following: Mission objectives and success criteria Mission requirements Performance requirements Programmatic requirements Functional requirements of mission segments (payload, spacecraft, launch vehicle, mission operations, ground system, mission design and navigation) Results of tradeoff studies Mission drivers Contract deliverables Open trades and issues

(Continues)

Table 5.7: Cont'd

Review	Description
	Lessons learned In some cases, the SRR and the conceptual design review are combined into a single review.
Conceptual Design Review (CoDR)	The purpose of the CoDR is to assure that the proposed implementation will meet the mission requirements. The focus is on the proposed design concept and major interfaces. The CoDR should be held early enough that changes can be made without major impacts to the project. Topics covered in the CoDR include the following: Mission overview, including objectives and mission success criteria Changes since the proposal Action item closure status System performance requirements System constraints System drivers Proposed design approach for mission segments (payload, spacecraft, launch vehicle, mission operations, ground system, mission design and navigation) Major system trades Technical and programmatic interfaces Project risks and mitigations System redundancy System margins System heritage Integration and test plan Ground support equipment Mission assurance and system safety plans Lessons learned The results from the CoDR form the basis for the preliminary design (Phase B). In some cases the concept study oral site-visit review is substituted for the CoDR.
Preliminary Design Review (PDR)	The PDR is the first detailed review of the system design. The focus of the review is on demonstrating that the design meets all system requirements with an acceptable level of risk and within the cost and schedule constraints. The results from the PDR form the basis for proceeding with the detailed (i.e., critical) design (Phase C). The topics covered in the PDR include the following: Mission overview/project overview, including mission objectives, mission success criteria, launch dates, mission phases, key mission milestones) Project status, schedule, management metrics Project risks and mitigations Action item closure status

(Continues)

Table 5.7: Cont'd

Review	Description
	Changes since the last review System performance requirements and specifications System interface specifications (with preliminary interface control documents [ICDs]) System budgets and margins (mass, power, communication link, data storage, processor throughput, processor memory usage, attitude control, alignments, etc.) Spacecraft subsystem descriptions Structural design and analysis Electrical systems design and analysis Software requirements and conceptual design Fault management design Design heritage Verification plans Integration and test flow Ground system and ground system equipment (GSE) design (mechanical and electrical GSE) Launch vehicle interfaces Mission operations plans Electromagnetic interference (EMI) control plans Contamination control plans EEE parts processes, quality control, quality processes, and inspections Plans for failure modes, effects, and criticality analysis (FMECA), fault tree analysis (FTA), probabilistic risk assessment (PRA), reliability/redundancy analyses Preliminary safety analysis (hazard identification) Orbital debris assessment Documentation status Lessons learned
Confirmation Review/Non Advocate Review (NAR)	The NAR is often conducted as part of the PDR. This part of the review provides the customer with an independent assessment of the readiness of the project to proceed to the next phase of development (Phase C).
Critical Design Review (CDR)	The CDR is the most comprehensive project-level review of the detailed system design. The focus is on demonstrating that the detailed design will meet the final performance and interface specifications. At the time the CDR takes place, all actions from previous reviews should be closed, or there is a good rationale for why they remain open. Most drawings should be ready for release. The material covered should revisit all items listed for the PDR plus the following additional items:

(Continues)

Table 5.7: Cont'd

Review	Description
	Final implementation plans Detailed hardware block diagrams showing signal and power interfaces and flow Detailed software diagrams showing logic, task communication, and timing Completed design analyses (loads, stress, torque, thermal, radiation) Design and expected lifetime Engineering model/breadboard hardware/software test results Released ICDs (spacecraft, payload, launch vehicle) Test verification matrix System functional, performance, environmental test plans Ground operations (including during launch campaign) Transportation plans, shipping container design Results of the planned analyses such as FMECA, FTA, PRA, reliability/redundancy analyses Hazard analyses and safety control measures Single-point failures list Spares philosophy The results from CDR form the basis for fabrication, assembly, integration, test and launch of the flight system (rest of Phase C, Phase D).
Mission Operations Review (MOR)	The purpose of the MOR is to demonstrate that the mission operations plans are sufficient to conduct the required flight operations to achieve the mission objectives. This review is the first of two major prelaunch reviews focused primarily on the ground system and mission operations (the second is the operations readiness review, described below). The MOR is held before the major integration and test activities, such as environmental test. The MOR should include the following topics: Mission overview, including mission objectives, mission success criteria, launch dates, mission phases, key mission milestones Spacecraft overview, including subsystem descriptions Schedule status, including status of documentation and procedures Mission operations risks and mitigations Mission operations and ground system action item closure status Mission operations objectives, including launch and early operations Mission operations plans and status (payload and spacecraft) Mission operations staffing, training, and facilities Onboard data management and data flow diagrams Health and safety monitoring, contingency operations

(Continues)

Table 5.7: Cont'd

Review	Description
	Data trending
	Configuration management
	Operations functional process flow (planning, control, and assessment)
	Ground system requirements, design, and status
	Ground software
	Operational interfaces
	Flight constraints
	Prelaunch tests plans and status (tests, simulations, exercises)
	Payload operations and data analysis
	Documentation status
	Launch critical facilities
	Lessons learned
	Issues/concerns
System Integration Review (SIR)	The SIR assesses the readiness to begin integration and test of the system. This review is held before the start of integration and test. The material covered in this review include integration and test plans, verification and validation plans, status of flight system components and associated procedures, ground equipment status, and staffing plans.
Pre-Environmental Review (PER)	The purpose of the PER is to show the flight system and ground support equipment are ready for environmental test and that the plans and procedures are complete and comprehensive. This review is held before environmental test and includes the following:
	Project overview and status
	Action item closure status
	Changes since the last review
	Closure status of problems, anomaly reports, deviations, and waivers
	Test objectives, descriptions, plans, procedures, and flow
	Test facilities, equipment, instrumentation
	Pass/fail criteria
	Test verification matrix
	Contamination control safety
	Thermal vacuum profile with test
	Component-level test history/results
	Baseline comprehensive performance test results
	Failure-free operating hours
	Documentation status
	Calibration plans
	Open issues and concerns

(Continues)

Table 5.7: Cont'd

Review	Description
Preship Review (PSR)	The PSR serves to confirm that the spacecraft has successfully completed all environmental testing and is ready to be shipped to the launch site. The PSR takes place after environmental test and before shipment. The PSR should include the following: Project overview and status Action item closure status Closure status of problems, anomaly reports, deviations, and waivers Risk assessment of any open items Compliance with test verification matrix Comparison of measured margins to estimates Trending data Failure-free operating hours Transportation plans, shipping container, ground support equipment Launch site operations Status of safety approvals for launch site operations
Launch Vehicle Readiness Review (LVRR)	The LVRR certifies that the project is ready to proceed with integration of the launch vehicle with the spacecraft. Typically, the LVRR is held before the mission readiness review and is chaired by the launch services program manager.
Mission Readiness Review (MRR)	The MRR certifies that the project is ready to move the spacecraft to the launch pad.
Flight Readiness Review/Mission Briefing (FRR)	The FRR is held to close out any issues or actions from the LVRR and to certify that the launch vehicle is ready to start the launch countdown. This review is held about 3 days before the opening of the launch window and is chaired by the NASA launch manager.
Launch Readiness Review (LRR)	The LRR is held to close out any issues or actions from the FRR and certify that the project is ready to start the launch countdown. At this review the certification of flight readiness is signed. The LRR is held no later than 1 day before launch and is chaired by the space operations assistant associate administrator or may be delegated to the launch services program manager.
Operations Readiness Review (ORR)	The purpose of the OOR is to demonstrate the flight system, ground system, and operations team are ready for launch. In some cases this review may be called the mission operations readiness review (MORR) or the flight operations review (FOR). This review is held near the completion of the pre-flight testing and include the following information: Final launch, operations, and commissioning plans

(Continues)

Table 5.7:   Cont'd

Review	Description
	Mission operations procedures (nominal and contingency operations) Flight and ground system hardware and software characteristics Personnel and staffing User documentation
Lessons Learned Exercise (LLE)	LLEs are held to capture and document knowledge from experiences of the project team. This activity helps preserve knowledge and improve processes. These exercises are held periodically or after major events.
Post Launch Assessment Reviews (PLAR)	PLARs are held to assess project performance after significant events or accomplishments. For example, a commissioning review, where results of the spacecraft and instrument commissioning activities are presented, is an example of a PLAR.
Critical Event Requirements Review (CERR)	Prior to critical events, such as a trajectory correction maneuver (TCM) or other major operation, a CERR, is held to confirm that the requirements are well understood and the event can be accomplished with an acceptable level of risk.

In addition to the major project-level reviews described in Table 5.7, peer reviews of components and subassembly designs are also a key part of the technical review process. In these peer reviews, component test plans are checked to ensure that all requirements are met. At these lower levels of assembly, more of the details of the designs and tests are evaluated.

Implementing good review practices help make any review more productive. Every review should have a review chair, who is responsible for selecting the review committee and conducting the review. The review chair should be experienced in the subject matter and independent from the members of the team presenting the review. For project-level reviews, the review chair is selected from individuals outside the project and often outside the organization. For some projects, the review chair of project-level review may be a member of the sponsor organization. For subsystem-level peer reviews, the chair may be part of the project, but must be independent from the team implementing the design under review. In general, the objective of the review is to evaluate whether the design meets the requirements. To achieve that objective, the requirements must be documented and the team must have selected a single design. In advance of the meeting, the agenda and presentation material must be supplied to the review committee with sufficient time to prepare for the review. This time may vary depending on the amount of material involved in the review. Typically, a week or more is required for a project-level review. During the review, meeting minutes and

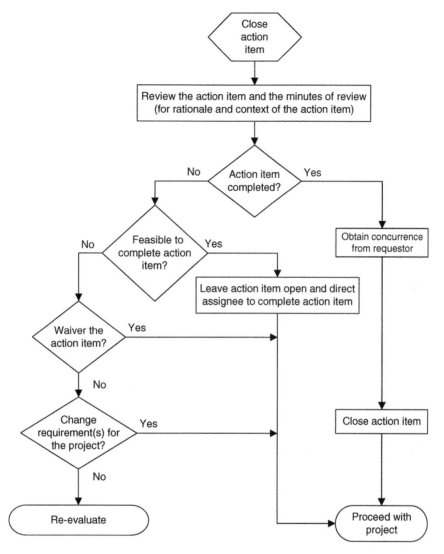

Figure 5.7: One example for closing out an action item.

action items must be recorded. After the meeting, the minutes should be distributed and the action items tracked to closure.

Closing an action item requires a procedure or process; Fig. 5.7 illustrates one example for closing an action item. An action time may lead to a minor correction or it may force a major design change or it may prove intractable and force a major change in the mission or re-evaluation. A record of all action items and their disposition should be archived in the project depository or database.

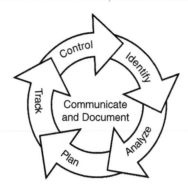

Figure 5.8: Risk management process from NASA NPR 8000.4.

2.1.7. Risk Management

The purpose of risk management is to identify potential problems early so that mitigation strategies may be implemented in time to minimize the impact to the project. The project manager is responsible for developing and implementing the risk management plan, but all members of the project team are involved in the risk management process. Risk management is a continuous process throughout all phases of the project, and it spans all aspects of the project. The risk management process, as defined in NASA NPR 8000.4, involves a number of steps in a cycle: identification, analysis, planning, tracking, controlling, communication and documentation. Figure 5.8 illustrates the risk management cycle.

Risks in production have much greater variability than operational risks. Human involvement in production introduces and maintains that variability. The risks occur in all stages of the life cycle of a product. The obvious stages are design, fabrication, assembly, test, shipping, installation, and maintenance. Some of the problems might include (but this is not necessarily a complete list):

- Late ordering and shipping of components and materials can delay delivery of the final product.

- Incorrect design and development of circuits can lead to incorrect operation of the final product.

- Incorrect design and development of software can lead to incorrect operation of the final product.

- Incorrect fabrication of materials can lead to incorrect operation of the final product.

- Incorrect assembly of components can lead to incorrect operation of the final product.

- Incorrect test procedures that miss desired or true measured values or that damage the equipment from incorrect power or signal levels.

- Mechanical damage from dropping, hitting, cutting, over flexing, bending, poking, or splashing with corrosive substances.

- Electrical short, over-voltage, over-current, or electrostatic discharge (ESD) from receiving personnel or from other subsystems connected to ecliptic subsystems.

Risk identification, the first step in the process, involves the identification and documentation of potential undesirable events and associated impacts to the project. The documentation of the risks should include all necessary information to provide a complete understanding by someone who has no prior knowledge of context of the risk. The identification of risks can come from many sources, including team members, reviewers, the lessons learned from other projects, and safety and reliability analyses.

Risk analysis (also called risk assessment) is the assignment of risk factors, on the basis of established criteria, to the *likelihood* that a risk will occur and to its *consequence* if it does occur. These criteria may vary from project to project and may depend on the type of risk being analyzed. For example, safety-related risks are often ranked with a lower tolerance on the probability of occurrence (i.e., they must be less likely to occur) than are technical or cost related risks. In addition to the likelihood and consequence, the time frame within which action is required is also documented for use in prioritizing the risks. Examples of likelihood and consequence guidance are given in Tables 5.8 and 5.9. A method for ranking risks is shown in Fig. 5.9. Risks whose likelihood and consequence put them in the far right corner (red) of the matrix in Fig. 5.9 are given the highest priority, followed by those in the center of the matrix (yellow), and finally those in the far left corner (green). Some methods for assigning risk factors include engineering judgment, statistical analysis, and probabilistic risk assessment.

Risk planning involves the development of strategies to mitigate risks and identify trigger dates when decisions must be made. For example, a new design may offer the promise of better performance, but it may also be a greater risk to the project schedule than an existing design. A mitigation strategy for the late delivery of the product could be to incorporate a decision point in the product schedule that would require a particular milestone to be met or the older, less-risky design would be built instead. In some cases a mitigation strategy might include multiple options with different trigger dates.

Table 5.8: Example of Risk Likelihood Assessment Guidance

Likelihood	Safety	Technical	Cost/Schedule
5 Very high	$(P > 10^{-1})$	$(P > 50\%)$	$(P > 75\%)$
4 High	$(10^{-2} < P \leq 10^{-1})$	$(25\% < P \leq 50\%)$	$(50\% < P \leq 75\%)$
3 Moderate	$(10^{-3} < P \leq 10^{-2})$	$(15\% < P \leq 25\%)$	$(25\% < P \leq 50\%)$
2 Low	$(10^{-4} < P \leq 10^{-3})$	$(2\% < P \leq 15\%)$	$(10\% < P \leq 25\%)$
1 Very low	$(10^{-5} < P \leq 10^{-4})$	$(0.1\% < P \leq 2\%)$	$(P \leq 10\%)$

Table 5.9: Example of Risk Consequence Assessment Guidance

Consequence	Safety	Technical	Cost	Schedule
5 Very high	Death or permanent disabling injury	Loss of spacecraft, instrument, or payload	Potential project cost overrun greater than 20%	Schedule slip greater than 3 months
4 Major	Severe injury	Loss of one or more Level 1 science requirements; major loss of capability of spacecraft, instrument, or payload	Potential project cost overrun greater than 10%	Schedule slip of 1 to 3 months
3 Medium	Injury with lost work time	Major loss of capability of spacecraft or payload	Potential project cost overrun from 3% to 10%	Schedule slip affecting critical path but not launch or postlaunch critical event
2 Minor	Minor injury with no lost work time	Decrease in spacecraft or payload capability/margin but all mission requirements met; need for requirement definition or design/ implementation workaround	Potential cost overrun less than 3%	Non–critical-path schedule slip
1 Minimal	None of the above	None of the above	None of the above	None of the above

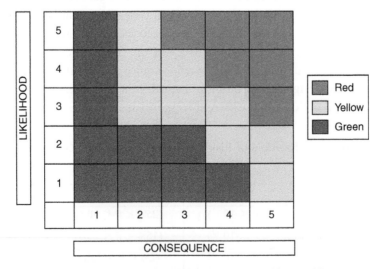

Figure 5.9: An example of risk severity ranking guidance.

Risk tracking involves the monitoring of risks and gathering of new information that could affect risk plans or other aspects of the risk management process. This new information is used to evaluate the risk management performance and is incorporated into updated reports of risk status and trends.

Risk control involves the decisions made based upon existing and new risk-related information. In some cases the decision may be to continue following the planned strategy for a particular risk, which may lead to the execution of the associated contingency plan. In other cases, the decision may be to re-assess the risk and plan for alternative mitigations strategies. Finally, if a risk has been successfully mitigated, the residual risk can be accepted and the risk closed.

Risk communication and documentation, as Fig. 5.8 shows, is the core of the risk management process. The importance of disseminating the risk information throughout the project so that the entire team is aware of potential issues cannot be understated. In some cases, an individual on the project team may think of a mitigation strategy that has gone unnoticed. In other cases, knowledge of a particular risk may lead a team member to discover a previously unidentified risk. By communicating risks to the sponsor or customer, a project may open options that had been considered unavailable. Documenting and communicating risk information also provides a record of the rationale for risk-related decisions made during the project. A risk management plan fulfills this need for documenting and some of the need for communication.

A project's spares philosophy is an example of an important risk mitigation strategy. Purchasing spares is necessary in any spaceflight development effort. Determining how much is enough is a complex tradeoff between cost and benefit. One of the factors in this trade involves the amount of downtime that can be tolerated in the event of a failure. If the mission is designed for an operational capability with little or no downtime, it may be necessary to launch multiple spacecraft so that spare spacecraft are waiting on orbit in the event of a failure. This approach would be a costly solution. Alternatively, a mission that doesn't need on-orbit spares but that has a very narrow launch window without a backup opportunity may need to be able to recover from a fault on the launch pad within a short period of time. This could drive the need to have fully qualified spare components ready and waiting for installation on the spacecraft. Finally, if a particular mission has many launch opportunities throughout the year, the spares philosophy may be simply to have minimal set spare parts so that repairs can be made to existing boards in the event of a failure. An organization's ability to build or repair certain types of parts may also affect the project's spares philosophy. An organization with a complete manufacturing facility may be able to divert resources during a crisis to expedite a particular repair or develop a replacement part. It would likely take longer for an organization without on-site facilities to obtain the necessary repairs quickly. An organization without on-site facilities might choose to develop spares at higher levels of assembly, such as at a board- or box-level.

2.2. Systems Engineering

Systems engineering on space missions embodies the management and oversight of all technical aspects on the project. The mission system engineer (MSE) is responsible for the requirements engineering and performance verification of the overall system. The MSE relies on the systems engineering team for support in many of the technical responsibilities. Other responsibilities include performing trade studies and administering the lessons-learned process. The MSE works closely with the program manager (PM) with regard to trade studies or other issues that affect cost and schedule. All of these plans for processes used on the project are documented in the systems engineering plan (SEP) and in the systems engineering management plan (SEMP) including a description of the requirements engineering and verification processes, the technical review plans, and the resource and configuration management plans. Some of the details of these processes are described in the following. An outline of the table of contents for a PMP that is a combined SEP and SEMP is in Chapter 1 of this volume. Appendix A has an outline of an example SEP. A template of an example SEMP can be found in NPR 7123.1 [13].

2.2.1. Requirements Engineering

Requirements engineering is the area of systems engineering that deals with the process of developing and verifying the system requirements. Following good requirements engineering practices helps achieve the primary objective of making sure that the delivered system meets the customer's needs. The requirements engineering process starts with requirements formulation, which is followed by validation, and then by verification.

Requirements formulation involves the collecting, organizing, communicating, and managing requirements. Requirements come from many sources: requirements can come from the customer, from environmental factors, from other system interfaces, from company facilities, from personnel safety, and from government regulations.

The requirements are organized into levels on the basis of the amount of decomposition needed (i.e., how far they flow down and how much detail this flow-down entails) and on the group of stakeholders they involve. Figure 5.10 illustrates the various levels of requirements and flow-down. The highest level (Level 1) represents the agreement with the customer as to the requirements of the mission. Level 2 documents identify the allocation of requirements to the mission segments. Levels 3 and 4 provide detail of subsystem requirements and component specifications. The lower-level requirements are derived by performing the functional decomposition and flow-down of the requirements.

Part of the requirements formulation process involves iterative communication of the requirements with the stakeholders to confirm that requirements have been accurately documented. The use of standard document templates and well-defined nomenclature helps

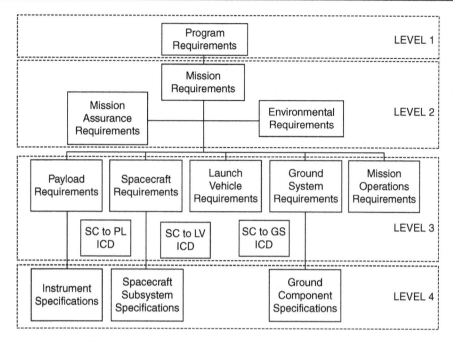

Figure 5.10: Example of requirements flow-down. SC, spacecraft; PL, payload; LV, launch vehicle; GS, ground system; ICD, interface control document.

develop documents that are readable and easily understood. An example of a requirements document template can be found in Appendix B. Examples of commonly used nomenclature include the terms "shall," "should," and "will," each of which has a specific and distinct meaning. The word *shall* is used to indicate a mandatory requirement; for example, "the spacecraft wet mass shall be less than 478 kg." The word *should* is used to indicate a goal that is desired but not mandatory. The terms "is" and "will" indicate a statement of fact or provide descriptive information. Such terms should be defined early in any requirements document. Well-written requirements should be clear, concise, understandable, unambiguous, complete, consistent (without contradictions), traceable, and verifiable. Well-written requirements usually involve an iterative process of documentation and review, with stakeholders involved. This iterative process illustrates the need for the next part of the formulation phase, requirements management.

Requirements management involves the processes of managing the configuration of the requirements and associated changes. Tools that are especially useful in the management of requirements are requirements databases, such as DOORS® by Telelogic®. This tool has the ability to provide a structured but flexible database schema for storing requirements and associated flow-down and traceability links. Fields for rationale and verification method can be added for each requirement. The edit history for each requirement is also preserved in the tool. These types of tools help facilitate the requirements management process.

After the requirements have been formulated and documented, they must be validated. The purpose of the requirements validation process is to confirm that the requirements accurately and completely describe what is needed. The validation of the requirements involves communicating them to the customer and other stakeholders for review and approval. Typically this validation process concludes with a technical review of the requirements being validated.

Finally, the requirements must be verified. Verification involves the testing or analysis of the system to confirm that it meets the specified requirements. Traceability between the requirements and the associated test plans is necessary to ensure that all requirements are verified. Establishing the traceability early during requirements formulation reduces the burden of the verification process late in the mission.

2.2.2. Interface Management

Interface management encompasses the definition, documentation, and control of system interfaces. As part of the functional decomposition of requirements, interfaces are defined and optimized. Interface boundaries depend on how the requirements are flowed down to subsystems. Once the interfaces are established, all aspects of the interface are documented in interface control documents. The types of information contained in interface control documents include both mechanical and electrical interfaces: physical footprint, thermal environment, vibration environment, field of view requirements, power requirements, command and telemetry formats, and electromagnetic compatibility requirements. Throughout the development, accurate documentation of interfaces is critical to the interface control function.

2.2.3. Resource Management

Resource management involves the budgeting and tracking of project resources and the controlled release of margin to address unanticipated changes in requirements or system performance. The resource management process is closely aligned with the risk management process. Resource trends can be used as a risk indicator, and resource margins can be used as a risk mitigation strategy.

In general, smaller margins tend to indicate greater risk, but determining appropriate guidelines for the amount of margin depends on a number of factors; the correlation between the amount of margin and the amount of risk is not precise. Many factors can be used to determine the appropriate margin for a project. One significant factor is the experience and capability of the project team and the organization implementing the project. As a result, an organization may develop company margin guidelines to indicate levels of risk, but application of these guidelines to other organizations may or may not be appropriate depending on the similarities of the organizations. Trends of estimates that repeatedly exceed allocations also can be used as an indication of other risks within the development.

Two commonly referenced and closely tracked resources are mass and power. During the initial phases of the project, allocations for each component are itemized to determine the system requirements. These allocations start with the *current best estimate* (CBE), which represents the best estimate of the system performance at the current time. The term *contingency* (sometimes called *reserve*) is an estimate of the uncertainty in the CBE for each component. The sum of the CBE and the contingency is the allocation for each component. Once the allocations are identified for all components of the system, they are added together to determine the *total system allocation*. The difference between the total system allocation and the capability is the *unallocated margin*. For example, the sum of all the mass allocations must be less than the lift capability of the launch vehicle. Any capability to lift more than the sum of all mass allocations is considered unallocated margin. To calculate the "margin + contingency" percentage, use the following equation:

$$\frac{(capability - CBE)}{CBE} \times 100\%$$

To calculate the unallocated margin percentage, use the following equation:

$$\frac{[capability - (total\_system\_allocation)]}{(total\_system\_allocation)} \times 100\%$$

The results of the previous calculations can be compared to the margin guidelines such as those shown Table 5.10 to determine a corresponding risk factor. Guidelines for recommended levels of margin vary by project phase and resource. While the mass margin is intended to trend to 0% at launch, the power margin should remain above 5% even after launch. Other system budgets related to resource management include link margins, data rates, and delta-V (i.e., the spacecraft capability to change velocity), as well as attitude knowledge, attitude control, alignment, navigation position/velocity, life-limited items, and radiation.

Guidelines for performance of central processing unit (CPU) resources, such as time, memory, and throughput, are usually specified as a percentage utilized rather than margin. To calculate the percent utilization, divide the CBE of usage divided by the capacity multiplied by 100%. Guidelines for CPU utilization resources are listed in Table 5.11.

2.2.4. Lessons Learned

Samuel Levenson (American author and humorist, 1911—1980) said, "You must learn from the mistakes of others. You can't possibly live long enough to make them all yourself." The same applies to space missions, which would be doomed if they repeated all the mistakes of previous missions. From a project's perspective, the lessons learned process can be broken into two functions: (1) it provides a resource for an ongoing project to incorporate lessons learned from other missions; and (2) it provides a process for capturing the lessons learned on the current project so they are not forgotten and can benefit other missions.

Table 5.10: Example of Margin Guidelines

Resource Margin	Risk Level	Concept	PDR	CDR	PER	PSR
Mass margin + contingency	Green	>30%	>20%	>15%	>7%	2%
	Yellow	20–30%	15–25%	10–15%	5–7%	1–2%
	Red	<20%	<15%	<10%	<5%	<1%
Unallocated mass margin	Red	<10%	<10%	<5%	<5%	<1%
Power margin + contingency	Green	>30%	>20%	>15%	>10%	10%
	Yellow	20–30%	15–25%	10–15%	5–10%	5–10%
	Red	<20%	<15%	<10%	<5%	<5%
Unallocated power margin	Red	<10%	<10%	<5%	<5%	<5%
Uplink (Earth orbiting)	Green	≥6 dB	≥6 dB	≥6 dB	≥6 dB	≥6 dB
	Red	<6 dB	<6 dB	<6 dB	<6 dB	<6 dB
Uplink (deep space)	Green	>6 dB	>6 dB	>6 dB	>6 dB	>6 dB
	Yellow	3–6 dB	3–6 dB	3–6 dB	3–6 dB	3–6 dB
	Red	<3 dB	<3 dB	<3 dB	<3 dB	<3 dB
Downlink (Earth orbiting)	Green	≥3 dB	≥3 dB	≥3 dB	>3 dB	>3 dB
	Yellow	—	—	—	2–3 dB	2–3 dB
	Red	<3 dB	<3 dB	<3 dB	<2 dB	<2 dB
Downlink (deep space)	Green	≥3 dB	≥3 dB	≥3 dB	≥3 dB	≥3 dB
	Red (Emergency)	<3 dB	<3 dB	<3 dB	<3 dB	<3 dB
	Red (Science)	<3 dB	<3 dB	<3 dB	<1.5dB	<1.5dB
Data rates	Green	≥30%	≥20%	≥15%	≥10%	≥10%
	Yellow	20–30%	10–20%	10–15%	0–10%	0–10%
	Red	<20%	<10%	<10%	<0%	<0%
Delta V (deterministic)	Green	>15%	>15%	>10%	>5%	>5%
	Yellow	10–15%	10–15%	5–10%	3–5%	2–5%
	Red	<10%	<10%	<5%	<3%	<2%

For ongoing projects, periodic review of lessons learned databases during the project life cycle is necessary to ensure that new lessons are not overlooked, even if they occurred after the beginning of the ongoing project. A method for incorporating this activity into standard practice is to include a review of the lessons learned databases as topic at each major mission-level review. By incorporating relevant process improvements, the project can avoid similar pitfalls.

Table 5.11: Example of CPU Utilization Guidelines

CPU Resource	Risk Level	Concept	PDR	CDR	PER	PSR
CPU time	Green	<40%	<45%	<60%	<65%	<70%
	Yellow	40–50%	45–55%	60–66%	65–70%	70–75%
	Red	>50%	>55%	>65%	<70%	<75%
CPU memory (RAM)	Green	<40%	<50%	<65%	<70%	<75%
	Yellow	40–50%	50–60%	65–70%	70–75%	75–80%
	Red	>50%	>60%	>70%	>75%	>80%
CPU memory (EEPROM, flash)	Green	<40%	<50%	<60%	<70%	<75%
	Yellow	40–50%	50–60%	65–70%	70–75%	75–80%
	Red	>50%	>60%	>70%	>75%	>80%
CPU memory (PROM)	Green	<50%	<60%	<75%	<80%	<85%
	Yellow	50–60%	60–70%	75–80%	80–85%	85–90%
	Red	>60%	>70%	>80%	>85%	>90%
Data bus throughput	Green	<40%	<50%	<65%	<70%	<75%
	Yellow	40–50%	50–60%	65–70%	70–75%	75–80%
	Red	>50%	>60%	>70%	>75%	>80%

The process for capturing lessons learned within an organization should include a vehicle for open communication of information within the project. Such communication can take the form of soliciting all members of the team for lessons learned at specific times during the project. Each lesson should include a description of the lesson, a proposed corrective action or rationale for no corrective action, the individual responsible for implementing the corrective action, and a date for completion. Sometimes the information related to a lesson may be sensitive or evoke emotion among individuals on the project team. To avoid such situations, lessons learned should be reviewed and distilled into a form that encourages process improvement rather than criticism of past performance. This task should be completed before lessons are distributed to the team for training.

Databases for tracking anomalies and nonconformances also provide useful resources for identifying lessons. Making sure that the project team has easy access to a system for recording issues will help improve the knowledge retention and accuracy in the documentation of lessons learned.

2.2.5. Architecture/Design Trades

The design of any architecture always involves tradeoffs. Some are simple decisions made by a single engineer, such as which part to use in a circuit. Other trades may involve input from an entire team of engineers, such as what occurs during the definition of the operational modes of the spacecraft. The responsibility for conducting a particular trade depends on the elements affected by the results of the trade. In general, when a trade affects multiple systems, the decision-making responsibility falls on the next level of authority in the project organization structure. The process involves comparison of alternatives using quantifiable parameters where possible. As alternatives are selected, the requirements must be folded back into the system to assess the impacts. This iterative process continues as the system concept is refined.

There are four steps to the trade study process:

- Analysis—Identify alternatives and establish objective evaluation criteria.

- Parameterization—Define alternatives in terms of parameters for performance, risk, reliability, cost, schedule, mass, power, and other aspects of the evaluation criteria.

- Selection—Evaluate the alternatives and select the best option, document the trade study results and the rationale for selection in a report.

- Reallocation—Reallocate the resources and requirements on the basis of the trade study results.

2.3. Fault Protection

The development process for fault protection is a systems engineering approach to evaluate and improve the robustness of the overall system. The fault protection engineer (FPE) is responsible for leading this effort. The process begins during the requirements development phase, which involves identifying potential faults and responses, as well as working with the MSE to allocate the requirements to the appropriate mission segments. After the development of the fault protection requirements, the process follows with the design, implementation, and verification phases similar to other system engineering processes. As with many development processes, the end of one phase sometimes overlaps the beginning of the next. Details of each phase are identified below.

2.3.1. Fault Protection Requirements

The fault protection requirements development process begins with the flow-down of the requirements from the mission level. These requirements define the framework of the fault protection system, and include aspects such as safe modes, system availability, hardware redundancy, and capabilities of the autonomy rule engine. Since different missions can have widely different levels of acceptable mission risk, the scope and complexity of the fault protection system can vary widely from mission to mission. For example, a fully redundant, fully cross-strapped system can require a significant development effort to ensure that power cycles appropriately and components are swapped in response to failures. A single-string spacecraft, on the other hand, may require only a particular component to be power cycled when a failure is detected.

In general, the fault protection requirements are derived from the need to detect a particular failure and respond with an action to mitigate the consequence. To determine the fault protection requirements, the FPE must first identify the potential faults. Identification of potential faults includes both formal and informal methods (see Table 5.12). Although the majority of the analysis involves identifying potential faults with the flight system, the FPE is responsible for addressing critical faults in any part of the system (in flight and on the ground) that could affect the success of the mission.

Once a potential fault has been identified, the FPE must identify the method for detecting the fault and the method for mitigating the consequence. In some cases, the fault may require continuous monitoring of spacecraft telemetry with an immediate onboard response if a fault is detected. This requirement is usually implemented by the autonomy rule system and represents a major effort in the development of the fault protection system. In other cases, an immediate response may not be necessary; it may be sufficient to wait until the next ground contact, when the mission operations team can diagnose and remedy the problem. Another method for addressing a potential fault may be to levy operational constraints that eliminate the need for an immediate detection and response. For example, a propulsion system typically has redundant catalyst bed heaters on each thruster. By requiring both redundant heaters to be powered whenever the propulsion system is in use, the consequence of one heater failing is mitigated by the redundant heater. The FPE works closely with the MSE in the evaluation of fault responses, since fault responses affect the requirements flow-down. In all cases, no matter what fault response option is selected, the requirements must be documented with traceability to the source of the requirement.

The autonomy rule system is one of the most substantial efforts in the area of fault protection. The responsibility for developing the autonomy rule system may be assigned to a separate autonomy subsystem lead engineer. An autonomy rule is an "if, then" statement that provides the spacecraft with the capability of reacting autonomously to events onboard. An example of such a rule is "if an instrument is drawing too much power, then turn it off." The FPE is responsible for defining the requirements of the entire autonomy rule system, from the performance capabilities of the rule engine and commands to the actual rules and responses that are loaded into the system. Examples of the performance capabilities of a rule engine include rule capacity, frequency with which rules are evaluated, telemetry that can be

Table 5.12: Sources for Fault Protection Requirements

Source	Description
Fault tree analysis (FTA)	Fault tree analysis (FTA) is an important analytical technique for identifying potential causes of an undesirable system-level event or fault. A fault tree is a graphical representation of the logical combination of series and parallel occurrences that could lead to a particular undesirable outcome. FTA is something of a "top-down" analysis when compared to FMEA or FMECA. See reference [10].
Failure modes and effects analysis (FMEA)	Failure modes and effects analysis is a method for analyzing the potential outcomes and associated probabilities resulting from the various failure modes of a specific component in the system. For example, a harness connector may fail in an open-circuit or shorted condition. FMEA would provide an estimate of the consequence and probability of the various outcomes caused by these two failure modes. FMEA is something of a "bottom-up" analysis when compared to FTA.
Failure mode effect and criticality analysis (FMECA)	Failure mode effect and criticality analysis is similar to FMEA with additional focus on the criticality of consequences of the effects and controls to minimize the probability of critical consequences. FMECA is something of a "bottom-up" analysis when compared to FTA.
Probabilistic risk assessment (PRA)	Probabilistic risk assessment is a comprehensive method for assessing risk in complex systems. The PRA involves quantitative evaluation of the probability of events in a sequence leading to an undesirable outcome with severe consequences. A PRA can also be helpful in comparing the benefits of various risk mitigation strategies. See reference [11].
Scenario analysis	Scenario analysis starts with the mission-critical operations and examines the consequences of various faults during each operation, thereby identifying high-priority requirements early in the development. In some cases a fault protection response may need to be different depending on the ongoing activity. For example, a typical fault protection requirement would be to enter safe mode in the event of a fault, but if the fault were to occur during the one and only planetary flyby, entering safe mode might prevent the spacecraft from gathering critical data, resulting in a mission failure. For this reason, fault protection requirements must be analyzed for critical events, as well as nominal operations. Examples of critical operations include launch, deployments, maneuvers, planetary flybys, and critical science observations.

(Continues)

Table 5.12: Cont'd

Source	Description
Event tree analysis (ETA)	Event tree analysis (ETA) helps you understand how your product responds to each potential, possible circumstance. An event tree gives the sequence of hardware, software, and operator functions that make up a scenario. ETA gives both failed and successful paths through scenarios; it can also handle multiple failures simultaneously. ETA is a combination of graphical and tabular techniques; it can be more powerful for analyzing than FMECA. ETA should capture all single-point failures and many multiple-point failures in the system, indicate marginal operations, and indicate proper operation. ETA can be part of scenario analysis (list above). ETA is something of an "outside-in" type of analysis. See reference [12].
Meetings with subsystem lead engineers	Meetings with subsystem lead engineers are an informal but effective way to identify potential faults and consequences. Subsystem engineers are most familiar with the subsystem behavior and can offer insight into potential faults. By collecting information from all subsystems, the FPE may uncover unforeseen interactions that could lead to unanticipated consequences. Meetings with subsystem engineers should occur during all phases of the fault protection process.
Peer reviews	Peer reviews with independent reviewers offer a fresh perspective on the potential faults and serve to identify concerns that may have been overlooked or initially discounted. Typically the FPE will perform at least three major reviews. First, a fault protection requirements review is held to make sure all necessary requirements are understood, are consistent, and have been accurately documented. Next, the fault protection PDR is held before the mission PDR to make sure the preliminary design meets the fault protection requirements. Finally, the fault protection CDR is typically held before the mission CDR to make sure the detailed design meets the performance specifications. In addition, a code walkthrough of the autonomy rules and responses is recommended near the completion of the implementation phase and prior to the verification phase.

examined by a rule, and commands that can be executed as responses. These types of requirements must be defined early so that the functionality is ready for the fault protection implementation phase.

Fault protection requirements that are common among most space missions include redundancy management, command loss timeout protection, processor reset (recovery), low bus voltage and high current load shedding, component communication failure mitigation, and anomalous attitude (recovery). A common technique for simplifying the responses to potential faults is to define a "safe mode" of operation and enter this state when something goes wrong. Characteristics of safe mode usually include maintaining the spacecraft indefinitely in a thermally stable, power-positive state with communications capability with the Earth. By defining a safe-mode response that addresses multiple fault conditions, the FPE can simplify the development and testing of the autonomy rule system.

The application of risk management techniques is useful for prioritizing the fault protection requirements. By assigning severity and likelihood factors to each fault, the FPE can address the most critical faults first. In general, the complexity of the fault protection requirements increases substantially with the number of independent faults. For most missions a single-fault tolerant system is sufficient (i.e., continue safe operation in the event of any single fault); however, range safety regulations require dual-fault tolerance for catastrophic failures. Even for spacecraft that are described as "fully redundant," there is often a short list of single-point failures in a design. The purpose of this list is to document the accepted risks for the project. Common single-point failures include the spacecraft mechanical structure, the propulsion tank, and common software designs in redundant components.

2.3.2. Fault Protection Design

During the design phase, the FPE works with subsystem lead engineers to determine how each requirement will be implemented. Interfaces between each subsystem of the fault protection architecture are clearly defined, so that each interface is understood by all involved. Strict control of interfaces is necessary to avoid unintended interactions. For example, if the onboard autonomy system is designed to switch the spacecraft antenna every 4 hours in response to a command-loss timeout, the mission operations staff must be aware of this timing so that they do not send commands that would adversely affect the antenna switching operation.

As part of the design phase, specific autonomy rules along with their logical expressions and command responses are defined. Similarly for fault protection hardware, schematics are drawn or VHDL (very-high-speed integrated circuit [VHSIC] hardware description language) code is developed. Those hardware designs are simulated and worst case analysis and other reliability analyses are performed. As the design is refined and issues are uncovered, it may be necessary to update requirements. This iterative cycle is a normal part of the systems engineering process.

Whenever possible, the detection method for a fault should be unique. Sometimes the indication of a fault can have multiple possible causes. In such cases, it is important for the response to address all potential causes of the fault indication. For example, the guidance and control system on a spacecraft indicates that it is not meeting its pointing specification. Possible causes could include an invalid setting in the processor, a failed star tracker, or a failed thruster. The response to the condition must address all potential problems. A possible response would be to switch to the backup processor, start tracker, and thrusters all at the same time. Another possible approach, often called a "tiered approach," would be to switch the processor first, and then wait to see if the condition is resolved. If not, the second tier would be to switch the star tracker, and so on. If the response is urgent, the first approach (addressing all possible sources at once) may be preferred, because it takes less time. If the response is not urgent and the spacecraft spends long periods of time unattended, the tiered approach may be preferred, because it addresses some cases of multiple failures, such as a failure of the primary processor followed by a failure of the redundant star tracker.

In some cases the fault protection may drive other subsystem requirements, especially in the area of command and data handling (C&DH) flight software, where the capabilities of the autonomy rule engine are implemented. Details about these capabilities are described in the next section. Also related to flight software are the requirements on the command structure, especially with regard to redundancy management. To reduce the number of autonomy rules it is advantageous to keep track of the active and inactive components onboard and provide the capability of toggling from one to the other. With this capability a single rule can be used to manage two redundant components. For example, if the "active" star tracker has failed, then power off the active unit, power on the inactive unit, and toggle it to make it active.

2.3.3. Rule-Based Autonomy

Rule-based autonomy is the most powerful aspect of the fault protection system. The autonomy rule engine, running in the C&DH system, provides a telemetry monitoring capability that responds to onboard triggers without interaction from the mission operations team. Each autonomy rule can be thought of as a simple if/then statement. For example, "If an instrument draws too much power, then turn it off." Since other factors can affect the autonomy rule performance (such as start-up transients, measurement noise, or intermittent failures), the actual rule syntax is a bit more complex. Table 5.13 lists examples of some of the advanced capabilities of autonomy rules. These capabilities form the infrastructure that the fault protection autonomy rules are based on.

The autonomy rule engine evaluates the set of rules once per second. If a rule is enabled, and its premise has evaluated true for M of the last N seconds, and the fire count has not reached its maximum value, then the command macro is executed. Command macros may consist of just a few commands or a long chain of commands. Command macros can execute other macros and enable or disable other rules.

2.3.4. Fault Protection Implementation

The fault protection implementation occurs concurrently with the overall project implementation. As the hardware and software of the overall system are being developed, the associated fault protection aspects are also implemented. The FPE is responsible for overseeing the implementation of these fault protections designs and ensuring the compatibility of these fault protection system components.

Table 5.13: Examples of Autonomy Rule Capabilities

Capability	Description
Mathematical comparison	Compares telemetry values to constants, storage variables, or other telemetry values (e.g., $x > 10$, $x = 0$).
Logical expression	Defines a Boolean function involving logical operators (e.g., AND, OR).
Computed telemetry	Provides a method for evaluating floating point mathematical functions using telemetry values.
M of N count	Specifies the number of times (M) in a given number of samples (N) that the premise must be true for the rule to fire.
Enable/disable	Designates which rules are enabled or disabled for evaluation
Write protection	Protects individual rules from being overwritten. A rule must be disabled and not write-protected before it can be reprogrammed.
Initial state	Defines the initial enable/disable state of a rule after a processor reset or power cycle.
Maximum fire count	Limits the number of times a rule can fire. The fire count can be cleared by command.
Priority	Identifies which rule response is executed first if multiple rules fire at the same time.
Storage variable	Stores a current value of a telemetry point for subsequent comparison to a future value. Storage variables may also be used to count events by incrementing and decrementing the value.
Command macro	Executed in response to a rule firing. A command macro contains a sequence of commands necessary to accomplish a specific task.
Stale bypass	Controls whether the premise is evaluated when the telemetry is stale.

In addition to the oversight responsibilities, the FPE is also responsible for the implementation of the fault protection rules and command macros. As autonomy rules and macros are implemented, they are also tested by the developers in a standalone configuration. Once a set of autonomy rules has been tested by the developers, the set can be released for test. A set of release notes is provided indicating the content of the release and any changes from the previous version. At this point the set is loaded to the spacecraft and programmed into non-volatile memory for verification during spacecraft-level integration and test. If issues are discovered that necessitate updates to the set of rules, the processes of requirements definition, design, implementation, and verification are repeated.

2.3.5. Fault Protection Verification

The fault protection system is verified during spacecraft integration and test in the form of the comprehensive performance test, mission simulations, and special fault protection scenario testing. During the comprehensive performance test, specific hardware features are tested to verify performance. For example, thermostats that trip when the spacecraft is too cold are verified during a cold cycle of the thermal vacuum testing. For the mission simulations, the purpose is to verify that the autonomy rules do not trigger during nominal operations, when no fault condition exists. For scenario testing, the purpose is to verify the proper responses by injecting simulated faults into the system while monitoring the autonomous responses of the fault protection system.

When implementing scenario tests certain capabilities can improve the testing efficiency. First, it is important to have a high-fidelity simulator for testing the fault protection system in ways that may not be possible on the flight spacecraft. Also, it is important to automate the spacecraft state setup and spacecraft state checking to facilitate unattended monitoring of scenario tests. This approach allows testing to be conducted overnight when the facilities are in less demand. Another important goal is to develop modularity between the initial spacecraft state setup script and the script that initiates the fault. This approach simplifies the process of checking the same fault in multiple spacecraft states and allows multiple scenarios to be chained into a single test.

2.4. Mission Assurance and Safety

Every company, program, and project wants to safely succeed at their mission. Consequently, most companies have a form of mission assurance and safety built into their modes of operation; either it can be implicit in their processes for doing business or it can be explicitly stated in company policies. Those that have an explicit policy often have a "boilerplate" template for preparing and stating mission assurance and safety for every project. Regardless, if you are involved in developing a spacecraft or a subsystem, you will need to state a program for mission assurance and safety. That program typically has the elements listed in this section.

NASA has a defined policy in NPD 8700.1C, *NASA Policy for Safety and Mission Success,* for safety and mission success [14]. Companies building space instruments and spacecraft should address the points in NASA's policy; a good place for a company to start is in the corporate quality policies, which should address the following aspects of quality:

- Assessment

- Assurance

- Audits

- Engineering

- Management

- Component parts

- Product identification and traceability

- Reviews

- Software

- Surveillance

- Workmanship

- Metrology and calibration

Companies should also address safety issues in each program or project; this can be done in a separate safety program. The risk management plan, mentioned earlier in this chapter, can serve as a complementary document or as input to the safety and mission assurance program. That program should address appropriate concerns from the following list:

- Management

- Safety engineering

- Aviation

- Confined spaces

- Cryogenic

- Electrical

- Explosives, propellants, and pyrotechnics

- Extravehicular activity

- Facility

- Fall protection

- Fire

- Hazardous materials

- Hazardous operations

- Hydrogen

- Inert gas

- Ionizing radiation

- Lockout/tagout

- Lifting devices

- Motor vehicle

- Nitrogen

- Nonionizing radiation

- Nuclear (viz., the launching of radioactive materials)

- Oxygen

- Payload

- Pressure vessel

- Promotion and motivation

- Range

- Software

- System

- Test operations

The following subheadings treat specific concerns that are typical for all projects. Prepare a plan for safety and mission assurance (SMA); some example areas that deserve attention to assure the mission and safety in the plan are establishing and maintaining independent lines of communications, defining and documenting SMA requirements, certifying safety and operational readiness, risk assessment techniques, training, and monitoring failure-free hours of operation. The plan for SMA and its execution require approval from the cognizant technical and SMA authorities.

2.4.1. Custom or Unique Design

Whenever a subsystem has a custom or unique design, it has a higher probability of failure. It simply does not have the time and history reveals many modes of failure. This is particularly true of mechanical design where fatigue from varying loads, for example, takes time to complete a fully developed fracture. The concept of failure-free hours of operation for a system before the operational mission will help develop some of the desired history for the system. A properly prepared and followed design plan, such as the mechanical design plan, will lay the groundwork to address the concerns over safety and mission assurance.

"Studies have shown that mechanical failures are more frequent and more likely to significantly affect mission success than are electronic failures. Spaceflight mechanisms are usually unique designs that lack the years of testing and usage common to electrical devices and in general cannot be designed with the same level of block of functional redundancy, or graceful degradation, common with electronic circuit design. The lesson provides a recommended checklist that contains 92 measures for enhancing the reliability, ease of fabrication, testability, maintainability, and robustness of mechanical designs" [18].

2.4.2. Software

Software is another area that is always a custom or unique design and prone to incorrect operation. A properly prepared and followed Software Design Plan will lay the groundwork to address the concerns over safety and mission assurance.

Over the years some heuristics have arisen to help plan for the rigorous and correct development of software. The following rules of thumb are examples that might help a particular development (but this is not necessarily a complete list):

1. Keep it simple (rationale: a 10% increase in *problem* complexity can double the *software* complexity) [15].

2. Use the right people (rationale: good programmers can be up to 30 times more effective than mediocre ones) [15].

3. Efficiency results from good design not necessarily as much from good coding [15].

4. Have serious design reviews and code walkthroughs (rationale: rigorous review can remove up to 90% of the errors before testing) [15].

5. Design, test, encapsulate, and execute critical functions independently of control operations code (rationale: critical functions tend to be the least flexible; get them right and the easier code can take up the "slack").

6. Practice defensive programming, design critical functions first (rationale: same as 5).

   - Begin with a carefully thought out RA/HA matrix that includes hardware, environmental, software, and user associated risks.

   - Verify the critical backbone subsystems independently before focusing on integration with operational code.

   - Meet specific needs with specific, traceable code.

7. Avoid open-loop designs (rationale: open loops make measurement more difficult).

8. Avoid mirroring or modeling hardware when the physical state is directly measurable (rationale: don't create extra work that ultimately can give false results).

Note: Items 5 through 8 are summarized from Chapter 2 of this volume.

2.4.3. Reliability

Reliability is an important part of safety and mission assurance. As part of the development plans that you prepare, you can include the following topics:

- Reliability engineering

- Human reliability

- Reliability management

- Software reliability

- Reliability-centered maintenance

2.4.4. Electrical, Electronic, and Electromechanical Part Procurement

Your project plan needs to establish the baseline criteria for selecting, screening, qualifying, and derating EEE parts. Screening tests are intended to remove nonconforming parts (parts with random defects that are likely to result in early failures, known as infant mortality) from an otherwise acceptable lot and thus increase confidence in the reliability of the parts selected for use. Qualification testing consists of mechanical, electrical, and environmental inspections, and is intended to verify that materials, design, performance, and long-term reliability of the part are consistent with the specification and intended application, and to assure that manufacturer processes are consistent from lot to lot. Derating is the reduction of electrical and thermal stresses applied to a part during normal operation in order to decrease the degradation rate and prolong its expected life.

Be aware that NASA defines three reliability levels for EEE parts requirements based on mission needs. The levels of part reliability decrease by reliability level; Level 1 has the highest reliability and Level 3 the lowest. A part at Level 1 has the greatest manufacturing control and testing per military or Defense Supply Center Columbus (DSCC) specifications. Level 2 parts have reduced manufacturing control and testing. Level 3 parts have no guaranteed reliability controls in the manufacturing process and no standardized testing requirements. The reliability of Level 3 parts can vary significantly with each manufacturer, part type and lot date code (LDC) due to unreported and frequent changes in design, construction and materials [16].

Sometimes you may have to perform destructive parts analysis (DPA) to qualify individual components. This might be called for when the project purchases modules, such as hybrids and DC/DC converter modules. Hybrids and modules do require a thorough knowledge of their design and fabrication, which incurs a substantial additional cost in acquiring them.

Another area of concern for EEE parts is for counterfeit or known bad lots of components. Your project should consider signing up for alerts from the Government-Industry Data Exchange Program (GIDEP) to expedite identification of counterfeit or bad components.

One last area of concern that is gaining momentum in space missions is the use of commercial off-the-shelf (COTS) components. The problem is that COTS parts, particularly integrated circuits, do not necessarily have the test data or documentation to show that they can withstand the rigors of spaceflight. Often the project needs an explicit contract or a waiver to use COTS components. Even then additional tests such as DPA, inspection, environmental tests, and highly accelerated stress screening (HASS) may be required to ensure that the COTS parts are acceptable to the program. (See the section on COTS earlier in this chapter.)

2.4.5. Radiation Hardness

Radiation hardness is a part of SMA. A number of issues for radiation hardness, particularly the NASA guidelines are listed and summarized earlier in this chapter. Designing for radiation hardness is tied directly to the mission and its trajectory or orbit. Even the areas, such as single-event effect (SEE), single-event fault indication (SEFI), single-event latchup (SEL), single-event upset (SEU), and total dose ionizing (TDI), have particular definitions and margins that depend on the mission parameters.

2.4.6. Materials and Process Control

A project establishes a program-approved materials and processes list (PAMPL) to help control materials and processes. The PAMPL can address some or all of the following sections:

- Control of nonconforming materials.

- Prohibited materials to verify that prohibited materials, such as tin, zinc, and cadmium, are not used.

- The configuration management plan and version control system should provide the necessary traceability.

- Corrosion for mechanical parts should be resistant to corrosion per MSFC-SPEC-250 and to stress corrosion per NASA MFSC-STD-3029.

- Incompatible metals should be identified and protected per MSFC-SPEC-250, NASA MFSC-STD-3029, and MIL-STD-889; an example might be using aluminum alloy 6061 with A286 stainless steel alloy inserts and stainless steel washers in a metal component. If the aluminum has a chemfilm coating applied before inserting the A286 stainless steel alloy inserts, this reduces the concerns for corrosion between dissimilar metals.

- Mechanical fasteners and parts—An example might be requirements for "metallic components shall be designed using 'A' values in MIL-HDBK-5."

- Nonmetallic structures—An example might be requirements for "using 'B' values listed in MIL-HDBK-17/2."

- Absorption and outgassing—NASA's outgassing data for selected spacecraft materials (http://outgassing.nasa.gov); an example of following this direction might involve using conformal coating in manufacturing and assembling circuit boards and controlling the water vapor exchange and outgassing with an alcohol rinse of circuits and components followed by a bake cycle before conformal coat. This also means that the materials will not absorb nor release corrosive by-products.

- Atomic oxygen erosion—Polymeric/organic materials used on spacecraft external surfaces shall be resistant to atomic oxygen erosion.

- Flammability—An example might be requirements for "materials shall be selected to the flammability requirements of NASA-STD-6001."

- Non-nutrient and fungal control—An example might be requirements for "materials shall be non-nutrient and/or non-incubating to fungi per MIL-HDBK-454, Guideline 4."

- Structural bonding per MIL-HDBK-83377 or MIL-PRF-534 or 535.

- Nonstructural bonding per MIL-A-83376.

- Protective finishes—An example might be requirements for "metals requiring protective finishing systems shall be selected consistent with the requirements for electrical grounding as defined by MIL-HDBK-464."

One area of process control is for electrostatic discharge (ESD). The company and project need facilities, such as static mats and appropriately grounded equipment benches, to reduce and control ESD. The company also must institute regular ESD training for the production staff and anyone who handles electrical components and integrated circuits and subsystems.

One part of the controlling materials and processes is to establish a parts, materials, and processes control authority (PMPCA) for a particular project. The PMPCA can regulate parts, materials, and processes for safety and mission assurance. The project can also establish a parts, materials, and processes control board (PMPCB) to meet regularly to review parts, materials, and processes.

2.4.7. Workmanship

Planning for safety and mission assurance needs to address workmanship on a project. Some workmanship issues include (but this is not necessarily a complete list):

- Soldering of components, PWBs, wires, cable, and connectors
- Conformal coat coverage of EEE parts
- Staking components
- Crimping wires into connectors
- Installation of bolts, nuts, and screws
- Inspection of components, subsystems, materials, and assemblies

2.4.8. Configuration Management

Chapter 1 introduced the need and rationale for configuration management. For your convenience some of the areas that a configuration management system should monitor are repeated here:

- Software
- Hardware
- Products and intermediate stages of production
- Identification
- Database and network

A configuration management system archives the records, schematics, source code, documents, memos, meeting and review minutes, and communications on the project. Configuration management is very tightly intertwined with data management; the only difference is that configuration management also accounts for state of physical product.

2.4.9. Data Management

Data management, as a part of configuration management, has a design repository that is version controlled as mentioned in Chapter 1. It stores and maintains requirements, descriptions, source listings, schematics, manufacturing and fabrication instructions, records of production, and records. The system needs security as it stores the intellectual property for the project. File transfers should be effortless and well understood.

A significant data package accompanies the delivery of any system. The content of a data package might contain the following information; an example of how it might be divided and arranged into sections follows:

- Title page: Contains item description, buyer's part number, dash number, revision number, serial number, product specification number and revision number, purchase order or subcontract number, and change notice number.

- Certification of conformance: Approval page from seller with signatures from engineering, program management representative, and QA; this certificate contains the item name, part number, and serial number; it also states whether it is for acceptance, proto-qualification, or qualification test and whether it is the first test sequence or a retest sequence.

- Summary page: Describes all events affecting each serial number delivered. Include pertinent events from both a manufacturing and test perspective.

- Test procedure and data: State whether it is for acceptance, qualification, or proto-qualification.

- Test flow block diagram: Includes each test performed a block identifying the deliverable hardware name, serial number, the date the test started, the date the test completed, name of the test, test specification number, and revision number.

- Operating time record: This is the individual time record for each limited life critical item; this is where you demonstrate the failure free hours of operation.

- Summary of test discrepancies occurring during acceptance, qualification, or proto-qualification test (description, root cause, disposition, corrective action) and copies of all failure review board (FRB) reports.

- Any deviation or waiver documentation.

- Summary data parameter sheets with evidence of seller's QA approval.

2.4.10. Nonconformance Processes

The goal of nonconformance processes is to prevent defective materials and components from entering or leaving fabrication, manufacturing, assembly, and production. Nonconformance processes address the following:

- Inspection report

- Material review disposition form

- Material review board

- Failure review board

- Deviations and waivers

Inspection is visual examination of a material or component or assembly to determine if the component type or the assembly workmanship is correct according to design instructions or the bill-of-materials (BOM). Inspection may occur in a number of different situations during the course of product development and production. One area is receiving inspection where incoming or received components, subsystems, or materials are checked for compliance with the purchase specification in the purchase order (PO). This is where the markings and identification of the received material is checked for correctness. If a functionality test is called for in the design instructions for receiving, then that test takes place during inspection. If the material fails the inspection it is either returned or scrapped. You should maintain a log of all incoming inspection activities. Sometimes a customer receiving inspection conforms to MIL-STD-105.

Workmanship inspection occurs during production (fabrication, manufacturing, and assembly) and usually occurs in several points along the way. Inspection for workmanship follows the instructions on the manufacturing traveler or router. You should maintain a log of all workmanship inspection activities.

When an inspection or a test fails a component or subsystem, then the project must have a way of recording the failure and what happens to the problem material. Often that record is in a form such as an inspection report or a failure report or a problem report. Then it needs to be reviewed by a team assembled by the project management. That review team often is called something like a failure review board or a material control board. That team conducts a root cause analysis and takes appropriate corrective action regarding the disposition of failed components or materials and other affected hardware. Key project disciplines should participate in the investigation, such as the responsible design engineer, quality, and manufacturing. Typically, senior program management chairs the failure review board to review root cause analysis and corrective action and to authorize break-of-configuration, troubleshooting and disposition of the hardware. Sometimes when a failure cannot be repeated or where most probable root cause cannot be determined, operational constraints or actions should be defined and documented.

Once a discrepancy or problem is found or a failure occurs, the project needs to start a corrective action to fix the concern. The corrective action needs to be a closed loop system with clearly defined methods and responsibilities. The system must maintain a status of all the corrective actions in the project and someone must periodically report the problems and corrective actions to the customer for review. The report should include a statement of the problem, an analysis of the cause of the problem, a statement of the action taken to prevent the recurrence, and the effectiveness of the corrective action. Often a response time is contractually defined (e.g., 2 weeks unless otherwise authorized by the customer). A corrective action is not closed out until it is reviewed and verified to be successful.

The nonconformance process requires that disposal of the rejected raw materials or products for. That disposal may be either to scrap the materials or products which are determined to be unfit for use or rework the product. Rework only takes place for nonconforming products that can be corrected to existing work instructions without adverse effects and still meet the original specifications.

The project needs a deviation/waiver process for nonconforming materials or product that customer review deems acceptable either "as-is" or with rework. This process is a set of procedures to determine, record, and approve the use of nonconforming materials or product.

2.4.11. Contamination Control

Mission assurance typically requires a contamination control plan for fabricating, assembling, and producing components and systems. Often it focuses on proper clean room practices. The facilities and procedures must address the following to control contamination by particulates and from outgassing:

- Design
- Proper materials
- Component cleaning
- Conformal coating to control outgassing
- Clean room facility
- Clothing apparel
- Packaging for storage and shipping

2.4.12. Safety

Since this book does not address human-rated spacecraft, the written material here addresses the safety for personnel and for the equipment during production and launch. Concerns include hard hats in facilities where a satellite or large subsystems are moved, procedures for moving large or heavy structures, protection for extreme temperatures, such as cryogenics, protection from hazardous fuels, such as rocket propellant, and protection from electrical shock and currents.

2.4.13. System Safety

System safety refers to the proper care of flight hardware situations; its goal is to avoid situations where the project or equipment can be damaged or the mission degraded. These sorts of problems occur because of mistakes, incorrect procedures, or unforeseen circumstances and consequences. An example of damage that illustrates a lapse in system safety is the NOAA N-Prime mishap in September 2003; this was a large, expensive satellite that fell off its support cradle because all 24 bolts that secured the satellite to the cradle were removed before technicians tilted it. The satellite tipped over onto the floor and completely destroyed itself. The bolts had been removed without communication to the technicians before they tilted the satellite to work on it. Ultimately, incorrect test procedures, poor communications, and corporate culture were all blamed for the accident [17].

2.4.14. Personnel Safety

The system can affect the health and lives of a development team, ground personnel, and the local population. It has to do with assembly and test of the launch vehicle and satellite, prelaunch integration tests, and range safety during launch.

Fire, hazardous materials, and falling loads are primary concerns, as all are potential safety problems on the ground during production and assembly. Range safety includes situations such as clearing aircraft from the airspace and water craft from the ocean near the launch area and time. If a launch fails catastrophically and pieces of the launch vehicle fall to Earth, the SMA plan needs to address those sorts of potential scenarios.

If there is potential for injury or harm to human life, NASA policy requires:

> *Formal consent to take any human safety risk by the actual risk taker and an appropriate member of his/her supervisory chain ... are two elements to the consent to take risk. The first element is that the risk takers themselves volunteer to take the risk. The second element is that the appropriate member of the supervisory chain also consents to the risk-taking. The first element focuses on the willingness of the risk taker to volunteer while the second element provides for a check and balance on the risk taker to alleviate situations where a risk taker might be reluctant to decline taking inappropriate risk [14].*

Your SMA plan needs to address the list from the NASA policy noted at the beginning of this section, which has areas of concern such as cryogenics, electrical issues, explosives, propellants, pyrotechnics, fire, and hazardous materials. You may need to address two primary areas: emergency preparedness and non–health-associated personal protective equipment. Emergency preparedness has to do with worker health and safety aspects while personal protective equipment has to do with items such as fall protection restraints and hard hats. Examples of these areas in the SMA plan include hard hats in facilities where a satellite or large subsystems are moved, procedures for moving large or heavy structures, protection for extreme temperatures, such as cryogenics, protection from hazardous fuels, such as rocket propellant, and protection from electrical shock and currents.

2.4.15. Hazard Analysis

A hazard analysis identifies potential hazards to personnel and product that could occur during manufacture and test of the product. A hazard should be defined as a condition where significant risk of injury or damage exists because of the nature of the operation being performed or of the material being handled. Typical hazards include electrical shock, explosion, fire, contamination, radiation, temperature extremes, corrosion, and collision. Each hazard should be listed as either catastrophic or critical. A catastrophic hazard can cause loss of life or damage to adjacent hardware. A critical hazard can cause significant personnel injury or damage to deliverable product.

2.4.16. National Environmental Protection Act Approval Process

One of the system safety concerns is launch approval process that considers protection of the environment. Figure 5.11 details that process.

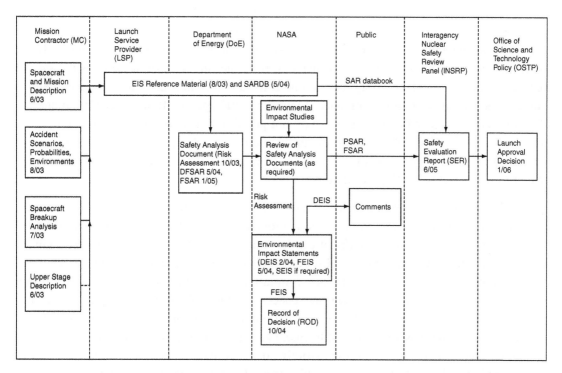

Figure 5.11: An example launch approval process that satisfies the National Environmental Protection Act.

2.5. Integration and Test

The two main objectives of the integration and test (I&T) process are the safe assembly of the flight system and the comprehensive validation and verification of the system performance. This section describes a hierarchical I&T method that has consistently led to successful spaceflight missions. Throughout project duration, plan and conduct tests in the following sequence:

- Requirements analysis

- Test planning

- Test development

- Test execution

- Test reporting

- Retesting resolved defects

- Regression testing

- Test closure

The process begins with early interface testing of breadboard circuits and encompasses numerous I&T activities up to launch. Requiring the most rigorous testing of hardware as early as possible and at the lowest levels of assembly uncovers potential problems early, when they are both easier and less expensive to fix. Later in the development, as the system reaches higher levels of assembly, end-to-end tests, and mission scenario tests; they help validate the overall system performance. Since scientific spacecraft are typically unique designs, and since schedule constraints may not permit the development of a separate qualification unit, a protoflight test program is often used, which is the test program described in this section. A protoflight test program means that the actual flight hardware is tested to flight levels for flight durations.

The I&T process is like the anchor leg of a long distance relay race. This phase of the project often inherits schedule delays from previous problems and the team must make up time by working long hours or multiple shifts. It can be tempting to cut corners or skip steps to make up schedule, but following established processes can mean the difference between a successful mission and a complete failure. Similarly, a gap in test coverage could result in a problem slipping through and leaving the spacecraft vulnerable to a catastrophic failure in flight. See references [19] and [20] for a discussion of the importance of validation. These examples show why it is so important to plan carefully, establish I&T processes, ensure comprehensive test coverage, and then strictly adhere to the planned processes. Every I&T plan should include a list of "incompressible tests." These tests represent the minimum set of tests that must successfully be performed before the spacecraft can be launched.

2.5.1. Requirements and Test

A process for ensuring comprehensive coverage of the tests was introduced in the Systems Engineering Section earlier in this chapter. Proper flow-down of requirements ensures that the spacecraft design meets the mission requirements. Identification of the verification method and verification event for each of the requirements forms the basis for the test plan and associated analysis. This information is often portrayed in the form of a verification matrix. Every test program should use a system verification matrix to confirm that there are no unverified requirements that could leave the door open for a problem to slip through unnoticed.

When identifying a method for requirements verification, the preferred method is by test. For all tests a "test-as-you-fly" philosophy should be used whenever practical, because it ensures the system is verified in the flight configuration according to the mission operations plans. When planning a test, there are often trades between the fidelity of a test and the associated resources required to accomplish the test. To increase confidence that the project is making the correct choices with respect to this difficult balance, independent review of test plans should take place early in the planning stages to allow for changes or enhancements if necessary.

In some cases a test-as-you-fly approach is not practical or possible. As an alternative, a "fly-as-you-test" philosophy may be appropriate. For example, if there is not enough time in the schedule to complete failure mode testing in every configuration of the spacecraft, it may be acceptable to constrain operations to only those configurations that have been tested. Similarly a communications system may be designed with a large number of data rates to provide operational flexibility. But with many rates, it may be impractical to test all rates prior to launch. An alternative approach is to constrain the operations team to only use the rates that have been tested for nominal operations. In the case of an unforeseen event requiring additional data rates, the new rates could be carefully tested at that time and made available for use in normal operations.

The remainder of this section describes a multi-layered test approach starting with the component-level tests. Component-level testing requirements are followed by spacecraft-level testing requirements. The processes and documentation ensure that the tests are repeatable and accurately reported and explained; guidelines for handling unexpected events and troubleshooting are included.

The first step is to make sure that no requirements are missed in the evaluation of the system during I&T. Next, start with a review of the list of requirements and make sure that all objectives are achieved in a test. The goal is for every requirement to have a corresponding metric and test that provides a clear result and indication. This is not always possible; consequently, requirements may be evaluated by one of four methods—assessment (or analysis), inspection, demonstration, or test. Generally, the strongest and most desirable criterion is test. Finally, in addition to verifying the functionality of the requirements, it is also important to test the performance of the design.

2.5.2. Data Trending

Trending of data and performance parameters is important because it can provide an early indication of a problem before a failure actually occurs. Don't wait until integration and test to think about important telemetry points that should be trended. Require each lead engineer to provide a list of trending points in advance of the component delivery, so that the I&T team is ready to monitor and trend these telemetry points. Once the points are identified, start trending early and often. By trending early, a baseline can be established by which other measurements can be benchmarked. Begin with the component-level performance test parameters to compare with the parameters after integration. Trending often improves the likelihood that a sudden shift in performance can be correlated to a root cause or external event.

2.5.3. Independent Data Review

An independent data review is warranted after a major spacecraft-level environmental test, such as thermal vacuum testing. During this long-duration, continuous test, the volume of data is likely too large to adequately review during the test. Without a careful review of the data after the test is complete, a subtle problem could go unnoticed until after launch, when it is discovered that there was an indication of the problem prior to launch. The benefit of the independent data review is that it gives an opportunity for probing questions and scrutiny of the results with a fresh set of eyes. This data review is one of the criteria in the test closure process.

An example of a subtle problem that might have been found through an independent data review is the WIRE Mishap in 1999. "The Wide-Field Infrared Explorer Mission objective was to conduct a deep infrared, extra galactic science survey. The Wide-Field Infrared Explorer was launched on March 4, 1999, and was observed to be initially tumbling at a rate higher than expected during its initial pass over the Poker Flat, Alaska, ground station. After significant recovery efforts, WIRE was declared a loss on March 8, 1999.... The root cause of the WIRE mission loss is a digital logic design error in the instrument pyroelectronics box. The transient performance of components was not adequately considered in the box design. The failure was caused by two distinct mechanisms that, either singly or in concert, result in inadvertent pyrotechnic device firing during the initial pyroelectronics box power-up. The control logic design utilized a synchronous reset to force the logic into a safe state. However, the start-up time of the Vectron crystal clock oscillator was not taken into consideration, leaving the circuit in a non-deterministic state for a time sufficient for pyrotechnic actuation. Likewise, the startup characteristics of the

Actel A1020 FPGA were not considered. These devices are not guaranteed to follow their 'truth table' until an internal charge pump 'starts' the part. These uncontrolled outputs were not blocked from the pyrotechnic devices' driver circuitry" [21].

2.5.4. Test Closure Process

Once all requirements have been successfully verified (or waivered) and the validation suite of tests are accepted, the test phase may be closed. The closure process requires completion of the tests and customer approval of the results after data review. Test closure also requires successful retest of corrections and fixes to the system under test and to the test procedures. Test closure requires approval from the I&T manager, mission system engineer, and mission assurance engineer.

2.5.5. Elements of a Successful Integration-and-Test Program

Clearly, well-planned tests are critical. In addition, there are certain other considerations that are typical of a successful I&T program: appropriate approval authority, mission systems engineer involvement, good leadership, communications, and balance control with flexibility.

Three key roles provide the approval authority for all integration and testing: the I&T manager, the mission system engineer, and the mission assurance engineer.

It is important to have the mission systems engineer (MSE) actively involved in the daily activities during I&T. The I&T phase provides the most comprehensive verification of the system requirements, more than any other phase of the project. For this reason, it is imperative that the MSE be involved in the formulation and review of all test plans and procedures, as well as in the closure of the as-run test documentation. In addition, the MSE offers technical oversight with regard to process execution.

Good leadership fosters open communication and effective coordination. During the I&T phase, morning I&T meetings serve to coordinate the activities for the day, review upcoming events, and identify any required preparations or prerequisites.

There's a delicate balance between effective process control and the flexibility to deal with the uncertainties that occur during integration. First, consider process control through the test directive, which is the agreement between the project developer, the test facility, and the customer. It defines the technical and schedule requirements of the project and the degree of customer on-site participation. Its goal is to optimize testing and minimize costs. Next there is the need for flexibility and to respond to changes; requests for a change in design usually leads to a change in a drawing to create a new design; this leads to the red-line change process, which marks up a design drawing to indicate the new design before it is committed to a final drawing. Red-line changes require less rigor in the approval process with the understanding that the final design will be reviewed before final approval. Black-line changes are the full design cycle with all approvals in place and represent full control.

2.5.6. Outline of Test Activities by Outside Organization

Sometimes an organization or company outside the project management might have to test a subsystem or the system. This outline represents an example of the sequence of events and activities that might occur during test by an outside organization. It is based on a brochure from NASA's White Sands Test Facility [22]:

1. Cost estimate—Generated by the test facility and based on customer stated requirements.

2. Funding.

3. Test requirements—Generated by the customer.

4. Test directive.

5. Design reviews and facility buildup.

6. System verification and validation.

7. Test readiness review (TRR)—Management and customer review to ensure the systems and planned operations will satisfy customer requirements in a safe and efficient manner.

8. Open paper review—A review conducted before testing to ensure that there are no outstanding tasks or problems which would preclude a safe and effective test.

9. Pretest briefing—Review of pretest operations, countdown timeline, station assignments, and credible emergency conditions and responses.

10. Test conduct—By procedures approved by the test facility and the customer; all changes must be documented and approved; completed procedures are signed off by the appropriate authorities and verified by QA.

11. Post-test briefing—Test team reports included in official records.

12. Data reduction.

13. Data analysis.

14. Final report.

2.5.7. Component-Level Tests

Component-level testing is done early, because problems are easier and less expensive to solve earlier in the development. At higher levels of assembly, it may not be possible or practical to expose every component to the predicted environment, so this type of testing must be done at the component level. The test program must include verification of every interface. The component-level test serves as the baseline performance against which all future tests are measured.

As early as is practical, interface testing should be conducted to demonstrate the compatibility of each interface with other components and subsystems. Usually this test is performed with non-flight breadboards and is not intended as a formal verification of the interface; rather, it helps mitigate the risk of discovering a problem after the flight hardware is built.

After the flight hardware is built and functionally tested, performance tests characterize the behavior of the component in a flight-like environment. The specifications for these environmental tests are typically documented in the project's environmental test specification, sometimes referred to as the component environmental specification. Vibration, thermal vacuum, acoustic, shock and EMI/EMC tests are typical requirements that are specified in this document. In addition to the test requirements, this document also specifies the ambient requirements such as temperature and humidity during integration and transportation. The document may also include for reference a description of spacecraft level tests that will be conducted after spacecraft-level integration.

Depending on the mission objectives, contamination control may be another important part of the component-level preparations prior to delivery. For missions with sensitive optical imaging payloads, it is usually required to bake out components prior to delivery to reduce spacecraft outgassing after launch and minimize contamination of the imaging instruments. Often times this bake-out is conducted as an unpowered long-duration hot cycle at the beginning of thermal-vacuum testing.

Before delivery to the spacecraft, the I&T team should coordinate with the component (or subsystem) development team to write the integration procedure. This coordination helps insure a safe component installation on the spacecraft. Sometimes it is necessary to incorporate verification events into the integration procedure, because certain tests can only be done before connecting the spacecraft harness. For example, there are often grounding and isolation requirements between primary and secondary power returns. If these are not verified during the installation, they cannot be measured later without removing connections to the component. To help facilitate verification and validation, it is useful to include a cross-reference table with all requirements that are verified in the procedure and the step where the requirement is verified. Finally, you should hold a component-level preship review (PSR) to review the data results and confirm that the components and subsystems are indeed ready for integration on the spacecraft.

2.5.8. Spacecraft-Level Tests

Because each spacecraft development is unique, there is no such thing as a "typical" test flow. Depending on the location of the test facilities and the characteristics of the spacecraft, activities or tests may be required for some missions that are not needed for others. The New Horizons spacecraft, for example, was too large to fit in the thermal vacuum chambers at

The Johns Hopkins University Applied Physics Laboratory, where it was built, so the spacecraft was packed and shipped to NASA's Goddard Space Flight Center after vibration testing. A spacecraft with a sensitive magnetometer may need a magnetic survey as one of the tests in the I&T flow, but a spacecraft without this type of instrument will not need magnetic tests. This section describes many of the tests that make up the I&T flow for integration with the spacecraft and explains the development of an example flow.

As mentioned previously, the I&T team often inherits the problems experienced earlier in development. It is important for the team to be prepared with contingency plans in case the component deliveries do not occur as planned, which is often the case. One may begin the project with an integration flow that starts with the assembly of the structure and is followed by the installation of the power system and then the central computer, but when the structure is delayed, the I&T lead must be prepared with creative alternatives to keep the program on schedule. One option could be to integrate components on a flat plate until the structure arrives. This approach enables the team to work out any electrical or cable harness issues prior to the arrival of the structure. The success of integration and testing depends on the team's ability to adapt to changes as the project progresses.

A goal of the I&T flow is to accomplish efficiently the required set of I&T activities within the project constraints. As a general rule, the test flow typically follows the sequence of events in the order they would occur during launch and operations. As shown in the example flow of Fig. 5.12, the environmental testing starts with the vibration and acoustics tests, then the shock test, then the thermal vacuum tests, including a comprehensive performance test at a hot and cold plateau. This order covers the launch, separation, and on-orbit operations of the spacecraft. Some tests, such as the solar array deployment, are not possible while inside the thermal vacuum chamber. In such cases we rely on tests of the mechanisms at lower levels of assembly, where the performance was tested in thermal vacuum at various temperatures, and a test in room-temperature air before and after the vibration and thermal vacuum testing to determine if any changes in performance occurred.

Each of these activities is described in detail either in the System Test Plan or in a separate standalone test plan. The I&T plan lays out the overall flow of the I&T activities. It reaffirms the standard processes by which the activities will be conducted and, if necessary, identifies any exceptions to those standard processes. For reference, Table 5.14 describes each activity in the test flow. Many of the activities make up the set of spacecraft-level verification tests. In addition to these standard verification tests, other performance tests and special tests are performed to satisfy specific verification requirements.

Appendix C to this chapter has an example test plan, which may serve to give something of a description of what you might expect in a test plan. It does not provide a schedule, which should be an integral part of a test plan. Figure 5.13 describes an example set of test

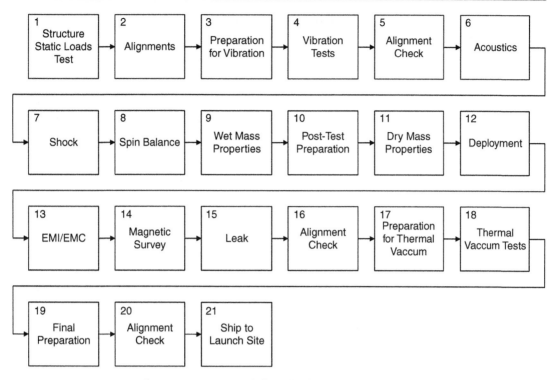

Figure 5.12: General flow for integration and test.

Table 5.14: Integration and Test Activities

Activity	Description
Structure Static Loads Test	This test provides the verification of the strength of the structure. Before integrating the valuable components onto the structure and performing a vibration test, you want to be reasonably sure that the structure will support the predicted loads during the vibration tests. Prior to the test the structure is configured with mass load simulators and strain gauges. During the test the stress and strain loads are monitored through the strain gauges.
Alignments	These tests use theodolites to measure alignment of the subsystems and later the payload module and the spacecraft. Tests for spacecraft alignment stability verify the mechanical stability of the spacecraft. Photographing the set-up is good for later review.
Preparations for vibration	You generally perform vibration tests on subsystems and then the entire spacecraft. Design the procedures and vibration profiles to cover what is expected in the mission and to prove the survivability of the spacecraft. Align each article (subsystem or spacecraft) on the shake table for the axis called for by the test, attach accelerometers, and tape down wires. Photographing the set-up is good for later review.

(Continues)

Table 5.14: Cont'd

Activity	Description
Mass Properties Test	These tests measure and infer the moments of inertia, the total mass, and the center of mass. You may use a load scale to suspend the subsystem or instrument to determine the mass. A fixture with gimbals in three axes may be used to measure center of gravity on all three axes. Sometimes a spin balance test can be used to determine the moments of inertia.
Dry mass properties	Tests for dry mass properties measure the spacecraft without water, propellant, or cryogens that may be later required in the mission. Dry mass properties are an initial form of test to project what the final mass properties of the spacecraft will be.
Wet mass properties	Tests for wet mass properties measure the spacecraft mass properties with very clean deionized water. As a final determination, you might use propellant or cryogens as they will be loaded into the spacecraft.
Deployment Test	This test is a functional exercise to extend an instrument boom or communications dish or solar panels. It helps avoid mechanical interference problems. Deployment tests can sometimes simulate a 0 g environment by suspending the boom subsystem either with counterbalancing springs or in a water tank [23].
EMI/EMC Test	These tests look for interference and susceptibility in instruments, subsystems, and the entire spacecraft. A number of standards, such as MIL-STD-461E, and guidelines can provide the necessary direction to prepare and run these tests. The concern is that RF radiating communications do not "jam" sensitive instruments or that power system noise does not affect subsystems or that spacecraft electrostatic fields do not affect sensors.
Magnetic Survey	Strong sources on spacecraft, such as thrusters, reaction wheels, and solar panels, sometimes generate significant, local magnetic fields. To test for these, magnetic measurements can use a gradiometer to probe the spacecraft during the different test phases and map its magnetic fields and calibrate sensors for them.
Modal Survey	A modal survey verifies the finite element model for coupled loads analysis of the spacecraft. Modal surveys use a vibration table and generate multipoint random, stepped-sine, and sine-dwell mechanical vibrations to characterize the dynamics of the spacecraft.
Sine Vibration Test	Vibration tests help characterize the instruments and spacecraft for the mechanical environment, particularly during launch. Sine waves are the easiest to generate, analyze, and interpret. Tests use a sine wave whose frequency is swept from very low values, such as 5 or 10 Hz, up to 2000 Hz in each of the three axes. Accelerometers firmly attached to the instrument or spacecraft record the mechanical response of the article under test. Review of the data compares the mechanical response and resonant peaks to the analytical model.

(Continues)

Table 5.14:   Cont'd

Activity	Description
Random Vibration Test	Vibration tests help characterize the instruments and spacecraft for the mechanical environment, particularly during launch. Random vibrations are the most realistic in simulating the environment. Tests use arbitrary waveforms whose frequency spectrum range from very low values, such as 5 or 10 Hz up to 2000 Hz, in each of the three axes. Accelerometers firmly attached to the instrument or spacecraft record the mechanical response of the article under test. Review of the data compares the mechanical response and resonant peaks to the analytical model. Some recent studies may indicate that random vibration tests in all three axes simultaneously may be more effective [24].
Acoustic Test	This test is very similar to the vibration tests—you are attempting to simulate the launch mechanoacoustic environment to verify the mechanical integrity of the spacecraft. Arrays of speakers point at and bombard the spacecraft with acoustic energy.
Shock Test	This test is a mechanical impulse test of the instrument and later of the spacecraft, again to verify the mechanical integrity of the spacecraft. A hammer with a calibrated mass can strike a plate bearing the instrument or spacecraft or a vibration table can simulate an impulse with a very broad spectrum of mechanical energy.
Spin Balance Test	For spacecraft or subsystems that spin during the mission, this test measures and adjusts the mass rotational symmetry of the spacecraft. You add masses to the structure so that it aligns with the geometrical axis of symmetry of the structure. This is like a car tire balance machine, only bigger and more accurate. The test uses a turntable with the spacecraft attached to balance it.
Thermal Vacuum Cycling Test	Thermal vacuum tests help characterize how well the instruments and spacecraft will function, particularly during the mission. These tests occur in large steel chambers that can accommodate a spacecraft or large subsystem and be evacuated to simulate the vacuum of space. Heating cords or high-wattage lamps provide the heat and cooling coils in the chamber walls produce the cold environment. These tests verify operation in a space environment or help find weaknesses in joints and materials that fatigue under thermoelastic expansion and contraction or outgassing problems. These tests cycle between extreme temperatures with dwell times at the extremes to allow temperature equilibration and time to run operational tests on the test article. These tests can run for several days up to weeks in duration.
Thermal Vacuum Balance Test	Thermal balance testing verifies the thermal model of the instrument, subsystem, or spacecraft; it also verifies the ability of the instrument, subsystem, or spacecraft to operate in vacuum and at the temperature extremes, both upper and lower, that are expected for the system. The test uses an environmental vacuum chamber to test the powered-on instrument, subsystem, or spacecraft, which has thermocouples attached to measure temperature. Once a temperature extreme or plateau is reached and the instrument, subsystem, or

(Continues)

Table 5.14: Cont'd

Activity	Description
	spacecraft has equilibrated in temperature, you measure the temperature of the various sensors to determine the thermal balance and then verify the accuracy of the thermal model.
Leak Test	This test looks for leaks in the propellant and cryogenic systems, particularly in the service valves. A helium leak checker attached to the system can easily quantify the flow of pressurized helium in the plumbing to indicate the quality of the tubing and valves.

Req. #	Description	Type of test	Initials of tester	Results (pass/fail)	Comments or resulting data output
3.4.1	The design of Instrument "A" shall conform to good electromagnetic compatibility (EMC) design practice.	a			
3.4.2	Circuit boards (PCBs or PWBs) within Instrument "A" shall contain ground (signal return) planes that are continuous and uninterrupted with slots or large, open regions with no metal plating.	i			
...	....				
3.4.6	Paired signal and return lines within cabling between the vehicle and Instrument "A" shall be twisted at 6 turns per meter or 2 turns per foot.	i			
3.4.7	The cabling between the vehicle and Instrument "A" shall have a continuous, electrically conductive capacitive shield over all wire conductors.	i, t			
...	....				
4.2.1	The software development team(s) shall maintain a record of lines of code (LOC) generated, debugged, tested, and integrated into the vehicular portion of Instrument "A".	i, d			
4.2.2	The software development team(s) shall maintain a record of the time/effort expended to generate, debug, test, and integrate each software module in the vehicular portion of Instrument "A".	i, d			
...	....				
5.4.2	Input power on the supply bus will be supplied to Instrument "A" at a nominal voltage of 28 VDC. The variation in voltage could be ± 20%. Instrument "A" shall operate throughout this range of voltage input	d, t			
5.4.4	Instrument "A" shall consume a maximum of 90 W.	d, t			max. recorded voltage = _____ and current = _____

Figure 5.13: Example portion of a test procedure. a, assessment (or analysis); t, test; d, demonstration; i, inspection.

procedures. The life cycle of a test should follow this order: after the test plan is released, a procedure is developed to capture each step of the activity. List requirements that are verified at the top of the procedure to help facilitate verification. The process life cycle of a test procedure follows:

- Procedure development

- Procedure review

- Procedure approval

- Activity execution

- Data review

- Activity closure

During the I&T activities, it may be necessary to alter the test directives (the agreement between the project developer, the test facility, and the customer to define the technical and schedule requirements of the project with the goal to optimize testing and minimize costs). These changes may occur through red-line and black-line changes. Test closure requires verification that the procedure was followed and any red-line or black-line changes to the procedure are appropriate.

Several cautions seem to be in order. You should have "safe to mate" procedures for plugging cables into instruments and into the spacecraft. You should have a procedure to always discharge harnesses before mating them together or to connectors (long cables can build significant static charge through triboelectric effects during flexing).

2.5.9. Performance Tests

Performance tests are conducted in multiple places during integration and test and in multiple ways. Here are four common performance tests:

- Aliveness

- Functional

- Comprehensive performance

- Mission simulation

The aliveness test powers up an instrument or the spacecraft to perform an elementary power-on self-test. The aliveness test is not comprehensive but it gives some assurance that instruments and the spacecraft may be working as expected. It usually is the first performance test run and the last when the spacecraft is on the launch pad. Aliveness tests can take between 10 and 30 minutes, such as the aliveness test for the New Horizons spacecraft in the case study.

The functional test is a full test of all interfaces and operational modes. It is far more detailed than the aliveness test and can take a day or more, as for the New Horizons spacecraft in the case study. This test is only for the engineering team in the engineering mode; it is not for the mission operations staff, although there may be many similarities.

The comprehensive performance test puts every instrument and subsystem through a full test at both hot and cold plateaus at least. This test can take from 1 to 7 days as for the New Horizons spacecraft in the case study.

The mission simulation test, which is sometimes called "a day in the life," operates the spacecraft as it will be flown. Only those inputs that cannot be directly generated by the spacecraft will be simulated. The mission simulation tests help train the mission operations staff and are similar to the functional test.

2.5.10. Special Tests

Special tests are additional tests specific to a particular subsystem; usually you conduct each test only once during the integration and test period. They typically verify a subsystem characteristic.

One such test is the plugs-out test, which means operating the spacecraft with no (or minimal) electrical cable connectors, suspended on a crane with an isolating block. Communication with the spacecraft is only via the RF link. This practice ensures that test facility problems are not distorting the test results, such as might be caused by illicit ground loops.

Another type of test is the polarity/phasing test, which tests the guidance and control system to ensure that spacecraft commands actually work in the correct orientation and direction. These sorts of tests particularly aim at actuators and sensors. An example might be to test that the torque rods respond correctly and are not installed backwards. Another example would be to test that thrust valves open correctly upon command.

2.5.11. Conducting Tests

Every test and test procedure should have a formatted data collection form for recording the results that allow later data review to reconstruct the events. The data collection should have clearly marked places for recording the following:

- Start times
- Test configuration
- Equipment used
- Calibration dates
- Tester's signature
- Data and results

Sometimes explanatory notes or rationale might be included for a particular test procedure. Some test procedures might also have the expected results as a separate column to allow immediate comparison with the recorded results. Finally, some test procedures might need references to specific documents.

Every test procedure should indicate what to do if a problem occurs—that is, a deviation from the expected. The first step is to file an anomaly report, described in the next section. From there, a number of actions can arise, also described below.

2.5.12. Anomaly Reporting

During testing, it is important to have a multi-level system for documenting, tracking, and resolving unexpected behavior. Here are the majority of recognized forms to report problems:

- Anomaly reports

- Problem/failure reports

- Software change requests

- Electronic media change requests—database

- Engineering change requests (ECRs)

- Engineering change notices (ECNs)

Each of these forms has a level of significance and effort associated with it.

At the first level are anomaly reports. It is important to provide all individuals on the project immediate accessibility to reports of anomalies. A networked database tool can facilitate these. The closure process at this level should be fairly simple and straightforward. The main purpose of this level of anomaly reporting is to log any unexpected behavior that might be an indication of a potential problem or failure. Closure of anomalies at this level requires the I&T manager, mission system engineer, and mission assurance engineer.

In the event the anomaly requires rework, the previous testing is no longer valid. The next section outlines what testing can be done to revalidate the testing.

If an anomaly proves to be a significant problem that is not waived or closed out, it then goes into a problem or failure report. These reports reach the next level of importance; they trigger a greater level of review, either by a failure (or control) board or by a designated group. This review often requires a root-cause analysis or ancillary analysis to be performed, as well. Once the work has been done and a course of action to resolve the problem or failure has been recommended, then the failure (or control) board signs off. Closure of a problem or failure report requires the I&T manager, mission system engineer, and mission assurance engineer.

A problem or failure report can lead to a software change request or an electronic media change request or an ECR. Each of these requests initiates an effort to investigate a change. If analysis of the problem, such as root-cause, has been performed, then the appropriate team or expertise proposes a solution. Any solution then requires a tradeoff analysis to confirm its suitability. Generally a control board signs off the request. If approved, the team performs work on the request or generates an ECN to prepare to do the work. Sometimes the work is quickly completed or waived and does not need the effort for an ECN. Closure of a request requires the I&T manager, mission system engineer, and mission assurance engineer.

An ECR or a change in requirements or a change driven by the customer may alter the system design. When this happens, the developing team prepares an ECN to red-line and then black-line the design. The ECN not only describes the change in design, it also describes the effort required. If a tradeoff analysis has not been performed, then that needs to be included to provide another level of assurance for the ECN's suitability. The engineering or program manager uses the ECN and the results of any analyses to estimate the impact on the project's schedule and budget. An ECN is the highest level of importance when a problem occurs. Generally a control board signs off the ECN when it has been both completed and reviewed. Closure of an ECN, after review and control board sign-off, requires the I&T manager, mission system engineer, and mission assurance engineer.

2.5.13. Retest Requirements

If an instrument or subsystem becomes damaged while bolted to the spacecraft, it must be fixed and requalified through an appropriate set of tests. The goal is to avoid an entire requalification, which sometimes just is not possible on the final system. Requalification should be a minimal set test. Often that set can be a hot/cold thermal cycle, reduced vibration profile (it sometimes is called a workmanship vibration), functional test of the instrument, and then finish with a thermal cycle test. Regardless, retest requirements have to be defined for each particular project and system.

2.5.14. Test Closure

Every test procedure may be closed only after data analysis and signed off by the lead engineer. Finishing out and closing all testing requires an independent inspection; often the customer is involved in the final review. Test closure must be approved by the I&T manager, mission system engineer, and mission assurance engineer.

2.5.15. Test Equipment Calibration/Certification

Any tool, meter, or instrument that generates a value recorded on a qualification (or test) report to a customer must be calibrated. Examples of such tools and instruments are tools such as torque wrenches and calipers, meters such as mass scales and digital multimeters, laboratory equipment such as oscilloscopes, logic analyzers, power supplies, waveform generators, network analyzers, and filters, environmental recording instruments such as hygrometers, humidity controllers, particulate counters and filters, and temperature recorders. The type and period of calibration is set by each manufacturer or by government standards, whichever takes precedence in the particular type of calibration or project. Often the calibration and certification of equipment is already a part of the quality system of the producing company. The I&T manager, mission system engineer, and mission assurance engineer should be cognizant of these calibration procedures.

2.6. Mission Operations

Mission operations are control procedures generated on the ground and sent to the spacecraft. Even though they are unique to each spacecraft and its mission, they all have a common core of basic operations [25]:

1. Telemetry, commanding, and data archive.

2. Spacecraft load management—This is the process of building command loads (lists of commands) that are to be uplinked to the satellites. Example inputs to the command load set include ground station schedules, products from flight dynamics, and science observation timelines.

3. Automation rules and operator notification—Automation rules diagnose problems or anomalies as they occur and react according to predefined system rules.

4. Flight dynamics system—Flight dynamics estimates need to be generated on a routine basis to assist with mission planning activities. The primary prerequisite to running a flight dynamics product generation session is an ephemeris file for the satellite.

5. Spacecraft health and safety monitoring.

Lead engineers review all command loads going to the spacecraft for correct operation. The time duration covered by a command load can vary widely; for New Horizons, it can be for 1 week or 9 months. Another important component is that careful exercise of configuration management makes sure that command versions are correct.

3. Documentation

Documentation is an essential part of a spacecraft development project. In addition to serving as a record of the system configuration, it also serves as a method for communication and training within the team. Documentation used on a project can be divided into two categories: corporate documentation, such as corporate-wide quality processes and standards, and project documentation, such as assembly drawings, requirements documents, and verification/test plans. This section provides examples and descriptions of typical project documentation and typical documents in corporate libraries that are relevant to the project and are likely to be referenced in project documentation. Following these document descriptions is a list of tools that are useful in the management of corporate and project documentation.

3.1. *Project Documentation*

Every development project must have a documentation tree. This tree provides a hierarchical description of assembly of the spacecraft from the tiny screw that holds a connector bracket in place on a circuit board to the top-level assembly drawing for the entire spacecraft. Every document produced on the project has a place in this tree. In addition to the assembly documentation, other process and configuration documentation are included in the tree as well. For example, the test plan of a component should be located in the tree under the branch that corresponds to that component. This approach provides quick access to all documents associated with a component by simply extracting all documents on a particular branch of the tree. Table 5.15 lists examples of documents typically developed for space missions. In some cases, different organizations may use different titles for these documents, but each of the items in the descriptions should be documented within the project documentation tree.

Table 5.15: Examples of Project Documentation

Document Title	Document Description
Project plan	The project plan is a high-level description of the overall project. It documents the mission objectives, management approach, technical implementation, and project commitments. Performance goals including cost and schedule are also included in the project plan. The WBS is also part of the project plan. A template for a project plan is included in NPR 7120.5D.
Systems engineering management plan (SEMP)	The SEMP describes the project's approach for systems engineering. This document explains the plan for requirements development and verification/validation of those requirements. A template for the SEMP is included as part of NPR 7123.1.
Mission requirements document	This document provides the necessary requirements to describe the overall goals and purpose and the mission profile and schedule. It supplies the framework that guides the preparation of the system requirements documents.
System requirements document	The spacecraft, ground system, and each instrument generally have their own set of requirements in a separate document. These must be an integrated set that fulfills the mission requirements.
Mission assurance documents	Safety and mission assurance plan (also referred to as a product assurance implementation plan [PAIP] in some organizations), contamination control plan, ESD control plan, configuration management plan (these can be company processes and procedures)
Validation/verification (V&V) or test plan	V&V plan is the overview document for setting the philosophy of test. A V&V matrix correlates each requirement with an appropriate test; the matrix should address every technical requirement for the

(Continues)

Table 5.15:   Cont'd

Document Title	Document Description
	system. Many times this is called the test plan, which gives a detailed description and instructions for all the tests necessary or required within a project. A test plan performs the two basic functions of verification and validation. Closure documentation confirms that V&V has completed satisfactorily.
Documentation plan	This plan strives to provide consistent control over all the documents generated in a project. It defines what documents will be generated and what their format will be.
Trade studies	Trade studies occur at several levels from architectural concerns and the integration of major subsystems down to individual circuits and components. Architectural trade studies might consider, as examples, whether implementation of a particular function should be in hardware or software, whether communications should be serial or parallel, and whether power regulation should be a centralized function or at the point of load. Trade studies continue until the analyses and requirements have settled to a stable point and the project management signs off on the approval of the results.
Risk management plan	Risk is the potential inability to stay within defined cost, schedule, performance, or safety constraints. Risk analysis establishes the likelihood of problems and their severity. Risk management uses these metrics to then manage the system margins. Risk Management identifies potential problems (e.g., mission, technical, personnel, and funding concerns), categorize their effects, or criticality, and then plan risk-handling activities across the life of the spacecraft to mitigate adverse impacts on achieving objectives. For smaller projects, the hazard analysis might be a part of this plan.
Interface control documents (ICDs)	Payload to Spacecraft ICD, Spacecraft to Launch Vehicle, Spacecraft to Ground System ICD. Each of these documents provides the necessary characteristics for each type of interface (e.g., electrical, software/data/information, mechanical and thermal) so that the system or subsystems may connect seamlessly during integration. An electrical ICD specifies items such as the signal names, connector pin numbers, polarities, frequencies, shielding, and signal levels. The software/data/information ICD specifies items such as data rates and formats.
Development plans	Each of these cover disciplines, such as system architecture, software, electronics, mechanics, and structure and materials, that may proceed in parallel with the other disciplines. A development plan gives detailed coverage, description, and instructions of the particular system or subsystem or component. The level of detail

(Continues)

Table 5.15:   Cont'd

Document Title	Document Description
	needs to be sufficient to allow it to be either replicated in production and manufacturing or available for future modifications.
Software management plan	This plan indicates the methods, techniques, tools, certifications, and guidelines that developers will use to create software. It does not directly relate to the actual design requirements. Sometimes this management plan is rolled in with the software development plan.
Component and material specifications	This is a general design requirements specification for components and parts. It may require a program approved parts list (PAPL). It might contribute to a program approved materials and processes list (PAMPL), which covers the following concerns: corrosion, incompatible metals, mechanical fasteners and parts, nonmetallic structures, absorption and outgassing, atomic oxygen erosion, flammability, non-nutrient and fungal control, structural bonding, nonstructural bonding, and protective finishes.
EMC/EMI control plan	This plan should provide guidance for preparing circuits, modules, and subsystems for EMC; examples might include, "Always pair a signal line with a return path that is unvarying in separation or dielectric constant to minimize inductive impedance," or "Eliminate slots or holes in the return path (i.e., slots in the ground plane) to minimize inductive impedance." The plan should also have guidance for EMC tests, which should include instructions for testing for conducted susceptibility, conducted interference, radiated susceptibility, and radiated interference.
Mission concept of operations (CONOPS)	Concept of operations, flight constraints, standard operating procedures, contingency operating procedures
Contamination control plan	A plan for facilities and procedures to control contamination from both particulates and outgassing during design, manufacturing/ production, and packaging for storage and shipping. It addresses the use of proper materials, component cleaning, conformal coating to control outgassing, clean room facility, and clothing apparel.
Spacecraft disposal plan	This plan directs the disposal when the spacecraft finishes its mission. It may a plan to de-orbit if the spacecraft is to come down in 25 years or boost into an orbital graveyard to stay up for >1000 years.
Test documentation	System test plan, instrument test plans, system test procedure, instrument and subsystem test procedures, test directives, test conductor logs.

(Continues)

Table 5.15: Cont'd

Document Title	Document Description
Configuration management plan	This plan is often a company process; it provides direction for consistent control over the status of the project and state of the developing spacecraft.
Acquisition plan (subcontracting plan)	This plan indicates how vendors are selected and qualified and what work they will perform for the project.
Design description documentation	A design description for a subsystem gives detailed coverage, description, and instructions. The level of detail needs to be sufficient to allow it to be either replicated in production and manufacturing or available for future modifications. Design descriptions should be written for the electronic and mechanical hardware, the structure and materials, and for the software.
Technology development plan	This plan describes the technologies that will be used, such as electronic components, mechatronics, new materials, and new software tools and techniques. This plan is particularly important for new technologies; it becomes an important adjunct to the risk management plan.
Presentations communications informal documentation	Design reviews—plans, materials, slides, and communications.
	Progress/status reports, memos.
	Engineering notebooks, e-mail messages.

3.1.1. Safety Analysis

The missile system prelaunch safety package (MSPSP) provides a detailed description of the hazards and safety-critical systems of the flight and ground hardware and software associated with the launch of the spacecraft. It is one aspect of obtaining launch approval. Instructions for the preparation of the MSPSP are contained in the EWR 127-1 [26]. The contents of the MSPSP are listed in Table 5.16. In some cases there could be a separate MSPSP for separate systems. Table 5.17 gives examples of other documents that you may encounter or need to prepare but it is not an exhaustive list.

3.2. Corporate Documentation

Corporate documentation comprises the enterprise-wide standards and processes that are referenced in the project documentation. These documents include standard fabrication procedures that describe a process so that it need not be repeated on every document. Corporate documentation also includes the quality management system, such as described in

Table 5.16: Missile System Prelaunch Safety Package Contents

MSPSP Section	Description
Table of Contents and Glossary	The MSPSP must have a table of contents and glossary of terms.
Introduction	The introduction addresses the scope and purpose of the MSPSP.
General Description	The MSPSP contains technical information concerning hazardous and safety-critical equipment, systems, and materials and their interfaces used in the launch of vehicles and payloads. Where applicable, previously approved documentation shall be referenced throughout the package. The MSPSP is a detailed description of the design, test, and inspection requirements for all ground support equipment and flight hardware and materials and their interfaces used in the launch of launch vehicles and payloads [26]. The detailed description of each of the following categories shall be provided. Each description shall include the following information: a. Nomenclature of system b. Function of the system c. Location of the system d. Operation of the system e. System design parameters f. System test parameters g. System operating parameters h. Summaries of any range-safety-required hazard analyses conducted [26]
Flight Hardware Subsystems	Overview of the flight hardware subsystems with specific data relevant to hazards in the appropriate subsection.
Structure and Mechanisms	Details of flight hardware used in lifting critical loads and details of each flight hardware mechanism that could pose a hazard.
Pressure/Propulsion Subsystems	Details of hazards and safety devices associated with pressure vessels or propulsion subsystems and components.
Electrical/Electronic Subsystems	Details of the electrical subsystem hazards, including flight hardware batteries.
Ordinance Subsystems	Details of hazards associated with flight hardware ordinances and explosive actuators.
Nonionizing Radiation Sources	Details of sources of nonionizing radiation such as RF transmitters and lasers and associated radiation protection plans. Must include location of the source, hazard areas, and local facilities.
Ionizing Radiation Sources	Details of sources of ionizing radiation such as radio-isotopes and associated radiation protection plans. Information to be included are manufacturer and model number, a description of the system and its operation, a description of the interlocks, inhibits and other safety features, a diagram showing the location, a description of the radiation levels, in millirems per hour, and a copy

(Continues)

Table 5.16:   Cont'd

MSPSP Section	Description
	of the RPO-approved 45 SWI 40-201 Radiation Protection Program Radiation Use Request Authorization.
Acoustical Subsystems	Details of flight hardware acoustic noise sources that could generate sound levels that are hazardous to personnel, the location of all sources generating noise levels that may result in hazardous noise exposure for personnel and the sound level (in dBA) for that noise, the anticipated operating schedules of these noise sources, methods of protection for personnel who may exposed to sound pressure levels above 85 dBA (8-hour time-weighted average), and copy of the bioenvironmental engineering approval stating the equipment and controls used are satisfactory.
Hazardous Material Subsystems	Details of any hazardous materials including flammability, toxicity, and compatibility with other materials along with any associated safety plans.
Computing Systems	Details of computing hardware, data flow, hardware/software interfaces, test plans, and hazard analysis.
Ground Support Equipment	Details of the hardware, data flow, hardware/software interfaces, test plans, and hazard analysis for the ground support equipment.
Material Handling Equipment	Details of the hardware, test plans, and hazard analysis for the material handling equipment.
Pressure Propellant Subsystems	Details of the hardware, test plans, and hazard analysis for the propellant subsystems and hazardous materials including flammability, toxicity, and compatibility with other materials along with any associated safety plans.
Electrical/Electronic Subsystems	Details of the hardware, data flow, hardware/software interfaces, test plans, and hazard analysis for the electrical and electronic subsystems.
Ordinance Subsystems	Details of the hardware, test plans, and hazard analysis for any ordinance subsystems and hazardous materials including detonation, explosion, flammability, toxicity, and compatibility with other materials along with any associated safety plans.
Operations Safety Console (OSC)	Details the schematic of the OSC and outside interfaces and a narrative of each of the features of the OSC, including the following: function, operation, outside interface, and operating limits. It also provides a summary of test plans, test procedures, and test results in accordance with the OSC Validation and Test Requirements section.
Vehicles	Documentation certifying that vehicles used to transport bulk hazardous material on the range comply with DoT requirements or are formally exempted by DoT. If DoT certification or exemption documentation is not available, the following information is required: —Design, test, inspection requirements —Stress analysis —SFP analysis

(Continues)

Table 5.16: Cont'd

MSPSP Section	Description
	—FMECA —Comparison analysis with similar DoT-approved vehicle —"Equivalent safety" (meets DoT intent) analysis
Seismic Data Requirements	Identify equipment that has the potential, directly or by propagation, to cause the following seismic hazards: severe personnel injury, a catastrophic event, significant impact on space vehicle or missile processing and launch capability, damage to high value flight hardware. Identify the expected G forces, the level of G forces the equipment can withstand, the magnitude of potential damage, and the method of restraint used.
Compliance Checklist	A compliance checklist of all design, test, analysis, and data submittal requirements in this chapter shall be provided. The checklist shall indicate for each requirement if the proposed design is compliant, non compliant but meets intent, noncompliant (waiver required) or nonapplicable. An example of a compliance checklist can be found in Appendix E of the Ranger User Handbook.
Change List	Summary of changes since the last revision.

Table 5.17: Examples of Other Safety Documents That May Be Required as Part of Project Documentation

Document Title	Document Description
Launch Site Support Plan (LSSP)	Reference [27] provides a template of an LSSP; reference [28] provides a flowchart.
Test Interface Requirements Document (TIRDOC)	The TIRDOC can have example entries such as: Initial Power On Measurements—test interface for stray voltages, check signal pins for levels that could cause damage Power—conformance to voltage specification (operation at nominal, high, low), in-rush current, power consumption over input voltage range, noise on power lines Signaling characteristics—check voltage range, jitter, and noise on signaling lines Electrical interface continuity and isolation test—verify power and ground isolation, signal isolation from power and ground, expected impedances between signal pairs Mate-demate, workmanship standard—inspect and verify that connector savers are being used, both halves of a mate for reference designators, cleanliness, pin/socket characteristics, discharge cables, and prepare paperwork for mate-demate cycles by filling in mate-demate log

(Continues)

Table 5.17:   Cont'd

Document Title	Document Description
	Applies to subassemblies (CCA, unit, cable) Performed when item has undergone any re-work or when I&T receives an item from a subsystem supplier or when item has not been used for some period of time
Process Waste Questionnaires (PWQs)	Identifies solid wastes are defined as the materials and equipment that are non-hazardous, non-liquid items that may be sold for scrap or as a recyclable product; or be disposed of in a landfill.
Orbital Debris Analysis	Analysis that considers particle size, velocity, initial altitude, and flight angle relative to the surface.
Planetary Protection Analysis	Analysis that considers planetary protection requirements that protect solar system bodies from biological contamination.
Environmental Impact Statement (EIS)	The EIS presents the analysis for the launch site and globally for concerns over exhaust plume pollutants and the effects on atmospheric ozone; also covers the potential effects of an RTG accident.
Preliminary Safety Analysis Report (PSAR)	The PSAR appraises the risk to public health and safety resulting from the handling, transportation, emplacement, operation and recovery of hazardous or radioactive materials such as an RTG system.
Final Safety Analysis Report (FSAR)	The FSAR is the final report that appraises the risk to public health and safety resulting from the handling, transportation, emplacement, operation and recovery of hazardous or radioactive materials such as an RTG system.

a quality standard such as AS9100. In most cases the project does not have the authority to directly modify these processes. Table 5.18 gives some examples of corporate documentation that may be important to a project.

3.2.1. Documentation Revision and Approval Processes

Depending on the project, documentation for spacecraft or instrument subsystems can require several different levels of rigor and configuration control. Some companies might have these levels defined as Level 0, 1, and 2. Level 2 is complete configuration control where no change is made unless all changes have been approved and have been signed off, and all design processes iterated before implementing the change—basically red-line changes are not tolerated in manufacturing. Level 1 is similar except that red-line changes may institute a change in production but the change is not approved until the design drawings have been through the change process to black-line changes. Level 0 has no such controls and is seldom ever used in development of spacecraft.

Table 5.18: Corporate Documentation Likely To Be Referenced in Project Documentation

System Title	Description
Quality Management System	Good quality is a necessary ingredient to being profitable and even to survival. An important component of quality is to have consistent application across the enterprise, which means instituting appropriate processes that involve people, tools of industry and business, and procedures. ISO 9001, AS9100, and CMMI are all standards efforts to develop quality management systems (QMSs) that arose from different industries but can be applicable to developing spacecraft. A quality management system (QMS) must identify processes, controls processes, document the effort through record keeping, and control the records. A QMS has the following components and concerns: Management and staff responsibility Resource management Product realization Measurement, analysis, and improvement QMS implementation
Configuration Management System	Configuration management maintains the current state of the project. Its database keeps records of the requirements, the documents, the design descriptions, the minutes of reviews and meetings, the purchase orders (POs), the work orders (WOs), the test results, and the communications with the customer, subcontractors, and employees.
Documentation System	Documentation is integral to product development; it works in concert with the configuration management system. Documentation provides the necessary explanation for the "who, what, when, where, why, and how" within that development. Documentation records many aspects of development and the product, including the requirements, the development plans, the design descriptions, the minutes of reviews and meetings, the purchase orders (POs), the work orders (WOs), the test results, and the communications with the customer, subcontractors, and employees.
Engineering Processes	Company process provides the guidelines or procedures for engineering design in electrical, mechanical, material, and software development. Often these processes and procedures provide instructions and flowcharts for engineering design. They can also specify the procedures, checklists, and documents needed in a spacecraft or subsystem.
Software Style Guide	This document describes how the team will prepare software modules. It will help make review of software easier and more likely to catch mistakes through peer review.
Production and Fabrication Processes	Company process provides the guidelines or procedures for manufacturing, fabrication, assembly, and test. Often these processes and procedures provide instructions and flowcharts for the production. They can also

(Continues)

Table 5.18: Cont'd

System Title	Description
	specify the procedures, checklists, and documents needed in manufacturing, fabricating, assembling, and testing a spacecraft or subsystem.
Printed Wiring Board (PWB) Fabrication	A company often has procedures for procuring and assembling printed wiring boards to order, fabricate, and receive printed wiring boards. An example may be to procure printed wiring boards (PWBs) that have been fabricated to MIL-PRF-31032.
Metrology	Company process provides the guidelines for the metrology procedures and instructions. It also can specify the procedures, checklists, and documents necessary to calibrate equipment.
Prohibited Materials Verification	A company often has controls to prevent the inadvertent introduction of prohibited material into hardware. The procedures guide the evaluation of materials and components for prohibited materials, such as tin, zinc, and cadmium.
Contamination Control	A company often has facilities and procedures in place to control particulate contamination and hydrocarbon outgassing.
Vendor Qualification	A company often specifies the qualification process for vendors who supply components and services.
Thermal Analysis	Company process provides the guidelines or procedures for analysis, simulation, and assessment of the thermal performance of components, subsystems, and spacecraft.
Vibration Analysis	Company process provides the guidelines or procedures for analysis, simulation, and assessment of the vibration performance of components, subsystems, and spacecraft.
Structural Analysis	Company process provides the guidelines or procedures for analysis, simulation, and assessment of the structure of components, subsystems, and spacecraft.
Risk Analysis	Company process provides the guidelines or procedures for analysis, simulation, and assessment of risk. It may include FMECA, FTA, ETA, and hazard analysis.
Worst Case Analysis	Company process provides the guidelines or procedures for analysis, simulation, and assessment of worst case and derating.
Dependability Analysis	Company process provides the guidelines or procedures for analysis, simulation, and assessment of dependability, which includes reliability, availability, maintainability, fault tolerance, and testability.
Audit and Review	Company process provides the guidelines for reviewing processes and procedures. It also describes how frequently and what types reviews may occur.

3.3. Documentation Tools

Readily accessible information results in continuous review and improvement. Web-based access to autonomy rules provides quick access to current spacecraft configurations as well as to new builds under development. Having information readily accessible to the team helps enhance quality and speeds troubleshooting during I&T and during mission operations.

A number of Web-based tools exist to aid your documentation efforts. There are tools to maintain a database for requirements; they make sure to flag each requirement for test and verification and then record the results. There are tools for version control of documents that also provide for electronic approval processes. These Web-based tools can have interfaces to configuration management tools that provide immediate access from remote locations to hardware design documentation. There are also tools for managing action items and risk management.

4. Case Study—New Horizons

The unique requirements of NASA's New Horizons mission to Pluto and the Kuiper Belt make it a particularly suitable example of mission critical systems. The long 9.5-year journey to Pluto requires a robust and reliable spacecraft design, while the importance of the data gathered during the final days before the closest approach to Pluto highlights the mission-critical nature of the mission operations planning and test processes. In addition to the mission critical attributes, New Horizons also includes safety-critical aspects. The radioisotope thermoelectric generator (RTG), as well as the hydrazine propulsion and RF communication systems provide examples of fault tolerant designs for safety-critical systems. With New Horizons on its way to a close encounter with Pluto and its moons Charon, Nix, and Hydra, it's beneficial to assess how well certain processes worked and where improvements could be made. This section describes the application of the principles given previously in this chapter in a real life application for the New Horizons spacecraft.

The two authors of this chapter had involvement with the development of the New Horizons spacecraft. Chris Hersman was the spacecraft systems engineer and is now the mission systems engineer for New Horizons. Kim Fowler monitored and facilitated board fabrication and assembly for the LORRI and PEPSSI instruments.

4.1. Pluto-Kuiper Belt Announcement of Opportunity

Missions to Pluto had been proposed many times before the Announcement of Opportunity (AO) for the Pluto-Kuiper Belt (PKB) mission [29, 30]. A number of mission incarnations were short lived, but NASA's announcement of plans for the PKB AO kicked off the beginning of the New Horizons project. The PKB AO was modeled after the successful

PI-led mission paradigm. Though this approach was commonly used for smaller solar-powered missions, the PKB AO was the first application to a mission of this type. As with other PI-led missions, the selection process was conducted in two steps, beginning with the Step-1 proposals.

The New Horizons Step-1 proposal demanded a substantial effort by the entire New Horizons team to produce the document within the two-month schedule. To add to the anguish, in the middle of the proposal preparation, the AO was cancelled based on the new administration's future budgetary plans not to fund it. Within a week, the decision was made to allow the process to proceed despite the fact that there were no plans by the administration to fund the mission. Near the end of the effort, NASA granted an extension of several weeks. On April 5, 2001, the New Horizons Step-1 proposal was delivered to NASA, joining four others in the evaluation.

On June 6, 2001, two proposals were awarded funding for the second step of the selection process, the development of the Phase A concept study report. Both New Horizons and the Pluto Outer Solar System Explorer were given 3 months to refine their mission concepts and submit their reports. Following the delivery of the report, each team had nearly a month to prepare oral presentations to the evaluation team. Formal notification of the selection of New Horizons was given on November 29, 2001.

After a brief celebration, preparation quickly began for the many design and development obstacles to come. Some of these challenges, which included both technical and programmatic issue, are described later in this section.

4.2. Mission Concept Overview

The beginning of any mission starts with a concept. For New Horizons that concept began with the science objectives including the PKB AO. Of those objectives, the New Horizons mission concept addressed all but one. Table 5.19 lists the New Horizons science objectives. The objectives are achieved through science investigations of surfaces, atmospheres, interiors, and space environments through imaging, visible and infrared spectral mapping, ultraviolet spectroscopy, radio science, and in-situ plasma measurements. Following its primary mission, New Horizons was designed to achieve similar discoveries at one or more Kuiper Belt objects (asteroid and planet-like objects beyond Pluto).

The New Horizons suite of instruments is well matched to the science objectives. Efforts early in the concept study phase (Phase A) documented detailed traceability from the science objectives to the required observations to the instrument performance specifications.
In addition, the team produced a straw-man timeline for the Pluto encounter to show that all observations could be completed within the time available. Table 5.20 describes the instrument suite.

Table 5.19: New Horizons Science Objectives

	Objective
Group 1 (Primary Objectives)	Characterize the global geology and morphology of Pluto and Charon.
	Map surface composition of Pluto and Charon.
	Characterize the neutral atmosphere of Pluto and its escape rate.
Group 2 (Secondary Objectives)	Characterize the time variability of Pluto's surface and atmosphere.
	Image Pluto and Charon in stereo.
	Map the terminators of Pluto and Charon with high resolution.
	Map the surface composition of selected areas of Pluto and Charon with high resolution.
	Characterize Pluto's ionosphere and solar wind interaction.
	Search for neutral species including H, H_2, HCN, and C_xH_y, and other hydrocarbons and nitriles in Pluto's upper atmosphere, and obtain isotopic discrimination where possible.
	Search for an atmosphere around Charon.
	Determine bolometric Bond albedos for Pluto and Charon.
	Map the surface temperatures of Pluto and Charon.
Group 3 (Tertiary Objectives)	Characterize the energetic particle environment of Pluto and Charon.
	Refine bulk parameters (radii, masses, densities) and orbits of Pluto and Charon
	Search for additional satellites and rings.

The New Horizons Spacecraft was specifically designed to accommodate the instruments' requirements, including fields-of-view, power, thermal dissipation, and mass. Figure 5.14 shows the mounting positions of the instruments and view directions of each instrument. Additional spacecraft design characteristics were driven by the unique constraints of the mission. Many of these constraints are identified in the system engineering section under the requirements flow-down and design tradeoffs subsections [31–34].

The New Horizons mission has a carefully designed trajectory that uses planetary gravity "boosts" to gain velocity and course changes. An early launch window, such as the actual launch in January 2006, allows the travel time to Pluto to be 9.5 years. Later launch windows would have different mission trajectories and longer travel times. A very late launch would not have any gravity "boosts" and a 14.5-year travel time. Delaying a launch does not give a proportionally longer travel time, it actually causes step function delays in travel with later and later launch windows. Another consequence of longer travel times is that less power is available and less data are collected. Since the power from the RTG relies on radioactive decay, power declines with mission time. Although not a linear decline, power drops

Table 5.20: New Horizons Instrument Suite

Instrument	Description	Function
Ralph	Visible and infrared imager/spectrometer	Produces color maps, surface composition and thermal maps
Alice	Ultraviolet imaging spectrometer	Provides atmospheric measurements of the composition and structure at higher altitudes
REX	Uplink radio science experiment	Measures atmospheric composition and temperature at lower altitudes
LORRI	Telescopic imager with panchromatic framing CCD	Obtains high resolution geological images and surface maps of the back side during the departure from the encounter
SWAP	Solar wind and plasma spectrometer	Measures atmospheric escape rate and interaction with solar wind
PEPSSI	Energetic particle spectrometer	Measures composition and density of ions escaping from the atmosphere
VB-SDC	Venetia Burney-Student Dust Counter	Student built instrument detects dust particle impacts during the cruise through the solar system to map the dust-density distribution

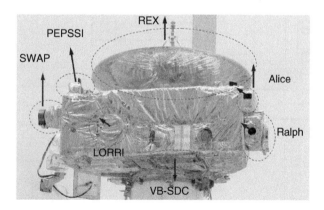

Figure 5.14: New Horizons spacecraft.

approximately 3 W per year; therefore, an early launch has more available power than a later, longer mission. Reduced power capacity would force the mission operations to change significantly to handle the instruments and data collection for more power conservation. New Horizons' early launch allows both computers to run simultaneously for better fault tolerance and to record more data redundantly on both computers. The trajectory design is discussed in the references [35–38].

4.3. Project Management

The Johns Hopkins University Applied Physics Laboratory (JHU/APL), in collaboration with Southwest Research Institute (SwRI), was responsible for building the New Horizons spacecraft and most of the instruments. JHU/APL has a long history in building satellites, systems engineering, and project management.

Figure 5.15 illustrates the project organization. Figure 5.16 summarizes the project's milestone schedule.

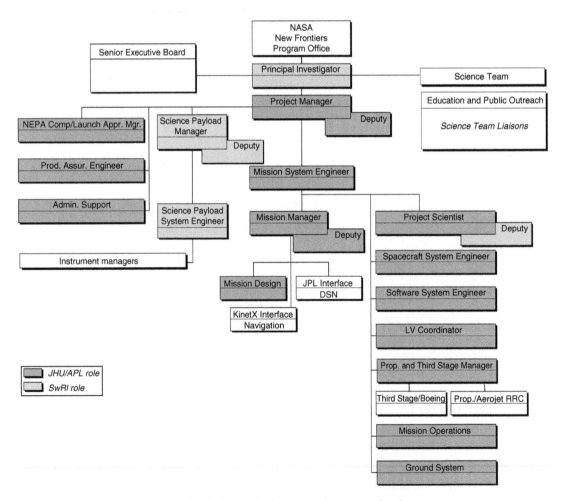

Figure 5.15: New Horizons project organization.

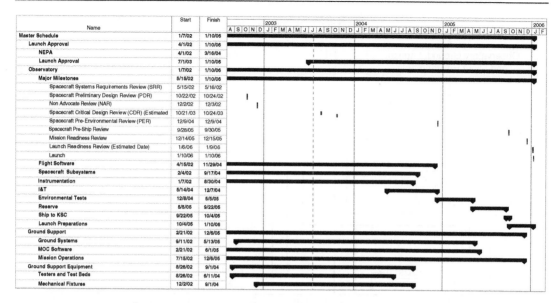

Name	Start	Finish
Master Schedule	1/7/02	1/10/06
Launch Approval	4/1/02	1/10/06
NEPA	4/1/02	3/16/04
Launch Approval	7/1/03	1/10/06
Observatory	1/7/02	1/10/06
Major Milestones	5/15/02	1/10/06
Spacecraft System Requirements Review (SRR)	5/15/02	5/16/02
Spacecraft Preliminary Design Review (PDR)	10/22/02	10/24/02
Non Advocate Review (NAR)	12/2/02	12/3/02
Spacecraft Critical Design Review (CDR) (Estimated)	10/21/03	10/24/03
Spacecraft Pre-Environmental Review (PER)	12/9/04	12/9/04
Spacecraft Pre-Ship Review	9/28/05	9/30/05
Mission Readiness Review	12/14/05	12/15/05
Launch Readiness Review (Estimated Date)	1/6/06	1/9/06
Launch	1/10/06	1/10/06
Flight Software	4/15/02	11/29/04
Spacecraft Subsystems	2/4/02	9/17/04
Instrumentation	1/7/02	8/30/04
I&T	5/14/04	12/7/04
Environmental Tests	12/8/04	5/5/05
Reserve	5/5/05	9/22/05
Ship to KSC	9/22/05	10/4/05
Launch Preparations	10/4/05	1/10/06
Ground Support	2/21/02	12/6/05
Ground Systems	9/11/02	5/13/05
MOC Software	2/21/02	6/1/05
Mission Operations	7/15/02	12/6/05
Ground Support Equipment	8/26/02	9/1/04
Testers and Test Beds	8/26/02	6/11/04
Mechanical Fixtures	12/2/02	9/1/04

Figure 5.16: New Horizons project milestone schedule.

4.4. Systems Engineering

4.4.1. Requirements Flow-Down

Figure 5.17 illustrates the requirements flow-down for the New Horizons project. The long distance from the sun precluded the use of solar power; consequently, a single RTG power source was specified. Mass constraints prevented the using multiple RTGs. The power was one of the most constrained resources on the spacecraft—the power limitations and long cruise duration (9.5 years), in turn, drove the need for low-power operations. Requiring low-power operations had further ramifications in the design tradeoffs for computing power. Many requirements flow-down and design tradeoffs deal with mass versus power available for the payload.

4.4.2. Examples of Design Tradeoffs

A number of trade studies were evaluated during the development. As already mentioned, many design tradeoffs were between mass and available power (or power efficiency).

Originally, the functions of the separate Alice and Ralph Instruments were to be combined in a single instrument called PERSI. The intent was to add several new functions to a proven function that had already flown on another spacecraft. Separating the instrument functions into Alice and Ralph provided separate instrument interfaces, which were less complex and simpler to test than the single interface of the more complex PERSI instrument. The two separate instruments allowed greater flexibility in the schedule even though they added extra

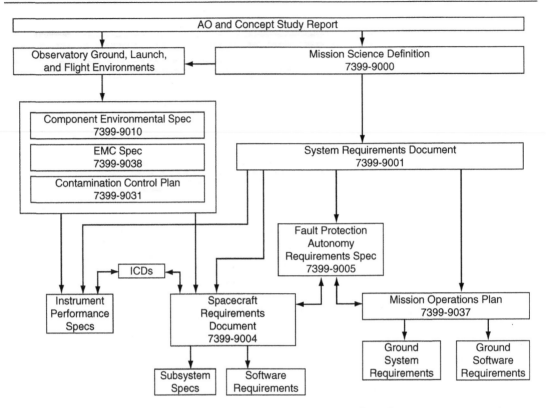

Figure 5.17: New Horizons requirements flow-down.

work with two separate instrument ICDs and two sets of thermal-vacuum tests, whereas PERSI would have had only one ICD and one set of thermal-vacuum tests. The advantages of schedule flexibility with two instruments over one were that the more capable instrument outweighed the additional effort with two sets of ICDs and environmental tests for the two instruments.

Another design tradeoff involved a radiation shield for the Alice instrument. Alice has a sensor that is sensitive to radiation, which can easily interfere with the functioning of the sensor and swamp its data. The project team decided to go ahead and build a tantalum radiation shield to protect the sensor from potential radiation from the RTG. Building of the shield proceeded in parallel with the effort to analyze and test the spacecraft layout for radiation interference of the Alice instrument sensor. Eventually, the analysis and testing proved that the spacecraft did not need the tantalum shield on Alice. The programmatic decision was that the parallel effort cost extra but significantly reduced risk in the face of uncertainty for the potential problem until the structure and spacecraft configuration solidified.

In general, shielding the electronics total ionizing dose (TID) from the RTG is always a concern. The tradeoff made here was between a boron-impregnated polymer shield and counterbalancing the RTG on one side of the spacecraft and the electronics placed on the other side. Early analysis proved that the RTG did not need a shield; placing the electronics on the other side of the spacecraft was sufficient—doing this also helped with the spin balance of the spacecraft.

In any spacecraft development, there is always the trade between how much development is done in hardware and how much is done in software. In the New Horizons power system the decision was to go all hardware, putting any computations in FPGAs; this had the advantage of better results over software in the safety analysis. Software is generally more difficult to prove "safe," whereas hardware has physical limitations that are more readily understood.

The New Horizons team also made tradeoffs for the main computer between processor speed and processor power consumption. They used a Mongoose V processor, which had adequate capability but low-power consumption, rather than using a Rad6000 processor, which is much more computationally capable but too power hungry for this particular mission.

Spacecraft always have tradeoffs between payload layout and clear fields of view for the instruments. Much effort went into the layout of the New Horizons spacecraft to avoid glint from the structure of the spacecraft into instruments.

Often a tradeoff between mass and power in a spacecraft is not a single-point optimization but rather a "chain" of considerations. One such example tradeoff was for the specific form of amplifier for the RF communications. A traveling wave tube amplifier (TWTA) is more inefficient than a solid-state power amplifier (SSPA) in an absolute sense. The TWTA dissipates 18 W of heat—it required 30 W of input power to generate 12 W of radiated energy. The SSPA has a smaller size at lower power levels but it is less efficient as well—it requires 20 W of input power to generate 6 W of radiated energy, which means that it dissipates 12 W of heat. The "chain" of consequences was that even though the TWTA dissipated more heat, it also generated more RF energy, which means that it can use a smaller high-gain antenna (HGA). The SSPA dissipate less heat than the TWTA, it also generates less RF energy, which then requires a larger HGA to attain the desired communications link. In this tradeoff, greater power consumed means less mass required to maintain sufficient RF communications.

An early architectural tradeoff involved dual redundancy and cross-strapping between modules. Cross-strapping, in essence, allows one subsystem to "shadow" the operation of its redundant counterpart; the spacecraft's autonomy system switches out a block subsystem if it is not operating correctly. The design trade was whether to cross-strap every subsystem or not. The final decision was to not cross-strap between radios, which means that operations cannot command the computer directly from the other "shadow" receiver; the computer can still get commands from the other computer. The development team did cross-strap the antennas; separately, they cross-strapped the power systems with both the 1553 bus and the serial interface.

In another architectural consideration, the New Horizons' design called for dual redundant inertial management units (IMUs); the desire was to operate only one at a time to conserve power. The team had to select an IMU from one of two different vendors. One vendor had a single component with two internal, redundant IMUs; the problem was that one IMU could not be selected while the other was switched off. This meant that both internal IMUs were powered simultaneously; overall this component had less mass but consumed more power. The other vendor supplied a single IMU component, therefore the spacecraft would need two units resulting in greater mass, but that it could select and power a single IMU, consuming less power. Programmatic caution about the delivery capability of a vendor ultimately drove the tradeoff decision for the single IMU; it proved to be fortuitous from a technical viewpoint because the IMU consumed less power.

4.4.3. Risk Management

Table 5.21 lists the top 10 risks at CDR during the project. Another way to manage risk is through its spares and repair philosophy. The spares philosophy for New Horizons was to have a complete flight parts kit for each board type.

Table 5.21: Ten Top Risks for New Horizons Project at CDR

Rank	Risk	Mitigation
1	Third-stage anti-rotation cord stress	Modify design (Boeing) or change LV ascent maneuvers (LM).
2	Launch approval process to support 2006 launch	2007 launch.
3	Inadequate staffing resources	Monitor continuously; use team members from other institutions, subcontract, hire new staff.
4	Cold sparing of receiver, USO	Multiple layers of protection in spacecraft design; thorough testing.
5	Reserve liens, FY04	Monitor vigilantly; not a major concern presently.
6	House FY04 funding cut	2007 launch.
7	Power margin for 2007 launch	Manage power allocations during implementation phase; power management during flight.
8	Increased actel current	Manage power allocations during implementation phase; power management during flight.
9	Late testbed delivery	Deputy program manager is managing testbed needs dates against expected deliveries. Establishes priorities as necessary.
10	Thruster valve temperatures	Revisit conservative assumptions in initial thermal analysis and reanalyze.

4.4.4. Margin Management

One example of margin management was for the mass of the spacecraft. The lift capability for the New Horizons spacecraft changed somewhat over the course of the project as the provider of the launch vehicle considered various performance enhancements for the vehicle. This affected the percentage trend of the mass margin. There were other allocations involving the third stage that affected the margin, too. In general the mass of the spacecraft gradually increased over time resulting in the decrease in margin as the project progressed. (See Fig. 5.18.) The mass margin for the New Horizons spacecraft was "green" most of the time during the development of the spacecraft.

4.5. Fault Protection

The fault protection system has autonomy rules to put the spacecraft into safe modes. A typical course of action when the system encounters a problem is to transition to a normal mode operation; if that does not solve the problem, then the system transitions to a safe mode. One safe mode is called earth safe; the fault protection system points the communications antenna at Earth. The other safe mode is called sun-safe; the fault protection system points the antenna toward the Sun. The earth safe mode requires more functionality to operate and generally occurs while the trajectory is closer to Earth. The sun-safe mode requires less functionality to operate; it is used when the trajectory is farther from the Earth—at a great distance, the Earth is always in the same direction as the Sun.

Another example of the operation of the fault protection system and its autonomy rules is during a trajectory correction maneuver (TCM). Whenever mission operation commands a TCM, the autonomy rules watch over the spacecraft. Figure 5.19 illustrates the state diagram of the operation of the fault protection system during a TCM.

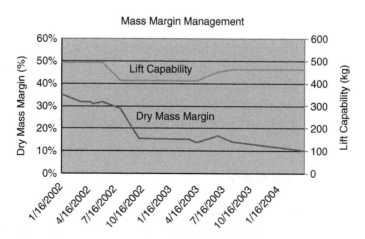

Figure 5.18: The mass margin changes over the course of development of the New Horizons spacecraft.

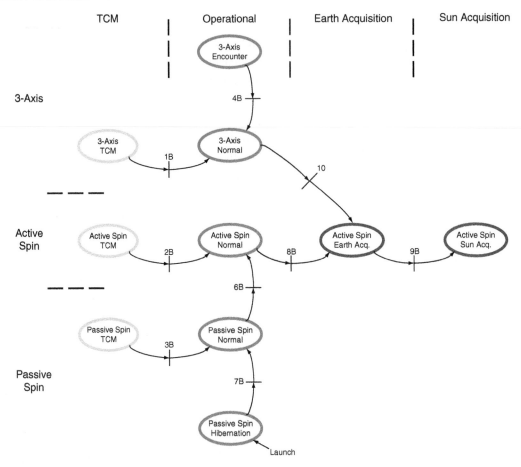

Figure 5.19: Spacecraft mode state diagram for trajectory correction maneuvers.

The fault protection system has many autonomy rules. Table 5.22 lists some sample requirements for fault protection and a high-level view of how autonomy rules operate. Table 5.23 lists all of the New Horizons autonomy rules; some autonomy rules are interconnected with other autonomy rules to cover situations completely. This complete listing of the rules is only to give a flavor of the type of autonomy rules that can be in a fault protection system.

4.6. Mission Assurance and Safety

The engineer conducting the reliability analysis caught an architectural problem through a skew in the probability distribution of the mission success. He found multiple connections to one power switching board that decreased the probability of mission success.

Table 5.22: List of Sample Requirements for Fault Protection

Fault	Response	Category
C&DH processor reboot	If the C&DH processor reboots, go to safe mode.	Health and safety responses
C&DH processor failure	If the primary C&DH processor fails, switch to the backup and go safe mode.	Health and safety responses
Emergency over temperature	If the spacecraft is too hot, go into Earth-safe mode and turn on a preset number of loads.	Health and safety responses
Emergency under temperature	If the spacecraft is too cold, then reset thermal software control to a default and then turn on the heaters.	Health and safety responses
Command loss timeout	If the command loss timeout expires indicating commands have not been received by the spacecraft, then reconfigure the communication system and go to sun-safe mode.	Health and safety responses
Ultra-stable oscillator side-A failure	If side-A fails, then listen to side-B.	Redundancy management
Guidance and control (G&C) system	If failure, then try a series of different actions and if none of these work, then go into sun-safe mode.	Some redundancy management, and health and safety responses
Instrument health and safety	If science instrument has a problem, power it off.	Health and safety responses
Clear SSR over-current indication	If false set of over-current, then clear indication.	Convenience/maintenance
Update last thruster count	If too many thrusters used, then something must be wrong and update the last thruster count.	Convenience/maintenance

The mission assurance had a number of objectives; one objective stated that no single failure would turn on a whole branch of the propulsion system; another objective stated that no single failure would prevent operation of a branch of the propulsion system. Clearly there existed a dynamic tension between the mission assurance objectives. The propulsion system had four separate thruster branches while the spacecraft had eight power-switching boards. The reliability engineer found a single power-switching board wired to multiple propulsion branches in such a way that losing the power-switching board would have brought down two branches, instead of just one. Consequently, a design change to the architecture and ultimately the wiring harness improved the system reliability.

Table 5.23: New Horizons Autonomy Rules—Some Are Interconnected with Other Autonomy Rules

C&DH Processor Has Failed	Active Sun Sensor Powered Off	Low Voltage Sense in Encounter Mode
Encounter Mode Entry	PEPSSI Instrument Boot Mode in Encounter Mode	Still Not In Spec in AS-SA
Bus Controller Reboot Not 3-Axis Normal	Target Ephemeris Is Invalid	Power Distribution Unit 1553 Board Unhealthy
C&DH Processor Command Loss Timer Expired	Not Meeting Pointing Specification in Encounter Mode	Active Power Distribution Unit Command Decoder Off
Other Command Loss Timer Expired	Ephemeris is Still Invalid	Power Distribution Unit FET Board Telemetry 1553 Board Unhealthy
Encounter Mode Handover	Encounter 7-Day Timeout	Low-Voltage Sense Recovery in Active Modes
Generic Safe Mode Entry	C&DH Processor Reboot in 3A-N	Last Ditch AS-SA
Safe Mode from PS-H	C&DH Processor Reboot in 3A-N Go Safe Timer	Continuous Latch Valve Current A
Safe Mode from PS-N	Safe Mode from AS-TCM	Continuous Latch Valve Current
Safe Mode from AS-N	Safe Mode from 3A-TCM	Propulsion Leak Cruise
Safe Mode from 3A	Unforeseen Bus Controller Reboot in Any Mode	Propulsion Tank Pressure Temperature PS-H
MIL-STD-1553 Bus Failure	Previous Fault during Encounter or Command Loss Timeout	Propulsion Tank Pressure Temperature PS-N
Encounter Switch-all Complete	60-Day Safety Net	Active Safety Relay Is Open
Safe Mode from PS-TCM	Low Voltage Sense Recover Backup Timer	Active Propulsion Relay Is Open
Aborted TCM	Primary Makeup Heaters Selected	Emergency Over Temperature
	Redundant Makeup Heaters Selected	Emergency Under Temperature
Catalyst Bed Warm-up Timer	Excessive Thruster Use in AS-N	Emergency Under Temperature Swap C&DH Processors
14 Day Wait for PS-H	Excessive Thruster Use in 3A-N	Primary Makeup Heater Failed Shorted

(Continues)

Table 5.23: **Cont'd**

Last Ditch AS-EA	Persistent Thruster Current	Primary Makeup Heater Failed Open
Configure Downlink AS-EA	Primary Thrusters Selected	Primary Catalyst Bed Heater Failed Open
Configure Beacon AS-SA	Redundant Thrusters Selected	Primary Catalyst Bed Heater Failed Shorted
Catalyst Bed Countdown	Inappropriate Thruster Mask Setting	Primary Star Tracker Heater Failed
Ultra-stable Oscillator B Failure	Enforce Redundant Thruster Selection	Star Tracker Heater Not On
Loss of Inertial Reference AS-EA	Enforce Primary Thruster Selection	Star Tracker Heater Failed
Low Voltage Sense Recover in Encounter Complete	Alice Instrument Side A Turn On	Power Distribution Unit Command Decoder A Active
Primary Catalyst Bed Heaters Selected	Alice Instrument Side B Turn On	Power Distribution Unit Command Decoder B Active
Redundant Catalyst Bed Heaters Selected	Red-1 Beacon	Star Tracker in Standby Mode in Spin Mode
Hardware MET Stuck at 0	Red-3 Beacon	Star Tracker in Initialization Mode in Spin Mode
Hardware MET Stuck at 1	Red-5 Beacon	Mask Excessive Thruster Use
Load Storage Variable Defaults	PEPSSI Instrument Turn On	Star Tracker Unhealthy in 3A Mode-1
Synthesizer Lock Off	G&C Processor Heartbeat Stuck at 1	Pointing Back in Spec
Receiver Latchup Protection	G&C Processor Heartbeat Stuck at 0	Clear SSR Over-Current Indication
SWAP Instrument Heartbeat Stuck at 1	Sun Sensor Unhealthy	Clear SSR Over-Current Indication Quickly
SWAP Instrument Heartbeat Stuck at 0	Sun Sensor Still Unhealthy	Flash Corruption Protection
LORRI Instrument Boot Mode in Encounter Mode	Enforce Sun Sensor Selection	Solid State Recorder Setup for Safe Mode
Alice Instrument Heartbeat Stuck at 1	Ephemeris Is Invalid	Check PDU A Critical Heartbeat Missing
Alice Instrument Heartbeat Stuck at 0	Pointing Is Not in Spec Tier 1	Check PDU B Critical Heartbeat Missing

(Continues)

Table 5.23:   Cont'd

Alice Self Turn Off Request	Pointing Is Not in Spec Tier 2	Check Tank Pressure Temperature Ratio
LORRI Instrument Heartbeat Stuck at 1	Pointing Is Not in Spec Tier 3	Check Star Tracker Heater
LORRI Instrument Heartbeat Stuck at 0	Need To Go To AS-SA	Check Internal Timing
LORRI Instrument Boot Command Not Encounter Mode	G&C Processor Heartbeat Stuck at 1 Quick Response	Check Receiver Phase-Lock-Loop
PEPPSI Instrument Heartbeat Stuck at 1	G&C Processor Loss of Inertial Reference Quick Response 1	Check Receiver Phase-Lock-Loop Latchup Counter
PEPPSI Instrument Heartbeat Stuck at 0	G&C Processor Loss of Inertial Reference Quick Response 2	Check Flight Software to Uplink Card Communication
PEPPSI Instrument Boot Command Not Encounter Mode	Star Tracker Unhealthy in AS Mode	Check PDU Data
Ralph Instrument Over-current	Star Tracker Unhealthy in 3A Mode	Check PDU FET Data A
Ralph Instrument Heartbeat Stuck at 1	Inertial Measurement Unit Unhealthy	Check PDU FET Data B
Ralph Instrument Heartbeat Stuck at 0	Star Tracker in Standby Mode in 3A Mode	Update Last Thruster Count (Maintenance)
REX Instrument Heartbeat Stuck	Low-Voltage Sense Recovery in Progress	G&C Processor Parameter Load Support for Segment 24
SDC Instrument Heartbeat Stuck at 1	Pre-Low-Voltage Sense Detection	G&C Processor Parameter Load Support for Segment 8
SDC Instrument Heartbeat Stuck at 0	Low-Voltage Sense during Cruise	C&DH Processor Reboot during Flash Programming
SWAP Anomalous Event	Repeated Low-Voltage Sense	

Lessons learned from the Cassini spacecraft development were incorporated into the risk communication planning for developing the New Horizons spacecraft. Those lessons follow:

- Respond in a coordinated, well-defined way.

- Prepare risk communication materials well before they are needed.

- Make risk communication materials available on the Internet.

- Train project spokespersons in the importance of communication.

The guiding principles for communications are listed below [40].

Principle 1: Be transparent.

- Be honest, candid, and open.

- Make information available and easily accessible, as early as possible.

- Use plain language.

- Ensure the transparency to the public of the process by which missions are chosen, designed, and operated.

- Ensure that communications channels to the public easily provide information about safety, mission objectives and benefits, programmatic changes, and successes and failures [40].

Principle 2: Be inclusive.

- Seek as many perspectives as possible.

- Be sensitive to cultural differences.

Principle 3: Be interactive.

- Listen respectfully and respond constructively to colleagues, critics, and supporters.

- Be clear in establishing where NASA can and is willing to accept input.

- Based on input, be open to modifications or new options.

4.7. Assembly, Integration, and Test—Fabrication and Assembly of Circuit Boards

An interesting phenomenon in spacecraft projects, which was not peculiar to the New Horizons project, is the "churn" in designing, fabricating, and assembling circuit boards. This is where designs are submitted to CAD and fabrication before being fully analyzed and reviewed; a correction in the design or even a change in another instrument or subsystem ripples through the effort to design, fabricate, and assemble and causes considerable delay. Fowler details the effect of "churn"—iterating through PWB fabrication and inspection, assembly and inspection, and final circuit test and verification—see reference [39]. "Churn" causes real problems when the production staff and facilities can only take a limited amount of work at any one time; the reference details how a typical circuit board can average up to 10 months from initial design entry to finished and tested circuit board. "Churn" can be reduced if engineers analyze and complete designs before initiating production. A common problem that exacerbates "churn" is the desire to show linear progress during design reviews, when that is not the actual way of production. Mapping production progress tends to follow a series of step function changes rather than a monotonically increasing indication.

4.8. Subsystem Tests and Testing—Notable Anomalies and Lessons Learned

Every spacecraft development has unique challenges; the development within New Horizons was no different. Here are three examples of some of those challenges that arose during the development of the New Horizons spacecraft. In the first two cases, extra checks designed into the circuits saved the equipment from harm during subsystem test.

Shunt Installation Anomaly—A shunt regulator (i.e., big resistor) must load the RTG to draw down the voltage from 50 V to 30 V during subsystem tests. The team designed the GSE to limit over-voltage at 35V and they specified and designed the instruments to survive any input voltage from 0 V to 40 V. During I&T a shunt was accidentally not connected and the RTG voltage went up to 35 V, at which point the voltage was limited. This action prevented the over-voltage from destroying any circuits and saved the team from having to do extensive, post-test failure analysis.

Probing Voltage with the "Amp" Setting—The team designed the circuitry with circuit breakers to prevent over-currents during spaceflight. During testing of the PEPPSI instrument, a technician probed a voltage point with a multimeter but the multimeter was set on the "Amp" setting. When the multimeter probe touched the voltage point, it drew a momentary surge of current. The circuit breakers in the PEPPSI instrument tripped and saved the circuitry.

IEM DC/DC Converters Wrongly Installed—The integrated electronics module (IEM) had two DC to DC converters, each with its own unique position. During development, engineers on the development team put them into the engineering model of the IEM; they knew the orientation and keying for the converters to fit into the module. Unfortunately, three separate events conspired during assembly to install the converters incorrectly. First, the assembly drawing was wrong. Second, ESD concerns in the assembly area prevented the engineer who knew the correct position of each converter from getting close enough to see the exact location as the technician installed the converters in the IEM. Third, the keying mechanism for orienting the converters was weak and if pushed hard enough, it would bend out of the way. Consequently, the two DC/DC converters were pushed into the wrong slot and powered the IEM incorrectly. Tests found the problem and the ensuing analysis revealed the sequence of events that led to the problem.

4.9. Launch and Mission Operations

The New Horizons spacecraft launched from Earth as the fastest manufactured object in January 2006. Five months before the launch, however, tests indicated a potential catastrophic failure of the main tank in the launch vehicle. Examination following a pressure test of a tank, identical to the one on the launch vehicle, revealed cracks due to the extra stress during pressurization. Review of the inventory of seven launch vehicle tanks showed cracks or corrosion problems in five tanks. The tank on the New Horizons launch vehicle was one of the two without cracks; furthermore, the New Horizons tank had been previously tested beyond the stress planned for the New Horizons flight. A high-level review of the problem resulted in the decision to proceed with launch. A good read about this problem can be found in the reference for the New Horizons RP-1 tank decision [41].

The launch and initial mission operations went smoothly. Following is an excerpt of a press conference given by Glen Fountain and Alan Stern; it gives a taste of some of the issues during the initial portion of the mission [42]. The questions are in italics followed by the response.

> **What are risks and dangers?** *Fountain: I was remarking through the countdown, that once it launches, we'll finally get it out there where it will be safe. It's safest on its way to Pluto. As we get beyond Pluto, the power margin will drop, and we'll have to get smarter about how we operate the spacecraft.*

> **Any problems?** *Stern: We're not working any problems now. The onboard fault protection system did not return any faults whatsoever. So it's now our job to be good stewards of the spacecraft, and to learn to fly it in the environment it was built for....*

> **What's next on the agenda?** *Fountain: First we will start exercising the propulsion system. We will spin down the spacecraft. We'll exercise the other pieces of the guidance control system. In about 2 hours we'll get the precise information about what our trajectory is. It looks like we're on a very nominal trajectory. We'll then start planning a TCM that will take place in about 10 days. The second one will take place in about 20 days. We're going at something like 16 km per second. We have about 300 meters per second of Delta Vee to control the spacecraft. We'll then do an initial checkout of the instruments. Next summer, we'll go through an in-flight calibration. Then we'll start planning for the Jupiter encounter, which will start in the fall. The Jupiter encounter will take place in late February.*

5. Future Directions

Hopefully the development of spacecraft will become more routine and the processes better understood. Using these best practices and judicious use of thoughtful program management and systems engineering may improve the efficiency, effectiveness, and speed of development. A modified approach to COTS, where products are largely off-the-shelf but have the flexibility for some changes may also reduce cost and development time. Fault-tolerant computing should become more accessible and cost-effective [43].

6. Summary of Good Practices

Good project management processes are critical to mission success:

- The team takes a vested interest in the project.

- The team needs a talented project manager to coach them.

- The team requires clearly defined lines of authority to operate efficiently and effectively.

- Communication and teamwork are essential to the successful implementation of a project, in particular, regularly scheduled team status meetings.

- Maintain an up-to-date contact list, including roles and work, home, and mobile phone numbers.

- Open dialog with the customer—keeping the customer or sponsor informed of current progress, even when the news is bad, helps build a trusting relationship.

- Have standard scheduling software to improve the efficiency of schedule updates.

- Implement good review practices to make reviews more productive.

- Supply the agenda and presentation material to the review committee a week or more before the review.

- Involve all members of the project team in risk management.

- Disseminate risk information throughout the project so that the entire team is aware of potential issues.

Systems engineering is critical to all technical aspects of the project:

- Establish effective plans for configuration management, data management, and quality assurance.

- Adding rationale and verification method for each requirement can be very useful in clarifying intent.

- Accurate documentation of interfaces is critical to understanding and maintaining the interface control function.

- Smaller resource margins tend to indicate greater risk.

- Two commonly tracked resources are mass and power.

- Lessons learned captures important experiences from the current project so that they can benefit future projects.

Software engineering and design has some basic guidelines:

- Keep it simple.

- Use the right people.

- Efficiency results from good design not from good coding.

- Hold serious design reviews and code walkthroughs.

- Design, test, encapsulate, and execute critical functions independently of control operations code.

- Practice defensive programming, design critical functions first.

- Avoid open loop designs.

- Avoid mirroring or modeling hardware when the physical state is directly measurable.

Establish plans for integration and test (I&T) early in the project:

- Following established processes can mean the difference between a success and complete failure.

- Have a list of "incompressible tests," which represent the minimum set of tests that must complete successfully before the spacecraft can be launched.

- Develop a system verification matrix to insure that all requirements are tested.

- "Test-as-you-fly"; if that is not possible, then "fly-as-you-test."

- Of the four evaluation methods—assessment (or analysis), inspection, demonstration, and test—the strongest and most desirable criterion is test.

- Perform trending of data and performance parameters to get an early indication of any problems before failures actually occur.

- Require each lead engineer to provide a list of trending points in advance of the component delivery.

- Start data trending early and do it often.

- Begin interface testing as soon as possible in the development.

- Have a multilevel system for documenting, tracking, and resolving unexpected behavior.

- Documentation is an essential part of a spacecraft development project.

The use of COTS products is appropriate when:

- The intended application is close to the original design intent of the COTS product.

- There is independent verification and validation of the technical requirements.

- There are open and frequent communications with the vendor of the COTS product.

- Modification of a COTS unit for a space application is treated as a development project.

Acknowledgments

We would like to acknowledge individuals at the Johns Hopkins University Applied Physics Laboratory for their contributions to the company practices described herein. We would also like to thank the organizations involved in the formulation of the aerospace industry standards and the valuable resources they have published that guide the best practices throughout the industry, specifically NASA, ANSI, PMI, and IEEE. Finally, we would like to thank the numerous individuals who have contributed to the success of the New Horizons mission to Pluto, which served as a case study for this chapter.

References

[1] Siewiorek DP, Swarz RS. Reliable computer systems: design and evaluation. 3rd ed. London: AK Peters, Ltd.; 1998.

[2] Jackson B. A robust fault protection strategy for a COTS-based spacecraft. 2007 IEEE Aerospace Conference, 2007. Available at: http://ieeexplore.ieee.org/xpls/abs_all.jsp?tp=&arnumber=4161525&isnumber=4144550.

[3] Williams T. EMC for product designers: meeting the European directive. 3rd ed. New York: Newnes; 2001; 47–80.

[4] LCROSS quick facts. Spaceflight Now, June 2009. Available at: http://spaceflightnow.com/atlas/av020/090610lcross.html.

[5] Ray J. Delta 2 rocket puts military experiment into space. Spaceflight Now; June 21, 2006.

[6] Barnes C. New radiation issues for spacecraft microelectronics—commercial off-the-shelf (COTS). JPL briefing, 1998. Available at: http://parts.jpl.nasa.gov/cots/external/cots_rad_iss.pdf.

[7] Goodman JL. Lessons learned from flights of "off the shelf" aviation navigation units on the space shuttle. Joint Navigation Conference, Orlando, Florida; 2002.

[8] Committee on Principal-Investigator–Led Missions in the Space Sciences, National Research Council. Principal-Investigator–Led Missions in the Space Sciences. Washington, DC: National Academies Press; 2006. Available at: http://books.nap.edu/openbook.php?record_id=11530&page=101.

[9] Verzuh E. The fast forward MBA in project management. 2nd ed. New York: John Wiley & Sons; 2005.

[10] Stamatelatos M, et al. Fault tree handbook with aerospace applications. Version 1.1. Washington, DC: NASA; 2002. Available at: http://www.hq.nasa.gov/office/codeq/doctree/fthb.pdf.

[11] Stamatelatos M, et al. Probabilistic risk assessment procedures guide for NASA managers and practitioners. Version 1.1. Washington, DC: NASA; 2002. Available at: http://www.hq.nasa.gov/office/codeq/doctree/praguide.pdf.

[12] Dunn WR. Practical design of safety-critical computer systems. Solvang, CA: Reliability Press; 2002.

[13] NPR 7123.1, NASA systems engineering processes and requirements. Available at: http://nodis3.gsfc.nasa.gov/.

[14] NPD 8700.1C, NASA policy for safety and mission success. Available at: http://nodis.hq.nasa.gov/displayDir.cfm?t=NPD&c=8700&s=1C.

[15] Moore RC, Hoffman E. Hot tips for flight software. Slide 17. Available at: http://www.aticourses.com/sampler/Satellite_RF_Comms.pdf.

[16] EEE-INST-002, Instructions for EEE parts selection, screening, qualification, and derating, 2003 Available at: http://nepp.nasa.gov/DocUploads/FFB52B88-36AE-4378-A05B2C084B5EE2CC/EEE-INST-002_add1.pdf.

[17] NOAA N-PRIME Mishap investigation final report, September 13, 2004. Available at: http://www.nasa.gov/pdf/65776main_noaa_np_mishap.pdf.

[18] Space mechanisms reliability, 2000. Available at: http://mynasa.nasa.gov/offices/oce/llis/0913.html.

[19] Duren RM. Validation (not just verification) of deep space missions. 2006 IEEE Aerospace Conference, Big Sky, MT.

[20] Duren RM. Validation and verification of deep-space missions, 2004. Available at: http://trs-new.jpl.nasa.gov/dspace/bitstream/2014/37253/1/03-0887.pdf.

[21] WIRE Mishap investigation board report. June 8, 1999. Available at: http://www.aoe.vt.edu/~cdhall/courses/aoe4065/NASADesignSPs/wire.pdf.

[22] NASA, White Sands Test Facility. Propulsion test office method of doing business. Available at: http://www.nasa.gov/centers/wstf/pdf/269491main_prop_test_office_method_of_doing_bus.pdf.

[23] Freeman MT. Spacecraft on-orbit deployment anomalies: what can be done? IEEE AES Syst Mag 1993;3–15.

[24] Tustin W. Movers and shakers: simultaneous multiaxis testing tops sequential axis methods. Mil Embedded Syst 2009;20–23.

[25] Surka DM, et al. Integrating automation into a multi-mission operations center. American Institute of Aeronautics and Astronautics 2007. Available at: http://www.emergentspace.com/pubs/AIAA-2007-2870.pdf.

[26] Eastern and Western Range (EWR) 127-1, Range safety requirements, 1997. Available at: http://snebulos.mit.edu/projects/reference/NASA-Generic/EWR/EWR-127-1.html.

[27] Expendable launch vehicle boilerplate launch site support plan. K-ELV-14.000-BL, Rev. B, 2000. Available at: http://www.ksc.nasa.gov/procurement/elvis/docs/boilerplatessp.doc.

[28] NASA. Launch site support plan (LSSP) development. Available at: http://www.ksc.nasa.gov/procurement/elvis/docs/launsitesupplandev.pdf.

[29] Stern SA. Journey to the farthest planet. Sci Am 2002;56–63.

[30] Stern SA. The New Horizons Pluto Kuiper Belt mission: an overview with historical context. Space Sci Rev 2008;140:3–21.

[31] Fountain GH, et al. The New Horizons spacecraft. Space Sci Rev 2008. Available at: http://www.boulder.swri.edu/pkb/ssr/ssr-fountain.pdf.

[32] Weaver HA, et al. Overview of the New Horizons science payload. Space Sci Rev 2008;140(1):75–91.

[33] Kusnierkiewicz DY, et al. A description of the Pluto-bound New Horizons spacecraft. Acta Astronautica 2005;57:135–144.

[34] Vernon SR, et al. Launch vehicle integration of the New Horizons/Pluto mission. Chantilly, VA; 2006, 6th NRO/AIAA Space Launch Integration Forum.

[35] Guo Y, Farquhar RW. New Horizons Pluto-Kuiper Belt mission: design and simulation of the Pluto-Charon encounter. IAC Paper, IAC-02-Q.2.07, 53rd International Astronautical Congress, The World Space Congress–2002, Houston, TX, October 10–19, 2002.

[36] Guo Y, Farquhar RW. New Horizons mission design for the Pluto-Kuiper Belt mission. AIAA/AAS Paper, AIAA-2002-4722, AIAA/AAS Astrodynamics Specialist Conference, Monterey, CA; August 5–8, 2002.

[37] Hersman CB, et al. Optimization of the New Horizons spacecraft power demand for a single radioisotope thermoelectric generator. IAC-06-C3.4.05, 57th International Astronautical Congress, Valencia, Spain; October 2–6, 2006.

[38] Ottman GK, et al. The Pluto-New Horizons RTG and power system early mission performance. AIAA 4th International Energy Conversion Engineering Conference and Exhibit (IECEC), San Diego, CA; June 26–29, 2006.

[39] Fowler K. What every engineer should know about developing real-time embedded products. Boca Raton, FL: CRC Press; 2008.

[40] JPL. New Horizons risk communication strategy, planning, implementation and lessons learned, 2006. Available at: http://trs-new.jpl.nasa.gov/dspace/bitstream/2014/40677/1/06-1169.pdf.

[41] Launching New Horizons: the RP-1 tank decision. NASA Academy of program/project & engineering leadership. Available at: http://www.nasa.gov/offices/oce/appel/pdf/New_Horizons_Case_Study.pdf.

[42] Lakdawalla E. New Horizons post-launch press conference. The Planetary Society Blog, 2006. Available at: http://www.planetary.org/blog/article/00000435/.

[43] Fault-tolerant spaceborne computing employing new technologies, Workshop report draft. Albuquerque, NM; 2008. Available at: http://www.zettaflops.org/spc08/Vision-And-Findings-Draft.pdf.

Appendix A: Example of a Systems Engineering Plan

This is an example of a systems engineering plan (SEP). It is not necessarily complete but should give some idea what might be expected content in a SEP that is separate from a SEMP.

1. Introduction

    1.1 Purpose

    1.2 Scope

    1.3 Definitions, Acronyms, and Abbreviations

    1.4 References

    1.5 Overview

2. Project Overview

    2.1 Project Purpose, Scope, and Objectives

    2.2 Strategy

    2.3 Technology Evolution

    2.4 Assumptions and Constraints

    2.5 Project Deliverables

    2.6 Evolution of the Project Plan

    2.7 Certifications

    2.8 Systems Engineering Process

    2.9 Systems Engineering Process Improvement—cite SEMP or QA Plan

    2.10 Legal Requirements and Contracts

3. Project Organization

    3.1 Program Structure

    3.2 Organizational Structures

    3.3 External Interfaces and Organizations

    3.4 Roles and Responsibilities—cite responsibility matrix in SEMP

    3.5 Integrated Product Teams (IPTs)

4. Project Monitoring and Control—cite SEMP (including Earned Value, Risk Management, Configuration Management, Quality Management and Quality Assurance)

5. System Specification and Performance Verification

    5.1 Requirements and Flow-down

    5.2 Technical Performance Standards

    5.3 Trade Analyses

    5.4 Interface Definition and Control—ICDs

        5.4.1 Electrical

        5.4.2 Mechanical

        5.4.3 Optics or sensor

        5.4.4 Data or software

        5.4.5 Support equipment

        5.4.6 Miscellaneous interfaces

    5.5 System Validation

    5.6 Performance Verification

    5.7 Technical Performance Trending

    5.8 System-Level Design Guidelines

    5.9 Design Considerations

6. Reviews—cite SEMP

7. Documentation Plan—can cite Documentation Plan or a checklist derived from it if already a separate document

8. Development Plans

    8.1 System Architecture

    8.2 Software

    8.3 Electronics

    8.4 Mechanical or Packaging

    8.5 Sensor or Optics

    8.6 Support Equipment

9. Close-out Plan

10. Supporting Process Plans—cite SEMP

    10.1 Test Plan—often a separate document

        10.1.1 EMC/EMI Test Plan

        10.1.2 Shock and Vibration Test Plan

        10.1.3 Thermal-Vacuum Test Plan

        10.1.4 Extreme Environment Test Plan

    10.2 Problem Resolution Plan—cite SEMP

    10.3 Infrastructure Plan

    10.4 Tools and Resources

11. Glossary

12. Technical Appendices

Appendix B: Example of a Small Requirements Document for a Subsystem

This is an example of a requirements document. Some requirements documents have a rationale statement after each requirement. Other requirements documents have a matrix that contains each requirement, its ID number, and a business rule or document reference. Both statements of rationale and document references can be useful for helping with understanding the intent of the requirements.

1. Introduction

   1.1 Purpose

   This document supplies the information to comply with XXXXX (e.g., contract or customer document). This document describes the product requirements for the YYYYYY (project's title) and the metrics for measuring project performance and verifying the system components.

   1.2 Scope

   This document provides all the requirements to complete the YYYYYY (project's title). It covers the subsystem and the ground support equipment (GSE) supplied by (our company) to the (XXXXXX—customer's name). It describes six main sections: system, mechanical packaging and cabling, electronic hardware, mechatronic hardware, software, and EGSE.

   1.3 Objectives

   These requirements should direct development of (xxxxxxx) in such a way that it has the following objectives:

   > (e.g., support the success of YYYYYYY mission)

   > (e.g., support the success of YYYYYYY customer)

   > (e.g., cause the ZZZZZ experiment to function in a specified manner)

   > . . . .

   1.4 Keywords and Notations

   The following keywords have special significance herein.

   Shall. A keyword indicating a mandatory requirement that must be implemented.

   Should. A keyword indicating flexibility of choice with a preferred alternative that shall be considered.

May. A keyword indicating a flexibility of choice with no implied preference. It can be interpreted as permission.

Will. A keyword expressing a commitment by some party to provide something.
All sentences containing the keywords "shall" or "should" shall be interpreted by designers and implementers as instructions; they should be expected to be contractually binding. Any sentence not containing one of these keywords may be interpreted as informational.

1.5 Acronyms

CCSDS	Consultative Committee for Space Data Standards
ECN	Engineering Change Notice
ECR	Engineering Change Request
ECU	Experiment Control Unit
EICD	Electrical ICD
I^2C	Inter-Integrated Circuit (bus)
ICD	Interface Control Document
IICD	Information ICD
JTAG	Joint Test Association Group
MICD	Mechanical ICD
PPS	Pulse per Second
PRCA	Problem Reporting Corrective Action
PRD	Product Requirements Document
PWB	Printed Wiring Board
SOW	Statement of work
WBS	Work breakdown structure

2. References

2.1 Standards

MIL. STD. 461E, Electromagnetic Compatibility (TBD)
MIL-HDBK-338B Electronic Reliability Design Handbook
Std. identification, Document title.

2.2 Processes and Documents

1. (our company) Contract for the (YYYYY project's title)
2. Dyyyxxx, (our company) Statement of Work (SOW) for the (YYYYY project's title)
3. Dyyyxxx, (our company) Schedule for the (YYYYY project's title)
4. Dyyyxxx, (our company) Budget for the (YYYYY project's title)
5. Dyyyxxx, (our company) Quality Management System (QMS)
6. Dyyyxxx, (our company) Quality Assurance Plan for Product Realization

7. Dyyyxxx, (our company) Project Overview and Checklist
8. Dyyyxxx, (our company) Quality Assurance Procedures for Project Management
9. Dyyyxxx, (our company) Configuration Management Plan
10. Dyyyxxx, (our company) Documentation Plan
11. Dyyyxxx, (our company) Quality Assurance Procedures for Design
12. Dyyyxxx, (our company) Software Style Guide
13. Dyyyxxx, (our company) Quality Assurance Procedures for Production and Manufacturing
14. Dyyyxxx, (our company) Plan and Procedures for Audit and Process Review
15. Dyyyxxx, (our company) Quality Assurance Procedures for Metrology
16. Dyyyxxx, (our company) Quality Assurance Procedures for Test
17. Dyyyxxx, Contamination Control Plan
18. Dyyyxxx, Soldering Procedures
19. Dyyyxxx, Conformal Coating Procedure
20. Dyyyxxx, Staking Procedure
21. Dyyyxxx, Prohibited Materials Verification Plan
22. Dyyyxxx, Qualification Test Plan
23. Dyyyxxx, Vendor Qualification Plan
24. Dyyyxxx, Production and Manufacturing Plan
25. Dyyyxxx, Product Identification and Traceability Procedure
26. Dyyyxxx, Identification and Marking of Details and Assemblies Procedure
27. Dyyyxxx, (our company) ESD Control Plan
28. Dyyyxxx, (our company) Product Assurance Manual
29. Dyyyxxx, Worst Case Stress Analysis
30. Dyyyxxx, Structural-Dynamic Analysis
31. Dyyyxxx, (our company) Material Handling
32. Dyyyxxx, (our company) Manufacturing Flow Procedure
33. xxxxxxx, (our company) EICD for the (YYYYY project's title)
34. xxxxxxx, (our company) MICD for the (YYYYY project's title)
35. xxxxxxx, (our company) IICD for the (YYYYY project's title)
36. xxxxxxx, (our company) SICD for the (YYYYY project's title)
37. xxxxxxx, (our company) GICD for the (YYYYY project's title)

2.3 Customer Documents

1. xxxxxxx, Document title
2. <TBD>

2.4 Other

1. <TBD>

3. Project Overview

3.1. Customer's Project

Put in a brief overview (usually 4 to 8 sentences) to concisely describe what the main mission has been designated to do and what the customer has done to prepare.

3.2 Assumptions and Constraints

List the major constraints that will likely bound or limit (our company)'s execution of the development of the specified subsystem.

- An example might be the dynamic range of light levels that sensors are designed to tolerate.
- Another example may be the maximum number of frames per second that can be transmitted through telemetry or the maximum number of sensors supported by the subsystem.
- A final example might be that the system can only survive within a specified radiation environment for a specified time, outside this and the system is not guaranteed to operate without failure or latchup or upset.
- . . . .

4. System Requirements

The System Requirements divide into three main areas:

- Function—describes basic operation of the system without directing how it is done.

- Performance—gives metrics for acceptable operation of the system.

- Environment—describes what the system must withstand for shock, vibration, cooling, and heating.

4.1 Function

4.1.1 The (product) shall format data received from sensors on the vehicle and transmit the data to the ground for display, storage, and analysis.

4.1.2 The (product) shall have two distinct operational partitions—a vehicle package and a ground system.

4.1.3 The vehicle package in the (product) shall comprise sensors, circuit boards, mechanical enclosures, and cabling.

4.1.4 The GSE shall comprise desktop computer(s).

4.1.5 The (product) shall accept data from sensors, these sensors include: two (2) cameras for the visible spectrum, a camera for the infrared (IR) spectrum, a radiometer, and a spectrograph.

4.1.6　The (product) shall accept housekeeping signals, which derive from the following: a temperature sensor on each visible camera, three temperature sensors on the IR camera, a temperature sensor on the radiometer, and a temperature sensor on the spectrograph.

4.1.7　The (product) shall accept a GPS timestamp from the vehicle and incorporate it in the data stream.

4.1.8　The (product) shall multiplex the data from the sensors and the housekeeping signals into a single bit stream for transmission from the vehicle.

4.1.9　The (product) shall perform compression algorithms and processing on data from the cameras and instruments.

4.1.10　The (product) shall have the following priority in transmitting data:

　　4.1.10.1　First priority is compressed data from the radiometer.

　　4.1.10.2　Second priority is compressed data from the radiometer.

　　4.1.10.3　Third priority is housekeeping from the analog board.

　　4.1.10.4　Fourth priority is housekeeping from the power supply board.

　　4.1.10.5　Fifth priority is host communications and GPS time.

　　4.1.10.6　Sixth priority is housekeeping data from IR camera #2.

　　4.1.10.7　Seventh priority is housekeeping data from visible camera #1.

　　4.1.10.8　Eighth priority is compressed data from IR camera #2.

　　4.1.10.9　Ninth priority is compressed data from visible camera #1.

4.1.11　The (product) shall receive power from the vehicle to operate its systems.

4.1.12　The (product) shall receive commands from the vehicle and respond to them. The (product) operates within a system context and executes a mission sequence provided by or approved by (facility).

4.1.13　The (product) shall eject instrument covers only upon command from the vehicle as part of the mission sequence.

4.1.14　The GSE shall display, store, and analyze data sent from the vehicle.

4.1.15　The GSE shall have the capability to recall stored data and display it.

4.1.16　The GSE shall drive a separate S-Video monitor display.

4.1.17　The GSE shall incorporate a removable hard disk drive.

4.2　Performance

4.2.1　The (product) shall support a fixed bit rate of 10 Mbits/sec for transmitting data.

4.2.2　The (product) shall accept data from the radiometer at its signaling rate of 10 Mbits/sec.

4.2.3　The (product) shall accept data from the radiometer at its nominal raw data rate of 7.2 Mbits/sec.

4.2.4 The (product) shall accept data from the spectrograph at its signaling rate of 10 Mbits/sec.

4.2.5 The (product) shall accept data from the spectrograph at its nominal raw data rate of 3.6 Mbits/sec.

4.2.6 The (product) shall accept data from the IR camera at its signaling rate of 25 Mbits/sec.

4.2.7 The (product) shall accept data from the IR camera at its nominal raw data rate of 8.1 Mbits/sec.

4.2.8 The (product) shall accept data from two visible cameras at its signaling rate of 25 Mbits/sec.

4.2.9 The (product) shall accept data from two visible cameras at its nominal raw data rate of 8.1 Mbits/sec.

4.2.10 The (product) shall use lossless compression processing that supports the radiometer with algorithm(s) and methods similar to the APL "FAST" algorithm.

4.2.11 The (product) shall use lossless compression processing that supports the spectrograph with algorithm(s) and methods similar to the APL "FAST" algorithm.

4.2.12 The (product) shall use lossless compression processing that supports the IR camera with algorithm(s) and methods similar to the APL "FAST" algorithm.

4.2.13 The (product) shall use compression processing that supports the visible cameras with algorithm(s) and methods similar to wavelet compression.

4.2.14 The compression ratio for the radiometer shall be 3:1.

4.2.15 The compression ratio for the spectrograph shall be 3:1.

4.2.16 The compression ratio for the IR camera shall be 3:1.

4.2.17 The compression ratio for the visible cameras shall be TBD.

4.2.18 The housekeeping shall produce a bit stream rate not greater than 300 Kbits/sec.

4.2.19 The GSE shall display, store, and analyze data sent from the vehicle in real-time.

4.3 Environment

4.3.1 The (product) meets the environmental specifications as called out in the reference: (ZZZZZZZZZ).

4.3.2 The vehicular portion of the (product) shall operate with only conductive baseplate cooling.

4.3.3 The GSE shall operate in a typical office environment for shock, vibration, and temperature.

4.4 GSE Software Performance

    4.4.1 The GSE software shall allow for the selection and viewing of each sensor:

        4.4.1.1 IR camera.

        4.4.1.2 Visible camera #1.

        4.4.1.3 Visible camera #2.

        4.4.1.4 Radiometer.

        4.4.1.5 Spectrometer.

        4.4.1.6 Sensor temperature measurements.

    4.4.2 The GSE software shall allow for viewing, editing, and initiating the pre-programmed mission sequence including cover ejections.

    4.4.3 The GSE software shall allow for showing the sensor system status.

    4.4.4 The GSE software shall display the visible video with no perceptible delay.

    4.4.5 The GSE software shall display the IR video with no perceptible delay.

    4.4.6 The GSE software shall allow user selection of filenames for storing data.

    4.4.7 The GSE software shall allow verification of data storage on the local hard-drive of the GSE after a test.

    4.4.8 The GSE software shall be capable of turning on or off the power to each sensor separately.

    4.4.9 The GSE software shall be capable of displaying the test modes of the IR camera.

    4.4.10 The GSE software shall be capable of sending the FFC command to the IR camera.

    4.4.11 The GSE software shall display the APL logo somewhere on screen.

5. Structural Enclosure and Cabling

The Mechanical Interface Control Document (MICD) contains important requirements for the interface with the (larger system or vehicle).

The structural and cabling requirements will break into five main areas:

- Size, volume, and weight.

- Dissipation and cooling.

- Shielding—describes the electromagnetic environment and the guidelines for shielding.

- Connector policies—guidelines for keying connectors.

- Cabling policies—guidelines for labeling wires and cables.

5.1 Size, volume, and weight

5.1.1 The (product), which includes circuit board enclosures and cabling, shall not exceed a mass of 2.5 kg.

5.1.2 The (product), which includes circuit board enclosures and cabling, shall not exceed a volume of 0.005 m3.

5.1.3 The enclosure for circuit boards in the (product) shall have the following linear dimensions per (our company) document Dyyyxxx.

5.1.4 The enclosure for circuit boards in the (product) shall be capable of attaching to (TBD) in the vehicle.

5.1.5 The GSE shall be of a size, volume, and mass that is typical of desktop computers.

5.2 Dissipation and cooling

5.2.1 The (product) enclosure shall be capable of conducting 90 W of thermal energy with a maximum temperature increase of 20°C above ambient.

5.2.2 The (product) enclosure shall conduct the thermal energy into the baseplate on the spacecraft.

5.3 Shielding

5.3.1 The (product) shall conform to good electromagnetic compatibility (EMC) design practice by providing conductive gaskets and shields for connectors.

5.3.2 The cabling between the vehicle and the (product) shall pair each signal line with a return line to form a continuous circuit.

5.3.3 Paired signal and return lines within cabling between the vehicle and the (product) shall be twisted at 6 turns per meter or 2 turns per foot.

5.3.4 The cabling between the vehicle and the (product) shall have a continuous, electrically conductive capacitive shield over all wire conductors.

5.3.5 Capacitive shields surrounding the cabling between the vehicle and the (product) shall follow (customer)'s directions for electrical connection to the vehicle; if no directions from (customer), then capacitive shields shall be electrically connected to metallic enclosure that houses the circuit boards.

5.3.6 The GSE shall have EMC and shielding policies typical of desktop computers.

5.4 Connector policies

5.4.1 The (product) shall have an external connector to allow software uploads without opening its enclosure.

5.4.2 The connectors within the (product) shall be configured or keyed in a manner to reduce the likelihood of accidental connection in the wrong orientation.

5.4.3 The connectors within the (product) shall be configured or keyed in a manner to reduce the likelihood of accidental connection in the wrong location.

5.4.4 The connectors within the (product) shall be strained relieved in a manner to survive the specified shock, vibration, and temperature environment.

5.4.5 The GSE shall have connector policies typical of desktop computers.

5.5 Cabling policies

5.5.1 The cabling in the (product) shall be clearly labeled with (TBD) cable tags that attach to cables at (TBD location).

5.5.2 The cabling in the (product) shall be attached to the vehicle at (TBD location).

5.5.3 The cabling, if any, within the (product) shall be strained relieved in a manner to survive the specified shock, vibration, and temperature environment.

5.5.4 The GSE shall have cabling policies typical of desktop computers.

6. Software Requirements

The Information Interface Control Document (IICD) contains important requirements for the data protocols with the (larger system or vehicle).

The Software Requirements will break into five main areas:

- Development processes.

- Development metrics and rates.

- Error rates and defect records.

- Types of processing.

- Data throughput.

6.1 Development processes

6.1.1 The software development team(s) shall follow good industry practices to develop, generate, debug, test, and integrate software for the vehicular portion of the (product).

6.1.2 The software development team(s) shall follow established software style guidelines when developing code.

6.1.3 The software development team(s) shall document their practices and procedures.

6.2 Development metrics and rates

    6.2.1 The software development team(s) shall maintain a record of lines of code (LOC) generated, debugged, tested, and integrated into the vehicular portion of the (product).

    6.2.2 The software development team(s) shall maintain a record of the time/effort expended to generate, debug, test, and integrate each software module in the vehicular portion of the (product).

    6.2.3 The software development team(s) shall calculate the rate of development by dividing the LOC for each module by the time/effort to develop the module.

    6.2.4 The GSE software developed for the GSE shall make use of a pre-existing interface that is tailored to the KASP project; development records shall be kept available for audit.

6.3 Error rates and defect records

    6.3.1 The software development team(s) shall maintain a record of errors/defects in the code found during development (debugging, testing, and integration) of the vehicular portion in the (product).

    6.3.2 The software development team(s) shall calculate the rate of errors by dividing the errors/defects discovered in each module by the time/effort to develop the module.

    6.3.3 The GSE software developed for the GSE shall make use of a pre-existing interface that is tailored to the KASP project; records of defects and errors shall be kept available for audit.

6.4 Types of processing

    6.4.1 The software shall support algorithms for image compression of the data from the sensors in the (product).

    6.4.2 The software shall have a built-in-test with coverage to (TBD)% coverage. The coverage shall include:

        6.4.2.1 Data from video image sensors.

        6.4.2.2 Data from radiometers.

        6.4.2.3 Data from housekeeping sensors.

        6.4.2.4 Memory functionality and size.

        6.4.2.5 Internal communications between processor and FPGA.

    6.4.3 The software shall perform CRC error detection and correction on all commands received.

    6.4.4 The software shall perform CRC error detection and correction on all data transmitted.

    6.4.5 The CRC error detection and correction shall have a 128 bit code.

6.4.6 The software shall have a safe state for most faults that allows reporting the faults.

6.4.7 The software shall recover from a fault (TBD).

6.5 Data throughput

6.5.1 The software shall support a data throughput of xxxxx KBytes (or samples) per second.

6.5.2 The software shall support the xxxxxx transmission protocol.

6.5.3 The software shall be able to store up to xxxxx kbytes.

7. Electronic Hardware Requirements

The Electrical Interface Control Document (EICD) contains important requirements for the interface with the (larger system or vehicle).

The Electronic Hardware Requirements will break into seven main areas:

- Performance—describes types of logic families and component blocks

- EMC and signal integrity concerns

- Dependability

- Memory size

- Download and test ports

- Power—describes the input power, power dissipation, and supply voltages

- Radiation tolerance

7.1 Performance

7.1.1 The electronic circuitry within the (product) shall have sufficient capability to run the necessary software to compress and multiplex the data at the performance specified above for the system. Instrumentation profiling shall demonstrate the throughput margin of the (product).

7.1.2 The electronic circuitry within the (product) may use a variety of components, such as digital signal processors and field programmable gate arrays, to accomplish performance with power consumption at or below the specification below.

7.1.3 Components shall be derated according to the following:

7.1.3.1 Part Parameters and Deratings—Each parameter must be derated from the data book value for the intended environment to compensate for the effects of temperature, age, voltage, and radiation.

7.1.3.2 Timing Analysis—Set-up and hold times at all clocked inputs, pulse widths of clocks, and asynchronous set, clear, and load inputs, all clock inputs and asynchronous inputs such as sets, clears, and loads must be shown to be free from both static and dynamic hazards.

7.1.3.3 Gate Output Loading—Show that no gate output drive capacities have been exceeded.

7.1.3.4 Interface Margins—Show that all of the gates have their input logic level thresholds met.

7.1.3.5 State machines must be analyzed to assure that they will not exhibit anomalous behavior, such as system lock-up.

7.1.3.6 Asynchronous Interfaces—Must show either that asynchronous signals are properly synchronized to the appropriate clock or that the circuitry receiving asynchronous signals will function correctly if set-up and hold times are not met.

7.1.3.7 Reset Conditions and Generation—All circuitry must be shown to be placed into a known state during reset.

7.1.3.8 Part Safety Conditions—The analysis must prove that the circuit is designed so as to prevent its parts from being damaged.

7.1.3.9 Cross-Strap Signals between Redundant Modules—Show that isolation between boxes is actually achieved.

7.1.3.10 Circuit Interconnections—Show that circuit interconnection requirements are met from the standpoint of signal quality as affected by edge rates, loading and noise.

7.1.3.11 Bypass Capacitance Analysis—Show that the amount of on-board bulk and bypass capacitance is appropriate for the circuitry.

7.1.4 The GSE shall have performance typical of a desktop computer with at least 1 g byte of RAM and a removable hard disk drive.

7.2 EMC and signal integrity

7.2.1 Circuit boards (PCBs or PWBs) within the (product) shall contain ground (signal return) planes that are continuous and shall not be interrupted by slots or large, open regions with no plating.

7.2.2 Electrical connections between circuit boards (PCBs or PWBs) within the (product) shall use either multilayer backplanes with continuous planes for signal return or short, wire conductors that pair each signal with a return line connected to the ground (signal return) plane.

7.2.3 Within the (product), electrical connections between the circuit boards (PCBs or PWBs) and the connectors shall be either directly soldered or short, wire conductors that pair each signal with a return line connected to the ground (signal return) plane.

7.2.4 Conducted susceptibility of the product shall meet or improve upon MIL. STD. 461E.

7.2.5 Radiated susceptibility of the product shall meet or improve upon MIL. STD. 461E.

7.2.6 Conducted interference of the product shall meet or improve upon MIL. STD. 461E.

7.2.7 Radiated interference of the product shall meet or improve upon MIL. STD. 461E.

7.3 Dependability

7.3.1 The (product) system shall have a calculated reliability of 8000 hours. Calculations shall follow MIL-HDBK-338B Electronic Reliability Design Handbook.

7.3.2 The system shall have JTAG boundary scan capability that allows test coverage to 95% of all estimated faults.

7.3.3 The system shall recover from 95% of all estimated nonfatal faults. Nonfatal means that components and circuitry remain physically undamaged and not latched up.

7.3.4 System maintainability shall be TBD.

7.3.5 The system shall have a shelf life that exceeds 20,000 hours.

7.4 Memory size

7.4.1 The memory within the (product) shall have sufficient memory capacity to run all the code without reducing the performance to below those already specified.

7.4.2 The memory within the (product) shall have 30% or greater additional margin in memory capacity at the time of final integration at (facility).

7.4.3 The GSE shall have memory capacity typical of desktop computers and sufficient to run the software.

7.5 Download and test ports

7.5.1 The electronic circuitry within the vehicle package in the (product) shall have connections or a port available to download software code into the circuitry.

7.5.2 The (product) shall have the mechanical and electrical means to support software uploads without opening its enclosure.

7.5.3 The electronic circuitry within the vehicle package in the (product) shall have connections or a port available to test the circuitry.

7.6 Power

    7.6.1 The (product) shall draw input power from the vehicle.

    7.6.2 Input power on the supply bus will be supplied to the (product) at a nominal voltage of 28 VDC. The variation in voltage could be ±20%.

    7.6.3 The maximum input current on the supply bus supplied to the (product) will be 3 A.

    7.6.4 The (product) shall consume a maximum of 90 W.

    7.6.5 The GSE shall draw residential power typical of a desktop computer at 120 VAC.

    7.6.6 The GSE shall consume power typical of a desktop computer.

7.7 Radiation tolerance

    7.7.1 The (product) shall operate tolerate a total ionizing dose during the mission of 100,000 rad Si.

    7.7.2 The (product) shall survive 1 SEU every 20 hours.

    7.7.3 The (product) shall not be expected to survive an SEL.

8. Mechatronics Requirements

The Mechanical Interface Control Document (MICD) contains important requirements for the interface with the (larger system or vehicle).

The Mechatronics Requirements will break into two main areas:

- Function—describes what it does.

- Performance—describes how well it does it.

8.1 Functional

    8.1.1 The system shall open the covers over the optical lenses upon command.

    8.1.2 The system shall unfurl solar panels upon command.

    8.1.3 The system shall operate a filter wheel in front of the video imagers.

8.2 Performance

    8.2.1 The covers over the optical lenses shall open within 0.2 seconds after a command to open.

    8.2.2 The solar panels shall take between 6 and 10 minutes to unfurl.

    8.2.3 The solar panels shall accelerate between 0.1 rad/minute^2 and 0.2 rad/minute^2.

    8.2.4 The filter wheels shall rotate between 0.1 rad/second and 0.2 rad/second.

8.3 Dependability

8.3.1 The (product) system shall have a calculated reliability of 8000 hours. Calculations shall follow MIL-HDBK-338B *Electronic Reliability Design Handbook*.

8.3.2 The filter wheel mechanism shall operate up to 3000 rotations.

9. Sensor Requirements

The Sensor Interface Control Document (MICD) contains important requirements for the interface with the (larger system or vehicle).

The Sensor Requirements will break into four main areas:

- Performance—describes types of logic families and component blocks

- Memory size

- Download and test ports

- Power—describes the input power, power dissipation, and supply voltages

9.1 Function

9.1.1 The sensors shall view the visible, UV, and IR light spectrum.

9.1.2 The sensors shall provide video images and calibrated spectrums:
9.1.2.1 One video imager shall view the visible spectrum.
9.1.2.2 One video imager shall view the UV spectrum.
9.1.2.3 One video imager shall view the IR spectrum.
9.1.2.4 One spectrometer shall view the visible spectrum.
9.1.2.5 One spectrometer shall view the UV spectrum.
9.1.2.6 One spectrometer shall view the IR spectrum.

9.1.3 The sensors shall operate upon command.

9.2 Performance

9.2.1 The video imagers shall operate at command frame rates of 1.5, 2, 5, 7.5, 15, and 30 frames/second.

9.2.2 The spectrometers shall operate at command frame rates of 2, 5, and 7.5 frames/second.

9.2.3 The video imager parameters shall be as follows:
9.2.3.1 Measurand
9.2.3.2 Speed of transduction—samples per second
9.2.3.3 Span
9.2.3.4 Full scale output
9.2.3.5 Linearity—%, SNR

9.2.3.6 Threshold

9.2.3.7 Resolution—ENOB

9.2.3.8 Accuracy—SNR

9.2.3.9 Precision

9.2.3.10 Sensitivity—%

9.2.3.11 Hysteresis

9.2.3.12 Specificity

9.2.3.13 Noise—SNR, % budget

9.2.3.14 Stability

9.2.4 The sensors, individually, shall not consume more than 0.5 W during operation.

10. Technical Performance Standards

The project contract and SOW provide the basis for a product's technical performance. They may cite standards, which would be referenced in section 2.1 or 2.3 of this document.

Eventually, the project's PRD documents the technical performance requirements and their flow-down.

11. Glossary

ADC	analog-to-digital converter
APID	application process identifier (CCSDS)
C	Celsius
CCSDS	Consultative Committee for Space Data Systems
CDRL	contract deliverable requirements list
CM	configuration management
CMD	command
CRC	cylical redundancy code
DSP	digital signal processor
ECN	engineering change notice
ECR	engineering change request
ECU	experiment control unit
GSE	ground support equipment
EICD	electrical ICD
ELV	expendable launch vehicle
ENOB	effective number of bits
ETA	event tree analysis
FMECA	failure modes effects and criticality analysis
FPGA	field programmable gate array
FTA	fault tree analysis
GPS	global positioning system

HA	hazards analysis
Haywire	a wired correction on a PCB
I^{2}C	inter-integrated circuit (bus)
ICD	interface control document
IICD	information ICD
JTAG	Joint Test Association Group
LOC	lines of code
mA	milliamps
MICD	mechanical ICD
mW	milliwatts
NTP	network time protocol
PCB	printed circuit board
PPS	pulse per second
PRCA	problem reporting corrective action
PRD	product requirements document
PROM	programmable read only memory
PWB	printed wiring board
PWR	power
QoS	quality of service
RTN	return
RX	receive line
SEE	single-event effects
SEL	single-event latchup
SEU	single-event upset
SDIO	serial digital input/output
SOW	statement of work
STS	system timing signal
TCXO	temperature controlled crystal oscillator
TLM	telemetry
TX	transmit line
TBD	to be determined
TBR	to be resolved
UART	universal, asynchronous receiver/transmitter
V&V	validation and verification
V	volts
W	watts
WBS	work breakdown structure

Appendix C: Example of a Small Test Plan

1. Introduction

    1.1 Purpose

    This Test Plan will guide work performed by Company A1 in the test and integration of Instruments "A" and "B" with the Project XYZ System. Company A1 does not control the Project XYZ Payload integration or test, which is under control of the customer.

    1.2 Scope

    This Test Plan describes the processes, procedures, reviews, and documents that will guide and document the test and integration of Instruments "A" and "B" with the Project XYZ System. It outlines or references all test and integration activities necessary to complete the project. It describes the control of configuration, roles and responsibilities of the development team at Company A1, review processes, documentation, and schedule milestones.
    (*Please Note*: This Test Plan is for a relatively small subsystem. It is representative only—an entire spacecraft has a much larger and far more detailed test plan.)

    1.3 Definitions, Acronyms and Abbreviations

    The glossary defines the acronyms and abbreviations used on the Project XYZ System.

    Validation attempts to show that the system works as intended. It does not necessarily confirm that the software performs according to the requirements.

    Verification confirms that the software performs according to the requirements. It does not prove that the system works as intended.

    1.4 References

    Documents that describe the contractual deliverables for the Project XYZ Instrument "A" project and govern this Test Plan are:

    - Document xxx1, Project XYZ Instrument "A" Functional Specification
    - Document xxx2, Project XYZ Instrument "A" Design Description
    - Document xxx3, Project XYZ Instrument "A" Mechanical ICD
    - Document xxx4, Project XYZ Instrument "A" Electrical ICD
    - Document xxx5, Project XYZ Instrument "A" Data (or Software) ICD
    - Document xxx6, Project XYZ Instrument "A" Users' Guide
    - Document yyy1, Project XYZ Instrument "B" Functional Specification
    - Document yyy 2, Project XYZ Instrument "B" Design Description
    - Document yyy 3, Project XYZ Instrument "B" Mechanical ICD

- Document yyy 4, Project XYZ Instrument "B" Electrical ICD
- Document yyy 5, Project XYZ Instrument "B" Data (or Software) ICD
- Document xxx6, Project XYZ Instrument "B" Users' Guide
- Document zzz1, Project XYZ System Project Plan
- Document zzz2, Project XYZ System Users' Guide

Documents that provide guidelines for this Test Plan are:

- Company A1's "Quality Assurance Plan"
- Company A1's "Design Documentation Manual"
- ED012048, *Company A1 Software Implementation Standards*

This Test Plan directly specifies a number of plans and other documents. They include the following:

- Test Procedures, Project XYZ Instrument "A" Functional Test Procedures
- Test Procedures, Project XYZ Instrument "A" Environmental Test Procedures
- Test Procedures, Project XYZ Instrument "B" Functional Test Procedures
- Test Procedures, Project XYZ Instrument "B" Environmental Test Procedures
- Test Procedures, Project XYZ System Functional Test Procedures
- Test Procedures, Project XYZ System Environmental Test Procedures

1.5 Overview

This Test Plan is the primary document for testing Instrument "A" and the Instrument "B" for Project XYZ. It outlines the documents and activities needed to test and integrate the project.

Verification can be accomplished by one or a combination of the following: assessment (or analysis), test, demonstration, inspection.

Verification by assessment (or analysis) can be further divided into verification by analysis and verification by similarity. Verification by analysis uses calculations or modeling to verify compliance with specifications. Analysis may be used when it can be performed rigorously and accurately, when testing is not cost effective or is high risk (resulting in probable damage or contamination), or when verification by inspection is not adequate. Some amount of testing may be required to supplement or confirm part of the analysis. Assessment could also result in an operational or procedural constraints which preclude entry into a hazardous condition, which itself would require testing or analysis to verify.

Verification by similarity is a process of item comparisons taking into account configuration, test data, application, and environment. Engineering evaluations are required to verify that 1) differences in design between the item proposed for verification and the previously verified item are insignificant, 2) environmental stress shall not be greater in the new application, 3) the manufacturer and manufacturing methods are the same, and 4) there are no significant differences in use or application. Similarity does not eliminate the need for workmanship tests.

Verification by test consists of "proof by doing" to ensure that functional or environmental specifications for an item are met. Environmental test verification may be performed on prototype or flight hardware in conjunction with verifying functional performance. Environmental testing shall provide assurance that the hardware shall perform satisfactorily under conditions simulating the extremes of ground handling, launch, and flight operations.

Verification by demonstration is a method which denotes the qualitative determination properties of an item by observation. Verification by inspection is a method of visually determining an item's qualitative or quantitative properties such as tolerances, finishes, identification, specific dimension, envelopes, or other measurable properties.

2. Controls for Validation and Verification (V&V)

2.1 Controls

The controls to track and ensure that the verification program is being carried out according to the test plan shall consist of having approved verification test procedures, verification test reviews and verification test and analysis reports. All relevant plans, procedures, reports, waivers and liens shall be logged into the Verification Matrix/ Database as described below.

2.1.1 Verification Test Procedures
Test Procedures (TPs) shall be prepared based on this test plan and shall address the specifications for the system and instruments. Testing may include any supporting analyses, inspections, calibrations and checkout operations. The TPs shall specify the qualification or acceptance testing with critical pass/fail criteria and functional checkout requirements.

Procedures shall be developed to describe each test activity. Each procedure shall identify the requirement(s) that it is addressing. The TP shall identify the configuration of the hardware and software and test setup for the assembly, component, subsystem to which it is applicable. All procedures at the Instrument "A" and Instrument "A" levels shall be submitted to program management for approval 30 days prior to usage. Procedures for lower levels of assembly will be controlled and approved by the organization providing that component or subsystem.

2.1.2 Control of System Procedures
System level test procedures shall be controlled by the customer's Engineering Manager. All test procedures and all analysis procedures must be placed under control prior to the actual test or analysis being performed.

2.1.3 Test Reviews

Pre-test reviews shall precede system level tests of flight hardware/software. These reviews shall verify the readiness of the flight hardware, facility and test equipment, and procedures. For minor tests, reviews may be conducted by key test personnel. For major tests, formal reviews shall be conducted by the customer's Engineering Manager or a designated representative.

2.1.4 Monitoring of Procedures

The customer representative may review and monitor test activities. A schedule of verification activities provided in advance of the testing may serve as notification. Updates to this schedule should be provided as mutually agreed to by the customer representative and Company A1. All procedures, test data, analysis data, and reports will be available on Company A1's site to the customer representative for review and audit.

2.1.5 Verification Test Analysis Reports

A report shall be prepared after the completion of each qualification/acceptance test or analysis. Each report shall summarize the test results, correlate the results with the applicable TP requirements and note any nonconformities and re-verifications required. Company A1 will retain copies of all test procedures, test reports, and their updates. Company A1 will retain all the original signature sheets. These reports will be available for review by the customer representative.

Immediately following any verification test, a quick post test review should be performed. This should occur prior to breaking down the test set-up and moving hardware. This review can be informal; for instance, it could consist of a discussion among available engineers, managers, and quality assurance people. The purpose of the review is to determine if the test objectives were met, if there are any unanswered questions with respect to the test, and to determine if any additional testing is required.

2.1.6 Malfunction Reporting

All problems encountered are to be recorded on a Problem Report. If a major known or suspected failure occurs, the Test Conductor shall document, review, control, authorize and disposition failures. These failures will be submitted to the customer. The Test Conductor is responsible for preparing the Problem Report.

2.1.7 Waivers and Liens

Any verification requirement that is not met at the test or analysis level indicated in this document or the Verification Matrix/Database will require a waiver or a lien. A waiver shall be used when a requirement is not met, and will not be corrected or subjected to retest (flown "as is"). A lien shall be issued for any requirement verification that is deferred until later in the integration and test flow than its indicated test or analysis level. Waivers and liens shall be maintained by Company A1 and subject to review by the customer.

2.2 Verification Matrix/Database

A database of all requirements will be maintained; an example is shown in the following table. This database will include the requirement, type of verification performed, brief description of verification methodology, name and document number of procedures used, name and number of reports generated, and listing of waivers and liens associated with each. The database shall be under version control and configuration management; it may be updated on a daily basis during integration and test. The database is subject to review by the customer.

Example of Database Table

Requirement	Verification Test	Methodology	Results
The software development team(s) shall maintain a record of errors/defects in the code found during development (debugging, testing, and integration) of the vehicular portion in the system.	Audit	Review of records	All records from code inspections found and examined.

3. Instrument "A" Software Validation and Verification

3.1 Scope

The verification and validation (V&V) of the portion of software for Instrument "A" and the GSE includes elements from software development activities, specification testing, engineering testing, and usability testing. Reports from the testing areas as called by this plan combine to form the verification and validation report. Reports follow the IMRAD scheme (Introduction, Methods, Results, and Discussion).

Elements of verification and validation for the Instrument "A" consist of code inspections, white box testing, ground support equipment (GSE) screen properties, events verification, GSE data file creation and display, and Instrument "A" to GSE communication testing.

3.2 Code Inspection

Code inspections are highly productive in catching errors in software and increasing the quality of the final product. Inspections can be up to 20 times more efficient than testing. Acode inspection has the following components:

Elements
- Non-confrontational environment
- Checklist of things to consider
- Action item list
- Document the proceedings and results—action item list feeds this and may replace it

Team members
- Moderator
- Reader
- Recorder (can be same as moderator)
- Author

Proceedings
- Planning by author and moderator
- Overview and preparation by all team members
- Inspection meeting
- Rework by author
- Follow-up by moderator

Statement:	Code review with software engineers uses both the code review checklist and the action item list. Record issues and concerns from each review into a code review checklist, along with the date. The action item list then assigns personnel to work each line item from the code review checklist, when to perform the follow-up review, and date of resolution. These lists provide the basis for the code review report.
Tools:	Code review checklist, action item list, QA review procedures.
Training:	Code review, QA System training record.
Timing:	Before final software freeze and delivery to integration testing.
Personnel:	Software engineers and developers
Deliverable:	Code review report lists review dates, issues from reviews, and statement of resolution and name of assignee. Use the IMRAD format and report results in metrics (number of issues, number resolved, time for resolution, risk levels, confidence levels, etc.) and discuss review methods, future release criteria and specs, and note risky areas of the code.

3.3 White-Box Testing

White box testing attempts to exercise each module of code thoroughly. It uses knowledge of the code to design tests that attempt to exercise every possible branch within the logical flow of the code. This testing should address and verify that the Instrument "A" meets every requirement in the specification.

Statement:	White box testing targets every requirement for verification.
Tools:	Emulator, GSE, Test Procedures, Instrument "A" Functional Test Procedures
Training:	QA training procedures.
Timing:	Before final software freeze and delivery to integration testing.
Personnel:	Software engineers and developers
Deliverable:	A report that covers the testing during software development. It should include metrics on the number of defects, their subsequent resolutions, and a report from the version control system. It should list the issues and code changes. It should discuss risk areas and future improvements. It follows the IMRAD format.

3.4 Command Sequence and Communication Testing

These tests exercise the Instrument "A" with command sequences and the communications channels of the Project XYZ GSE. They are a necessary part of software development and might be considered a part of white box testing. Testing command sequences through the communications channels for the Instrument "A" overlaps with integration of the system; it is end-to-end testing of the system.

Statement:	Command sequence and communications testing includes using the Instrument "A" and the Instrument "B" in a controlled, laboratory environment. Like white box testing, it verifies every requirement for command and communications in the system.
Tools:	GSE, laboratory support equipment such as power supplies, oscilloscope, and logic analyzer. Test Procedures for Project XYZ, Instrument "A" Functional Test Procedures
Training:	User manual, test procedures
Personnel:	Validation person, electronic and software engineers
Deliverable:	Command sequence and communications test report includes metrics on the number of defects, their subsequent resolutions, and a report from the version control system. It should list the issues and code changes. It should discuss risk areas and future improvements. It follows the IMRAD format.

3.5 GSE Data File Creation and Display

These tests exercise the system and the GSE for creating and storing data files and then displaying the data in real time. The tests follow after command sequences are sent to the Instrument "A" and data are received from the Instrument "A" via the communications channels. They are a necessary part of software development and might be considered a part of white box testing. Testing command sequences through the communications channels for the Instrument "A" overlaps with integration of the system; it is end-to-end testing of the system.

Statement:	Stores and displays real time data after command sequence and communications testing includes using the Instrument "A" and the GSE in a controlled, laboratory environment. Like white box testing, it verifies every requirement for command and communications in the system.
Tools:	GSE, laboratory support equipment such as power supplies, oscilloscope, and logic analyzer. Test Procedures for Project XYZ, Instrument "A" Functional Test Procedures
Training:	User manual, test procedures
Personnel:	Validation person, electronic and software engineers
Deliverable:	GSE storage and display test report includes metrics on the number of defects, their subsequent resolutions, and a report from the version control system. It should list the issues and code changes. It should discuss risk areas and future improvements. It follows the IMRAD format.

3.6 Integration Tests for Validation

Validation attempts to show that the system works as intended. It does not necessarily confirm that the software performs according to the requirements.

The Project XYZ System is intended to [mission]. . . . One immediate and obvious way to achieve the intent is to receive data from a number of sensors on … and make the data available for analysis on the ground; this is one form of validation.

Deeper layers of validation follow. The Instrument "A" must receive data from a number of sources, compress the video data from four different types of sensors, multiplex the data into a single, serial data stream, and send the data stream to the target vehicle's data transmission system. On the ground the GSE must receive the data stream from the launch facility, demultiplex the data from a single, serial data stream into separate files that represent the various sources and sensors, decompress the video data from the four different types of sensors, and store and display those data in real time.

The testing in sections 3.4 and 3.5 will verify operation of the Instrument "A" in each of these concerns. The final end-to-end tests of the Instrument "A" also accomplish tests for validation. They should adequately demonstrate the intent of the system.

4. Instrument "A" Test and Integration

4.1 Overview

The system integration includes the software, electronics, mechanical, and packaging of Instrument "A." This section describes the "who, what, when, and where" of the integration of the instrument into the system.

Each of these activities will have a detailed script and corresponding checklist and log to record the results.

4.2 Instrument "A" Integration

4.2.1 Package Mechanical Placement

The Instrument "A" mechanical packages are oriented, attached to the spacecraft frame, and confirmed in correct orientation. If any package does not orient or attach as designed or scripted, then the integration team will follow a review and correction procedure. An integration script, in the Test Procedure will detail each of these actions.

4.2.2 Harness Placement

The vehicle and sensor harnesses are oriented, attached to the Instrument "A" frame, and confirmed in correct orientation. Bend radius limits will be confirmed. Connector polarities will be confirmed. If any harness does not orient or attach as designed or scripted or has incorrect polarities, then the integration team will follow a review and correction procedure. An integration script, in the Test Procedure will detail each of these actions.

4.2.3 Electronic Board Placement

The Instrument "A" electronic boards are oriented, attached to their respective connectors within the Instrument "A" enclosure, and confirmed in correct orientation. The electronic boards will install in the following sequence:

- Power supply board
- Analog telemetry board
- Serial digital board
- Video compression board

4.2.4 Connection of the GSE

The GSE can imitate the host vehicle and exercise Instrument "A" for testing. The appropriate test cables are oriented, attached to the Instrument "A" frame, and confirmed in correct orientation. Connector polarities will be confirmed. If any cable does not orient or attach as designed or scripted or has incorrect polarities, then the integration team will follow a review and correction procedure. An integration script, in the Test Procedure will detail each of these actions.

4.3 Instrument "A" Test

4.3.1 Electrical Test

After installing all the electronic boards during integration, they are removed. The Instrument "A" electronic boards will then reinstall in the following sequence:

- Power supply board
- Analog telemetry board

- Serial digital board
- Video compression board

After installing each board, the GSE will apply and monitor power. If power does not operate as designed or scripted, then the integration team will follow a review and correction procedure. Otherwise, the GSE will exercise each board according to the test script. If an electronic board does not operate as designed or scripted, then the integration team will follow a review and correction procedure. A test script, in the Test Procedure will detail each of these actions.

4.3.2 Verification of the Software

Once the entire Instrument "A" passes the electrical test, the GSE will verify operation of the Instrument "A" in the following order:

- Power supply board in self-test and tested for supplied voltages
- Power supply and analog telemetry boards in self-test and tested for supplied voltages
- Power supply, analog telemetry, and serial digital boards in self-test
- Power supply, analog telemetry, and serial digital boards with commanded sequences
- Power supply, analog telemetry, serial digital, and video compression boards in self-test
- Power supply, analog telemetry, serial digital, and video compression boards with commanded sequences
- Full Instrument "A" (power supply, analog telemetry, serial digital, and video compression boards) with analog housekeeping (can be simulated signals)
- Full Instrument "A" (power supply, analog telemetry, serial digital, and video compression boards) with radiometer attached
- Full Instrument "A" (power supply, analog telemetry, serial digital, and video compression boards) with spectrograph attached
- Full Instrument "A" (power supply, analog telemetry, serial digital, and video compression boards) with IR camera attached
- Full Instrument "A" (power supply, analog telemetry, serial digital, and video compression boards) with visible camera attached
- Full Instrument "A" with the radiometer and spectrograph attached and full compression and data multiplexing turned on
- Full Instrument "A" with the radiometer, spectrograph, and IR camera attached and full compression and data multiplexing turned on
- Full Instrument "A" with the radiometer, spectrograph, IR camera, and visible camera attached and full compression and data multiplexing turned on
- Full Instrument "A" with all sensors attached and analog housekeeping signals

Each step in this sequence will have a full suite of tests described in the test script. If any electronic board or subsystem does not operate as designed or scripted, then the integration team will follow a review and correction procedure. A test script, in the Test Procedure, will detail each of these actions.

Note: Verification confirms that the software performs according to the requirements. It does not prove that the system works as intended.

4.3.3 Validation of the Software

Once the entire Instrument "A" passes the software verification, the test procedure will validate the operation of the system. A test script, in the Test Procedure, will detail the validation. If any operation does not go as designed or scripted, then the integration team will follow a review and correction procedure.

Note: Validation attempts to show that the system works as intended. It does not necessarily confirm that the software performs according to the requirements.

The final end-to-end tests of the Instrument "A" accomplish tests for validation. They should adequately demonstrate the intent of the system.

4.3.4 Vibration Test

The integrated flight Instrument "A" will undergo vibration testing. All connector savers, with the exception of the host interface to Instrument "B," will be removed and the Instrument "A" will be in final flight condition. It will be subjected to the following sequences of vibration patterns:

- Low-level vibration at TBD g, swept-frequency, between at TBD Hz and TBD Hz, for TBD duration. This testing will determine mechanical resonance peaks.
- High-level vibration at TBD g, at each of the first five resonance peaks, for TBD duration.

After completing each sequence, the Instrument "A" will undergo visual inspection, mechanical torque checks, and electrical testing.

Upon inspection, no components or harness connections will have fallen out or be loose. If any component or subsystem falls out or loosens beyond the designed or scripted limits, then the integration team will follow a review and correction procedure. A test script, in the Test Procedure, will detail each of these actions.

If no components or subsystems appear loose, then the GSE shall exercise the system as described above in test procedure. If any component or subsystem does not operate as designed or scripted, then the integration team will follow a review and correction procedure. A test script, in the Test Procedure, will detail each of these actions.

4.3.5   Thermal/Vacuum Test

The integrated flight Instrument "A" will undergo thermal/vacuum testing. All connector savers, with the exception of the host interface to GSE, will be removed and the Instrument "A" will be in final flight condition. It will be subjected to the following sequences of conditions:

- Thermal cycling between TBD and TBD °C, with a period of TBD. The Instrument "A" will endure TBD thermal cycles.
- Vacuum testing to TBD torr for TBD time.
- Thermal cycling between TBD and TBD °C, with a period of TBD at vacuum of TBD torr. The Instrument "A" will endure TBD thermal/vacuum cycles.

After completing each sequence, the Instrument "A" will undergo visual inspection, mechanical torque checks, and electrical testing.

Upon inspection no components or harness connections will have fallen out or be loose. If any component or subsystem falls out or loosens beyond the designed or scripted limits, then the integration team will follow a review and correction procedure. A test script, in the Test Procedure, will detail each of these actions.

If no components or subsystems appear loose, then the Instrument "B" shall exercise the systems as described above in section X.X. If any component or subsystem does not operate as designed or scripted, then the integration team will follow a review and correction procedure. A test script, in the Test Procedure, will detail each of these actions.

4.3.6   Electromagnetic Compatibility Test

The integrated flight Instrument "A" will undergo electromagnetic compatibility (EMC) testing. All connector savers, with the exception of the host interface to Instrument "B," will be removed and the Instrument "A" will be in final flight condition. It will be subjected to the following sequences of EMC patterns:

- MIL-STD-461E—TBD

After completing each sequence, the Instrument "A" will undergo visual inspection, mechanical torque

If the Instrument "A" does not operate as designed or scripted, then the integration team will follow a review and correction procedure. A test script, in the Test Procedure, will detail each of these actions.

4.4   Test Tolerances

Unless otherwise specified, the following tolerances on the environments shall be used during verification and test. The tolerances include measurement uncertainties, and were derived from MIL-STD-1540C (Test Requirements for Launch, Upper Stage, and Space Vehicles).

Acoustic Noise

Overall level: ±1 dB
One-third octave band center frequencies from:
- up to 40 Hz: +3 db and −6 db
- 40 to 3150 Hz: ±3 db
- 3150 Hz and up: +3 db and −6 db

Electromagnetic Compatibility

Voltage magnitude: ±5% of peak
Current magnitude: ±5% of peak
RF amplitudes: ±2 db
Frequency: ±2%
Distance: ±5% of the specified distance or ±5 centimeters (greater of the two)

Humidity

Humidity level: ±5% RH

Loads

Static: ±5%
Steady-state (acceleration): ±5%

Pressure/Vacuum

Greater than 100 mm Hg: ±5%
100 mm Hg to 1 mm Hg: ±10%
1 mm Hg to 1 micron: ±25%
Less than 1 micron: ±80%

Temperature

Environmental control ±1 °C

Vibration

Sinusoidal
- Amplitude: ±10%
- Frequency: ±2%
Random
- RMS level: ±10%
- Acceleration spectral density: ±3 db

Mechanical shock

> Response spectrum: +25%, −10%
> Time history: ±10%

Mass properties

> Weight: ±0.1% (per ICD)
> Center of gravity: ±0.40 inch/axis (per ICD)
> Moments of inertia: ±5% (per ICD)

4.5 Documented Outputs

The Project XYZ Instrument "A" documents required for integration and test are the Test Procedures (with integration and test scripts) and Test Results.

4.6 Problem Reporting and Corrective Action

If any component or system does not operate as designed or scripted during any of the described tests above, then the integration team will follow a review and correction procedure. The Test Procedures describe all scripts and procedures and remedial actions.

4.7 Facilities, Tools, Techniques, and Methodologies

The flight Instrument "A" will be integrated and tested in a clean room at Company A1 for particulates. Company A1 will carry out the vibration and thermal/vacuum testing at the contract test facilities run by Testing-R-Us.

The flight Instrument "A" will be tested for EMC. The Company A1 will carry out the EMC testing at the contract test facilities run by Testing-R-Us.

5.   Instrument "B" Integration and Test

5.1 Overview

The system integration includes the software, electronics, mechanical, and packaging of the Project XYZ Instrument "B". This section describes the "who, what, when, and where" of the integration of the Instrument "B" in the Project XYZ system.

Each of these activities will have a detailed script and corresponding checklist and log to record the results.

5.2 Instrument "B" Integration

5.2.1 Package Mechanical Placement
The Instrument "B" mechanical packages shall be oriented and confirmed in an appropriate position. If any package does not orient or attach as designed or

scripted, then the integration team will follow a review and correction procedure. An integration script, in the Test Procedure will detail each of these actions.

5.2.2 Harness and Cables Placement

The Instrument "B" harnesses and cables are oriented and confirmed in an appropriate position. Bend radius limits will be confirmed. Connector polarities will be confirmed. If any harness does not orient or attach as designed or scripted or has incorrect polarities, then the integration team will follow a review and correction procedure. An integration script, in the Test Procedure will detail each of these actions.

5.2.3 Connection of the Instrument "B"

The Instrument "B" harnesses and cables, that imitate the host vehicle, are oriented, attached to the Instrument "A" frame, and confirmed in correct orientation. Connector polarities will be confirmed. If any harness does not orient or attach as designed or scripted or has incorrect polarities, then the integration team will follow a review and correction procedure. An integration script, in the Test Procedure will detail each of these actions.

5.3 Instrument "B" Test

5.3.1 Verification of the Software

The team will verify operation of the Instrument "B" in concert with the Instrument "A" in the following order:

- TBD—xxxxx

Each step in this sequence will have a full suite of tests described in the test script. If any electronic board or subsystem does not operate as designed or scripted, then the integration team will follow a review and correction procedure. A test script, in the Test Procedure will detail each of these actions.

One note: verification confirms that the software performs according to the requirements. It does not prove that the system works as intended.

5.3.2 Validation of the Software

Once the entire Instrument "B" passes the software verification, the test procedure will validate the operation of the system with Instrument "B." A test script, in the Test Procedure, will detail the validation. If any operation does not go as designed or scripted, then the integration team will follow a review and correction procedure.

One note: validation attempts to show that the system works as intended. It does not necessarily confirm that the software performs according to the requirements. The final end-to-end tests of the Instrument "B" accomplish tests for validation. They should adequately demonstrate the intent of the system.

5.4 Documented Outputs

The Project XYZ Instrument "B" documents required for integration and test are the Test Procedures (with integration and test scripts) and Test Results.

5.5 Problem Reporting and Corrective Action

If any component or system does not operate as designed or scripted during any of the described tests above, then the integration team will follow a review and correction procedure. The Test Procedures describe all scripts and procedures and remedial actions.

5.6 Facilities, Tools, Techniques, and Methodologies

The flight Instrument "B" will be integrated and tested in a clean room at Company A1 for particulates. Company A1 will carry out the vibration and thermal/vacuum testing at the contract test facilities run by Testing-R-Us.

The flight Instrument "B" will be tested for EMC. The Company A1 will carry out the EMC testing at the contract test facilities run by Testing-R-Us.

CHAPTER 6

Systems Engineering in Military Projects

Timothy Cathcart, Kim Fowler, and David Tyler

1. Introduction

The purpose of this chapter is to present selected best practices in developing military projects. The chapter provides high-level guidance and then details, such as military specifications and standards, and actual developments in the case studies. You will see similar threads of continuity in military development as was done in spacecraft and medical instruments.

The chapter will focus on systems engineering. After a very brief historical introduction that will lay the background for the current "best practices," it will summarize the state of defense systems acquisition, and then review portions of the discipline of systems engineering in military practice, EIA-632, IEEE 1220, and ISO/IEC15288:2008(E). Next, we focus on MIL-STD-499, the now-canceled standard for systems engineering; it provides a starting point to review certain concerns for developing military projects. The chapter discussion moves on to cover other issues within systems engineering, including—legacy systems, obsolescence, software development, and testing, and current case studies for developing military projects.

2. Historical Background

Engineering within military projects has changed considerably over the past two decades. From the 1950s through the 1980s, the military had clear and mandated specifications, standards, and handbooks for developing military equipment. Military needs could dictate some of the specifications of electronic components during this era because military purchasing dominated the market. By the early 21st century, the influence of military engineering in electronics and electronic components had diminished to nothing; the consumption of electronic components in commercial business, industry, and consumer markets overwhelmed and surpassed the demands of the military market. Furthermore, generations of more capable commercial equipment were leapfrogging the custom designs of military equipment. These factors made custom military designs far more expensive to field and maintain.

To reduce both cost and delays in fielding and to improve technical capabilities, the military moved to purchasing and using commercial off-the-shelf components and equipment, commonly referred to as COTS. In 1994, Defense Secretary William Perry broadcast a memo recommending COTS insertion by the U.S. military. Following is Perry's memo,

> *On January 4, 1995, the Defense Electronics Supply Center (DESC) issued an initial draft of Revision C to convert MIL-I-38535 into a "performance-based" specification. The term "performance-based" is being used by DoD and DESC to describe a specification that will no longer dictate to the manufacturer how to produce or test their devices. DoD intends to allow vendors to be able to use their own "best commercial practices" for manufacturing and testing microcircuits. DoD views these "how-to" requirements as significant cost-drivers and hopes that by eliminating them, vendors will be able to continue to support the military market.... The Office of the Secretary of Defense (OSD) has threatened to deactivate any specification that tries to tell vendors how to manufacture and test their devices. Vendors are to be allowed the flexibility to use their own normal (commercial) manufacturing methods and techniques. In response, DESC is restructuring MIL-I-38535 and is moving many of the "how-to" requirements to appendices. Most of the appendices are now intended to serve as guidelines [1].*

As a consequence of this move to COTS insertion during the late 1990s, the military canceled many standards or reduced their influence to guidance. That process continues. As new technology develops, military guidance continues in the form of handbooks.

Should a project have very narrow and specific requirements, they are stated in the contract between the government and the providing organization or in the statement of work that is attached to the contract. Military specifications may be used as guidance but each specification usually has the following disclaimer, "It is for guidance only and cannot be cited as a requirement. If it is so cited, the contractor does not have to comply."

2.1. JCIDS

Relegating military specifications and standards to "guidance only" had consequences. While it freed up design development to fit applications more effectively and to reduce cost and delay in many situations, it sacrificed flexibility and interoperability between different types of missions. Service-specific systems for generating requirements flourished in this environment, resulting in redundant capabilities that could not individually address the combined needs of all the U.S. military services (Army, Navy, Marines, and Air Force). The Department of Defense (DoD) identified shortfalls in the generation of requirements:

- Not considering new programs in the context of other programs

- Not sufficiently considering combined service requirements and effectively prioritizing joint service requirements

- Not accomplishing sufficient analysis [2]

In 2003, the DoD unveiled the Joint Capabilities Integration Development System (JCIDS), which defined acquisition requirements and evaluation criteria for defense programs. The intent of JCIDS was "to guide the development of requirements for future acquisition systems to reflect the needs of all four services (Army, Navy, Marines, and Air Force) by focusing the requirements generation process on needed *capabilities* as requested or defined by one of the U.S. combat commanders. In the JCIDS process, regional and functional combatant commanders give feedback early in the development process to ensure that their requirements are met" [2].

The central focus of JCIDS is to address capability shortfalls, or gaps as defined by combatant commanders. Thus, JCIDS is said to provide a capabilities-based approach to requirements generation. The previous requirements generation system focused on addressing future threat scenarios. While understanding the risks associated with future threat postures is necessary to develop effective weapons systems, a sufficient methodology requires a joint perspective which can both prioritize the risk associated with future threats and consider operational gaps in the context of all the services. If requirements are developed in this joint context, there is simultaneously a smaller chance of developing superfluously overlapping systems and a greater probability that weapons systems would be operational with one another (i.e.[,] common communication systems, weapons interfaces, etc[.]) [2].

JCIDS analysis should produce three documents that define needed capabilities, guide materiel development, and direct the production of capabilities. Such analyses are rooted in the capabilities based assessment (CBA) process and formally articulated in an initial capabilities document (ICD). The analyses conducted in the CBA [3] are as follows:

- Capability needs analysis: The *mission area, military problem, and required capabilities* are mapped, analyzed, and developed under a defined set of operating conditions. Tasks and associated standards are developed to measure the ability of an organization to meet its assigned missions.

- Capability gap analysis: The *assessment of how well an organization meets identified needs and associated risks* evaluates current and programmed capabilities and identifies the gaps/risks in capability based on the tasks, conditions, and standards defined earlier.

- Non–materiel solutions analysis: The *assessment of non-materiel solutions and identification of materiel solutions* analyzes potential non-materiel solutions and identifies materiel approaches to mitigating the capability gaps [3].

Each of these documents supports a major design approval decision each with gradual improving design maturity A, B or C . . . . The initial capabilities document (ICD) defines the capability need and where it fits in broader concepts, ultimately supporting the milestone A decision. (The Milestone A decision approves or denies a concept demonstration to show that a proposed concept is feasible). When the technology development phase is complete, a capability development document (CDD) is produced which provides more detail on the materiel solution of the desired capability and supports Milestone B decisions. (The milestone B approval starts the Engineering and Manufacturing Development Phase.) Most important, the CDD also defines the thresholds and objectives against which the capability will be measured. After approval, the CDD guides the Engineering and Manufacturing Development Phase of the acquisition process. The capability production document (CPD) supports the Milestone C decision necessary to start the Production & Deployment Phase to include low-rate initial production and operational tests. The CPD potentially refines the thresholds from the CDD based on lessons learned during the Engineering and Manufacturing Development Phase [2].

The DoD published two documents to help developers understand the JCIDS:

- The *Joint Chiefs of Staff (CJCS) Instruction 3170.01* provides a top-level description and outlines organizational responsibilities.

- *CJCS Manual (CJCSM) 3170.01* defines performance attributes, key performance parameters, validation and approval processes, and associated document content.

2.2. Defense Acquisition

Moving up one more level, the acquisition of military systems is considered. The DoD has prepared the *Defense Acquisition Guidebook* for the process that incorporates JCIDS methodology and systems engineering. Figure 6.1 illustrates the various disciplines that drive military acquisition. Figure 6.2 provides an overview of the framework for the life cycle.

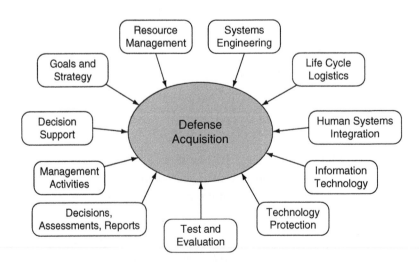

Figure 6.1: Functional view of defense acquisition. (Source: akss.dau.mil/dag/FWST/ acquisition-strategy.htm.)

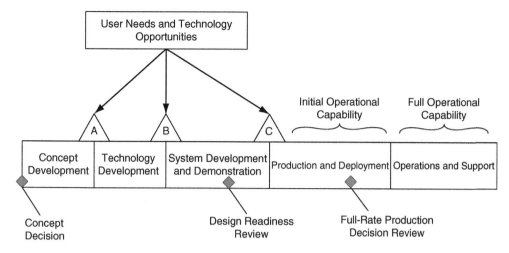

Figure 6.2: Life cycle framework of defense acquisition. (Source: akss.dau.mil/dag/FWST/ acq-framework.htm.)

The Defense Acquisition System exists to manage the Nation's investments in technologies, programs, and product support necessary to achieve the National Security Strategy and support the United States Armed Forces. In that context, our continued objective is to rapidly acquire quality products that satisfy user needs with measurable improvements to mission capability at a fair and reasonable price. The fundamental principles and procedures that the Department follows in achieving those objectives are described in DoD Directive 5000.1 and DoD Instruction 5000.2. The Defense Acquisition Guidebook is designed to complement those policy documents by providing the acquisition workforce with discretionary best practice that should be tailored to the needs of each program [4].

This *Guidebook* is readily available online. It contains the following 11 chapters:

Chapter 1, *Department of Defense Decision Support Systems*, presents an overview of the Defense Department's decision support systems for strategic planning and resource allocation, the determination of capability needs, and the acquisition of systems.

Chapter 2, *Defense Acquisition Program Goals and Strategy*, discusses acquisition program goals and the topics the program manager should consider in developing a strategy for the acquisition program. It addresses the required information associated with the acquisition program baseline and the program's acquisition strategy.

Chapter 3, *Affordability and Life cycle Resource Estimates*, addresses acquisition program affordability and resource estimation.

Chapter 4, *Systems Engineering*, covers the system design issues facing a program manager, and details the systems engineering processes that aid the program manager in designing an integrated system that results in a balanced capability solution.

Chapter 5, *Life cycle Logistics*, provides the program manager with a description of life cycle logistics and its application throughout the system life cycle, from concept to disposal.

Chapter 6, *Human Systems Integration*, addresses the human systems elements of the systems engineering process. It will help the program manager design and develop systems that effectively and affordably integrate with human capabilities and limitations; and it makes the program manager aware of the staff resources available to assist in this endeavor.

Chapter 7, *Acquiring Information Technology and National Security Systems*, explains how the DoD complies with statutory and regulatory requirements for acquiring IT and NSS systems and is using a network-centric strategy to transform DoD war-fighting, business, and intelligence capabilities. The chapter also provides descriptions and explanations of the Clinger-Cohen Act, the Business Management Modernization Program, and many other associated topics and concepts, and discusses many of the activities that enable the development of net-centric systems.

Chapter 8, *Intelligence, Counterintelligence, and Security Support*, describes program manager responsibilities regarding research and technology protection to prevent inadvertent technology transfer, and provides guidance for and describes the support available for protecting those technologies.

Chapter 9, *Integrated Test and Evaluation*, discusses many of the topics associated with test and evaluation, including oversight, developmental test and evaluation, operational test and evaluation, and live fire test and evaluation. The chapter enables the program manager to develop a robust, integrated test and evaluation strategy to assess operational effectiveness and suitability, and to support program decisions.

Chapter 10, *Decisions, Assessments, and Periodic Reporting*, prepares the program manager and Milestone Decision Authority to execute their respective oversight responsibilities.

Chapter 11, *Program Management Activities*, explains the additional activities and decisions required of the program manager that are not otherwise discussed in earlier chapters of this volume [4].

2.3. Where Is JCIDS Now?

A U.S. Government Accounting Office (GAO) report from September 2008 summarized the effectiveness of JCIDS as follows:

> *The JCIDS process has not yet been effective in identifying and prioritizing warfighting needs from a joint, department-wide perspective. GAO reviewed JCIDS documentation related to proposals for new capabilities and found that most—almost 70 percent—were sponsored by the military services, with little involvement from the joint community— including the combatant commands (COCOMs), which are largely responsible for*

planning and carrying out military operations. By continuing to rely on capability propo-sals that lack a joint perspective, DOD may be losing opportunities to improve joint war-fighting capabilities and reduce the duplication of capabilities in some areas. In addition, virtually all capability proposals that have gone through the JCIDS process since 2003 have been validated—or approved. DOD continues to have a portfolio with more pro-grams than available resources can support [5].

2.4. Recent History of Systems Engineering

Not only is JCIDS struggling within military projects, so is the practice of systems engineering. Concerning the state of systems engineering in military systems, one colleague said it best: "The whole industry retrograded [from] DoD contractor systems engineering capability when Mil Specs were removed [in the late 1990s]." In 2008, the National Research Council Air Force Studies Board released a report titled, "Pre-Milestone A and Early-Phase Systems Engineering: A Retrospective Review and Benefits for Future Air Force Acquisition" [6]. In part, the summary of the report stated the following:

Recent years have seen a serious erosion in the ability of U.S. forces to field new weapons systems quickly in response to changing threats, as well as a large increase in the cost of these weapons systems. Today the military's programs for developing weapons systems take two to three times longer to move from program initiation to system deployment than they did 30 years ago . . . . Many causes for this trend have been suggested, including the increased complexity of the tasks and the systems involved from both technological and human/organizational perspectives; funding instability; loss of "mission urgency" after the end of the Cold War; bureaucracy, which increases cost and schedule but not value; and the need to satisfy the demands of an increasingly diverse user community. The diffi-culty of focusing on a specific, homogeneous, post–Cold War threat made problems even worse. Yet although the suggested causal factors have merit, a common view is that better systems engineering (SE) could help shorten the time required for development, making it more like what it was 30 years ago [6].

Combine this Air Force report on systems engineering with the GAO's report on JCIDS and you can begin to see something of their pessimistic perspectives. So how did we get to this stage and what are the ramifications? The impetus behind Secretary Perry's 1994 memo was industrial persuasion to allow commercial companies to carry a greater proportion of engineering expertise and reduce the need for government expertise for oversight in engineering projects. The unfortunate consequence of this lack of oversight has contributed to delay and overruns in many recent military programs in addition to causes mentioned in the Air Force report.

Many people within the U.S. government and military have recognized the need for systems engineering oversight on programs. Military directorates have begun increasing systems engineering staff. Universities are developing curricula to teach SE.

The Air Force report summary says, "SE is the translation of a user's needs into a definition of a system and its architecture through an iterative process that results in an effective system design. SE applies over the entire program life cycle, from concept development to final disposal" [6]. The belief is that re-establishing SE will recover balance in developing military programs by melding different technical and operational disciplines, ensuring that each phase is properly conducted, and monitoring progress, schedule, and budget.

2.5. Evolution of Standards for Systems Engineering

Systems engineering overarches all design and development for a military project and works directly with the acquisition organization. The goal is to tie together many different disciplines, which include most of the same components in Fig. 6.1.

The government and private industry are still grappling with the problem of how best to define best practices within SE and which applications incorporate which practices. EIA 632 grew out of the canceled MIL-STD-499B effort in the late 1990s. An effort for the past decade resulted in the harmonization of IEEE-1220 and ISO/IEC15288 into a single standard—ISO/IEC 15288:2008 (E), which is also labeled IEEE Std 15288-2008. Along the way, organizations differentiated approaches within standards for SE into a life cycle approach or a process approach. Figure 6.3 illustrates some of the history of and connections between the standards.

The IEEE 1220 provided some basic definitions that are still useful in SE:

The customer is the organization(s) responsible for eight life cycle processes:

- Development
- Manufacturing
- Test
- Deployment
- Operations
- Training
- Support
- Disposal

System elements follow.

- Hardware
- Software

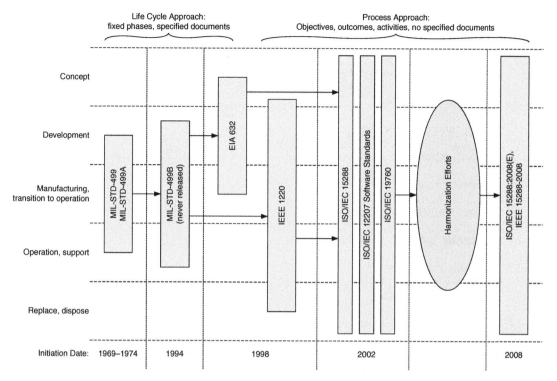

Figure 6.3: Influences and progression of standards for systems engineering. (© 2009 by Kim Fowler, used with permission. All rights reserved.)

- Personnel

- Facilities

- Data

- Material

- Services

- Techniques

IEEE 1220 is a "process approach" to SE. It focuses on the development process and produces the requirements baseline and its validation, the functional architecture and its verification, and the physical architecture and its verification. IEEE 1220 covers the following topics in Clause 4, General Requirements:

1. Apply the SEP (systems engineering process)

2. Policies and procedures

3. Plans and schedules

4. Strategies

5. Modeling and prototyping

6. Integrated repository

7. Integrated data package

8. Specification tree

9. Drawing tree

10. System breakdown structure

11. Integrate inputs

12. Technical reviews

13. Quality management

14. Product and process improvement [7]

Several presentations within the standards community have passed around a graphic that shows IEEE 1220 only focusing on development, but this is not accurate. In Clause 5, Life Cycle Processes, IEEE 1220 clearly covers processes for manufacturing, distribution and support, operations and training, and disposal (as shown earlier in the definitions and terms). The only thing it does not explicitly cover is conceptualization.

EIA 632 is sometimes considered broader in its coverage of the system life cycle than IEEE 1220 because it includes the phases for conceptualization and the transition to operation. EIA 632 does not explicitly cover manufacturing, support, or disposal. "The EIA 632 standard specifies accepted practices used for engineering systems but does not specify the details of 'how to' implement process requirements. Nor does it specify the methods or tools a developer would use to implement the process requirements. It is intended that the developer select or define methods and tools that are applicable to the development, and that are consistent with enterprise policies and procedures" [8]. EIA 632 began as a "life cycle approach" to SE while it was the interim standard in the mid-1990s; it migrated to a more "process approach" as a full standard in 1998. EIA 632 covers the following topics:

Acquisition and supply

-   Supply

-   Acquisition

Technical management

- Planning

- Assessment

- Control

System design

- Requirements definition

- Technology and architecture selection

- Solution definition

Product realization

- Implementation

- Transition to use

Technical evaluation

- Systems analysis

- Requirements validation

- System verification

- End products validation

ISO/IEC 15288:2008 (E):

Provides a common process framework covering the life cycle of man-made systems. This life cycle spans the conception of ideas through to the retirement of a system. It provides the processes for acquiring and supplying systems. In addition, this framework provides for the assessment and improvement of the life cycle processes.... This revised International Standard is an initial step in the SC7 harmonization strategy to achieve a fully integrated suite of system and software life cycle processes and guidance for their application. This revision aligns with the revision to ISO/IEC 12207 within the context of system life cycle processes and applies SC7 guidelines for process definition to support consistency, to improve usability and to align structure, terms, and corresponding organizational and project processes.... The processes in this International Standard form a comprehensive set from which an organization can construct system life cycle models appropriate to its products and services. An organization, depending on its purpose, can select and apply an appropriate subset to fulfill that purpose [9, 10].

The central importance of ISO/IEC 15288:2008 (E) is outlined as follows, starting with Section 5 and ending with Section 6:

5. Key Concepts

    5.1. System concepts—systems, system structure, enabling systems

    5.2. Life cycle concepts—model, stages

    5.3. Process concepts—description, application, tailoring

6. System Life Cycle Processes

    6.1. Agreement processes—acquisition, supply

    6.2. Organizational projection-enabling processes—management of life cycle model, infrastructure, project portfolio, human resource, and quality

    6.3. Project processes—project planning, project assessment and control, decision management, risk management, configuration management, information management, measurement

    6.4. Technical processes—stakeholder requirements definition, requirements analysis, architectural design, implementation, integration, verification, transition, validation, operation, maintenance, disposal

3. Processes, Procedures, and Tasks

Systems engineering defines the many important processes and tasks that apply throughout the system life cycle of a military program including new development, upgrade or modification, resolution of deficiencies, and exploitation of technology. SE brings balance to the design process by weighing function, performance, reliability, availability, maintainability, sustainment, environmental hazards, safety hazards, and manufacturing concerns against budget and schedule. SE establishes and uses both processes and procedures to bring projects to fruition.

A *process* is a group of interrelated activities and resources that transform inputs (typically specifications and requirements) into outputs (such as designs, schematics, and source code), often described by a high-level block or flow diagram of events. Each process may have a number of procedures within it. A *procedure* is specific implementation of the process for a single, focused area of concern; often it is a set of step-by-step instructions to accomplish a single kernel task.

PROCESS
STEPS

SUPPORTING
ACTIVITIES

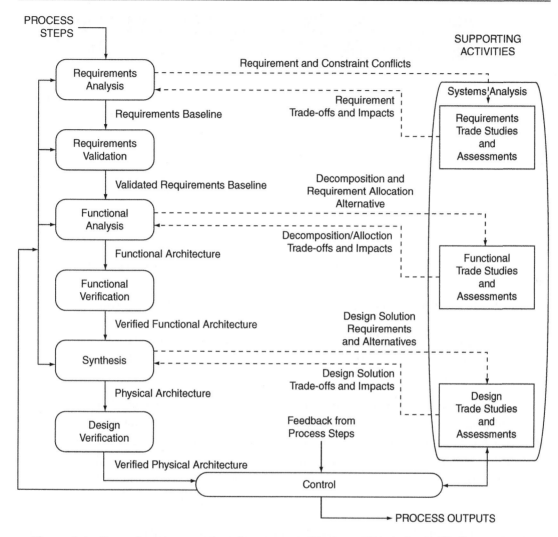

Figure 6.4: General systems engineering process. (Source: IEEE Std 1220™, © IEEE 2005, used with permission.)

Figure 6.4 illustrates the general SE process (SEP) from a flow perspective; this flow is derived from IEEE 1220. Figure 6.5 shows the traditional SEP from a top-level requirements perspective.

Thus far, the importance of processes for the SE of military projects has been established. Looking at some lower-level processes and procedures, MIL-STD-499B, although canceled as a requirement for military programs, can serve as guidance for tasks and procedures.

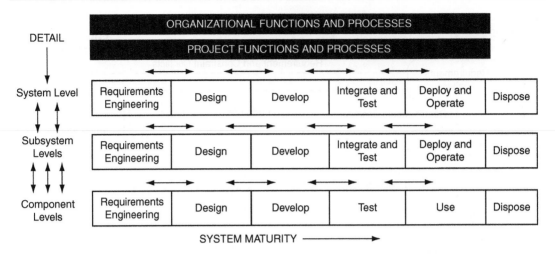

Figure 6.5: Systems engineering efforts in system life cycle. (Source: David Tyler.)

3.1. MIL-STD-499B: Systems Engineering Planning and Implementation

As mentioned, though MIL-STD-499B has been canceled as a requirement for military programs, it can still serve as guidance for SE tasks and procedures. The first major task is to plan and implement SE.

The technical manager ensures that technical execution and management efforts are integrated to conform with the systems engineering plan. The SE plan integrates technical tasks, which may include the task requirements from other standards that may be required by the customer or end user, to yield a single and complete process that focuses all activities on a common objective. Task planning and execution should demonstrate multi-disciplinary teamwork whereby the SE plan applies all the technical disciplines to satisfy the identified needs. The integrated technical effort is documented in the systems engineering tasking description (SETD), the integrated master plan (IMP), and the integrated master schedule (IMS).

The SETD contains the following planning components:

- Technical performance measurement (TPM)

- Technical review

- Technical integration

- Technology transition

The SETD should relate the TPM, IMP, and IMS to cost and schedule performance measurement.

In "How to Measure the Effects of Systems Engineering on the Outcomes of a Project," D. Smith and S. Russell write:

> *Technical Performance Measures (TPMs) express critical performance measures and these should be evaluated at regular intervals during a project life cycle to monitor progress toward meeting requirements, identify potential shortfalls in performance, and to identify technical risks (Ferraro, 2002, p. 14; Roedler & Rhodes, 2007, p. 7.11). . . . TPMs provide a means to balance cost and schedule by providing evidence that the design is adequate to meet requirements (MIL-STD-499A, p. 14). Without an adequate TPM program, a project can place too much emphasis on cost and schedule to the detriment of the technical objectives (Roedler & Rhodes, 2007, p. 7.11). Ferraro (2002, p. 16) suggests that TPMs should be integrated with Earned Value Management Systems (EVMS) to aid risk management and early detection of potential problems. Ferraro argues that the EVMS is a "lagging indicator" that reveals problems after they have occurred and that TPMs can act as a "leading indicator" to provide earlier warning of potential problems and facilitate preventative, strategic management [13].*

The Defense University Acquisition website provides a guide

> *For the preparation and implementation of a program's Integrated Master Plan (IMP) and Integrated Master Schedule (IMS). The IMP and IMS are fundamental management tools that are critical to performing effective planning, scheduling, and execution of work efforts. This Guide amplifies the event-based technical approach directed by policy in the February 20, 2004, USD(AT&L) Memorandum, "Policy for Systems Engineering in DoD," and October 22, 2004, USD(AT&L) Memorandum, "Policy Addendum for Systems Engineering; complies with the Earned Value Management (EVM) policy directed in the March 7, 2005, USD(AT&L) Memorandum, "Revision to DoD Earned Value Management Policy; and complements the guidance provided in the Defense Acquisition Guidebook [14].*

> *The IMP is an event-based plan that should provide sufficient definition to allow tracking the completion of required accomplishments for each event and to demonstrate satisfaction of the completion criteria for each accomplishment. In addition, the IMP demonstrates the maturation of the development of the product as it progresses through a disciplined systems engineering process. The IMP events are not tied to calendar dates; each event is completed when its supporting accomplishments are completed and when this is evidenced by the satisfaction of the criteria supporting each of those accomplishments. The IMP is generally placed on contract and becomes the baseline execution plan for the program or project. Although fairly detailed, the IMP is a relatively top-level document in comparison with the IMS . . . . The IMS flows directly from the IMP and supplements it with additional levels of detail. It incorporates all of the IMP events, accomplishments, and criteria; to these activities it adds the detailed tasks necessary to support the IMP criteria along with each task's duration and its relationships with other tasks. This network of integrated tasks, when tied to the start date, creates the task and calendar-based schedule that is the IMS. The IMS should*

be defined to the level of detail necessary for day-to-day execution of the program or project. The IMS is required on contracts that implement EVM [earned value management] in accordance with the EVM policy, and the delivery requirements are placed on contract via a CDRL [contract data requirements list] as a deliverable report [11].

The systems engineering management plan (SEMP) documents how the design organization will implement the SE plan and other SE activities. The chief engineer incorporates the expertise and advice of specialty engineers in the various disciplines to develop, communicate, and document the technical elements of the SEMP. The SEMP describes the systems engineering process for defining the system design, test requirements, system performance parameters, and system configuration; it describes the planning and controls of the technical program tasks; it also describes the management of the integrated effort of design engineering, test engineering, logistics engineering and production engineering. The *Systems Engineering Plan Preparation Guide* from the DoD has templates for SE plans [12].

A designated technical leader writes an integrated master schedule (IMS) and maintains it while detailed, time-dependent tasks evolve within the project. Normally, an initial version of the IMS accompanies the SEMP. During the execution of tasks, the IMS is maintained to track technical progress and associated risks—the IMS should follow the maintenance of the TPMs. This gives the program management insight into cost and schedule risks.

The design organization extends the program's work breakdown structure (WBS) developed by a designated technical leader to the level necessary to complete task requirements. The design organization has the flexibility to extend the task WBS below the reporting requirement to reflect how the work will be accomplished consistent with managing program risk. This has the added benefit of pinpointing where cost or schedule risk originates in the overall program. These risks are typically new and untested areas of technology or manufacturing techniques. In conjunction with the WBS, the TPMs, SETD, IMP, and IMS can be powerful additions to the overall performance, quality, or cost of a product and can help an organization keep the project state-of-the-art on schedule and on budget.

Please note that this level of detail and information is for large projects. Smaller projects might combine the SE plan, SEMP, SETD, IMP, and IMS into one or more documents.

3.2. Systems Engineering Input Information

As shown in Fig. 6.6, the process input has information necessary to support the continued technical effort, which includes results from technology validation and item verification. The process input begins at the initiation of a new phase of technical effort; it includes new or updated customer needs, technology base data, outputs from a previous phase, and program constraints. The systems developer(s) notifies the chief engineer that technical input

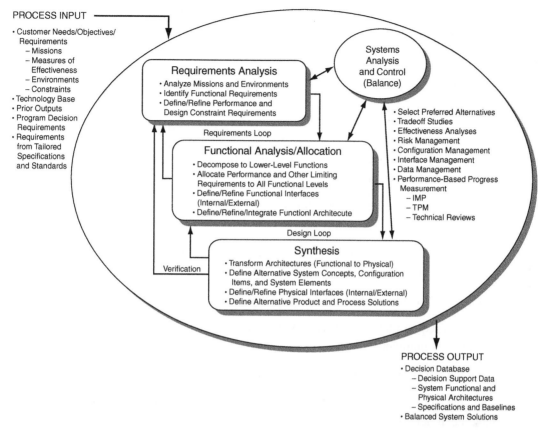

PROCESS INPUT

- Customer Needs/Objectives/
  Requirements
  – Missions
  – Measures of
    Effectiveness
  – Environments
  – Constraints
- Technology Base
- Prior Outputs
- Program Decision
  Requirements
- Requirements
  from Tailored
  Specifications
  and Standards

Requirements Analysis
- Analyze Missions and Environments
- Identify Functional Requirements
- Define/Refine Performance and
  Design Constraint Requirements

Requirements Loop

Functional Analysis/Allocation
- Decompose to Lower-Level Functions
- Allocate Performance and Other Limiting
  Requirements to All Functional Levels
- Define/Refine Functional Interfaces
  (Internal/External)
- Define/Refine/Integrate Functionl Architecute

Design Loop

Verification

Synthesis
- Transform Architectures (Functional to Physical)
- Define Alternative System Concepts, Configuration
  Items, and System Elements
- Define/Refine Physical Interfaces (Internal/External)
- Define Alternative Product and Process Solutions

**Systems
Analysis
and Control
(Balance)**

- Select Preferred Alternatives
- Tradeoff Studies
- Effectiveness Analyses
- Risk Management
- Configuration Management
- Interface Management
- Data Management
- Performance-Based Progress
  Measurement
  – IMP
  – TPM
  – Technical Reviews

PROCESS OUTPUT
- Decision Database
  – Decision Support Data
  – System Functional and
    Physical Architectures
  – Specifications and Baselines
- Balanced System Solutions

Figure 6.6: Systems engineering process requirements from Fig. 3 in MIL-STD-499B.

information is needed, why it is needed, and when it is needed. The chief engineer will respond as to whether the information can be provided or not. If the information will not be provided, then a defined task must generate the necessary information using documented research, analysis, and assumptions.

3.3. Technical Objectives

When sufficient data to establish requirements are not available, technical objectives provide a basis for defining and trading off relationships between need, urgency, costs, risks, and value. For some programs, technical objectives will be derived from key performance parameters (KPPs) in program requirements documentation; KPPs can be a subset of TPMs, and KPPs usually relate to the physical system while TPMs relate to progress of the project. Technical objectives are identified to assist in converging on a system solution, focus on factors critical to success, and offer substantial capability payoffs for the resources expended. The systems developer(s) identifies the necessary technical objectives and provides rationale; then develops metrics and success criteria to ensure that increases in system capabilities are cost effective when technical objectives exceed requirements. Finally, the systems developer(s) can use critical technical objectives in measuring technical performance.

3.3.1. Specifications and Baselines

The systems developer(s) generates unique documentation according to both the system and the configuration items (CIs). General criteria are necessary in the following areas:

1. Documentation, which establishes configuration baselines (functional, allocated, product), is developed progressively.

2. Specifications are formalized to establish configuration baselines commensurate with the program effort.

3. Configuration baselines are documented, controlled, and audited in accordance with program configuration management practices.

4. Essential requirements for processes are included in item specifications.

5. Specification requirements need to be verifiable. Traceability to the verification criteria and methods is maintained.

6. The systems developer(s) presents specifications for approval only when:

   • The cost, schedule, and performance risks associated with the item and its processes have been determined and the risk levels are acceptable.
   • Item costs have been determined and those costs satisfy established design-to-cost targets or other prescribed affordability limits.
   • Completeness and design attainability have been confirmed.

7. System functional and configuration item development specifications need to be performance-based.

3.3.2. Life Cycle Support Data

The systems developer identifies, annotates, and tracks those elements in the decision database necessary for life cycle management of the system for:

1. Product performance monitoring, analysis, problem identification, and corrective action recommendations.

2. Life cycle supportability analysis to identify operational and support resource requirements, to include any changes in requirements due to changes in the user community, missions, operational tempo, and operational strategy.

3. Identification of drivers of systems readiness degraders and excessive total ownership cost (TOC) contributors.

4. Analysis of alternative courses of actions and recommended actions to improve material readiness or reduce TOC.

5. Provide product support engineering services systems to user organizations.

These SE processes are DoD-adopted processes from the IEEE Standard 1220. "The systems engineering process (SEP) defines the interdisciplinary tasks that are required throughout a system life cycle to transform customer needs, requirements, and constraints into a system solution" [10]. The SEP is a problem-solving methodology to define and evolve system solutions. The SEP is conducted for each layer of the system.

The SEP contains eight processes that flow from requirements analysis to design validation. Each process contains a number of integrated tasks. Each process has clearly defined interfaces with known information inputs and outputs. The tasks within the SEP are individually tailored (streamlined) based on the complexity of the system requirements. The SEP ensures that all design activities are properly focused on all stakeholders/customers requirements and constraints. The SEP uses design tradeoff analysis to balance customer requirements against system development constraints. A general description of the eight processes is contained in Table 6.1 and illustrated in Fig. 6.4.

3.4. Systems Engineering Process Requirements

The project or program employs requirements analysis, functional analysis, functional allocation, synthesis, and systems analysis and control in progressive levels of detail according to the SEP. These activities continue throughout the effort both to achieve program objectives and to define requirements, designs, and solutions for the system life cycle.

3.4.1. Systems Engineering Output

Outputs of the SE effort vary by, and depend upon, the acquisition phase. The systems developer develops and implements a decision database that handles the following tasks:

- Document and organize data used and generated by the SE effort.

- Provide an audit trail of results and rationale from the identified needs to verified solutions for traceability of requirements, designs, decisions, and solutions.

Table 6.1: System Engineering Processes

Systems Engineering Process	Description
Requirements analysis	This process clarifies and defines the problem statement in verifiable quantitative terms. Requirements and constraints are identified and documented in the system requirements baseline.
Requirements validation	This process validates and resolves conflicting requirements and assumptions from all stakeholders.
Functional analysis	This process is used to identify and develop all functional tasks required to execute the requirements baseline.
Functional verification	This process validates the functional architecture to ensure that it meets the minimum requirements baseline objectives.
Synthesis	This process includes all of the design activities necessary to achieve specified functional architecture.
Design verification	This process is used to validate the system architecture against both functional and requirements baseline documentation.
Systems analysis	This problem-solving process is used throughout the SEP to make decision trade-offs.
Control process	This management process is used to coordinate, document, and track the systems engineering processes.

Source: Timothy Cathcart.

3.5. Requirements Analysis

The project or program analyzes customer needs, objectives, and requirements in the context of customer missions, environments, and identifies system characteristics to determine functional and performance requirements for each primary system function. Understanding customer needs, missions, and environments is something that is not always "done well," as pointed out by Paul Gartz in his case study at the end of this chapter. Commercial aerospace seems to have gained a much better understanding of this process than military contractors.

Prior analyses are reviewed and updated to refine the definitions for the mission and environment to support the ultimate system definition. Requirements analysis is conducted iteratively with functional analysis to develop requirements that depend on additional system definition (e.g., other system items, performance requirements for identified functions) and verify that people, product, and synthesized process solutions can satisfy customer requirements. In conducting requirements analysis, the organization should do the following:

1. Assist in refining customer objectives and requirements.

2. Define initial performance objectives and refines them into requirements.

3. Identifies and defines constraints that limit solutions (e.g., missions and environments with adverse impacts on natural and human environments).

4. Defines functional and performance requirements based on measures of effectiveness (MOEs) provided by the customer. When MOEs are not provided at the level of detail needed, the systems developer(s) develops and uses a set of MOEs relating to customer missions, environments, needs, requirements and objectives, and design constraints.

The functional analysis uses functional requirements, identified in the requirements analysis, and process inputs as the top-level functions. Performance requirements are developed interactively across all identified functions based on system life cycle factors and characterized in terms of the degree of certainty of the estimate, the degree of criticality to system success, and relationship to other requirements.

3.5.1. Requirements Analysis Process

The requirements analysis process collects and translates all system requirements and constraints. Table 6.2 summarizes the requirements analysis process description while Fig. 6.7 illustrates it. This process is critical to clearly define requirements early in system development. Clearly defined system requirements should prevent unnecessary analysis and design changes that could be expensive to implement once system development has progressed.

3.5.2. Requirements Validation

The requirements validation process helps ensure that the requirements baseline is correct. The baseline is examined for consistency, variance, and ambiguity; it is also compared to customer expectations and to enterprise and project constraints, to provide a "second check" before proceeding with the functional analysis process. Table 6.3 summarizes the requirements validation process while Fig. 6.8 illustrates it. This process often will

Table 6.2: Summary of Requirements Analysis Process

Description	Tasks
This process translates stakeholders, marketing, functional, performance, regulatory, and enterprise internal/external requirements and constraints into a requirements baseline. This process establishes the requirements baseline. The requirements baseline defines the system problems to be solved for the following areas. – Operational – Functional – Design	• Define customer expectations • Define project and enterprise constraints • Define life cycle process concepts • Define human factors, manpower, personnel, training, human engineering, and safety requirements • Define functional requirements • Define performance requirements • Define modes of operation • Define technical performance measures (TPMs) • Define design characteristics • Establish requirements baseline

Source: Timothy Cathcart.

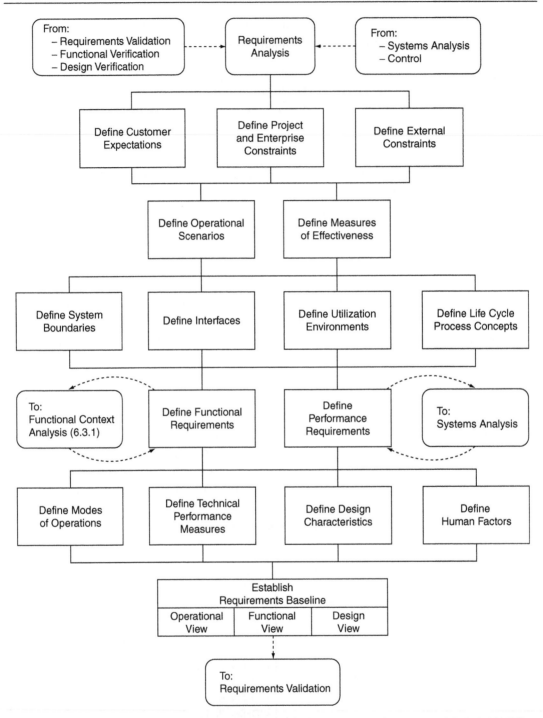

Figure 6.7: **System engineering requirements analysis processes. (Source: IEEE Std 1220™,**
© IEEE 2005, used with permission.)

Table 6.3: Requirements Validation Process

Description	Tasks
This process validates the requirements baseline to ensure that the baseline properly addresses the system requirements and constraints. The requirements baseline is an input to the functional analysis process.	• Compare requirements to customer expectations. • Compare requirements to enterprise and project constraints. • Compare requirements to external constraints. • Identify variances and conflicts. • Establish a validated requirements baseline.

Source: Timothy Cathcart.

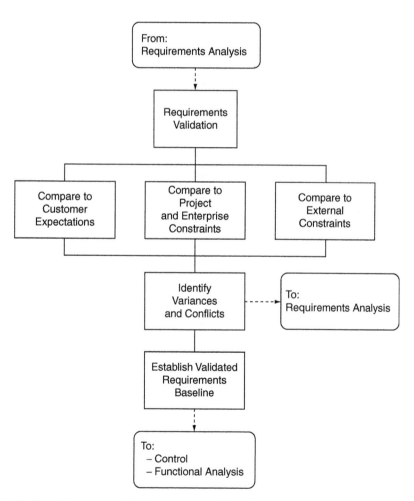

Figure 6.8: Requirements validation process. (Source: IEEE Std 1220™, © IEEE 2005, used with permission.)

include the final resolution of conflicting requirements that are resolved through the systems analysis process.

3.5.3. Functional Analysis Process

The functional analysis process translates requirements into system functions. Table 6.4 summarizes the functional analysis process while Fig. 6.9 illustrates it. System functions are continuously decomposed until enough details have been developed for the specified system level; that is, decomposition goes from system to subsystem to cell to equipment.

3.5.4. Functional Verification Process

The functional verification process verifies the functional baseline. The baseline is examined for traceability to the functional architecture and compared to defined requirements to verify decomposition and traceability. Table 6.5 summarizes the functional verification process while Fig. 6.10 illustrates it. This process resolves conflicting issues throughout the systems analysis process. Unresolved issues are sent to the functional analysis process for reassessment and correction. If the functional analysis process is unable to correct or resolve the issue, the problem is sent back into the requirements analysis process. The problem is then reprocessed from the beginning of the SEP. The requirements baseline and functional

Table 6.4: Functional Analysis Process

Description	Tasks
This process takes the requirements baseline and decomposes the system functions into lower level functions that must be performed by elements of the system design solutions. This process translates the requirements baseline into a functional architecture. The functional architecture defines the allocation of performance requirements to be solved during the synthesis process.	• Functional context analysis • Analyze functional behaviors • Define functional interfaces • Allocate performance requirements • Define external constraints • Define operational scenarios • Define measurement of effectiveness (MOE) • Define system boundaries • Define interfaces • Define utilization environments • Functional decomposition • Define sub-functions • Define sub-functions states and modes • Define functional timeline • Define data and control flows • Define functional failure modes and effects • Define safety-monitoring functions • Establish functional architecture

Source: Timothy Cathcart.

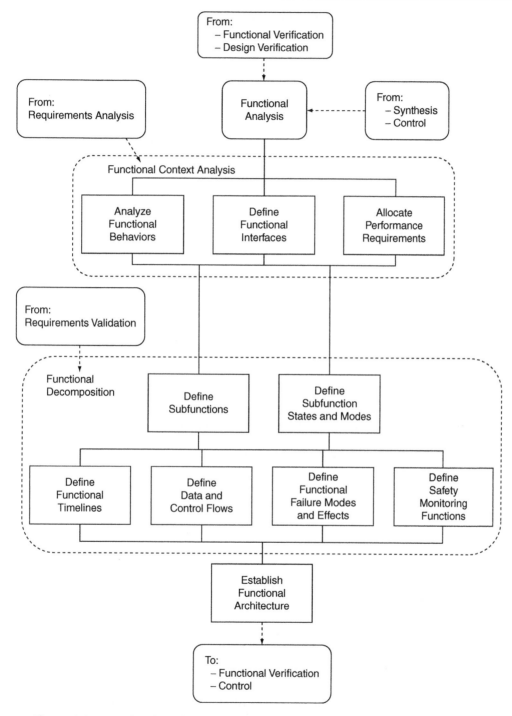

Figure 6.9: Functional analysis process. (Source: IEEE Std 1220™, © IEEE 2005, used with permission.)

Table 6.5: Functional Verification Process

Description	Tasks
This process verifies that the functional architecture satisfies the requirements baseline. This process ensures that all requirements and constraints can be traced to the functional architecture. The verified functional architecture is an input to the synthesis process.	• Define verification procedures • Conduct verification evaluation • Verify architecture completeness • Verify functional and performance measures • Verify satisfactory resolution of constraints • Identify variances and conflicts • Establish verified functional architecture

Source: Timorthy Cathcart.

architecture will be updated to capture any new changes. This method of processing unresolved problems illustrates the balancing and configuration control properties of the SEP. Often, "new solutions" are identified during the system analysis process, and these new solutions will require the same reprocessing cycle.

3.6. Functional Analysis and Functional Allocation

The systems developer(s) defines and integrates a functional architecture to the depth needed to support synthesis of solutions for people, products, processes, and risk management. The team iterates through functional analysis and allocation to achieve the following:

1. Definition of successively lower-level functions required to satisfy higher-level requirements and alternative sets of functional requirements.

2. Definition of mission- and environment-driven performance parameters through requirements analysis.

3. Determination that higher-level requirements are satisfied.

4. Flow-down of performance requirements and design constraints.

5. Definition of feasible alternatives that meet requirements through synthesis.

6. Fit the derived requirements into the functional architecture.

3.6.1. Functional Analysis

The systems developer(s) identifies and analyzes the functional requirements to determine the lower-level functions required to accomplish the parent requirement. The analysis includes all specified usage modes and arranges the functional requirements so that lower-level requirements are recognized as part of higher-level requirements. The team conducts a time-line analysis when performance or sequencing of functions are time critical.

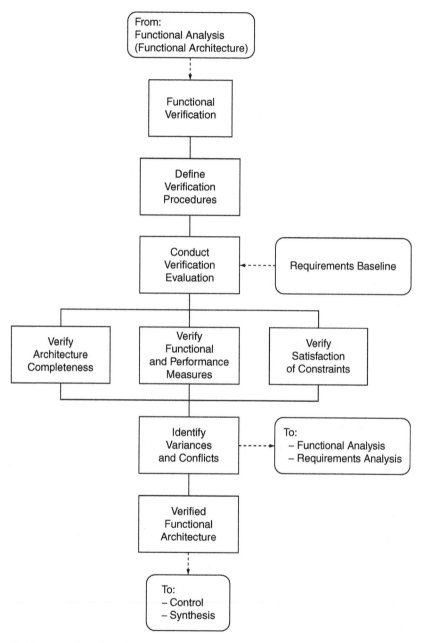

Figure 6.10: Functional verification process. (Source: IEEE Std 1220™, © IEEE 2005, used with permission.)

Functional requirements need to be logically sequenced, have the requirements defined for input, output, and functional interface (internal and external), and be traceable from beginning to end conditions and across interfaces.

3.6.2. Functional Allocation

The systems developer(s) establishes performance requirements for each functional requirement and interface successively from the highest to lowest. The systems developer(s) first determines and allocates the time requirements that are prerequisites for a function or a set of functions. The resulting set of requirements must be defined in measurable terms, applicable GO/NO-GO criteria, and in sufficient detail to be used as design criteria. Performance requirements need to be traceable throughout the functional architecture, through the analysis by which they were allocated, and to the higher-level requirement that they are intended to fulfill.

3.6.3. Synthesis

The systems developer(s) defines and designs solutions for each logical set of functional and performance requirements in the functional architecture and then integrates them into a physical architecture. The systems developer(s) iterates synthesis with functional analysis and allocation to define a complete set of functional and performance requirements necessary for the level of design output required. The systems developer(s) then uses requirements analysis to verify that the solution outputs can satisfy customer input requirements. In first defining the solution, the systems developer(s) needs to do the following:

1. Determine the completeness of functional and performance requirements for the design and identify the derived requirements needed for completeness in terms of function and performance.

2. Define internal and external physical interfaces including required function and performance and ensure that requirements are integrated and verifiable across interfaces.

3. Identify critical parameters, and then analyze parameter variability and solution sensitivity to the variability.

4. Define people, product, and processes (including the concepts, techniques, and procedural data applicable to each of the primary system functions), as well as required allowances for tolerances and variations for alternatives.

5. Define the system, configuration items, and system element solutions to a level of detail that enables verification for whether the required accomplishments have been met or not.

6. Translate the architecture into a work breakdown structure (WBS), specification tree, specifications, and configuration baselines.

Table 6.6: Synthesis Process

Description	Tasks
This process translates the functional architecture into a design architecture that provides an arrangement of system elements, the decomposition, and interfaces. The design architecture is developed for system breakdown structure level (i.e., system, subsystem, component, or cells). System analysis is used selectively to evaluate and manage risk, schedule, and performance impacts of alternative design options.	• Group and allocate functions • Identify design solution alternatives • Assess safety and environmental hazards • Assess technology requirements • Assess life cycle quality factors • Define design and performance characteristics • Define physical interfaces • Identify standardization opportunities • Identify make or buy alternatives • Develop models and prototypes • Assess failure modes, effects, and criticality • Assess testability needs • Assess design capability to evolve • Finalize design • Initiate evolutionary development • Produce integrated data package • Establish design architecture

Source: Timothy Cathcart.

3.6.4. Synthesis Process

The synthesis process develops design solutions from documented functional requirements. Table 6.6 summarizes the synthesis process while Fig. 6.11 illustrates it. The process requires the design to be assessed from multiple points of view (e.g., safety issues, environmental concerns).

3.7. Design

The outputs from synthesis describe the complete system, including the interfaces and relationships between internal and external items. For system design, the systems developer(s) must do the following:

1. Develop the information for establishing and updating applicable product baselines for function and allocation; develop system, CI, process, and material specifications including commercial item descriptions; create drawings and lists; generate interface control documentation; develop technical plans; define life cycle resource requirements; create procedural handbooks and instructional materials; and document personnel task loading.

2. Apply design simplicity concepts by evaluating alternatives with respect to factors such as ease of access, ready disassembly, common and non-complex tools, decreased parts

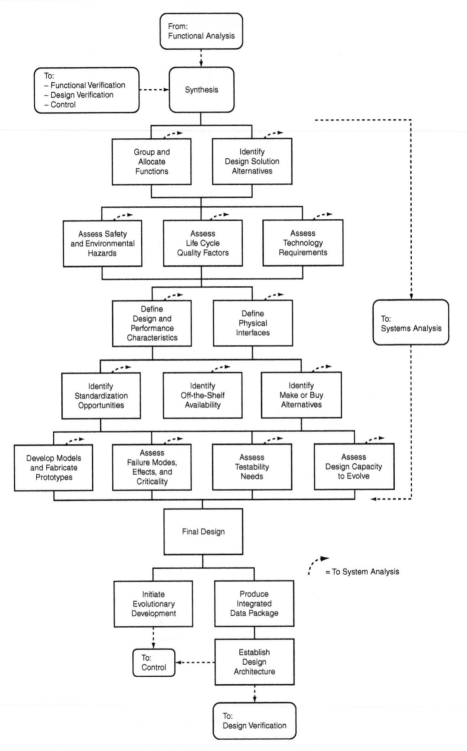

Figure 6.11: Synthesis process. (Source: IEEE Std 1220™, © IEEE 2005, used with permission.)

counts, modularity, producibility (e.g., ready assembly), standardization, and less demanding cognitive skills.

3. Demonstrate design consistency with results from risk reduction efforts.

4. Establish and control the correlation among interdependent and functionally related elements.

3.7.1. Design Verification Process

The systems developer(s) progressively verifies that the designs for product and processes satisfy the requirements (including interfaces) and that the requirements can be implemented. Verification goes from the lowest level of the current physical architecture up to the total system.

The design verification process ensures that the design architecture elements match the defined functional architecture elements. Table 6.7 summarizes the design verification process while Fig. 6.12 illustrates it. Design verification is a critical process because an official system baseline is established when design verification is complete. Design baselines are major gating points in the program management of the SEP. Incomplete data or analyses could result in costly overruns and redesign activities in the next iteration of the SEP for lower-level system design.

Table 6.7: Design Verification Process

Description	Tasks
This process verifies that the design architecture satisfies the functional architecture. This process ensures that all design architecture elements derived can be traced to the verified functional architecture. This process also ensures that the design architecture satisfies the requirements baseline.	• Select verification approach • Define inspection, analysis, demonstration, or test requirements • Define verification procedures • Establish verification environment • Conduct verification evaluation • Verify architecture completeness • Verify functional and performance measures • Verify satisfactory resolution of constraints • Identify variances and conflicts • Verify design architecture • Verify design architectures of the life cycle process • Verify system architecture • Establish specifications and configuration baselines • Develop system breakdown structure (SBS)

Source: Timothy Cathcart.

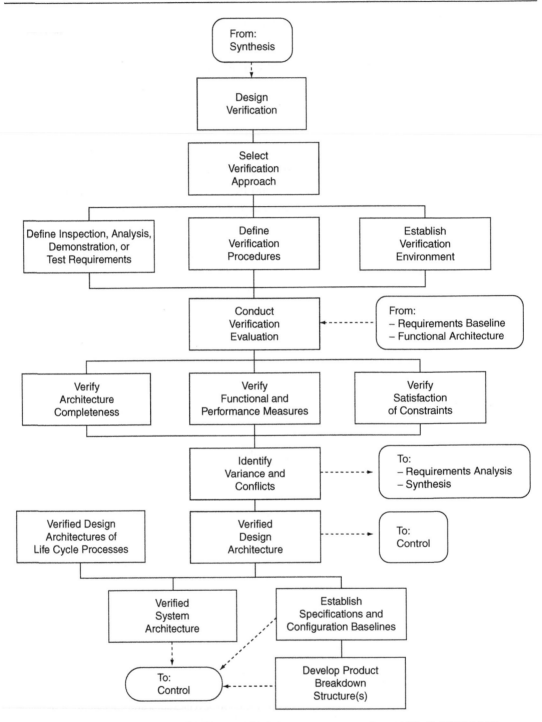

Figure 6.12: Design verification process. (Source: IEEE Std 1220™, © IEEE 2005, used with permission.)

3.8. Systems Analysis and Control

Systems developers measure progress, evaluate alternatives, select preferred alternatives, and document both the data and the decisions. They conduct systems analyses including tradeoff studies, effectiveness analyses and assessments, and design analyses to determine the progress toward satisfying technical requirements and program objectives; the systems analyses also provide a rigorous quantitative basis for performance, functional, and design requirements. Control mechanisms include risk management, configuration management, data management, and performance-based progress measurement including the IMP, TPM, and technical reviews. Systems developers implement systems analysis and control to ensure that the following areas are covered thoroughly:

1. Decisions on alternative solutions are made only after evaluating the impact of each solution on the system effectiveness, life cycle resources, risk, and customer requirements. The systems developer(s) identifies those solutions that will improve system effectiveness or costs when compared with those based on program requirements.

2. Technical decisions and system-unique specification requirements are based on systems engineering outputs and documented results of decisions.

3. Traceability from process inputs to outputs is maintained and includes changes in requirements.

4. Schedules for the development and delivery of products and processes are mutually supportive.

5. Technical disciplines are integrated into the SE effort.

6. Impacts of customer requirements on resulting functional and performance requirements are examined for validity, consistency, desirability, and attainability with respect to the available technology, physical and human resources, human performance capabilities, life cycle costs, schedule, risk, applicable statutes, designated hazardous material lists, and other identified constraints. This examination needs to either confirm existing requirements or determine that more appropriate requirements need to be defined for the system.

7. Product and process design requirements are directly traceable to the functional and performance requirements that the design requirements were designed to fulfill and vice versa.

3.8.1. System Analysis and Control Processes

The SEP has two common processes that are repeatedly used throughout the top–down flow of the overall SEP. The system analysis and control processes are used to integrate and coordinate the overall SEP. These processes are used to select and execute decisions derived from each of the previous processes. The system analysis process is used repetitively

Table 6.8: System Analysis Summary

Description	Tasks
This process is used to resolve conflicts among requirements analysis, functional analysis, and design synthesis. This process provides a rigorous quantitative basis to select a balanced set of requirements and design trade-offs and assessments. This process is part of a feedback network loop for the requirements analysis, functional analysis, and synthesis processes.	• Assess requirement conflicts • Assess functional alternatives • Assess design alternatives • Identify risk factors • Define trade-off analysis scope • Select methodology and success criteria • Identify alternatives • Establish trade-study environment • Conduct trade-off analysis • Analyze life cycle costs • Analyze system and cost effectiveness • Analyze safety and environmental impacts • Quantify risk factors • Select risk handling options • Select alternative recommendations • Consider trade-offs and impacts • Design effectiveness assessment

Source: Timothy Cathcart.

throughout the SEP to balance requirements and solution tradeoffs. The system analysis process is described in Tables 6.1 and 6.8 and is illustrated in Fig. 6.13. The SE tasks performed in each of these processes provide the necessary data inputs needed to provide a comprehensive evaluation of the available options.

3.8.2. Control Process

The control process provides a very structured technical management approach to assess, manage, and document the SEP. The control process is described in both Tables 6.1 and 6.9 and is illustrated in Fig. 6.14. The control process is the primary interface between the enterprise and the SEP. The control process is used to identify and allocate the resources needed to execute all SEP tasks. The SEP becomes a framework for the concurrent development and design of products and systems processes. The SEP framework synchronizes the design of customer products and the required system processes to produce the product, thereby creating a truly integrated product development environment.

3.9. Tradeoff Studies

The systems developer(s) identifies and conducts practical tradeoffs among user requirements, technical objectives, design, program schedule, functional and performance requirements, and life cycle costs. Tradeoff studies are defined, conducted, and documented

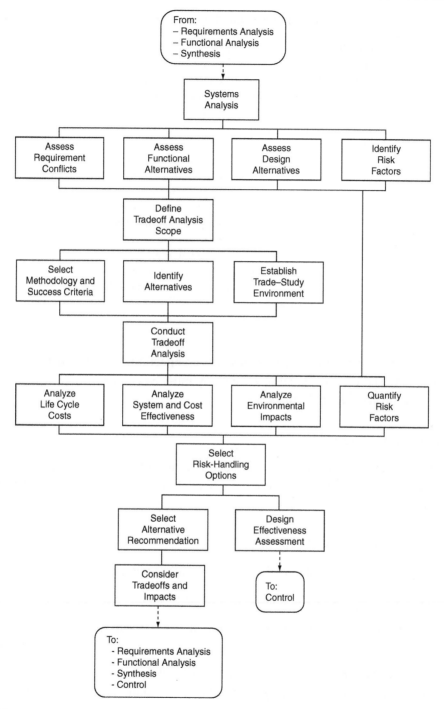

Figure 6.13: Systems analysis process. (Source: IEEE Std 1220™, © IEEE 2005, used with permission.)

Table 6.9: Systems Control Processes

Description	Tasks
This process identifies tasks required to manage and document the SEP. This process is the management interface to the enterprise. Resource allocation and planning is coordinated through this process.	• Technical management • Data management • Configuration management • Interface management • Risk management • Performance-based progress measurement • Track system analysis and test data • Track requirement and design changes • Track progress against project plans • Track product and process metrics • Update specifications and configuration baselines • Update requirements views and architectures • Update engineering plans • Update technical plans • Maintain technical databases

Source: Timothy Cathcart.

at the various levels of the functional or physical architecture in enough detail to support decision making. The level of detail of each study must be commensurate with cost, schedule, performance, and risk impacts.

3.9.1. Requirements Analysis

The systems developer(s) should conduct tradeoff studies of the requirements to establish alternative performance and functional requirements to resolve conflicts with customer requirements.

3.9.2. Functional Analysis and Allocation

The systems developer(s) should conduct tradeoff studies within and across functions to:

1. Support functional analyses and allocation of performance requirements.

2. Define a preferred set of performance requirements that satisfy identified functional interfaces.

3. Determine performance requirements for lower-level functions when higher-level performance and functional requirements cannot be readily resolved to a lower level.

4. Evaluate alternative functional architectures.

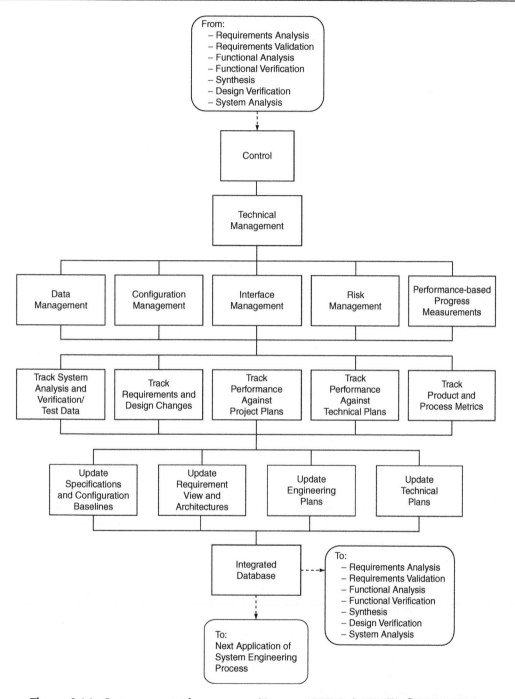

Figure 6.14: Systems control processes. (Source: IEEE Std 1220™, © IEEE 2005, used with permission.)

3.9.3. Synthesis

The systems developer(s) should conduct tradeoff studies on synthesis to:

1. Support decisions to build new products and process developments versus purchasing or acquiring nondevelopmental products and processes.

2. Establish system/CI configuration(s).

3. Assist in selecting system concepts, designs, and solutions (including availability of people, parts, and materials).

4. Support materials selection and make-or-buy, process, rate, and location decisions.

5. Examine proposed changes.

6. Examine alternative technologies to satisfy functional/design requirements including alternatives for moderate- to high-risk technologies.

7. Evaluate environmental and cost impacts of materials and processes.

8. Evaluate alternative physical architectures to select preferred products and processes.

9. Select standard components, techniques, services, and facilities that reduce system life cycle cost and meet system effectiveness requirements. Government and commercial databases should be used to provide historical information for evaluation decisions.

3.10. System/Cost-Effectiveness Analysis

The systems developer(s) plans and implements a systems analysis effort as an integral part of the SEP. The systems developer(s) develops, documents, implements, controls, and maintains a method to control analytic relationships and MOEs. It also identifies critical MOEs used for decision-making for technical performance measurement. It then uses system/cost-effectiveness assessments to support risk impact assessments. The systems developer(s) analyzes each primary system function to support the following areas:

1. Identification and definition of performance and functional requirements for the primary system functions to which system solutions must be responsive.

2. Selection of preferred product and process design requirements that satisfy performance and functional requirements.

3.11. Configuration Management

The systems developer(s) manages the configuration of identified system products and processes. This effort includes:

1. Configuration identification, which involves selecting the documents to compose the baseline for the system and CIs involved and the numbers and other identifiers affixed to the items and the documents.

2. Configuration control, which includes the systematic proposal, justification, evaluation, coordination, approval, or disapproval of all proposed changes to the CIs after the baseline(s) for the CIs has been established.

3. Configuration status accounting, which includes recording and reporting the information needed to manage configuration items.

4. Configuration audits, which include verification that the CI conforms to its current approved configuration documentation.

3.12. Interface Management

The systems developer(s) manages the internal interfaces within its program task responsibility. The systems developer(s) supports activities to ensure that external interfaces are managed and controlled. The systems developer(s) delineates the design compatibility of external and internal engineering interfaces as interface requirements in the specifications. The design compatibility for interfaces establishes, coordinates, and maintains controls for interface requirements, documents, and drawings; it includes all contract terms for the applicable systems developer(s), vendor, and subcontractor; it also includes controls for tasking-activity–furnished equipment, computer programs, facilities, and data. Interfaces need to be controlled to ensure accountability and timely dissemination of changes.

3.13. Data Management

The systems developer(s) establishes and maintains an integrated data management system for the decision database to:

1. Capture and organize all inputs as well as current, intermediate, and final outputs.

2. Provide data correlation and traceability among requirements, designs, solutions, decisions, and rationale.

3. Document engineering decisions, including procedures, methods, results, and analyses.

4. Be responsive to established configuration management procedures.

5. Function as a reference and support tool for the systems engineering effort.

6. Make data available and shareable as called out in the task.

3.14. Integrated Master Plan

The systems developer(s) implements the IMP for process control and progress measurement, at the top level, to ensure completion of the required accomplishments, to demonstrate progressive system and development achievements and maturity, to ensure that integrated, multidisciplinary information is available for decision and demonstration events, to provide an event-based, accomplishment-oriented framework for measuring progress, and to demonstrate control of cost, schedule, and performance risks in satisfying accomplishments, requirements, and objectives. IMP accomplishments with supporting criteria are devised and structured to ensure that:

1. Critical technical inputs and decision data are available for technical and program decision points, demonstrations, reviews, and other identified events.

2. Required progress and system maturity are demonstrated prior to continuing technical efforts dependent on that progress and maturity.

3. An IMP accomplishment is complete when all the associated criteria have been demonstrated.

3.15. Technical Performance Measurement

The systems developer(s) establishes and implements technical performance measurement (TPM) to evaluate the adequacy of evolving solutions and to identify deficiencies that impact the ability of the system to satisfy a performance requirement. Actions taken to redress deficiencies depend on whether the technical parameter is a requirement or an objective. The TPM level of detail and documentation needs to be commensurate with the impact on cost, schedule, performance, and risk.

3.15.1. Implementation of TPM

The systems developer(s) determines the achievement-to-date for each technical parameter. It assesses technical progress in terms of both allowed variation and the trend in achievement-to-date compared with the planned value profile. When progress in the technical effort supports revising the current estimate, it develops a new profile and current estimate. It updates risk assessments and analyses to reflect changes in planned-value profiles and current estimates and the impacts on related parameters.

3.15.2. TPM on Requirements

The systems developer(s) performs analyses for identified deficiencies to determine the cause(s) and to assess the impacts on higher-level parameters, interfaces, and system cost effectiveness. They develop alternative recovery plans that fully explore the impacts of cost, schedule, performance, and risk. For performance in excess of requirements, it assesses the marginal cost benefits and opportunities for reallocation of requirements and resources and defines an appropriate course of action.

3.16. Technical Reviews

The systems developer(s) plans and conducts the technical reviews necessary to demonstrate that required accomplishments have been successfully completed before proceeding beyond critical events and key program milestones. Technical reviews are conducted for the system- and program-identified CIs. Technical reviews occur at key events identified in the IMP when the systems developer(s) is ready to demonstrate completion of all of the IMP accomplishments associated with the event as measured by the associated criteria.

3.16.1. Technical Review Content

System and CI technical reviews need to be integrated reviews that include all disciplines, all primary system functions, and all products and processes of the item being reviewed. Reviews should be structured within the total system context to assess the following areas:

1. Confirm that the effects of technical risk on cost, schedule, and performance have been addressed, as well as risk reduction measures, rationale, and assumptions made in quantifying the risks.

2. Demonstrate that the relationships, interactions, interdependencies, and interfaces between required items and externally interfacing items, system functions, subsystems, CIs, and system elements have been addressed.

3. Ensure that performance, functional, design, cost, and schedule requirements and objectives and technical performance measurements and technical plans are being tracked, are on schedule, and are achievable within existing programmatic constraints.

4. Confirm that continued development is warranted, and when it is not, that executable alternatives have been defined.

3.17. Response to Change

The systems developer(s) defines the total program impact of identified changes to technical requirements with respect to cost, schedule, performance, and risk. It diagnoses the technical, cost, and schedule problems and determines the impacts. It determines the impacts of collateral effects induced by solutions and solution alternatives on the technical program, including interfaces. The systems developer(s) informs program management or the tasking activity of changes in cost, schedule, performance, and risk that impact the execution (on time, within budget, meets requirements) of the program. The systems developer(s) processes all resulting changes to program requirements and configuration baselines in accordance with established change control procedures. The systems developer(s) ensures that supporting data are accessible to the tasking activity and are documented in the decision database.

4. U.S. Department of Defense Resources

At least within the U.S. defense industry there are resources to help with specific concerns in a project. The DoD has the Defense Technical Information Center (DTIC®), which serves in the "transfer of information among DoD personnel, DoD contractors and potential contractors and other U.S. Government agency personnel and their contractors. DTIC is a DoD Field Activity under the Under Secretary of Defense for Acquisition, Technology and Logistics, reporting to the Director, Defense Research & Engineering (DDR&E)" [15].

DTIC® also operates the Information Analysis Centers (IACs). "IACs are research and analysis organizations chartered by the DoD and operated by DTIC ... to help researchers, engineers, scientists and program managers..." [15]. IACs maintain extensive databases, produce reports, handbooks, and data books, and perform technology assessments. IACs serve researchers, engineers, scientists, and program managers from DoD, government, industry, and academia.

IACs, apropos to this book, include the following:

- Advanced Materials, Manufacturing, and Testing Information Analysis Center (AMMTIAC) is operated for DoD by Alion Science and Technology Corporation and is supported by a multifaceted team of 24 industrial and academic organizations [16].

- Chemical, Biological, Radiological, & Nuclear Defense Information Analysis Center (CBRNIAC) generates, acquires, processes, analyzes, and disseminates CBRN Defense Science and Technology Information (STI) in support of the combatant commanders, war fighters, the reserve components, the CBRN defense research, development, and acquisition community, and other federal, state, and local government agencies [17].

- Chemical Propulsion Information Analysis Center (CPIAC) is a national clearinghouse and resource center for chemical rocket propulsion technical information and research services and provides archiving services for rocket, space, missile, and gun propulsion scientific and technical documents [18].

- Data & Analysis Center for Software (DACS) is managed and operated by the ITT Corporation; it serves as a centralized source for current, available data and information concerning software engineering and software technology [19].

- Information Assurance Technology Analysis Center (IATAC) provides the DoD with a central point of access for information on information assurance emerging technologies in system vulnerabilities, research and development, models, and analysis to support the development and implementation of effective defense against information warfare attacks [20].

- Modeling and Simulation Information Analysis Center (MSIAC) provides scientific, technical, and operational support information and services [21].

- Reliability Information Analysis Centers (RIAC) offers a collection of publications, tools, and articles in reliability, maintainability, quality, supportability and interoperability (RMQSI) [22].

- Military Sensing Information Analysis Center (SENSIAC) is operated by the Georgia Institute of Technology to facilitate the use of scientific and technical information in the Military Sensing Technology (MST) areas for the design, development, testing, evaluation, operation, and maintenance of DoD systems, military systems operated by allied and friendly nations, and the industrial and research base that provides and supports such systems. SENSIAC fosters communications within the MST community, creates standards, and collects, analyzes, synthesizes, maintains, and distributes critical information within the field [23].

- Survivability/Vulnerability Information Analysis Center (SURVIAC) collects, analyzes, and disseminates scientific and technical information (STI) related to all aspects of survivability and lethality for aircraft, ground vehicles, ships and spacecraft, to conventional homeland security threats including chemical, biological, directed energy, and non-lethal weapons [24].

- Weapon Systems Technology Information Analysis Center (WSTIAC) monitors the science, technology, and acquisition of conventional and directed energy weapons systems technology and related guidance, intelligence gathering systems, training, analyses, databases, model repositories, laboratory studies, testing, hardware, components, systems, and subsystems. The systems and subsystems include, but are not limited to, aircraft, ships, satellites, guns, ordnance, missiles, rockets, bombs, submunitions, projectiles, mines/countermines, munitions-dispersing canisters, lasers and high power microwaves, intelligence gathering systems, guidance and control, command guidance communications links, and undersea weapons. Technical areas of interest include military systems and supporting equipment; instrument and seeker development and test; manufacturing process development; system and subsystem simulation; development of computational techniques and hardware; control actuators and power sources; sensors for gathering and updating information; aerodynamic and reaction jet control devices; inertial components and system developments; GPS; guidance-aided fusing; energy management for navigation law profiles; special test equipment and techniques; theoretical performance computations; analytical test techniques; component design criteria; operational serviceability; maintenance and logistics equipment; training systems and specialized RDT&E systems; models, simulations, and basic science and technology activities; environmental protection; and materials areas specifically related to conventional and directed energy weapon systems technology [25].

- The Shock and Vibration Information Analysis Center (SAVIAC) serves as a central information resource for government activities, contractors, and academics concerned with structural dynamics design, analysis, and testing; and shock physics and weapons effects. SAVIAC focuses on the engineering tools and techniques needed for the design, analysis and testing of systems subjected to dynamic conditions and loading, including water shock, vibration, blast, impact, crash, and mechanical/climatic environments. The fields of application include ships, aircraft, spacecraft, ground vehicles, ground structures, machinery and humans; any area where Government engineers have a responsibility for the design, analysis or testing of mechanical or structural systems [26].

5. Military Standards and Handbooks

Beginning in 1995, the U.S. military canceled or downgraded many military standards and handbooks. Only 72 remain out of thousands of original military standards. The DoD now applies "best industry practice."

> *Currently, the majority of Military Handbooks have been replaced by Unified Facilities Criteria (UFC) published by a Tri-service committee with representatives from the three military services and the Department of Defense [27].*

So, what use are mounds of military standards and handbooks, many of which are canceled? They still provide guidance and can indicate good practices. Following this thread, MIL-STD-499 for Systems Engineering, which has been canceled, called out a number of standards and handbooks that can be instructive for guidance. Table 6.10 lists those standards and handbooks. Subsequent text introduces each of the standards and handbooks.

IEEE/EIA ISO 12207 *U.S. Software Life cycle Process* (www.stsc.hill.af.mil/crosstalk/1997/07/ lifecycle.asp). "The Electronic Industries Association (EIA) and the Institute of Electrical and Electronic Engineers (IEEE), both accredited standards-developing organizations, have been collaborating with the DoD on the development of software life cycle standards for use in the United States. Three primary objectives have motivated this work:

- The standards should represent the best commercial practice.

- The standards should be suitable for application to the complex requirements of defense acquisition.

- The standards should be compatible with those of the emerging global marketplace for software.

ISO/IEC 12207 was approved as an international standard in 1995. It provides a framework of software life cycle processes suitable for use in the acquisition, supply, development,

Table 6.10: Standards and Handbooks Called Out in MIL-STD-499

Topic	Military Standards/Reference Numbers
Configuration Management Guidance	MIL-HDBK-61
Corrosion Prevention and Control	MIL-HDBK-1250 MIL-HDBK-1568
Cost Engineering Policy and Procedure	MIL-HDBK-1010A*
Defense and Program-Unique Specifications Format and Content	MIL-STD-961
Electromagnetic Compatibility, Emissions, and Susceptibility	MIL-STD-461E MIL-STD-1541 MIL-STD-464 MIL-HDBK-237
Electronic Reliability Design	MIL-HDBK-338B
Electrostatic Discharge	MIL-STD-1686C
Engineering Drawing Practices	MIL-STD-100
Environmental Analysis	MIL-STD-810G
Human Factors Engineering	MIL-STD-1472F MIL-HDBK-46855
ID Markings	MIL-STD-130N
Logistic Support Analysis	MIL-STD-1388-2B
Logistics	MIL-HDBK-502 MIL-PRF-49506
Maintainability	MIL-HDBK-470 MIL-HDBK-791
Microelectronics Test Methods	MIL-STD-883
Military Training Programs	MIL-HDBK-1379.1, .2, .3, .4 MIL-PRF-29612
Nondestructive Inspection	MIL-HDBK-728 MIL-HDBK-731
Parts Control	MIL-HDBK-965*
Producibility	MIL-HDBK-727
Reliability Testing	MIL-HDBK-781
Reliability/Durability	MIL-HDBK-1530 MIL-HDBK-1798 MIL-HDBK-2164 MIL-HDBK-87244
Review and Audits; Software	MIL-STD-1521B

(Continues)

Table 6.10: Cont'd

Topic	Military Standards/Reference Numbers
Software	IEEE/EIA ISO 12207.0, .1, .2 DO-178C MIL-STD-2167* MIL-STD-2168*
Software Support Environment	MIL-HDBK-1467
Supportability	MIL-HDBK-502
Survivability	MIL-HDBK-336 MIL-HDBK-2069 MIL-STD-1799*
Systems Engineering	MIL-STD-499B*
System Safety Engineering	MIL-STD-882
System Security	MIL-HDBK-1785
Tech Manuals; Data Bases	MIL-PRF-87269
Technical Data Packages	MIL-DTL-31000C
Technical Manuals	MIL-STD-40051B
Testability	MIL-HDBK-2165*
Thermal Design/Analysis	MIL-HDBK-251
Training Requirements	MIL-HDBK-1379
Transportability	MIL-STD-1366
Vibrations	MIL-STD-167/2A
Work Breakdown Structure	MIL-HDBK-881

*Canceled or replaced.
Note: This table provides guidelines and is for reference only.

maintenance, and operation of software. ISO/IEC 12207 is intended to be adapted for application in specific situations."

MIL-DTL-31000C *Technical Data Packages* (assist.daps.dla.mil/quicksearch/basic_profile. cfm?ident_number=69492). This specification prescribes the requirements for preparing a technical data package (TDP), which is composed of one or more TDP elements and related TDP data management products.

MIL-HDBK-61 *Configuration Management Guidance* (www.product-lifecycle-management. com/download/MIL-HDBK-61B%20(Draft).pdf). This military handbook provides guidance and information to DoD acquisition managers, logistics managers, and other individual's assigned responsibility for configuration management. Its purpose is to assist them in planning for and implementing effective DoD configuration management activities and practices during all life cycle phases of defense systems and configuration items. It supports acquisition based on performance specifications, and the use of industry standards and methods to the greatest practicable extent. This handbook is closely related to the following Electronic Industries Alliance (EIA) Standards.

ANSI/EIA-649-1998, "National Consensus Standard for Configuration Management," EIA-836, "Consensus Standard for Configuration Management Data Exchange and Interoperability," and ANSI/EIA-632-1998, "Processes for Engineering a System."

The latter handbook provides the insight necessary to:

- Understand the application of the basic principles of CM articulated in ANSI/EIA-649 to the DoD acquisition and operational environment

- Plan for and make prudent and cost effective choices in effecting DoD configuration management activities throughout the life cycle of a material item

- Provide the necessary basis for CM in RFPs and contracts

- Evaluate contractor proposals and CM processes

- Acquire and process necessary CM information

- Use data models (schema), data dictionaries, and CM business object templates as a framework for translating and communicating configuration information among diverse, distributed databases in an integrated data environment

- Measure CM performance effectiveness of both government activities and contractors

MIL-HDBK-237D *Electromagnetic Environmental Effects and Spectrum Supportability Guidance for the Acquisition Process* (acc.dau.mil/GetAttachment.aspx?id=131917&pname=file&lang=en-US&aid=26637). Provides guidance for establishing and implementing effective electromagnetic environmental effects (E3) control and spectrum supportability (SS) for the design, development, and procurement of DoD platforms, systems, subsystems, or equipment.

MIL-HDBK-251 *Reliability/Design Thermal Applications* (www.everyspec.com/MIL-HDBK/MIL-HDBK+(0200+-+0299)/MIL_HDBK_251_154/ or www.weibull.com/knowledge/milhdbk.htm). This handbook recommends and presents electronic parts stress analysis methods that lead to the selection of maximum safe temperatures for parts so that the ensuing thermal design is consistent with the required equipment reliability. These maximum parts temperatures must be properly selected since they are the goals of the thermal design, a fact that is often overlooked. Many thermal designs are inadequate because improper maximum parts temperatures were selected as design goals. Consequently, the necessary parts stress analysis procedures have been emphasized.

MIL-HDBK-336 *Military Handbook, Survivability, Aircraft, Nonnuclear.* This military handbook is subsumed by *JTCG/AS Aerospace Systems Survivability Handbook Series* (www.bahdayton.com/surviac/PDF/SHBSVol1HDBKSeries.pdf). The handbook series is designed to provide its users with insight into the key activities performed by survivability personnel in support of aerospace systems acquisition. The series is not a specification or standard but rather a "how-to" guide for all survivability managers, engineers, and analysts associated with survivability activities likely to be needed on any program, government or commercial. The intent of this handbook series is to address the activities of the survivability community in response to the changing needs of the war fighter. Survivability personnel need to assist acquisition managers in establishing aggressive but realistic survivability objectives for all programs, as well as in considering tradeoffs of cost, performance, and schedule, beginning early in the program.

MIL-HDBK-338B *Electronic Reliability Design Handbook* (www.relex.com/resources/mil/ 338b.pdf). Reliability engineering is doing those things that insure that an item will perform its mission successfully. The discipline of reliability engineering consists of two fundamental aspects: (1) paying attention to detail, and (2) handling uncertainties.

The traditional, narrow definition of reliability is "the probability that an item can perform its intended function for a specified interval under stated conditions." This narrow definition applies largely to items that have simple missions, such as equipment, simple vehicles, or components of systems. For large complex systems, such as command and control systems, aircraft weapon systems, a squadron of tanks, naval vessels, it is more appropriate to use more sophisticated concepts such as "system effectiveness" to describe the worth of a system. System effectiveness relates to that property of a system output, carrying out of some intended function, which was the real reason for buying the system in the first place; if the system is effective, it functions well; if it is not effective, it does not function well and attention must be focused on those system attributes that are deficient.

MIL-HDBK-470A *Designing and Developing Maintainable Products and Systems* (www. sre.org/pubs/Mil-Hdbk-470A.pdf). Unlike previous handbooks that focused only on maintainability, this document provides information to help the reader view maintainability in the context of an overall SE effort. The handbook defines maintainability, describes its relationship to other disciplines, addresses the basic elements common to all sound maintainability programs, describes the tasks and activities associated with those elements, and provides guidance in selecting those tasks and activities.

MIL-HDBK-502 *Acquisition Logistics* (acc.dau.mil/CommunityBrowser.aspx?id=32543 or www.logsa.army.mil/lec/downloads/standards/milhdbk502.pdf). This handbook is for guidance only. This handbook cannot be cited as a requirement. If it is, the contractor does not have to comply. This handbook offers guidance on acquisition logistics as an integral part of the SE process.

MIL-HDBK-727 (NOTICE 1) *Military Handbook Design Guidance for Producibility* (aero-defense.ihs.com/document/abstract/FEAVDAAAAAAAAAAA).

This method was developed by the U.S. Army Material Command and published by the Naval Publications and Forms Center. The first edition was published in 1971, and the latest revision was published in April 1984. The primary principle is Taylor's motion economy and some other design tools, i.e., DOE. This method is not too popular. Not many people know about it, and it is not used very much outside the military. Some updates and revisions are needed to make it more practical to general manufacturing companies [28].

MIL-HDBK-728 Nondestructive Testing (www.everyspec.com/MIL-HDBK/MIL-HDBK+ (0700+-+0799)/download.php?spec=MIL_HDBK_728_1.2125.pdf). Although this handbook is provided as a guide to all those employed in nondestructive testing (NDT), it will be of specific interest to administrators, designers, production engineers, quality assurance personnel, and nondestructive test engineers and technicians. It has been formulated to cover both broad and specific applications of NDT, so as to satisfy individuality, as well as conformity, of interests and knowledge among the divisions of responsibility in NDT.

MIL-HDBK-731 (NOTICE 1) *Military Handbook Nondestructive Testing Methods of Composite Materials—Thermography* (aero-defense.ihs.com/document/abstract/ QXRWDAAAAAAAAAAA). This is the first volume of a nine-volume series on nondestructive testing of composite materials. Each volume is designed to provide an overview of the general method. This volume contains an overview of techniques available for the application of thermographic nondestructive testing and evaluation techniques. General information on contact and non-contact methods of temperature measurement are discussed as these pertain directly to thermography. Special emphasis is placed on non-contact, infrared methods as these can be used for real-time observations. Practical information on the procedures that have proven useful for application of thermography to composites is discussed in some detail.

MIL-HDBK-781 *Reliability Test Methods, Plans, and Environments for Engineering, Development Qualification, and Production* (aero-defense.ihs.com/document/abstract/

AKXZDAAAAAAAAAAA). This handbook explains techniques for use in reliability tests performed during integrated test programs. Procedures, plans, and environments that can be used in reliability development/growth tests (RD/GT), reliability qualification tests (RQT), production reliability acceptance tests (PRAT), environmental stress screening (ESS) methods, and durability/economic life test are provided.

MIL-HDBK-791 *Maintainability Design Techniques* (www.everyspec.com/MIL-HDBK/ MIL-HDBK+(0700+-+0799)/MIL-HDBK-791-MAINTAINABILITY_DESIGN_ TECHNIQUES_(METRIC)_2786/). Maintainability and reliability are the two major system characteristics that combine to form the commonly used effectiveness index availability. Although reliability and maintainability share importance, maintainability merits special consideration because of its influence on system maintenance activities, that is, the expenditure of man-hours and material, which represent significant budgetary costs over the life of the system. Maintenance activities also reduce the operational readiness of a system. Maintainability is a characteristic of design and installation. This characteristic is the measure of the ability of an item to be retained in or restored to a specified condition when maintenance is performed by personnel having specified skill levels and using prescribed procedures and resources at each prescribed level of repair.

MIL-HDBK-881 *Work Breakdown Structures for Defense Materiel Items* (www.acq.osd.mil/ pm/currentpolicy/wbs/MIL_HDBK-881A/MILHDBK881A/WebHelp3/MILHDBK881A. htm). This handbook presents guidelines for effectively preparing, understanding, and presenting a work breakdown structure (WBS). Its primary objective is consistent application of the WBS for all programmatic needs (including performance, cost, schedule, risk, budget, and contractual). It is intended to provide the framework for DoD program managers to define their program's WBS and guidance to defense contractors in their application and extension of the contract's WBS. Section 1 defines and describes the WBS. Section 2 provides instructions on how to develop a Program WBS in the pre-award timeframe. Section 3 offers guidance for developing and implementing a Contract WBS and Section 4 examines the role of the WBS in the post-award timeframe.

MIL-HDBK-965 *Acquisition Practices for Parts Management* (canceled without replacement).

MIL-HDBK-1010A *Cost Engineering Policy and Procedure* (canceled without replacement).

MIL-HDBK-1250 *Corrosion Prevention and Deterioration Control in Electronic Components and Assemblies* (www.everyspec.com/MIL-HDBK/MIL-HDBK+(1000+-+ 1299)/MIL_HDBK_1250A_2032/). This standard establishes minimum requirements for procedures, materials and systems for protecting electronic components in DoD systems and assemblies from adverse environments. Protective measures shall be sufficient to maintain performance characteristics within specified limits both during and after exposure to moisture, high and low temperatures, corrosive gases, chemicals and microbial (fungal) attack. This standard does not deal with protection against damage from stress, shock, vibration or nuclear, biological and chemical (NBC) contamination. It is not concerned with electrical or mechanical design except insofar as design details affect susceptibility to corrosion. Provisions for prevention of deterioration shall apply to housings, chassis, hardware and similar parts that are assembled into electronic equipment, as well as to electronic and electromechanical components. Unless specifically mentioned in the item specification or drawing, this standard does not apply to standard commercial equipment.

MIL-HDBK-1379-1 *Guidance for Acquisition of Training Data Products and Services.*

MIL-HDBK-1379-2 *Instructional Systems Development/Systems Approach to Training and Education.*

MIL-HDBK-1379-3 *Development of Interactive Multimedia Instruction.*

MIL-HDBK-1379-4 *Glossary for Training* (www.atsc.army.mil/itsd/imi/DODHandbooks.asp).

MIL-HDBK-1467 *Acquisition of Software Environments and Support Software* (www.dcs. gla.ac.uk/~johnson/teaching/safety/reports/0000_91_62_1467CO1.PD3). This handbook establishes guidelines for the acquisition of new COE items and ensuring the compatibility of these items with the designated COE. Further, it establishes guidelines for the contracting activity's use in contracting for SEE items, to ensure the compatibility of each software engineering environment (SEE) item with the contracting activity's designated life cycle SEE (LCSEE), and to ensure the existence of a complete life cycle support capability for application software that is to be delivered under the contract. This is for guidance only. This handbook cannot be cited as a requirement. If it is, the contractor does not have to comply.

MIL-HDBK-1530 *General Guidelines for Aircraft Structural Integrity Program* (engineering.wpafb.af.mil/corpusa/Handbook/Mh1530/MH1530.pdf). These guidelines describe the processes proven successful in the achievement of structural integrity in USAF air vehicles while the cost of ownership is minimized; and cost and schedule risks managed through a series of disciplined, time-phased tasks. This handbook is for guidance only. This handbook cannot be cited as a requirement. If it is, the contractor does not have to comply.

MIL-HDBK-1568 *Materials and Processes for Corrosion Prevention and Control in Aerospace Weapons Systems* (www.everyspec.com/MIL-HDBK/MIL-HDBK+(1500+-+1799)/MIL_HDBK_1568_2077/).

MIL-STD-1568B(USAF) has been redesignated as a handbook and is to be used for guidance purposes only. This document is no longer to be cited as a requirement. It establishes the requirements for materials, processes and techniques, and identifies the tasks required to implement an effective corrosion prevention and control program during the conceptual, validation, development and production phases of aerospace weapon systems. The intent is to minimize life cycle cost due to corrosion and to obtain improved reliability and to provide a mechanism for implementation of sound materials selection practices and finish treatments during the design, development, production and operational cycles of the aerospace weapon systems.

MIL-HDBK-1785 *System Security Engineering* (akss.dau.mil/Documents). This standard establishes the formats, contents and procedures for a contract SSE management program. The purpose of the System Security Engineering Management Program is to establish definitive guidance in the initial acquisition or modification of new or existing systems, equipment, and facilities to analyze security design and engineering vulnerabilities, and to develop recommendations for engineering changes to eliminate or mitigate vulnerabilities consistent with other design and operational considerations. SSE supports the development of programs and standards to provide life cycle security for critical defense resources no longer to be cited as a requirement. If cited as a requirement, contractors may disregard the requirement of this document and interpret its contents only as guidance.

MIL-HDBK-1798 *Mechanical Equipment and Subsystems Integrity Program* (www. barringer1.com/mil_files/MIL-HDBK-1798.pdf). This document describes the general process (program) to achieve and maintain the integrity of aerospace and ground mission mechanical systems, subsystems, and equipment. This standard allows for tailoring the process in a competitive environment to meet specific subsystem, equipment and/or system requirements. The Mechanical Equipment and Subsystems Integrity Program (MECSIP) is implemented into the procurement through the statement of work (SOW) and is to be used for guidance purposes only. This document is no longer to be cited as a requirement.

MIL-HDBK-2069 *Aircraft Survivability* (aero-defense.ihs.com/document/abstract/ IJPAEAAAAAAAAAAA). This handbook provides guidance and criteria for establishing survivability requirements and conducting survivability plans and programs throughout the system life cycle for fixed and variable wing aircraft, helicopters and remotely piloted vehicles. The survivability program will include a mix of threat avoidance, reconstitution and repairability, redundancy, and hardening techniques to enhance system survivability to the maximum practical extents. This handbook applies to combat and combat support aircraft expected to be exposed to non-nuclear (i.e., conventional, chemical, biological, and directed energy) and nuclear threat environments.

MIL-HDBK-2164 *Environmental Stress Screening Process for Electronic Equipment* (www. barringer1.com/mil_files/MIL-HDBK-2164A.pdf). This handbook provides guidelines for environmental stress screening (ESS) of electronic equipment, including environmental screening conditions, durations of exposure, procedures, equipment operation, actions taken upon detection of defects, and screening documentation. These guidelines provide for a uniform ESS process that may be utilized for effectively disclosing manufacturing defects in electronic equipment caused by poor workmanship and faulty or marginal parts. It will also identify design problems if the design is inherently fragile or if qualification and reliability growth tests were too benign or not accomplished. The most common stimuli used in ESS are temperature cycling and random vibration. A viable ESS program must be dynamic; the screening program must be actively managed, and tailored to the particular characteristics of the equipment being screened. It should be noted that there are no universal screens applicable to all equipment. This handbook cannot be cited as a requirement. If it is, the contractor does not have to comply.

MIL-HDBK-2165 (canceled).

MIL-HDBK-46855A *Human Engineering Program Process and Procedures* (www. everyspec.com/MIL-HDBK/MIL-HDBK+(9000+and+Up)/MIL_hnbk_46855A_142/). This handbook provides human engineering (HE) (1) program tasks, (2) procedures and preferred practices, and (3) methods for application to system acquisition. The program tasks outline the work to be accomplished by a contractor or subcontractor in conducting an HE effort integrated with the total system engineering and development effort. This handbook applies to the acquisition of military systems, equipment, and facilities; however, all guidelines and practices contained herein apply to every program or program phase. Section 4 of this handbook should be tailored to specific programs and the milestone phase of the program within the overall life cycle. This tailoring by requiring or performing organizations—contractual or otherwise—focus on attaining essential human performance requirements, consistent with avoiding unnecessary program costs.

MIL-HDBK-87244 *Avionics/Electronics Integrity* (aero-defense.ihs.com/document/abstract/ PAXYDAAAAAAAAAAA). This handbook provides rationale, guidance, and lessons learned for specific avionics/electronics applications, and parallels the sample performance specification in Appendix C. This handbook describes the Avionics/ Electronics Integrity Process (AVIP) and associated performance requirements that should be tailored and incorporated into appropriate contractual documents and the integrated engineering and manufacturing process to achieve integrity of airborne and ground-based electronics. The suggested performance requirements assist in the development of program performance and verification requirements that encompass equipment life, life cycle uses, environments, and supportability. This handbook is intended to complement guide specifications and handbooks in other specialty areas.

MIL-PRF-29612B *Training Data Products* (nawctsd.navair.navy.mil/Resources/Library/ Acqguide/29612b.htm). This document establishes data requirements to support the life cycle maintenance of training data products and is a source document for training.

MIL-PRF-49506 *Logistics Management Information* (acc.dau.mil/CommunityBrowser.aspx? id=33641). This document describes information required by the government to perform acquisition logistics management functions. Its principal focus "is on providing the DOD with a contractual method for acquiring support and support-related reengineering and logistics data from contractors."

MIL-PRF-87269 *Data Base, Revisable—Interactive Electronic Technical Manuals* (www.dt. navy.mil/tot-shi-sys/des-int-pro/tec-inf-sys/cal-std/index.html). This performance specification prescribes the requirements for an "interactive electronic technical manual data base" (IETMDB) to be constructed by a weapon-system contractor for the purpose of creating interactive electronic technical manuals (IETMs). The requirements herein cover the specification for the IETMDB and are intended to apply to one or both of two modes as specified in a contract: (1) the interchange format for the data base to be delivered to the government; or (2) the structure and the naming of the elements of the data base created and maintained by the contractor for purposes of creating IETMs that are in turn delivered to the government.

MIL-STD-100 *Standard Practice for Engineering Drawings* (elsmar.com/pdf_files/Military %20Standards/mil-std-100G.pdf). This standard, along with ASME Y14.100M, establishes the essential requirements and reference documents applicable to the preparation and revision of engineering drawings and associated lists for or by departments and agencies of the DoD.

MIL-STD-130N *Identification Marking of U.S. Military Property* (www.acq.osd.mil/dpap/ pdi/uid/attachments/MIL-STD-130N-20080111.pdf). This standard provides the item marking criteria for development of specific marking requirements and methods for identification of items of military property produced, stocked, stored, and issued by or for the DoD. This standard addresses criteria and data content for both free text and machine-readable information (MRI) applications of item identification marking.

MIL-STD-167/2A *Mechanical Vibrations of Shipboard Equipment (Reciprocating Machinery and Propulsion System and Shafting). Types III, IV, and V* (store.mil-standards.com/index.asp?PageAction=VIEWPROD&ProdID=1237). This standard covers the requirements of naval equipment including machinery as regarding both internally excited vibrations and externally imposed vibrations. In some special machinery, equipment, or installations, such as antennas, large machinery items, and certain "unique" designs it may be necessary to deviate from this standard. In those cases, special modifications shall be subject to approval by the command or agency concerned.

MIL-STD-461E *Control of Electromagnetic Interference (EMI) Characteristics of Subsystems and Equipment* (www.goes-r.gov/procurement/flight_documents/MIL-STD-461E.pdf). This standard covers electromagnetic effects that are both conducted and radiated. Each area addresses specific modes, either emissions or susceptibility, and bandwidths. This standard establishes interface and associated verification requirements for the control of the electromagnetic interference (emission and susceptibility) characteristics of electronic, electrical, and electromechanical equipment and subsystems designed or procured for use by activities and agencies of the DoD. Such equipment and subsystems may be used independently or as an integral part of other subsystems or systems. This standard is best suited for items that have the following features: electronic enclosures that are no larger than an equipment rack, electrical interconnections that are discrete wiring harnesses between enclosures, and electrical power input derived from prime power sources. This standard should not be directly applied to items such as modules located inside electronic enclosures or entire platforms. Application-specific environmental criteria may be derived from operational and engineering analyses on equipment or subsystems being procured for use in specific systems or platforms. When analyses reveal that the requirements in this standard are not appropriate for that procurement, the requirements may be tailored.

MIL-STD-464A *Electromagnetic Environmental Effects Requirements for Systems* (acc.dau.mil/CommunityBrowser.aspx?id=30513). This standard establishes electromagnetic environmental effects (E3) interface requirements and verification criteria for airborne, sea, space, and ground systems, including associated ordnance.

MIL-STD-499B *Systems Engineering* (canceled) (www.afit.edu/cse/docs/guidance/MS499BDr1.pdf). This standard defines a total systems approach for the development of defense systems. The standard requires establishing and implementing a structured, disciplined, and documented SE effort incorporating the SE process; multidisciplinary teamwork; and the simultaneous development of the products and processes needed to satisfy user needs. The SE process is defined generically to facilitate broad application. This standard defines the requirements for technical reviews. The tasks in this standard provide a methodology for evaluating progress in achieving system objectives.

MIL-STD-810G *Environmental Engineering Considerations and Laboratory Tests* (www.dtc.army.mil/publications/MIL-STD-810G.pdf). This standard contains materiel acquisition program planning and engineering direction for considering the influences that environmental stresses have on materiel throughout all phases of its service life. It is important to note that this document does not impose design or test specifications. Rather, it describes the environmental tailoring process that results in realistic materiel designs and test methods based on materiel system performance requirements.

MIL-STD-882 *Standard Practice for System Safety* (www.system-safety.org/Documents/ MIL-STD-882E-Feb05.doc). This document outlines a standard practice for conducting system safety. Mishap risk must be identified, evaluated, and mitigated to a level acceptable (as defined by the system user or customer) to the appropriate authority and compliant with federal (and state where applicable) laws and regulations, executive orders, treaties, and agreements. Program trade studies associated with mitigating mishap risk must consider total life cycle cost in any decision.

MIL-STD-961E (Change 1) *Defense and Program-Unique Specifications Format and Content* (assist.daps.dla.mil/quicksearch/basic_profile.cfm?ident_number=36063). This standard establishes the format and content requirements for the preparation of defense specifications and program-unique specifications prepared either by DoD activities or by contractors for the DoD. It also covers the format and content requirements for specification sheets, supplements, revisions, amendments, and notices.

MIL-STD-1366E *Transportability Criteria* (store.mil-standards.com/index.asp? PageAction=VIEWPROD&ProdID=1530). This standard establishes basic transportability interface criteria for use in the development and shipment of items of materiel. The standard covers dimensional and weight limitations for all modes of transport to ensure that new and modified systems meet the interface requirements of the Defense Transportation System (DTS) (highways, tunnels, bridges, railways, etc.) and the DTS lift assets (rotary and fixed wing aircraft, railcars, ships, barges, etc.) for unrestricted worldwide transport and deployment. It also covers lifting and tie-down provisions, containerization criteria, overloads, assembly/disassembly, air delivery, shelter criteria, transportability testing, and modeling and simulation of the transportation environment. This standard will allow materiel development and procurement activities to design military equipment to meet the transportability requirements of various modes.

MIL-STD-1388-2B *DoD Requirements for a Logistic Support Analysis Record* (www.navy. mi.th/logis/ils/data_files/MIL-STD.pdf). This standard prescribes the data element definitions (DED), data field lengths, and formats for logistic support analysis record (LSAR) data. It identifies the LSAR reports that are generated from the LSAR data and identifies the LSAR relational tables and automated data processing (ADP) specifications for transmittal and delivery of automated LSAR data. This standard applies to all system/equipment acquisition programs, major modification programs, and applicable research and development projects through all phases of the system/ equipment life cycle. This standard is for use by both contractor and government activities. As used in this standard, the requiring authority is generally a government activity but may be a contractor when LSA documentation requirements are levied on subcontractors.

MIL-STD-1472F *Human Engineering Design Criteria for Military Systems, Equipment and Facilities* (safetycenter.navy.mil/instructions/osh/MILSTD1472F.pdf). The purpose of this standard is to present human engineering design criteria, principles, and practices to achieve mission success through integration of the human into the system, subsystem, equipment, and facility, and achieve effectiveness, simplicity, efficiency, reliability, and safety of system operation, training, and maintenance.

MIL-STD-1521B *Technical Reviews and Audits for Systems, Equipments, and Computer Software* (acc.dau.mil/CommunityBrowser.aspx?id=110905). Although MIL-STD-1521 B is no longer in force, it remains a valuable source of information about the government and contractor roles in the establishment of agreement, a meeting of the minds as it were, on requirements, functional and allocated baselines, testing, and production.

MIL-STD-1541 *Electromagnetic Compatibility Requirements for Space Systems* (www. everyspec.com/MIL-STD/MIL-STD+(1500+-+1599)/download.php? spec=MIL_STD_1541A.1500.pdf). This standard establishes the electromagnetic compatibility requirements for space systems, including frequency management, and the related requirements for the electrical and electronic equipment used in space systems. It also includes requirements designed to establish an effective ground reference for the installed equipment and designed to inhibit adverse electrostatic effects. The purposes of this standard are:

1. To define minimum performance requirements for electromagnetic compatibility

2. To identify the system relationships pertinent to electromagnetic compatibility

3. To identify requirements for system and equipment engineering designed to enable achieving compatibility in a timely, predictable, and economical manner

4. To define requirements for equipment and system tests and analyses to demonstrate compliance with this standard

MIL-STD-1686C *Electrostatic Discharge Control Program for Protection of Electrical and Electronic Parts, Assemblies and Equipment* (www.dscc.dla.mil/downloads/packaging/ MS1686_C.pdf). The purpose of this standard is to establish comprehensive requirements for an ESD control program to minimize the effects of ESD on parts, assemblies, and equipment. An effective ESD control program will increase reliability and decrease both maintenance actions and lifetime costs. This standard shall be tailored for various types of acquisitions.

MIL-STD-1799 *Survivability, Aeronautical Systems (For Combat Mission Effectiveness).* This military handbook is subsumed by *JTCG/AS Aerospace Systems Survivability Handbook Series* (www.bahdayton.com/surviac/PDF/SHBSVol1HDBKSeries.pdf).

MIL-STD-40051B *Preparation of Digital Technical Information for Multi-Output Presentation of Technical Manuals* (aero-defense.ihs.com/document/abstract/ VUGIGBAAAAAAAAAA). This standard establishes the technical content, style and format requirements for all technical manuals (TMs) for major weapon systems, and their related systems, subsystems, equipment, weapons replacement assemblies (WRAs), and shop replacement assemblies (SRAs). The requirements are applicable for all maintenance levels throughout overhaul (depot) including depot maintenance work requirements (DMWRs) and national maintenance work requirements (NMWRs). The requirements can be used to develop TMs in a variety of output forms including interactive screen presentations (frame-based manuals) and paper page-based manuals.

6. Other Military Standards and Specifications

Table 6.11 and the descriptions that follow list some military standards that might be of guidance in developing military equipment. Even canceled standards might, in some cases, provide guidance or direction. This section finishes with some handbooks that provide guidance for the system architect and developer.

6.1. Specifications

MIL-E-4158 *General Specification for Ground Electronic Equipment* (www.everyspec.com/ MIL-SPECS/MIL+SPECS+(MIL-E)/MIL-E-4158E_8401/). Covers the general requirements of the design and manufacture of ground electronic equipment.

MIL-E-5400 *General Specifications for Aerospace Electronic Equipment* (aero-defense. ihs.com/document/abstract/JETWDAAAAAAAAAA). Covers the general requirements for airborne electronic equipment for operation primarily in piloted aircraft.

MIL-E-16400 *General Specification for Naval Ship and Shore: Electronic, Interior Communication and Navigation Equipment* (www.everyspec.com/MIL-SPECS/ MIL+SPECS+(MIL-E)/MIL-E-16400H_13914/). Covers naval ship and shore electronic, interior communication and navigation equipment, and those units of other naval ship and shore equipment that utilize electronic technology.

MIL-E-17555 *Packaging of Electronic and Electrical Equipment, Accessories, and Provisioned Items (Repair Parts)* (www.everyspec.com/MIL-SPECS/MIL+SPECS+ (MIL-E)/MIL-E-17555H_10374/). Covers the packaging (preservation, packing and marking) requirements for electronic equipment, accessories, auxiliary equipment, miscellaneous electrical equipment and provisioned items (repair parts).

Table 6.11: Other Military Standards and Specifications That May Be of Guidance in Military Projects

Designation	Topic
MIL-E-4158	General Specification for Ground Electronic Equipment
MIL-E-5400	General Specifications for Aerospace Electronic Equipment
MIL-E-16400	General Specification for Naval Ship and Shore: Electronic, Interior Communication and Navigation Equipment
MIL-E-17555	Packaging of Electronic and Electrical Equipment, Accessories, and Provisioned Items (Repair Parts)
MIL-M-28787	General Specification for Standard Electronic Modules
MIL-H-38534	General Specification for Hybrid Microcircuits
MIL-I-38535	General Specification for Manufacturing Integrated Circuits
MIL-H-46855	Human Engineering Requirements for Military Systems, Equipment and Facilities
MIL-PRF-19500K	General Specification for Semiconductor Devices
MIL-STD-210*	Climatic Extremes for Military Equipment
MIL-STD-414	Sampling Procedures and Tables for Inspection by Variables for Percent
MIL-STD-701*	Lists of Standard Semiconductor Devices
MIL-STD-721*	Definitions of Terms for Reliability, and Maintainability
MIL-STD-750	Tests Methods for Semiconductor Devices
MIL-STD-756*	Reliability Modeling and Prediction
MIL-STD-790	Reliability Assurance Program for Electronic Part Specifications
MIL-STD-975*	Standard Parts Derating Guidelines
MIL-STD-1540	Test Requirements
MIL-STD-1562*	Lists of Standard Microcircuits
MIL-STD-1670*	Environmental Criteria and Guidelines for Air Launched Weapons
MIL-STD-1772*	Certification Requirements for Hybrid Microcircuit Facility and Lines
MIL-STD-2155*	Failure Reporting, Analysis and Corrective Action System
MIL-HDBK-454	Standard General Requirements for Electronic Equipment
MIL-HDBK-471*	Maintainability Verification/Demonstration/Evaluation
MIL-HDBK-1547	Technical Requirements for Parts, Materials, and Processes for Space and Launch Vehicles
MIL-HDBK-2084	General Requirements for Maintainability

*Canceled or replaced.

MIL-M-28787 *General Specification for Standard Electronic Modules* (www.everyspec.com/ MIL-SPECS/MIL+SPECS+(MIL-M)/MIL-M-28787D_7402/). Establishes the requirements for standard electronic modules (SEM) as defined in MIL-STD-1378 for the Standard Hardware Acquisition and Reliability Program (SHARP) for use in military systems. The requirements herein serve to verify the design requirements of MIL-STD-1389.

MIL-H-38534 *General Specification for Hybrid Microcircuits* (ieeexplore.ieee.org/Xplore/ login.jsp?url=http%3A%2F%2Fieeexplore.ieee.org%2Fiel2%2F849%2F2572% 2F00077827.pdf&authDecision=-203). Provides hybrid microcircuit manufacturers with the first specification for hybrid microcircuit devices only. The basis for MIL-H-38534 is MIL-M-38510, MIL-STD-883, and MIL-STD-1772.

MIL-I-38535 *General Specification for Manufacturing Integrated Circuits* (www.dscc.dla. mil/Programs/MilSpec/ListDocs.asp?BasicDoc=MIL-PRF-38535). Establishes the general performance requirements for integrated circuits or microcircuits and the quality and reliability assurance requirements, which are to be met for their acquisition. The intent of this specification is to allow the device manufacturer the flexibility to implement best commercial practices to the maximum extent possible while still providing product that meets military performance needs.

MIL-H-46855 *Human Engineering Requirements for Military Systems, Equipment and Facilities* (www.everyspec.com/MIL-SPECS/MIL+SPECS+(MIL-H)/MIL-H-46855B_7563/). Establishes and defines the requirements for applying human engineering to the development and acquisition of military systems, equipment, and facilities.

MIL-PRF-19500M *General Specification for Semiconductor Devices* (www.snebulos.mit. edu/projects/reference/MIL-STD/MIL-PRF-19500-RevM.pdf). Establishes the general performance requirements for semiconductor devices.... Five quality levels for encapsulated devices are provided for in this specification, differentiated by the prefixes JAN, JANTX, JANTXV, JANJ, and JANS. Seven radiation hardness assurance (RHA) levels are provided for the JANTXV and JANS quality levels.

6.2. Standards

MIL-STD-210 *Climatic Extremes for Military Equipment* (www.everyspec.com/MIL-HDBK/ MIL-HDBK+(0300++0499)/download.php?spec=MIL_HDBK_310.1851.pdf). Replaced by guidance handbook, MIL-HDBK-310, which provides climatic data primarily for use in engineering analyses to develop and test military equipment and materiel.

MIL-STD-414 *Sampling Procedures and Tables for Inspection by Variables for Percent* (withdrawn February 1999). Future acquisitions should refer to acceptable non-government standard on sampling procedures and tables for inspection by variables for percent defective such as ANSI/ASQC Z1.9-1993.

MIL-STD-701 *Lists of Standard Semiconductor Devices* (replaced by MIL-HDBK-5961A, which was then canceled June 2005).

MIL-STD-750E *Tests Methods for Semiconductor Devices* (aero-defense.ihs.com/document/abstract/DDYUABAAAAAAAAAA). This standard establishes uniform methods for testing semiconductor devices, including basic environmental tests to determine resistance to deleterious effects of natural elements and conditions surrounding military operations, and physical and electrical tests.

MIL-STD-790 *Established Reliability and High Reliability Qualified Products List (QPL) Systems for Electrical, Electronic, and Fiber Optic Parts Specifications* (www.dscc.dla.mil/Programs/MilSpec/listdocs.asp?BasicDoc=MIL-STD-790). This standard is for direct reference to established reliability and high-reliability electrical, electronic, and fiber optic parts specifications and establishes the criteria for a manufacturer's qualified product system.

MIL-STD-1540 *Test Requirements* (see the chapter on spacecraft instruments and aerospace, Report No. TR-2004(8583)-1 Rev. A).

MIL-STD-1670A *Environmental Criteria and Guidelines for Air Launched Weapons* (store.mil-standards.com/index.asp?PageAction=VIEWPROD&ProdID=143). This standard establishes guidelines for the environmental engineering required in support of air-launched weapons developments. Air-launched weapons include the following: air-to-air weapons, air-to-surface weapons including free-fall weapons, and aircraft gun pods. The following environments are not considered in this standard: nuclear, electromagnetic, and short wavelength, coherent, and electromagnetic (laser effects).

MIL-STD-1772 *Certification Requirements for Hybrid Microcircuit Facility and Lines* (store.mil-standards.com/index.asp?PageAction=VIEWPROD&ProdID=136). This document establishes requirements governing certification and qualification of manufacturing construction techniques and materials for hybrid microcircuits. It is intended to standardize the documentation and testing for hybrid microcircuits for use in military and aerospace applications.

6.3. Handbooks

MIL-HDBK-454B *Standard General Requirements for Electronic Equipment* (www.dscc.dla.mil/programs/psmc/library.html). This handbook provides guidance and lessons learned in the selection of documentation for the design of electronic equipment.

MIL-HDBK-2084 *General Requirements for Maintainability* (store.mil-standards.com/index. asp?PageAction=VIEWPROD&ProdID=28). This standard covers the common maintainability design requirements to be used in military specifications for avionic and electronic systems and equipment.

6.4. Current Guidance

Considering the focus and direction of the DoD for "best industry practice" and guidance, handbooks provide guidance as opposed to mandating standards. Two examples are the *Naval Systems Engineering Guide* and ADS-13F-HDBK *Aeronautical Design Standard Handbook, Air Vehicle Materials and Processes*. Both carry the disclaimer that they are each for guidance only and cannot be cited as a requirement; even if the requisitioning body cites them, the contractor does not have to comply.

The *Naval Systems Engineering Guide* "is provided to help ensure the systems we develop for the fleet are affordable, operationally effective and suitable, and can be a timely solution to satisfy user needs at an acceptable level of risk. This Guide defines the systems engineering (SE) requirements and tasks, their implementation and products, and explains the tools and techniques used throughout a product life cycle. This Guide satisfies the DoD requirement for having a documented systems engineering process, and emphasizes the relationship between the technical management process and the systems engineering process" [8].

The ADS-13F-HDBK, *Aeronautical Design Standard Handbook, Air Vehicle Materials and Processes* embodies the general guidelines of the Army Aviation and Troop Command (ATCOM) for the materials and processes utilized in the design and construction of Army aircraft [29].

7. Avionics Standards: DO-178 and DO-254

Military avionics have adopted two commercial standards, DO-178B and DO-254, that have established safety while maintaining schedule, cost, and quality. Both standards required planning, detailed requirements, process control, and rigorous testing. "DO-178B and DO-254 mandate processes for safety, specification, design, implementation, correctness, data, and certification of airborne electronics" [45]. The avionics industry developed these two standards to establish hardware deployment guidelines for developers, verification engineers, quality managers, installers, and users [47].

The RTCA DO-178B document titled, *Software Considerations in Airborne Systems and Equipment Certification*, is "to provide guidelines for the production of software for airborne systems and equipment that performs its intended function with a level of confidence in safety that complies with airworthiness requirements" [46]. This document has a dual reference: DO-178B is the RTCA reference, and ED12B is the EUROCAE reference [47].

The RTCA DO-254 document titled, *Design Assurance Guidance for Airborne Electronic Hardware*, is "to help aircraft manufacturers and the suppliers of aircraft electronic systems assure that electronic airborne equipment safely performs its intended function" [48]. DO-254 is a recent standard aiming to bring the same rigor to development of airborne hardware that DO-178B did for software to airborne environments. This document has a dual reference: DO-254 is the RTCA reference and ED80 is the EUROCAE reference [47].

RTCA, Inc. (Radio Technical Commission for Aeronautics) is a private, not-for-profit corporation in the United States that develops consensus-based recommendations regarding communications, navigation, surveillance, and air traffic management (CNS/ATM) system issues. RTCA functions as a Federal Advisory Committee for the U.S. government. Its recommendations are used by the U.S. Federal Aviation Administration (FAA) as the basis for policy, program, and regulatory decisions and by the private sector as the basis for development, investment and other business decisions. You can find more about RTCA at www.rtca.org [47].

EUROCAE is the acronym for the European Organization for Civil Aviation Equipment. It is the European equivalent of RTCA. EUROCAE documents are considered by Joint Aviation Authorities as means of compliance to Joint Technical Standard Orders and other regulatory documents. EUROCAE has extended its activity from airborne equipment to complex CNS/ATM systems including their ground segment. The related documentation is also considered by Eurocontrol and by the European Commission. The main European administrations, aircraft manufacturers, equipment manufacturers and service providers are members of EUROCAE, and they actively participate in the working groups that prepare these documents. You can find more about EUROCAE at www.eurocae.org [47].

FAA is the acronym of the U.S. Federal Aviation Administration, the organization responsible for controlling air traffic safety in the United States. You can find more about the FAA at www.faa.gov [47].

The European Aviation Safety Agency is the centerpiece of the European Union's strategy for aviation safety. EASA's mission is to promote the highest common standards of safety and environmental protection in civil aviation. You can find more about EASA at www.easa.eu.int [47].

7.1. DO-178B/C

DO-178B scope encompasses "those aspects of airworthiness certification that pertain to the production of software for airborne systems and equipment used on aircraft or engines." The document "does not provide guidelines concerning the structure of the applicant's organization, the relationships between the applicant and its suppliers, or how the responsibilities are divided." DO-178B recognizes that you can develop software in many different ways. It states, "This document recognizes that the guidelines herein are not mandated by law, but represent a consensus of the aviation community. It also recognizes that alternative methods to the methods described herein may be available to the applicant" [46].

Definitions: DO-178B provides some working definitions for planning and designing avionic architectures. It does so by categorizing failure under five headings:

- Catastrophic

- Hazardous/severe–major

- Major

- Minor

- No effect

The level of the software is then defined according to its potential failure conditions:

- Level A for potential catastrophic failures

- Level B for potential hazardous/severe–major failures

- Level C for potential major failures

- Level D for potential minor failures

- Level E for potential no-effect failures [46]

These definitions will determine the process levels used under DO-178B for software development.

"DO-178B is primarily a process-oriented document. For each process, objectives are defined and a means of satisfying these objectives are described" [46]. The processes include the following:

- Software planning

- Software development

- Verification of outputs of software requirements

- Verification of outputs of software design

- Verification of outputs of software coding & integration

- Testing of outputs of integration

- Verification of verification process results

- Software configuration management

- Software quality assurance

- Certification liaison [46]

DO-178C is the first update to DO-178B in nearly 20 years. "By 2010, the updated DO-178C will be required on all military and commercial avionics. DO-178C will better address software technologies such as modeling, object-oriented software, formal methods, and COTS software; these technologies are increasingly applied with military avionics. DO-178C will also contain enhanced details on software tool qualification, an area of previous underutilization within military avionics" [45].

7.2. DO-254

DO-254 "is intended to help aircraft manufacturers and the suppliers of aircraft electronic systems assure that electronic airborne equipment safely performs its intended function. The document identifies design life cycle processes for hardware that include line replaceable units, circuit board assemblies, application specific integrated circuits (ASICs), programmable logic devices, etc. It also characterizes the objective of the design life cycle processes and offers a means of complying with certification requirements" [48].

"The DO-254 standard was formally recognized by the FAA in 2005 via AC 20-152 as a means of compliance for the design of complex electronic hardware in airborne systems. Complex electronic hardware includes devices like Field Programmable Gate Arrays (FPGAs), Programmable Logic Devices (PLDs), and Application Specific Integrated Circuits (ASICs)" [49]. DO-254 has the same five categories of failure and the five levels of compliance, A through E, which depend on the effect a failure of the hardware will have on the operation of the aircraft.

DO-254 has the following outline:

1. Introduction

2. System Aspects of Hardware Design Assurance—must capture and track requirements throughout the design and verification process. To substantiate the effort, supply either the FAA or the designated engineering representative (DER) representing the FAA the following:

   - Plan for hardware aspects of certification (PHAC)

   - Hardware verification plan (HVP)

   - Top-level drawing

   - Hardware accomplishment summary (HAS)

3. Hardware design life cycle—Hardware design and hardware verification need to be performed independently

4. Planning process

5. Hardware design processes

   - Requirements capture

   - Conceptual design

   - Detailed design

6. Validation and verification process

7. Configuration management process

8. Process assurance

9. Certification liaison process

10. Hardware design life cycle data

11. Additional considerations

   - Use of previously developed hardware

   - Commercial-off-the shelf (COTS) components usage

   - Product service experience

   - Tool assessment and qualification

DO-254 requires independent qualification of tools used for design and verification. Independent qualification ensures an acceptable level of confidence in each tool's capability.

> When it is possible to show that the tool has been previously used and has been found to produce acceptable results, then no further assessment is necessary. A discussion of the relevance of the previous tool usage versus the proposed usage of the tool should be included in the justification .... [I]f no such relevant history can be evidenced, then the tool must undergo "Basic Tool Qualification[,]" which includes tool configuration control, a tool problem reporting process, and a process "to confirm that the tool produces correct outputs for its intended application using analysis or testing" [48].

A tool qualification is performed for a given context and tool usage (i.e., for a particular project), and for a given version of the tool. Qualification has to be performed again for a different context or system. The users of the tool and their prime contractor are responsible for qualifying a tool with the certification authorities (e.g., FAA, EASA). Obviously, this is a process that requires the cooperation of the tool vendor, but it cannot be carried out by the tool vendor alone [47].

8. Test and Evaluation

Tests have a variety of formats and purposes. Each test usually performs under one, and only one, of the following formats:

- Confirm aspects of simulation and analysis

- Help debug problems

- Assess manufacturing quality

- Screen components and subsystems

- Validate the design intent

- Verify that the design meets requirements

Note the difference between validation and verification. *Validation* attempts to confirm that the product meets the customer's intent; validation is more qualitative in nature. *Verification* is quantitative and measures whether specifications are met.

All tests, whether inspections, subsystem tests, unit tests, integration, evaluation/shakedown, or commissioning must be recorded and maintained in a central data repository. These files are subject to audits from customers.

8.1. Inspection

Inspection is one form of test that has two purposes, either to find problems or to assure quality. Debugging a problem often starts with inspection that looks for a nonconformity in a circuit (e.g., missing components or backwards installation) or some source code (e.g., incorrect pointer or increment value). Inspection in manufacturing is a quality check for both workmanship and defects in components or assembly. Inspection is a form of verification.

8.2. Peer Review

Peer review is another formal test that strives to find problems in the logic and rationale of a design. It primarily addresses software development but can also be used to study the design and dynamics of circuits, mechanisms, or structures. Inspection is a form of verification.

8.3. Subsystem Tests

The subsystem tests are formal activities to confirm that each module (whether software or hardware) performs according to requirements; these tests are formal verifications. Generally a dedicated tester or test team might be responsible for running the tests. Generating or selecting tests should be a collaborative effort between the design team and the test team. These tests occur before integration of the full system.

Examples of hardware subsystem tests might include the following (this is not an exhaustive list):

- Functional specifications

  Correct operations

  Human factors

  Inspection for fit and form

- Performance specifications

  Mass and volume

  Power consumption

  Physical dynamics of mechanisms

  Data throughput

- Electromagnetic capability (EMC)

- Dependability

- Environmental

  Shock and vibration

  Radiation tolerance

  Condensation

  Thermal cycling

  Vacuum and outgassing

  Dust

  Immersion and leaks

The tests of software modules are often called unit tests. Examples of software subsystem tests might include the following (this is not an exhaustive list):

- Functional and logical specifications

  Correct operations

  Human factors—input and display

Proper function and calculation of parameters

Rejection of out-of-bounds values

Interface protocols

- Performance specifications

    Completion of tasks within deadlines

    Data throughput

- Fault tolerance

"Tests of software take two different forms, static and dynamic. Static tests are sometimes called glass box or white box tests because you can view the structure of the code. Dynamic tests are sometimes called black box or closed box tests because you don't see the structure of the code, rather the tests exercise the operating software . . . . Static tests examine and critique the structure of the code through either manual or automated means. Static, manual tests include code inspections and walkthroughs, which can be very effective in finding problems with the design and the translation of the original intent. Static, automated tests use software tools to check for statement coverage, branch coverage, and path coverage . . . . Dynamic tests are behavioral tests that exercise the behavior of the operating software. These tests rely on understanding the requirements of the design and testing the software to see if it meets each requirement" [30, 31].

8.4. Integration

Integration examines the multiple interactions between modules and subsystems. It combines modules, both hardware and software, in a planned and controlled fashion to reveal interactions, to understand consequences and potential consequences, and to limit undesirable operations. Good integration provides both verification of system requirements and validation of design intent. Integration requires the close cooperation of the design team, test team, manufacturing team, and the integration team.

Integration of hardware, including electrical, electronic, mechanical, and structural components, ensures that things fit and function together. Integration can include the tests listed above but it does so within the context of the system and not just a single module. Software integration ensures that software performs as specified and handles unexpected circumstances in an appropriate and predictable manner.

Integration should bring together a minimal number of subsystems and limit the number of variables. Too often people connect all the components together, in a "big bang" form of integration, and then hope for the best. This "big bang" integration is fatally flawed; even if it appears to work, there is seldom any way to find obscure faults and interactions; furthermore, if it works, it gives the false impression that all is well.

For more complex systems and projects, you will need simulators that imitate various subsystems. Each simulator provides a subset of the important functions and interfaces for the subsystem that it replaces. Simulators allow preliminary tests of various subsystem modules during integration; they can improve the parallel effort often needed to reduce time. When a subsystem completes component testing and moves to integration, it replaces its simulator. The closer the simulator comes to replicating the interface or coupled subsystem, the better the test coverage and the higher the final confidence in the integrated system. Simulators can fill the void while the various subsystems are being developed. The downside is that the closer a simulator gets to replicating the actual subsystem, the more expensive it becomes and the longer it takes to develop it [31].

In very large, complex systems, a test bed that replicates the final system allows simulators and subsystems to swap in and out as integration progresses. An example of an integration test bed was Boeing's 777 Systems Integration Laboratory (SIL). "The 777 SIL included all the electrical power systems, electromechanical systems, avionics, environment control systems, propulsion systems, and a portion of the payload electronics. The integration testing included realistic simulations of flight modes to support verification and validation of production equipment *before* the first flight of the aircraft. It also provided support for certification and validated the correct performance of both the physical and functional interfaces in the electrical and electronic systems during concurrent operation of multiple subsystems and failure" [32]. The Boeing 777 SIL had 40,000 airplane wires in 1000 bundles, 4000 signals and data buses to simulate, support for all the LRUs (line replaceable units) on the actual airplane, antennas for radios, GPS, and navigation systems, and generating capacity for 800 kW of power [32].

Several forms of tests can straddle either individual component tests or system integration. One set of these tests are environmental, another type is EMC.

8.5. Environmental

Environmental tests serve two purposes: to certify a system as capable of surviving the expected environments and to shake out problems.

For certification, environmental tests subject a module or a system to extremes, such as temperature, shock, vibration, humidity, pressure, and corrosion, and then confirm system operation; these tests are formal verifications. The specific application drives the certifying environmental tests. Military devices often do need to operate over wide temperature ranges, often specified between −40 and +80°C, and sometimes even wider. Military devices may also have to endure wide humidity ranges to 100% condensing. The power spectrum of the mechanical vibrations endured by the system varies widely from one application to the next. Some systems have to endure nearly continuous vibrations of less than 1 g over a frequency spectrum of fractions of a hertz to hundreds of hertz. Other systems must endure shock of 20,000 to 60,000 g while being fired out of a gun.

To shake out problems, environmental tests that stress test prototypes have a different purpose and format than certifying a product for extreme environments. Environmental stress tests are "looking to uncover incipient faults. Stress in the form of thermal or power cycling, vibration, shock, or condensation applied in extreme forms can reveal weaknesses. High temperature can precipitate or accelerate diffusion processes on silicon die, the oxidation of fractures, and collapse timing margins. Temperature cycling will expand and contract interconnections, such as solder joints and ball bonds. Elevated humidity that causes condensation can promote corrosion and breakdown electrical isolation" [31, 33].

> *These types of environmental stresses go by a variety of names: Accelerated Stress Test (AST), Accelerated Environmental Stress Screen (AESS), and Accelerated Life Test (ALT). They can also have the prefix of H to indicated "highly" accelerated, as in HAST, HASS, and HALT [31].*

Generally a dedicated tester or test team runs the tests. Generating or selecting tests should be a collaborative effort between the design team and the test team. Environmental testing usually requires a specialized test facility.

Some environmental tests might exercise selected subsystems before integration of the full system. Most of the time environmental tests take place after functional integration of all the subsystems.

8.6. EMC

Electromagnetic compatibility (EMC) tests serve two purposes: to certify a system (or module) as not susceptible to electromagnetic interference and to certify a system (or module) as not generating electromagnetic interference. Some EMC tests might exercise selected subsystems before integration of the full system. Generally, EMC tests take place after functional integration of all the subsystems.

Two military standards that are still very useful for testing systems and subsystems to EMC requirements are MIL-STD-461E, *Control of Electromagnetic Interference (EMI) Characteristics of Subsystems and Equipment*; and MIL-STD-464A, *Electromagnetic Environmental Effects Requirements for Systems* [34, 35].

MIL-STD-461E covers electromagnetic effects that are either conducted or radiated. Each area of the standard addresses specific modes (11 conducted and 6 radiated), either emissions or susceptibility, and bandwidths (somewhere between 30 Hz and 40 GHz).

This standard is best suited for items that have the following features: electronic enclosures that are no larger than an equipment rack, electrical interconnections that are discrete wiring harnesses between enclosures, and electrical power input derived from prime power sources.

MIL-STD-464A establishes electromagnetic environmental effects (E3) interface requirements and verification criteria for airborne, sea, space, and ground systems, including associated ordnance.

Clearly, testing a system for EMC requires specialized facilities, equipment, and trained technical personnel. The references contain several books that address EMC concerns in developing systems [36, 37].

8.7. Field Tests, Final Acceptance Tests, Builder's Trials, and Commissioning

A variety of test suites can look like integration or be a part of the integration plan for a system. They include field tests, final acceptance tests, customer evaluation tests, builder's trails, commissioning, and shakedown. These often are part of the development of a large system, such as a ship or aircraft.

Field testing can be a form of integration because it exercises the system in day-to-day circumstances with real users. It occurs either with prototypes to refine a design or after rigorous integration to perform validation, which confirms suitability to the intent of the customer's requirements. Field tests are particularly important for complex systems; functional integration testing cannot always cover all possible scenarios; sometimes the development team just overlooks some scenarios. Field testing often is a matter of recording and analyzing user responses to how the system operates.

Final acceptance tests are very similar to field tests in function but usually are carefully planned and prepared, much like integration tests for verification. Both the customer and development organization agree to both requirement and conduct of the final acceptance tests; usually this agreement is part of the contract. Typically the development organization turns over the system to the customer once the customer signs off on the results of the final acceptance tests.

Larger systems, like ships, undergo builder's trials, another form of tests very similar to field tests. They also are carefully planned and prepared, much like integration tests for verification or final acceptance tests. These take place before commissioning the final system. An example of builder's trials can be found in Michael S. Sanders book, *The Yard: Building a Destroyer at the Bath Iron Works*: "The first milestone . . . is ALO—Aegis Light-Off, when the approximately twenty-five component subsystems making up the Aegis combat system are brought on-line. Each and every part of this system—from the enormous SPY radar arrays at each corner of the main deckhouse to the smallest switch controlling a cooling fan down in the Combat Information Center—all must be rigorously tested individually and as they interact with each other. . . . Three months later comes GELO, or Generator Light-Off, the second post-launch milestone, when the electrical power generation, conversion, and distribution system comes to life. . . . MELO, Main Engine Light-Off, the last milestone before sea trials, takes place usually within a month of GELO. . . . All parts of the propulsion system—fuel pumps, air intake and exhaust, turbines, reduction gears, shafting, propeller blades—must be tested" [38].

The final act, after all the integration tests, field tests, and builders trials, is commissioning. Commissioning has various meanings and formats for different systems and different industries. For a ship, commissioning is the final act to bring her to life and begin her naval career. For fixed land installation, commissioning can look more like integration and builders trials but at the customer's site for the system.

Training is an important aspect to turning over a complex system to a customer. An outstanding practice is to bring users in early during design, construction, and test to learn the system and its operation. An example of practical training also derives from ship building as described in Sander's book: "[W]orkers and systems subcontractors, mostly electricians, technicians, and engineers, have been imparting to their Navy counterparts an invaluable trove of knowledge as the two groups work side by side to get their systems installed, tested, and up and running. This unique partnering is just the first step down the long road of melding crew and ship into a single highly functioning fighting unit" [38].

8.8. Manufacturing

Manufacturing test assures the quality of production and the quality of the fabricated item. Manufacturing tests are not verification of the design! Manufacturing tests only assure quality; they are not a thorough exercise of all modules. Examples of manufacturing tests might include the following (this is not an exhaustive list):

- Fit and finish
- Visual inspection of assembly
- Weight and volume
- Power consumption within limits
- Signal levels within limits
- Display function and brightness within limits
- Mechanism operation within limits [31]

8.9. BIT, BITE, and ATE

Often engineers need automated help to provide routine test coverage of subsystems under development. Built-in-test (BIT), built-in-test-equipment (BITE), and automatic-test-equipment (ATE) can provide automated help during test. BIT, BITE, and ATE become important adjuncts to a system and its support structure when high availability and reduced maintenance time are imperative to its operation. Usually a dedicated engineering team is responsible for specifying BIT, BITE, or ATE. This team must work with both the system design team and the system test team.

Each of these types of test systems measures, diagnoses, and estimates the state of health of the system. The definitions and differences between each is as follows:

- BIT is entirely self-contained within the module that it tests. Often it is a specialized software routine that exercises the module. It may also require some additional circuitry to be a permanent part of the module. "BIT can generally be described as a set of evaluation and diagnostic tests that uses resources that are an integral part of the system under test" [39].

- BITE may have software and circuitry contained within the system but it also requires external components to provide full coverage, such as wrap-around cables that connect outputs to inputs to allow the BIT to test those functions.

- ATE is test equipment entirely external to the system. It usually resides in the manufacturing plant or in the repair depot.

Specifications for BIT, BITE, and ATE are in terms of the amount of coverage to detect faults. An example from a military avionics program states that the BIT must "detect 95% of all faults, isolate 95% of any fault to one of over 50 replaceable assemblies, and allow false alarms for less than 2% of all detected faults" [40] for wide coverage, the development organization will expend significant engineering time and resources to design the BIT, BITE, or ATE.

BIT and BITE take snapshots of the system operation and often give pass/fail indications. BIT and BITE do not monitor its operation continuously—this would require redundant system architecture. ATE, on the other hand, can measure some portion of system operation continuously.

ATE has two very different purposes: one purpose of ATE is to confirm manufacturing quality; the other is to diagnose problems within the system. When compared to BIT and BITE, ATE has much more extensive coverage of the subsystems and modules, resides in centralized locations, requires high technical competence and skill to operate, and tends to be very expensive to acquire and use.

9. Obsolescence and Legacy Systems

Obsolescence is the state when a component is no longer useful, not current, or not available for manufacture [41]. A system may be obsolete but still useful—this is the case for legacy systems where a part or component or tool is no longer supported by industry, and which often disrupts the operation of a system and its maintenance and support.

Obsolescence within military systems is an increasing problem. One problem precipitating obsolescence is that military systems have long life spans of 20, 30, or 50 years or even more (the airframes on some B-52 bombers in the U.S. Air Force have exceeded 50 years), while commercial technology and COTS equipment have design cycles measured in months. This mismatch in design cycles is compounded by the falling demand from the military market, which dropped from 17% of the market for electronic components in 1975 to less than 0.5% in 2009. This means that military concerns in the first part of the 21st century play essentially no part in developing electronic components [42, 43].

Another problem in obsolescence is legacy software systems. This became most apparent to the general public at the turn of the millennium with the "Y2K" software problems. Less obvious but more significant are the costs to the military from the maintenance and upgrades to legacy software.

These problems with obsolescence and legacy systems are further exacerbated by the "lack of visibility in increasingly complex technology development, lengthy and costly redesign cycles, and fragmented communications channels" [43].

- Lack of visibility: "Microelectronics obsolescence is largely the result of commercial pressures: when it is no longer economical to produce a certain product, the manufacturer stops producing it" [44]. Manufacturers rarely provide adequate notice that they are ending the production of a particular component; often they transition rapidly from the original model to an upgraded or new model.

- Fragmented communications channels: This is another aspect of the problem with lack of visibility. Visibility is obscured by layers of vendors and suppliers between the supplier and the final customer. If a supplier issues an end-of-life notice for a component to its immediate customers; these customers, who had incorporated the component into a circuit board, might not pass the notice along to the next level of customer who incorporates the circuit board into a subsystem; this subsystem manufacturer might not pass the notice along to the next level of customer, even if they do get an end-of-life notice for the component. This builds delay into the production cycle.

- Costly design cycles: Bogdanski and Downey indicate that it can cost *"tens of millions* of dollars to replace, redesign, or requalify a product simply because a $30 part went obsolete" [43].

These problems cascade, leading to delays and cost overruns.

There are several, proactive avenues to reduce or avoid the cascading problems of obsolescence:

1. Design engineering. Preplan for product improvements, updates, and modifications.

    *   Use established bus architectures and enclosures to house circuit boards; when specific models of integrated circuits become obsolete and new components arrive on the scene, spin a new board design for that particular board; swap out the old circuit board and replace it with the new circuit board. This architecture allows for fast redesign of only a portion of the subsystem rather than completely redesigning the entire subsystem (or worse, the entire system).

    *   Use daughter boards for components such as memory chips and microprocessors that might become obsolete; when a component becomes obsolete and is replaced by a new component, modify the daughter board design to accept the new component (but the interface with the base circuit board stays the same); swap out the old daughter board with the newly designed daughter board. This architecture allows for fast redesign of only a portion of the circuit board, thus reducing the delay and cycle time for the entire system.

2. Production engineering. Preplan for components and modules going obsolete. Make a last-time buy of components that are in danger of becoming obsolete and put them into inventory or storage. "Bank" enough components to account for the estimated production, repair, and maintenance plus an additional number (as much as 30% of the estimated total).

3. Similar to the daughter board concept, "[c]ustom semiconductor packaging that leverages commercially available die technology that also combines the best of both techniques [i.e., design engineering and production engineering]. Leveraging volume-driven silicon provides flexibility by opening up effective alternatives for handling obsolescence concerns. For example, a design might call for four plastic COTS discretes in packages with lead-free terminations that are prone to 'tin whisker' growth. Instead, four die can be put in one package with lead-based terminations, thus saving cost by reducing the cost of failure. It would also conserve board space while providing a product in a package that has been proven mechanically reliable according to military requirements" [43].

4. The Defense MicroElectronics Activity (DMEA) offers obsolescence management services through the continual active monitoring of a system's parts to stay ahead of the obsolescence curve and plan ahead for mitigation [43].

5. Finally, you can build replicas of the obsolete components. "The DMEA's Advanced Reconfigurable Manufacturing for Semiconductors (ARMS) facility can produce any quantity, small or large, of microelectronic devices. Managers of weapons systems can use ARMS as a single solution or as part of a comprehensive supportability solution. If a

large number of components is approaching obsolescence, a program manager can use ARMS to develop these microelectronic replacement parts as a temporary measure while redesigning and fabricating an entirely new system or subsystem" [43, 44].

10. Case Studies

Case Study 6.1. The GNAT Pro Ada Compiler—A Case Study in Achieving High Quality in the Presence of Rapid Change

By Dr. Robert Dewar
President and CEO, AdaCore

Most applications either have very high-quality requirements and change slowly, or they are rapidly changing but can tolerate less than complete reliability. For example, the avionics software on a commercial airliner must satisfy safety certification requirements, but once fielded it stays stable. In contrast, many websites change rapidly, but we are used to temporary malfunctions, missing links, and so on. In this case study we look at AdaCore's GNAT Pro compiler and toolset as an example of an application requiring rapid change, such as adding new features needed by users while maintaining high reliability. Although not a safety-critical application itself, GNAT Pro is a vital part of the development of many safety-critical applications, such as the avionics for the new Boeing 787, and the iFacts component of the new UK air traffic control system.

For such applications, one might hope to treat the compiler as though it were safety-critical and carry out the equivalent of a complete certification/qualification. However this is not practical. First, certifying compilers is beyond the state of the art for all but simple toy languages, and second, traditional approaches for safety certification require a very static view of requirements and implementation, though as discussed in the context of the OpenDO initiative (www.open-do.org), this problem can perhaps be addressed.

For the GNAT Pro compiler we developed a dynamic methodology that allows rapid change but maintains a strong control over quality/reliability. An essential element in this approach is a team-based view of the software. Far too many projects enforce ownership of sections of the code by specific individuals, which works against the possibility of rapid change: any change must first receive buy-in from specific individuals, and these personnel need to be available to do the work. Furthermore if they leave, so does their critical knowledge. A very important component of achieving team ownership of the code is to have a rigorously enforced coding standard that everyone adheres to. Idiosyncratic coding styles are one way of "marking" code and achieving individual code ownership, and this must be avoided. You know you have achieved an important benchmark when you write a piece of code, and some time later someone else improves it or fixes a bug. If your reaction is, "What is that guy doing messing with my code," you have a problem. If your reaction is, "Great, now I don't have to do that," things are much better. In our case, the compiler itself has an option to enforce AdaCore's coding standard, and our version control system refuses to check in any code not meeting this standard.

(Continues)

Case Study 6.1.   Cont'd

In our situation where the technology changes rapidly (the compiler has been updated tens of thousands of times in the last 10 years, and we make new potential releases on over 40 target configurations every day), it is still of course critical to ensure quality. Our solution is to maintain an exhaustive test suite of over 10 million lines of code and run it as an intrinsic part of the development process rather than a separate quality assurance step. Anyone can change anything, but only if they first run the entire test suite on the proposed patch, no matter how small the change.

This approach has proved extremely successful in meeting the dual requirements of flexibility and reliability. Many of our customers developing highly critical applications update frequently to new versions of the compiler to exploit relatively minor new features, confident from experience that they will not be sacrificing reliability.

Case Study 6.2. DO-178C, DO-254, and Military Rotorcraft

By Vance Hilderman
HighRely Incorporated

DO-178C is the soon-to-be-released update to DO-178B standard and covers almost all avionics software development. DO-254 is the complex hardware corollary to DO-178, where complex hardware essentially means almost all silicon-based devices within avionics. Collectively, DO-178C and DO-254 increasingly govern airborne electronics as previously described within this book.

On a recent military rotorcraft (helicopter) project, U.S. Army suppliers were contractually obligated to comply with both DO-178 and DO-254 for all the new avionics systems they were deploying. The military recently began requiring DO-178 and DO-254 to emulate the commercial world's success in reducing avionics cost, risk, and accidents. The challenge of adopting DO-178 and DO-254 were daunting: upgrade field-proven systems to add new functionality while making the entire system compliant to DO-178 and DO-254. Given unlimited time and budget, success could be assured. But the real world was harsher: success needed to be assured within budget while adhering to an aggressive schedule. How did they proceed?

First, the suppliers procured DO-178 and DO-254 training via on-site instructors found through the 178 and 254 Industry Group website. The helicopter avionics suppliers ensured their managers and quality assurance personnel attended the training, not just the engineers, so that the entire program could understand the team-effort paradigm inherent with DO-178/254. They then commissioned a formal DO-178B and DO-254 Gap Analysis to analyze the "gaps" between their prior engineering activities and those required by DO-178/254. The gap analysis showed that they were already complying with 60 to 70% of the software and hardware processes required, but the challenge was in proving such compliance. The avionics suppliers procured various specific DO-178B and DO-254 tools specifically intended to reduce costs and risks associated with DO-178/254 namely a requirements management and traceability tool, structural coverage verification tool, DO-178/254 process plan templates and checklists, certifiable operating systems, and a project management dashboard system.

The new rotorcraft avionics projects made great initial progress but then ran into major "mental" stumbling blocks. Normally, gray hair and experience are highly beneficial in complex avionics systems, since the technical details are often best codified via the engineer's experience versus detailed engineering instructions. However, that very legacy experience handicapped the teams who benignly reverted to prior old processes every time a software or hardware process question arose. The younger, less-experienced engineers could see that the deterministic "prove it" rigor of DO-178 and DO-254 was often at odds with prior military standards. However, the older, more experienced staff was uncomfortable with the seemingly onerous, and unnecessary, requirements of 178/254. How did they reconcile the irreconcilable?

The military avionics suppliers teamed together and decided to elevate their quality assurance organizations to a role similar to that within commercial avionics: true arbiters of DO-178B and DO-254 compliance. In commercial systems, the Federal Aviation Administration (FAA) must oversee and approve all avionics systems and integration; the role of technical compliance oversight and initial approvals are performed by FAA registered designated engineering representatives (DERs). However, DERs have no formal mandate on military programs! Hence the suppliers procured the informal services of several consulting DERs to work with the military quality assurance personnel; together, they served as an "informal FAA approval team" to replicate the well-known commercial approval process. In the commercial world, companies are focused on avionics quality but reduced cost and schedule are inherent aspects recognized by all. Thus, the military suppliers were able to gain the benefit of prior engineering expertise while adapting to the new reality of DO-178B and DO-254 compliance. Initial systems were delivered on schedule, with remaining systems still under development but on-schedule as of the date of this case study.

This real-world synopsis of a recent military system shows that DO-178 and DO-254 can be successfully deployed but success is not automatic: careful attention must be paid. DO-178 and DO-254 are not cheap, and may even increase the cost of an initial deployment. Longer term, however, there is no question that DO-178 and DO-254 can greatly increase safety while reducing long-term costs and risks.

Vance Hilderman is co-founder of HighRely Incorporated who provides full training and technical services for DO-178B & DO-254 avionics and critical software. His e-mail is hilderman@sprynet.com. Additional free mission-critical software and blogs can be found at www.do178site.com and www.do254blog.com.

Case Study 6.3. Derating of a Modern Microprocessor Design at Saab for a Safety-Critical Application

By Torbjörn Månefjord and Håkan Forsberg
Saab Avitronics, Sweden

For many years, derating has been performed in safety critical military and civilian airborne applications. The reasons for derating in these applications are to extend the life time of the

(Continues)

Case Study 6.3.   Cont'd

component and to increase the robustness of the design. The derating rules and guidelines have over the years been outdated, especially for complex integrated circuits such as microprocessors. To correctly derate such circuits requires detailed knowledge of the internal design and manufacturing process. To arbitrarily derate such components by following old rules might lead to decreased lifetime and reliability instead.

Some of the non-obvious derating concerns are pointed out in the paper, "Derating concerns for microprocessors used in safety–critical applications," published in *IEEE A&E Systems Magazine*, March 2009 [50].

In the following case study, we point out some of the best practices that we have developed for non-obvious derating for a recent design within Saab aimed for safety critical airborne applications.

Power Derating by Frequency Derating

By observation of the plotted graph of power consumption versus frequency we note that the power increases linearly as a function of the frequency up to a certain point (we have a constant derivative). After the break point, the derivative suddenly increases (Fig. 6.15). The optimum balance between power consumption and performance is at this breakpoint. Above the breakpoint the stress of the component increases. The on-chip operating temperature increases leading to higher probability for electromigration and the time margin decreases in the critical path of the alternating current. The selected CPU has a working point lower than the breakpoint to extend the lifetime of the component. Sometimes it makes sense to derate power more than required. When comparing performance and power of microprocessors, this comparison should be at equal performance (performance is usually compared at normal operating frequencies). The reason for this is that a fast, but highly derated, microprocessor might consume less power when performing equal to a slower microprocessor that is normally derated.

Disabling and Spreading of Functionality to Decrease Hot Spots

The CPU chosen contains several integrated clock drivers for driving external memory components. These clock drivers are running at high frequency and are all phase aligned with low skew and compensated for the routing delay by a zero-delay buffer. They are also designed for high driving capability. These clock drivers introduce hotspots on the CPU. Therefore, external drivers with the same performance are used instead of built-in drivers in the CPU. In this way we derate the power consumption of the CPU.

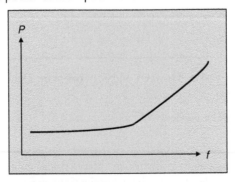

Figure 6.15: Frequency versus power.

Derating of Stabilizing Time Margin

The internal clock input to the CPU is fed by a low-frequency external clock oscillator. Inside the chip, this clock frequency is multiplied to a chosen higher clock frequency by means of a PLL (phase-locked loop). This is an analog circuit that is very sensitive and easily influenced by external disturbances and needs time to lock. The phase lock time is specified in the datasheet for the component. However this time is extremely important because if we release the reset signal before the PLL is locked the behavior of the CPU is unpredictable. Therefore, we extend the phase lock time to increase the safety margin.

Derating of Refresh Time for DRAM

The memory components are not part of the microprocessor but since the primary memory is so vital for the computer function, we mention them here as well.

In this particular design, we are using synchronous DRAM where each memory cell relies on a very small charge in a transistor. This small charge needs to be refreshed within a defined time period specified in the datasheet for the component in order to retain its value. This time is also extremely influenced by the ambient temperature. The chosen component is specified for 85°C and the equipment is designed for 70°C, which gives us a safety margin for self-heating inside the cabinet. To increase the margin on this sensitive component, we decrease the refresh time by dividing it by 2. This is also recommended by several manufacturers for operation beyond +85°C.

Power Supply Uprating for Increased Safety Margin to Noise and Decoupling Dips

The power supply to the CPU is specified to be 1.0 V with ±5% tolerance. Since ±50 mV is such a small margin to the recommended minimum operating limit, we may have unfortunate combinations of power supply tolerances, non-ideal decoupling capacitors, and power-intensive software—so-called hot code—that can cause the power supply to drop below its specification. This may lead to a failure in the CPU that might not be possible to measure and observe outside the chip by a supervisory circuit. To avoid this and gain extra margin we increased the typical value of the power supply to be 2.5% higher than the typical.

Case Study 6.4. Can Commercial Components Be Successfully Employed in Military Space Systems?

Contributed by Omar Facory
Director, Programs and Business Development, Aitech Space Systems, Inc.

It is natural for satellite equipment designers and users to aim for the same throughput and functionality that they find on their desktops, in their laboratories and in their complex ground systems. And, because space-qualified components do not always provide those levels of functionality, it is becoming more common that designers use commercial components to satisfy these design goals.

(Continues)

Case Study 6.4.   Cont'd

Experience has shown that starting with QML (qualified manufacturers list) components often yields the highest reliability. Yet, tight program schedules, increasing design complexity, and the need for improved performance and cost pressures have been driving the trend toward the use of non-QML, commercial parts.

For those systems where the use of commercial components are deemed an acceptable alternative, the selection process starts with commercial components that satisfy a minimal level of quality, and then parts are 100% screened. In fact, some parts are inherently radiation resistant, including:

- Silicon on insulator (SOI)
- Diodes (other than Zener)
- GaAs (gallium arsenide) technologies
- Bipolar devices with low-dose rate characterization
- Crystal oscillators
- Most passive devices

Verified by thorough and 100% testing, modern SOI microprocessors have an inherent transistor construction—eliminating the parasitic SCR—that resists total dose radiation effects and single event upsets found throughout space applications. This inherent survivability is equally applicable to the L1 instruction and data caches as well as L2 cache on the die, thus extending this radiation-tolerant robustness during full cache utilization. Error correction mechanisms are also built into the cache arrays to minimize data corruption from single-event upsets. Components of this technology type included in a space design are then typically de-rated and screened to provide added risk reduction by tightening design margins through additional up-screening and 100% unit qualification testing, as needed.

An added benefit of commercial microprocessors allows the use of the latest software tools and real-time operating systems that are commercially available, saving development time, improving the design and providing the added benefit of design re-use by lowering development time and costs.

Taking this concept to a higher level of organization, space systems engineers and mission managers are moving to space-qualified, commercial off-the-shelf (COTS) boards and subsystems. Single-board computers (SBCs), flash-based mass memory, and peripheral I/O cards designed and built for space applications allow design teams to complete complex, capable designs in a much shorter timeframe and with minimal expenditures (Fig. 6.16).

Manufacturers with this proven space experience, such as Aitech, provide a portfolio of space-qualified processors, peripheral cards and enclosures to give designers the freedom to implement flexible, modern architectures and the throughput to complete high-performance, highly complex systems. Companies well entrenched in the development of space-qualified, embedded computing products often offer extensive training to further reduce risk and improve their customers' time to market.

Commercial components pose challenges, such as frequently unannounced changes, lack of traceability and obsolescence. Through Aitech's extensive component engineering and

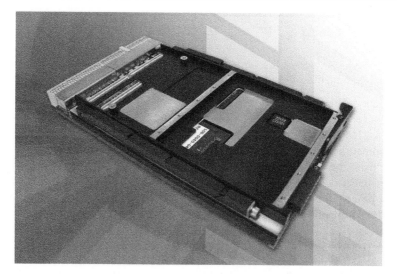

Figure 6.16: COTS circuit board from Aitech suitable for military applications. (© 2009 by Aitech, Inc. Used with permission.)

management processes, these and other issues are handled for our customers; therefore space electronics designs can be realized with much improved performance and functionality to meet demanding time to mission schedules with minimal risks.

Case Study 6.5. Joint Strike Fighter Alternate Mission Equipment Test Set

By Loofie Gutterman
Geotest-Marvin Test Systems, Inc.

Introduction/Overview

With the Joint Strike Fighter (JSF) program underway, many new airborne products will have to be tested and maintained for the next few decades. The JSF, also known as the F-35 Lightning II, is the future fighter aircraft for the U.S. armed forces and our allies.

Alternate mission equipment (AME) is the term used to describe military equipment that can be installed on, or removed from an aircraft to achieve specific mission requirements. AME covers military equipment such as aircraft pylons, missile launchers, bomb racks, and in some cases, auxiliary fuel tanks. AME is typically handled by armament technicians and test equipment is typically developed to address all AME associated with a specific aircraft. The test equipment is used either in the field or at the hanger where the AME is being maintained (back-shop). The Joint Strike Fighter is no different and will support a wide range of AME including pylons, launchers, and associated equipment and as a result, will require AME test equipment as well.

(Continues)

Case Study 6.5.   Cont'd

The Requirements

The following requirements were identified for the testing of AME for the JSF program:

- Portability—The test set had to be portable so that it can be easily carried by two soldiers (or war fighters) to the flight-line or from one shop to another.
- Ultra-rugged construction—The test set had to be rugged not only because of the environmental conditions it has to withstand, but also because of the anticipated abuse from the users. In other words, it had to be "soldierized." The environmental requirements of this test set are provided in Table 6.12.
- COTS-based—The new JSF AME tester had to be based on COTS technology to ensure that development costs are minimized and that it is not based on proprietary technology.
- Open architecture—The test set had to be based on an open-architecture to ensure that test instruments from different vendors could be used together and that the market could offer sufficient availability of products to prevent the need to design custom instruments.
- Small footprint—Other than being portable, the test set had to be small to allow placement on standard hanger benches and carts.

Table 6.12: Environmental Requirements

Environmental	Requirements
Operating low-temperature range	−40°C ± 5°C (−40°F ± 9°F)
Operating high-temperature range	+60°C ± 5°C (+140°F ± 9°F)
Storage low-temperature range	−45°C ± 5°C (−49°F ± 9°F)
Storage high-temperature range	70°C ± 5°C (+158°F ± 9°F)
Maximum relative humidity range	90% to 95%
Salt FOG	Per MIL-STD-810, Method 509.2
EMI	MIL-STD-461, Class TBD
Shock (half-sine)	Per MIL-T-28800, Paragraph TBD
Basic transportation	Per MIL-STD-810, Method 514.4
Loose cargo	Per MIL-STD-810, Method 514.4
Transit drop	Per MIL-STD-810, Method 516
Bench handling	Per MIL-T-28800, Paragraph TBD
Vibration (sine)	Per MIL-T-28800, Paragraph TBD
Explosive atmosphere	Per MIL-STD-810 (if applicable)
Max operating altitude range	10,000 ft ± 5000 ft
Max storage altitude range	35,000 ft ± 5000 ft

- Cost-effective—Like any other test system program, cost was a major consideration and the lower the tester's cost, the more widely used it will be, ensuring the increased readiness of JSF fleets worldwide.

Implementation

After reviewing available commercial T&M platforms, the PXI platform was selected since it met all six requirements. Additionally, the availability of an off-the-shelf generic PXI platform (Fig. 6.17) made the choice of PXI even more compelling. The MTS-207, forms the baseline test platform for the JSF AME tester or as it is officially known: the MTS-235-JSF AME STE (special test equipment).

The MTS-235 is based on the MTS-207, an ultrarugged PXI platform that features a 14-slot PXI chassis with seven 6U and seven 3U slots and room for custom electronics. of the 14 slots, slot #1 is used by the embedded single-slot PXI controller offering a Pentium 4–based 1.7 GHz CPU that operates over an extended temperature range. The CPU also includes a built-in flash disk, eliminating the need for a fragile mechanical hard disk drive that would not meet the environmental requirements.

Figure 6.17: The MTS-235 ultrarugged PXI platform. (© 2009 by Geotest—Marvin Test Systems, used with permission.)

The MTS-207 is field proven product and has been previously used on U.S. Air Force programs including the Maverick missile system and the AC-130 Gunship. As a Maverick Field Test Set (designated as the MTS-206, see Fig. 6.18), this ultra-rugged PXI platform has been qualified to operate at extreme environmental and EMI conditions that are more stringent than many other field testers used by our armed forces today.

(Continues)

Case Study 6.5.   Cont'd

Figure 6.18: The MTS-206 Maverick Field Test Set. (© 2009 by Geotest—Marvin Test Systems, used with permission.)

Test Set Configuration

The utilization of COTS products and in this case, PXI modules, was one of the requirements of this test set. This meant that PXI products had to be used in all cases where it was feasible. Indeed, most of the subsystems shown in Fig. 6.19 are based on COTS PXI products with very few unique requirements dictating custom circuitry.

As mentioned earlier, the PXI controller is an embedded single-slot, Pentium-based PXI controller. For a portable application, the fact that the controller occupies a single slot is an added benefit as the overall chassis solution is smaller and lighter (compared with the typical three-slot–wide PXI controllers).

Being a portable application, the use of a keyboard and a mouse was not feasible and the selected user interface was a remote control and display unit (RCDU), which is a combination of a rugged LCD display and a touch-screen. Using intuitive menu-driven displays created by the ATEasy Test Executive, the operators have full-control over test set functions and configurations and can access test data information when needed. The controller also provides standard peripheral interfaces including USB-2 and 1000BaseT Ethernet ports. Also provided

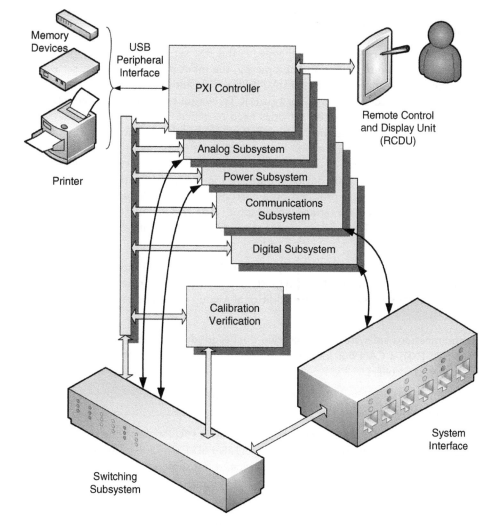

Figure 6.19: The JSF AME tester's block diagram. (© 2009 by Geotest—Marvin Test Systems, used with permission.)

are standard PS2 keyboard and mouse interface, which are only used for development purposes.

The USB interface allows connections of peripheral devices such as printers and USB thumb drives. This feature simplifies downloading of test data and supports updates to the test program(s) in the field via USB thumb drives. The 1000BaseT Ethernet (gigabit Ethernet) provides for the networking of the test set when required. One of the advantages of a networked test set is the use of remote diagnostics, which enhances the built-in test, and self-test capabilities of this JSF AME Tester.

(Continues)

Case Study 6.5.   Cont'd

The power subsystem represented a challenge, as this test set has to support MIL-STD-1760 requirements. To minimize weight and complexity, the input power to the test set was selected to be a three-phase source of 115V/400Hz. This allows the test set to switch the external power source to the units under test (UUTs) without the need to internally generate this power source. However, the test set does need to generate the 28VDC and 270VDC rails required by the UUTs as well as the power supplies required by the internal PXI chassis ($+5V$, $+3V$ and $\pm12V$ rails) and other internal custom circuitry ($\pm15V$ and $+28V$). Combined, these power supplies provide a total of over 1000W of power and while custom; they are still controlled by the embedded PXI controller that can monitor power supply voltage and current via additional custom circuitry provided on a custom PCB called the *power board*. The power board functions as the interface test adapter (ITA) and also includes all the high-power loads required by the UUT as well as special stray-voltage monitoring circuitry for the squib circuits.

The remaining instrumentation for this system employs COTS PXI modules. The analog subsystem includes a 6½ digit DMM (3U PXI) and a multifunction card with A/D and D/A channels (3U PXI). There are additional analog functions provided by the power board.

The communications subsystem includes two dual-redundant MIL-STD-1553 data bus interfaces (3U PXI), a CAN bus interface (3U PXI), and a RS-422 interface (supported by the embedded PXI controller).

The digital subsystem includes 128 digital I/O channels that are provided by the multifunction card (part of the Analog Subsystem).

The switching subsystem includes five switch matrix cards (6U PXI) and one high-current switch card capable of switching up to 7.5A per channel (also 6U PXI). One of the switch matrix cards is used for internal tester functions (BIT, self-test, and housekeeping) while the others are used for UUT testing. Additional high-current switching of up to 25A per channel is provided by the custom power board.

The system also includes an integrated calibration subsystem. Internal precision references are used to perform an automatic recertification/accuracy verification of the system. Traceability is maintained by verifying accuracy of the internal references on a 24-month cycle.

The COTS PXI instruments are accommodated by a 14-slot 3U/6U custom PXI chassis. The chassis employs a COTS backplane and a shock mounted, ruggedized mechanical assembly including heaters to ensure environmental compliance. Additionally, to ensure that the complete card cage assembly with instruments meets the system environmental requirements, all internal connections employ positive locking mechanisms or if necessary, the use of RTV (a silicon-based adhesive) is used to ensure reliable connections. Table 6.13 describes the 14-slot chassis and the instruments selected for the JSF AME tester.

Of the 14 available slots, all the 6U slots are in use, one by the embedded PXI controller and six by switching cards. The availability of the 6U slots played a major role in the switching density of the tester, providing a flexible matrix configuration up to 2 × 480 or 4 × 240 as well as 45

Table 6.13: The JSF AME Tester's PXI Chassis

Slot #	Description	Type of Slot	Function
1	Embedded PXI controller	6U	Computer that acts as system controller
2	GX6616 #1	6U	Switch matrix
3	GX6616 #2	6U	Switch matrix
4	GX6616 #3	6U	Switch matrix
5	GX6616 #4	6U	Switch matrix
6	GX6616 #5	6U	Switch matrix
7	GX6315	6U	High-current switch
8	DMM	3U	Digital multimeter
9	MIL-STD-1553	3U	Military serial bus
10	CAN Bus	3U	Commercial controller bus
11	Multifunction	3U	Analog and digital interfaces
12		3U	Available for future growth
13		3U	Available for future growth
14		3U	Available for future growth

Source: © 2009 by Geotest–Marvin Test Systems, used with permission.

high-current channels. The 3U side of the chassis is partially occupied by three 3U cards, leaving three 3U slots available for future growth.

The system's UUT interface had to be customized as well, as standard mass interconnect devices commonly used by testers cannot be used in the flight-line environment. Consequently, the system's interface consists of circular military connectors such as D38999 series III.

Test Set Maintenance

The JSF AME tester is provided with four maintenance capabilities:

- Built-in test (BIT)
- Self-test
- Remote diagnostics
- Calibration or recertification

The BIT is a power-up procedure which verifies that the test set's subsystems are operational, system power supplies output the correct voltages, and individual instruments are calibrated. It does not check any of the cables or the harnesses of the test set.

The system self-test utilizes a self-test adapter (stored in the test set's lid). The self-test TPS performs a complete comprehensive parametric test of the test set and all associated cables and is capable of providing troubleshooting information to the user in case of a failure. The test

(Continues)

Case Study 6.5.   Cont'd

set also provides the means for remote diagnostics via the Gigabit Ethernet port. This allows personnel in another location (or continent for that matter) to review parametric test data in real time and instructs the operators of the correct action to take.

The JSF AME tester, just like the Maverick field test set mentioned earlier (MTS-206) utilizes an innovative calibration approach. By employing a PXI card with high-precision calibration references, the test set's measurement instruments are checked (verified) against those references during the execution of BIT. A failure in any of these tests will prompt the execution of a calibration procedure that adjusts the test set's instruments to the required accuracy. The calibration verification card itself needs to be verified against NIST-traceable calibration equipment every 24 months. This period is based on current empirical data from other programs and will probably be extended in the future once data are available for the JSF AME Tester itself.

Summary

The JSF AME tester program is an example where COTS products and off-the-shelf testers are used to fulfill custom test requirements, thus reducing the development and life cycle costs of the test set and the time-to-market. It also demonstrates how innovative maintenance techniques can help reduce maintenance and life cycle costs of test sets.

Loofie Gutterman may be contacted at Geotest-Marvin Test Systems, Inc., 1770 Kettering, Irvine, CA 92614, (949) 263–2222, loofieg@geotestinc.com.

Case Study 6.6. U.S. Navy Mitigates Risk with Transformational Power and Propulsion Methods through Realistic Testing

By Edward Lundquist
Alion Science and Technology, Washington, DC

The Navy is testing new concepts in power generation, conversion, and distribution to make warships more efficient, economic, and combat effective. It's called the *integrated power system* (IPS), and it offers advantages of distribution and the availability of large amounts of momentary power for futuristic directed-energy weapons and rail guns. But as in the case of any new and highly beneficial technologies, they do not come without risk.

To meet the increased power demands for new sea-based weapon systems, next-generation surface combatants, such as the DDG 1000 *Zumwalt*-class of guided missile destroyers will feature IPS. On the DDG 1000, power will be generated by two large gas turbine generators and two smaller ones. Using efficient power management, that power is available to handle all of the electric loads throughout the ship, including potential future power-hungry weapons such as rail guns or directed energy weapons.

In a conventional configuration, the prime mover powers the propulsion drive train exclusively. About 90% of all the power generated on the DDG-51 guided missile destroyer in the fleet today is used for propulsion and cannot be transferred for other mission requirements. The remainder of the ships energy needs must be supplied by ship's service generators. IPS provides electric power to the total ship with an integrated plant. The prime mover powers a generator, and the drive train is powered by electric motors, with the flexibility to dedicate appropriate loads to the propulsion system, or other requirements. IPS enables a ship's electrical loads, such as pumps and lighting, to be powered from the same electrical source as the propulsion system (e.g., electric drive), eliminating the need for separate power generation capabilities for these loads.

The combat value of an electric ship goes well beyond weapon capability and capacity. There are significant efficiencies and redundancies. At full power, DDG 1000 will achieve speeds up to 30 knots. If one of the main turbines is lost, the plant can be isolated and still achieve 27 knots. Since a warship usually cruises at reduced power once it has arrived on station, normal station-keeping can be accommodated with the two small turbines to save fuel and reduce radiated noise. The power previously trapped in the propulsion train can now be directed to enhance combat capability and mission flexibility. At lower speeds, *Zumwalt* has a surplus of power that can be made available as needed. Further advantages include the elimination of maintenance-intensive and high-temperature auxiliary steam systems, reduced noise and vibration, and better fuel efficiency.

New Electric Motors

The key to IPS is new electric motors with high power density. The DDG 1000 will be powered by a Rolls Royce MT30, which is based upon the Rolls Royce "Trent" engine, which powers the Boeing 777 airliner. The aviation version of the engine has a demonstrated reliability of 99.98%. The "marinized" version of the MT30 has 80% commonality with the Trent 800, but is shock-mounted and has different blade coatings for operation in a saltwater environment. This engine is also serving today aboard the new Littoral Combat Ship, USS *Freedom* (LCS 1). *Zumwalt* will also have a smaller gas turbine, the Rolls Royce 4500.

DDG 1000 power generators produce 4,160 volts alternating current (AC), which is rectified to direct current (DC) and allows for ship service power distribution to be tailored to the ship's needs. There are three primary advantages to DC. First, DC uses solid-state power conversion that supplies loads which are converted back to AC, and is a cleaner way to supply power. Secondly, many of the combat systems' loads are DC. Finally, it enables power to be shared and auctioned. DC enables uninterrupted power even in the occurrence of a casualty.

Rigorous and Realistic Testing

The surface combatant IPS propulsion engineering development model (EDM) for DDG 1000 is being tested at the Land-Based Test Site (LBTS) at the Ships Systems Engineering Station in Philadelphia. The test site has been used to evaluate different configurations and motors, which mitigates the risk involved with newer technology applications. The test program validates key system metrics such as torque, speed, and power output, and specific fuel consumption for the various configurations, and runs them under realistic loads to assess reliability.

The Navy has tested the 18-megawatt (MW) advanced induction motor (AIM), which will be the baseline for DDG 1000, produced by Alstom, at the LBTS. This is essentially the same system installed on the Royal Navy's new Type 45 destroyer, HMS *Daring*, which has just been commissioned. The IPS features Integrated Fight through Power (IFTP), a fully automated DC

(Continues)

Case Study 6.6.   Cont'd

Zonal Electric Distribution System (DC ZEDS) that provides flexible, reliable, high-quality power to all shipboard loads. Other configurations are also being tested. The IPS system is fully automated with little operator intrusion. The testing at the LBTS will validate that the DDG 1000 IPS will automatically take appropriate corrective action if there is a malfunction or casualty without the input of an operator.

Engineers at the LBTS have also tested a 36-megawatt permanent magnet motor (PMM), developed by DRS Technologies. PMM has greater power density—which means it's smaller for the same amount of power and less weight—than the AIM, and may be used in future ships.

DRS Technologies and General Atomics Electromagnetic Systems are developing a hybrid electric drive that permits a smaller service gas turbine to power a permanent magnet motor that can power the ship at slow or "loiter" speeds. Using a smaller turbine can result in significant fuel savings. Furthermore, the motor can be reversed to function as a generator when propulsion gas turbines are online.

The Next Big Thing?

Since some materials are much better conductors at very cold temperatures, with virtually no electrical resistance, super cooled conductors make for much more efficient motors. Superconducting wire can carry more current and generate higher magnetic fields in very small areas, and can result in a significantly smaller motor.

American Superconductor and Northrop Grumman have recently tested a 36.5-megawatt, high-temperature superconductor (HTS) ship propulsion motor at the LBTS. The motor uses HTS wire that can carry 150 times more power than copper wire used in more conventional motors. The advantage is more compact propulsion systems that have greater power density. Superconducting wire can carry more current and generate higher magnetic fields in very small areas, and can result in a significantly smaller motor. In other words, more power is available from smaller, lighter motors. That means Navy ships can carry more fuel and munitions and have more room for crew's quarters and weapons systems.

General Atomics' (GA) superconducting DC homopolar motor for propulsion applications is small and light compared to traditional and superconducting AC motor systems. This motor uses low-temperature supercooling that employs gaseous helium to maintain the superconducting wire within the motor at 5 degrees Kelvin, which is almost absolute zero. Since some materials are much better conductors at very cold temperatures, with virtually no electrical resistance, super cooled conductors make for much more efficient motors. A comparable high-temperature super cooled system operates between 40 and 75 degrees Kelvin, depending upon the technology chosen. Refrigeration at higher temperatures is easier, but the high temperature superconducting material is not as easy to produce and is much more expensive than the superconducting niobium-titanium wire in the low-temperature motor. Niobium-titanium wire is the most widely used and available superconducting wire in worldwide commercial applications.

GA has built a 5000-HP motor that is 4.5 feet in diameter. This technology is slender, light, and fuel efficient and can be more readily adapted to propulsion pod applications.

Additionally, while superconducting AC motors have similar costs to the superconducting DC motor, there is no need for power inverters and the associated electronics to switch it to AC.

Edward Lundquist is a retired U.S. Navy captain and a senior science advisor with Alion Science and Technology.

Case Study 6.7. Integrating Systems Engineering with Earned Value Management

By Paul J. Solomon, PMP

Earned value management (EVM) is *capable* of integrating a project's cost, schedule, and technical performance. However, the EVM standard (EVMS) lacks guidance for integrating EV to technical performance. Fortunately, systems engineering (SE) standards fill the gap. EVM data will be reliable and accurate only if the right base measures of *technical* performance are selected and if progress is *objectively* assessed.

Standards

EVMS has a quality gap because it states that EV measures the quantity, not quality, of work performed. SE standards focus on product requirements and using technical performance measures (TPM) to determine progress. Table 6.14 shows pertinent guidance follows from:

- IEEE Std 1220 Standard for Application and Management of the SE Process
- ANSI/EIA-632 Processes for Engineering a System

Table 6.14: IEEE Std 1220 and EIA-632 Standards for Systems Engineering

IEEE 1220
6.8.1.5.b. TPMs are key to progressively assess technical progress. Track relative to time with dates established for when: Progress will be checked. Full conformance will be met. Key technical parameters are measured relative to lower-level elements of the breakdown structure by estimate, analysis, or test, and values are rolled up.
6.8.1.5.c. Cost and schedule performance measurements are integrated with TPMs to: Provide current schedule and performance impacts. Provide an integrated corrective action to variances identified.
6.8.6. Track product and process metrics Metrics are collected, tracked, and reported at pre-established control points
ANSI/EIA-632 Processes for Engineering a System
4.2.2, Requirement 10, Progress Against Requirements Identify product metrics and their expected values that affect the quality of the product and provide information toward satisfying requirements, as well as derived requirements.

(Continues)

Case Study 6.7.   Cont'd

Use of Product Requirements

Traditional EV measures only progress toward completing the set of enabling work products, such as drawings or code. Perform-based earned value adds progress toward meeting the product requirements in the technical baseline.

Example

In the example (Table 6.15), the work package is the design of a component of a subsystem, a set of wire harnesses. EV is based on completion of the enabling work products (drawings) and meeting requirements. EV, based on completed drawings, is reduced if the requirements are not met when planned (see Table 6.16). There are two TPMs with planned values (PV) as follows:

- Maximum weight: 200 lb.
- Maximum diameter: 1 inch

The allocated budget and EV penalty follow:

- Budget per drawing: 40 hours
- EV penalty if PV not met:

Table 6.15: Schedule for Meeting Requirements

Schedule	Jan	Feb	Mar	Apr	May	Total
Drawings	8	10	12	10	10	50
Requirements met:						
Weight				1		1
Diameter				1		1

Table 6.16: Net EV Based on Component Requirements

Design (drawings)	Jan	Feb	Mar	Apr	May	Total
Planned drawings	8	10	12	10	10	50
BCWS—current	320	400	480	400	400	2000
BCWS—cumulative	320	720	1200	1600	2000	2000
Actual drawings completed	9	10	10	12		
EV (drawings)—current	360	400	400	480		
EV (drawings)—cumulative	360	760	1160	1640		
Negative EV				−100		
Net EV (drawings and requirements)				1540		
Schedule variance (SV)	40	40	−40	−60		

Component weight requirement (req.): –100 hours
Diameter required: –200 hours

The schedule status at April month-end follows:

- Cumulative drawings completed: 41
- Diameter requirements met
- Component weight requirements not met

The time-phased budgeted cost for work scheduled (BCWS) is positive EV based on completed drawings, and negative EV when the design fails to meet PVs. (See Table 6.16.)

In this example, TPMs were used at the component or work package level. However, requirements are usually established at higher levels. When completion of the component design depends on achieving PVs at a higher level, EV at the component-level work package should be dependent on meeting both the component and higher-level technical performance.

Conclusion

Engineering standards fill the quality gap in EVMS by providing guidance to link EV with technical performance and meeting the product requirements.

References

This article is an abbreviation of the article published in the Project Management Institute's *Measurable News*, Fall 2008. The article and additional information on Performance-Based Earned Value® are available at www.PB-EV.com.

Case Study 6.8. Case Study of Case Studies: Lessons Learned in Systems Engineering in the Commercial Aviation and DoD Markets

By Paul E. Gartz

Why Do Programs and Projects Need Systems Engineering?

For decades people have struggled to define "systems development" methods in efforts to increase efficiency and effectiveness. Many programs and projects found benefits in what is called "systems engineering" (SE) and significant setbacks when they didn't use "systems engineering." Why? Large-scale systems projects are very expensive, are much more complex technically and managerially, are risky, take a lot of time, and often don't deliver as expected (i.e., both over-runs in schedule and budget and under-runs in performance).

When projects exceed the right combination of size and complexity, sophisticated, multilevel coordination methods are needed to ensure that the entire program or project concept comes together as expected in both technical and business aspects. Technical projects are analogous to music: while a string quartet can self-organize, a Wagnerian opera requires a conductor,

(Continues)

Case Study 6.8.   Cont'd

stage manager, choir leader, and set designer, among others, and they must all work together—so too with managing a technical project, as many different disciplines need coordination and must work together.

So why has this higher level of coordination—or systems development—been so hard to accept, deploy, and sustain? There are many reasons but when you cut through it all it is often found that the "stovepipe" systems or "disciplines" resist intrusion, or management feels that SE is wasted overhead, or management simply has no experience with the negative consequences of not coordinating the systems development.

The bottom line is that the discipline of SE and the developers (and managers) of large-scale systems need to communicate better in both directions. The process community needs facts and case studies to convince upper management of the value of SE. Technical organizations need to be more open to tailoring based on project characteristics and on team maturity.

"Best practices" that have been shown effective both in technical realms and in project management over a domain of systems can arise and evolve in organizations, companies, and industries. Since at one level "a system is a system is a system," it should be possible to develop some level of best practice across all system types and sizes. This has been the holy grail that I have sought for over 30 years. What I have found over my career is the tremendous opportunity for this relatively common process for system development over many types of systems. The two boundaries that proved difficult to work are (1) across "software/IT systems" and large systems employing these as subsystems, and (2) across defense and commercial systems at almost any scale level. It would still be very useful to have common processes, tools, and methods and a way to tailor processes.

Agility, Stability, and Control and Convergent Projects

The key "sales point" might be considered "stability and control" for the project (or program). In other words, "How does one exercise project control so that the program stabilizes and is controlled to deliver what it promises?" We can extend this concept a bit further by way of "Forty Second Boyd" (USAF Colonel John Boyd who added "agility" to stability and control for fighter jets as a key criteria for system design) by making the key to be a project's "agility, stability, and control." Too much stability in general and especially at the wrong point in the program limits creativity and proper closure. This is where the agility part is needed—the ability to be creative, do trade studies efficiently, and change the requirements and design. At another point in the program/project if there is too little stability, the program becomes chaotic with massive, positive feedback loops that self-amplify problems.

To rephrase the point, what is desired is a "convergent" project where unknowns and risk decrease with time to meet program cost, schedule, and technical targets. An overly constrained program converges too quickly to a less-than-desirable final state and cannot deliver the best solution because it limits creativity and flexibility. Alternatively, a program with too little stability and control mechanisms, especially in the mid-to-later stages of the program/project becomes unstable and diverges from the schedule, budget and technical envelopes—and does not deliver on commitments.

Two Key Skill Bases

The two skill bases that have evolved to increase the likelihood of convergence are "systems engineering" for the technical aspects and "project (or program) management" for the cost, schedule and sequencing aspects. The earliest efforts in SE were in submarines and fire control systems followed by aerospace projects where complexity and risk needed comprehensive organization. Today SE has spread to many other industry sectors and is an enabler of large increases in functionality and interconnectedness and shareholder value.

Project management can be traced back to the building of the pyramids. It has a long history of advancement by the construction industry including civil, mechanical, and architectural engineering. The Project Management Institute (PMI) has a "body of knowledge" (PMBOK Guide) on the subject (see www.pmi.org/Resources/Pages/Library-of-PMI-Global-Standards-projects.aspx).

Beginning 50 years ago, DoD programs actively promoted, required, and trained the services and defense contractors in SE excellence. Degrees and certifications in systems and software engineering and many tools and theories have emerged in recent years. Commercial programs have become much more complex in all domains due to extensive software, much cheaper computing power, and greater data bandwidth. Yet in spite all of this advancement the DoD has reported regression in program performance across the services and contractors; it has been traced in part to regression in SE usage. Commercial programs have also seen similar issues in the aviation sector. What is going on?

This "case study of case studies" does not look at a single project but rather discusses key concepts that underlie the field. It also recommends some proven, simple best practices, and briefs the applicability of "systems-of-systems" characteristics and principles that directly apply to mission-assured systems. This new field uses COTS (commercial off-the-shelf) systems.

So What Happened?

As mentioned above there were different trends and timing in the various market sectors. A few of the key events and occurrences follow.

DoD and Defense Contracting

As pointed out earlier in the chapter, the DoD had mandated processes for developing systems. These required certain steps and documents such as a SEMP and requirements. They used a "waterfall" (or top–down) process, with a Descarte-style decomposition, and had processes to ensure verification and validation. Many defense contractors complained about the onerous nature of the standards and that they could be a lot more efficient if these were lessened or removed. Also, as pointed out in this chapter, the DoD lost leverage, and therefore control of its supplier base, when the volumes of commercial applications far exceeded military volumes. This situation forced the move to COTS. In hopes of correcting the loss of military influence upon suppliers in producing needed components, Defense Secretary Perry removed many "mil-spec" requirements with the intention that defense contractors would embrace commercial best practices, products, and lessons learned and thereby increase their efficiencies. (See the beginning of this chapter for a good review.)

(Continues)

Case Study 6.8.   Cont'd

Unfortunately, the "counterintuitive behavior of social systems" occurred rather than best practices [51]. Instead of the defense contractors adopting commercial best practices, they generally went a different way with the new DoD freedoms. Most defense contractors were not and are not familiar with commercial sectors, their processes, their different market drivers, or their attitudes. The model was "the DoD says I must do 'x,' so I do 'x.'" It wasn't the commercial model that says, "I must invest my own money so I must find the most efficient methods and succeed in meeting customer expectations, or I will lose money and possibly go out of business." In both cases, however, upper management put pressure on lower project managers to reduce costs to meet a contract bid and increase profit margins. So as a rule and across all service contractors, SE often became viewed as a needless cost center versus a risk reducer and a way to have good program "agility, stability, and control." Project performance went down across all DoD services and one of the major contributing factors was the reduction or even abandonment of good SE principles, one example of which was system requirements.

Commercial Aviation

Commercial aviation began to embrace SE around 1980 as the complexity of going from analog to digital avionics allowed the networking of subsystems into larger, higher-level systems such as flight management systems. Some companies created whole new SE and integration organizations and sought to capture and extend best practices. Progressively, each new program absorbed the lessons learned and extended the methods beyond avionics into other major airplane systems including the use of common best practices. Why? Simple! They found that it contributed value. Systems engineering and best practices reduced the program risk by forcing risk to surface earlier and become visible across the airplane's entire development versus only locally in small subsystems.

By the beginning of the 21st century, SE practices were used quite widely for the technical side. The development of the Boeing 777 served as an excellent example that was filmed in a PBS series; it was an excellent success story that brought the concept of "working together" to a higher level [52]. In SE terms this was the people and process side of working the interfaces early, often, and systemically. Since this progress was based upon proof of business value and had historical antecedents over more than two decades, it continued in certain commercial aviation sectors. Defense aviation contractors who had been told they had to do it, conversely had much less motivation in their culture when they were no longer told they had to do it. All of this is a bit ironic as the defense community most proactively pushed the concept of SE practices; furthermore, the Defense Systems Management College was known as a chief proponent of SE practices.

However, another event occurred in the commercial sector. Transport jets were not producing the desired profit margins and, one could argue, that they did not have the margins needed for such a high-risk business. Margins varied between 5 and 9% for commercial transports versus drug companies, which had profit margins in the 35% range. The risk-reward ratio was off for commercial aviation companies, and they found that they could make more money from investments in U.S. Treasury notes, than in launching a new airplane. Again this is due to the very different business model of commercial markets versus defense markets in the same technology field. Commercial markets are driven by the airlines, which are driven by the consumers, who largely want the least expensive airfare, and by the economy. Commercial markets are highly cyclical based upon a lot of discretionary spending.

This drove endless bankruptcies, mergers, and acquisitions after governmental deregulation of the airline industry began in the late 1970s and the 1980s. To stay in business transport manufacturers had to look at both reducing costs and new partnering arrangements; in other words the "project management" side of the equation. This resulted in reducing or "leaning out" some of the oversight functions that ensured project success. Boeing and Airbus, the two main players, did this "leaning out" differently but both manufacturers had unprecedented problems and delays.

What Are "Systems" and the Discipline of Systems Engineering?

The development, operations, and analysis of any system infer the study of "systems." It is the field of "wholes" versus parts. A systems engineer first puts a black box around all the contents of the system and then asks about total performance. The engineer then progressively opens the box in layers to see if it will deliver. When parts come together in special ways they give rise to completely new and often unpredictable "emergent behaviors." This is what characterizes them as a "system" rather than a group of parts and determines the need for a specialized discipline. For example, who could have predicted that an assemblage of fabric, wood, and metal could defy gravity, lift humans into the air, and wing skyward before it was done in 1903? Furthermore, predict that the flimsy contraption would then transform the world of transportation and intercultural awareness in the 20th century?

Since similar activities often share some levels of commonality we can therefore talk of "common and best practices for processes, methods, deliverables and tools"—in other words, a discipline. Such a discipline can be codified into a body of knowledge and trained through degrees, training courses, and certifications in well-known, traditional methods.

On the other side of commonality is uniqueness. Each market, system, and team for a "mission-assured" system is different. The demands of a particular market require developers to tailor the discipline and best practices of SE. Excellent "best practices" allow for and specify not only a norm but also variations to the discipline. Some project characteristics that vary are the domain (e.g., health care, aerospace, energy and power, Earth observation and prediction), the system type (e.g., control, sensing, display, IT), the funding structure (e.g., industrial, government, academic), and the size of the project (small and self-contained to large, multiorganizational to global, multinational). In each combination there is almost always a difference in the language used and sometimes a difference in the "language" used within the team and in the ways of thinking about their activities. These differences raise the need for tailoring processes. Here "language" can mean transnational cultural differences or bringing together people from totally different backgrounds who do not understand each other's ways of thinking about the world.

All this being said, I have found the key characteristics around which tailoring is required are size, complexity, risk, regulation, and whose money is involved.

(Continues)

Case Study 6.8.   Cont'd

Program Variables Affecting Best Practices and Tailoring Needs

Size

Growth in a system's size and scope requires proportionately more rigor in SE. In commercial aviation, for example, a single airplane can require millions of parts, which come from more than 10,000 global suppliers; the aircraft development can require tens of billions of dollars in development cost spread globally; its development will bring together a staggering diversity of perspectives and global standards. The aircraft development can also risk the very existence of the development company and many of its suppliers if the product fails. Finally, product failures can risk hundreds of lives per event and, more recently, risk thousands of lives and hundreds of billions of dollars of loss if the product is intentionally used wrongly, such as in suicide attacks. This type of development requires attention to layers upon layers of SE and project management in all life cycle phases from concept to development to manufacturing to operations to eco-appropriate retirement.

Contrast this with the development of a pilot-static sensor system used on the same airplane and developed by a single organization in a single company. This system is still subjected to the same certification processes and must fit into the overall larger, airplane program, but it is a much simpler project requiring less orchestration and therefore a lower level of SE. But still an appropriate level of SE is needed.

Regulation

Regulation is the government's way to ensure safety of citizens and reduce risk to the economy. Regulation includes safety of life but also, as seen in 2008 and 2009, regulation can also include the safety of finances. Title 14 is the U.S. Federal Aviation Regulations; it covers commercial aviation design, manufacturing, and operations. Simply put it says that the greater the risk, the greater is the government oversight and the greater the burden of process, proof, record keeping, and liability of both the system developer and the operator. The main program or project "control" for DoD systems is different. But in all cases these regulations and controls are essential to public safety and the public's trust in products and services. Public needs can be complicated and must be addressed by good SE processes.

Funding Source

There are many funding models, but one of the determinants that affects the system development approach is whether it is an RFP-driven-and-funded system development or a privately funded one. RFP-driven projects are typical of U.S. DoD projects and most government projects. Development of an iPod or 777 is typically privately funded.

For a RFP-driven government project the winning company usually has a solution at some level specified to them. The company usually makes money at every phase of the project by meeting specified deliverables. Usually, the risk to the company is relatively limited when compared to the massive losses possible in privately funded projects. The contractor almost never loses much of the value of the contract as often this is partially accepted by the government contracting agency. These comments may not be true if the RFP is in the commercial-to-commercial sector.

For private projects, the company funds the entire development and can lose all of its investment if the product is unsuccessful or if performance is below expectations. In commercial aviation, customer expectations are often more than what is understood by the buyer or seller, which causes enormous additional risk that influences all of the development processes of the project. Proven and best practices can address each of these issues. When compared to an RFP-driven project, this much higher level of risk in a market, which can evaporate, vastly complicates the SE processes by an order of magnitude. The trick is to link together the market process, the technical process, the business process, and the uncertainties and then do iterative trade studies across all three process areas simultaneously.

Macro Program and Project Uncertainties (Risks and Opportunities)

Risk is the negative side in the larger subject of uncertainties. The positive side is opportunity. Here again the difference between the approach to uncertainties for either commercial or government RFP-driven programs is quite dramatically different from severe risk to great opportunity. The reason again is the different types of money and the different types of markets they address—each must adapt to its environment. The commercial side pays much more attention to both since they are risking their own money at large risk to gain large market opportunities if the timing is correct. While defense projects always look at both, their opportunities take the view of additional sales of the same product rather than being innovative game changers.

Managing uncertainties is an essential part of both good SE and good project management. Risk awareness, analysis, and mitigation skills are essential to any endeavor whether done intuitively in a small business or with sophisticated techniques in a large organization. Techniques from "seat of the pants" to "risk cubes" to sophisticated "decision analysis" add billions of dollars of value to some commercial businesses. Because their investment and even their company's future existence are at stake in a commercial, privately funded case, this subject is extraordinarily important to them.

If a commercial product has never been done before or never proven or if it is radically different, then the risk increases exponentially over more conventional efforts. There are many reasons including the market, the technology, and the team's capabilities. If the market's requirements or expectations or timing are not known, a company could be developing the wrong product or have the wrong timing.

A simplified way to think of top program risks is that it is the sum of the independent categories of risks raised to an exponential power; the set of interactions of the risks is itself a higher risk. Examples include:

- The market is not known by the team.
- The technology is not known by the team.
- The system or the service has never been done before anywhere.
- The team employs multiple "cultures" and these cultures have never worked together before in this system context.

If all of these are *yes*, then the risk may easily be 4^2 or 16 times higher than if all the answers are *no*. Naturally, one can argue whether it is a linear or exponential function, but this brief calculation

(Continues)

Case Study 6.8.   Cont'd

should be a wake up call to senior management and senior technical leaders to perform thorough risk analysis and to support continuing risk monitoring and mitigation. Large, complex projects should quantitatively examine risk and opportunities for the elements of decision analysis through methods such as "tornado charts." The example below (Fig. 6.20) shows that in a hypothetical service business the most critical variable is the initial fee charged followed by five more or less equal but far less important parameters of the service business. This quickly informs the team that the fee needs a deeper and more thorough analysis. Although this is a business example the same can be applied to technical risk sensitivities.

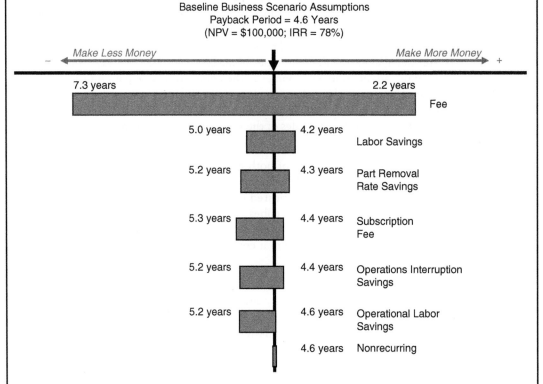

Figure 6.20: A decision analysis "tornado chart" that assesses sensitivity to uncertainties.

Enormous risk is often much less for RFP-driven business because the customer has said what they are committed to in the contract. Yet the DoD by its nature is often pushing the technology envelope and often in such large steps that the risks in this category may be high. Many commercial sectors will not take these risks but instead will perform more incremental development to test the market and ensure cash flow by funding the future from the sale of early units.

Best Practices

People, Processes, and Performance

It is important that people, processes, and performance all come together to achieve program success. The product or system must perform just as must as the business success of the venture. Any integrated system requires processes to ensure that everything in a system integrates together. This means that the organizations and people must work together in the same way as the system or service. People are the ones who make both happen. They are human, not machines. The right program skills are needed and they need the right type of incentives, management direction, and motivation. Sometimes "stretch goals" can be helpful to team spirit and to individuals' sense of achievement. Silly goals such as providing impossible schedules that violate all known benchmarks usually only serve to discourage everyone and cause people neither to stretch nor to work together.

Some Best Practices

Good SE and "systems approaches" do not view tasks as isolated; they work across task boundaries. Many lessons have been learned and oversights addressed in aerospace best practices. Unfortunately, they are often forgotten by the next program or next generations of program managers, project managers, and senior technical people.

First, understand the different "levels" of "systems" that exist in the final system from largest to smallest. Second understand the various "life cycles." Third, understand which parts or subsystems already exist (e.g., COTS) and the interrelationship between them; this gives rise to the "integration" of the system. If done properly and early, understanding these will dramatically and positively affect how a program is organized and its effectiveness and efficiency in the costs and in the time to develop the resulting system.

Life Cycles

The life cycles of a program can be many. A life cycle is a sequence of tasks, activities or operations that repeat over time; it is not exactly a schedule that attaches specific dates. A life cycle has a number of phases within it and each phase has tasks. The larger the project, the greater the levels of these tasks within the life cycles; there can be "sub–life cycles" within the larger ones. A typical example of the various life cycles of a product or service can include:

- Research life cycle
- System development life cycle
- Manufacturing life cycle
- Customer introduction life cycle (including training)
- Operations life cycle
- Customer support life cycle
- Retirement

DoD, NASA, and commercial ventures use life cycles because they provide integration efficiency for repeated tasks. However, organizations use life cycles very differently. One way is for each life cycle and phase to be worked independently of the others. This can cause much heartache, cost, and deficient system performance when viewed across all of the life cycles. A classic example is to not consider nuclear waste disposal before designing a system that uses nuclear fuel. Another is for engineering to not work together early with manufacturing resulting in extra costs from an unbuildable product. Aerospace manufacturing complains about engineering not talking to them

(Continues)

Case Study 6.8.   Cont'd

about requiring a milling operation when a forging would do. Another classic example of this type of program failure is the engineering community not understanding the operational environment and the pilots thereby leaving a cockpit design with high workload during an intense phase of operations when pilots' lives are at stake. A more local example within engineering is the failure to coordinate among various disciplines for the optimal design. The worst oversight, especially for a commercial entity risking its own money, is to not understand the market, customers, and variability resulting in developing the wrong product goals or timing.

A more subtle way to use the life cycles is with incremental development and "spiral" development, which can deliver a product to the customer earlier for test, review, or even sales. Field use then feeds back for product and process improvements. Spiral development is essentially identical but produces a final and complete system. First, it takes a high-level look across the various life cycles to understand the gross needs, which include requirements, designs, test, training, customer introduction, and operations. This way the whole team understands where the constraints are very early. Then the team should understand the integration requirements and priorities of what is most important and then it can plan better where to focus resources. This sequence of events can get the most bang for the buck in project stability and early risk reduction.

COTS and Pre-existing Subsystems

Another key is to understand which parts of the end system already exist (e.g., COTS or reuse of other designs) as the SE of each needs to be handled slightly differently. For example, if COTS are handled like a top–down design where the supplier is under your control, the project will likely be heading toward massive rework and tensions or failure.

As mentioned, best practices applied to each of the above elements need to be defined in "layers" of significance instead of a rigid set where all are implied to have equal significance for all program and project types. One such set was developed over more than two decades in the commercial aviation field on a variety of large-scale, integrated systems and service programs. Most of these elements apply to small projects, too, but on a smaller scale.

The most important concerns are to assess the scope, complexity, and size of the project relative the organization's experience in the required technologies, in the market, and in the specific application. If there is a large mismatch in any of these, a serious and honest risk assessment and mitigation plan is needed. One convenient tool is CMMI on a scale from 1 to 5. An example is that if a program or project requires a sophistication of CMMI Level 4.5 but the organization only has a CMMI level 2.5, this should serve as an immediate wake-up call to management.

Program goals must be made clear—particularly the top-level completion criteria, in both a technical and a business sense. This seems obvious but many program and projects trip up on this. What this means is that all team members should know what the "final exam" questions are early in the project (e.g., main test criteria for acceptance). A development team needs the skills and experience that match the scope, size, and risks of the project. This may seem so very obvious, yet I have seen major programs in major companies staff a large-scale systems integration project with no systems integration people or no systems integration tasks when this is the main purpose of the organization. The team should have somewhat matching capabilities in both technical and management skills.

The project needs a single, integrated schedule for the whole broken down by levels and important "events." It also needs a single, integrated, top-level technical definition of the product or service that is understandable to all. An example of being understandable is the geometry of a product in a CAD/CAM data set; one can zoom in or out of "viewing detail" from the whole to the smallest design detail. The same is needed for all the technical elements of total system—requirements, designs, trades and test/verification.

The "Vee" Model and Company Intellectual Capital

Historically, organizations needed a practical, simple yet integrated process and model to pull together the above concepts. Next, they needed to deploy it and get it accepted by both management and technical staff. Process models began separately in the DoD and commercial aviation in the late 1970s and early 1980s. Both came up with a "vee" model, but there were differences. While both were useful, the commercial model seems to have more staying power. The commercial model had at its core the concept of "intellectual capital" of the company, which the DoD model did not. The key question answered by the commercial model can be summed by asking a corporation's VP of Engineering what his job is and how is he rated by the CEO. The answer should be "to create intellectual capital through timely, effective, and efficient market-focused designs that increase shareholder value." This means a producible portfolio of "design data sets." So a process was needed to do this.

The commercial "vee" model was a cornerstone. Each level of the "vee" and each element was simply a "view" into this data set with it being seamlessly integrated so all elements were traceable. To do this, each level of the "vee" was a "system" unto itself and therefore each level could use common SE processes and deliverables. This meant identical document and data set templates so traceability was not a separate many-to-many matrix but a simple same section-to-same-section leveling of the information. The analog is a CADCAM data set or more recently Google–Earth that allows one to zoom in and out.

Since projects are more complicated than geometry, the project needs to facilitate this "zooming" by sorting out early the different levels and types of systems that make up the whole and those of the systems with which it interfaces. The levels are not defined randomly, but rather *each level must be a system unto itself* meeting the definition of a "system" versus just a collection of parts. Once this is done then each level of development *is* a system and one can apply the same SE life cycle processes, deliverables, tools, and documentation schema to each level. If the documents are structured to a common template, and then each document section at each level is a "zoom-in" or "zoom-out" of the same document section at a higher or lower level, thus enabling an integrated data set for the project. This hugely improves integration and traceability and eliminates the need for traceability matrices. Note: the term "document" is used for convenience. The correct term would be "data set" with a "document" being the simplest version of this. The "vee" spreads these layers out into deliverables at each level with the left side being "systems definition" (requirements, designs, and trade studies) and the right side being "systems verification." The "vee" then simply became a convenient way to divide the responsibilities. It did not matter whether the sequencing was done top-down as in the DoD "waterfall" or bottom-up or inside-out or outside-in or "sandwich." This "vee" was neutral to this and could therefore be used and add value even if the program didn't start in the optimal sequence. The key is to stabilize the upper left levels of the "vee" before going

(Continues)

Case Study 6.8.   Cont'd

to the next level. This means configuration control and change management must be instituted (see below).

Second, each level the "vee" must have a complete but simple and integrated set of deliverables (Fig. 6.21). A proven method uses "R," "D," and "V" deliverables for capturing the requirements, designs, and verification means at each level. "Verification" here is used somewhat differently than traditional Mil Spec 499B usage. It means to "verify" that the requirements and designs were implemented properly. This is done with walkthroughs, tests, and simulations. In the simplest form, these deliverables can be a single document each. On more complex programs and projects each of the three may be an entire set of deliverables. The quality of each deliverable is rated by the user(s) of the document, that is, the "customer" performs this rating and not the developer.

RAAs

For project management it is essential to use the same life cycle deliverables ("R," "D," and "V") to develop the statements of work (SOWs), work breakdown structures (WBSs) and the roles, responsibilities, accountabilities, and authorities (R-RAAs) for these tasks. Project management can throw a program into chaos by developing WBSs and SOWs on a schedule that does not consider or understand the technical levels nor develops clear RAAs.

Program Stability via Configuration and Change Management

To ensure stability of the program, the key deliverables must themselves be stabilized. A good way to do this is through configuration management and change management (CM^2). The first step is to identify the key deliverables "R," "D," and "V." These determine the main technical structure of the product or service development life cycle. Next is to determine at which program or project "gates" or "events" these deliverables are put under CM^2. Typical gates are critical requirements review (CRR) and critical design review (CDR) for "R" and "D" deliverables, respectively. Once under configuration control they can only be changed by a change management board where all key stakeholders get visibility and perhaps a vote. At some

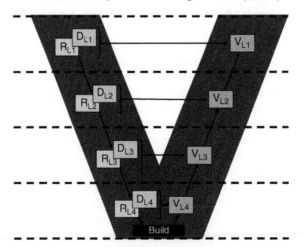

Figure 6.21: Commercial aviation "vee" model for intellectual capital. L1–L4, system levels; R, requirements; D, design; V, verification.

point many projects also "freeze" the deliverables so they cannot be changed unless by a very high authority (e.g., the project manager) to assure what is to be delivered is known. These processes are designed to gain inertia to minimize changes late in the project.

Systems-of-Systems

An emergent concept is systems-of-systems (SoSs). Why should you care about SoSs? Because they will affect the architecture of the mission-assured and other systems on which you may work by setting new contexts and requirements; they will also change and provide new funding sources. Your company may be large enough to take advantage of the concepts to participate in these. Finally, the concepts, processes, and skills used for complex SoSs have applicability to smaller-scale programs and projects.

The term was coined by the DoD in the late 1990s as a means to get more war-winning effectiveness from their investment dollars with the right combination of equipment and field tactics. The DoD did this by looking at the whole mission before settling on what investments give the biggest bang for the buck rather than simply buying another good-sounding platform— in other words, to globally optimize rather than locally optimize.

An SoS has certain characteristics that apply to some of the smaller, mission-assured systems discussed in this book, especially with regard to DoD COTS. An SoS consists of a number of stand-alone, self-contained, major systems (e.g., platforms, C4ISR, or a commercial transport linked into airline operations and air traffic management) that together constitute a higher-level system. The goal of SoSs is to "re-imagine" and then "reconfigure" the interfaces of all these systems of a SoS such that the total performance in operations and costs is improved. SoS examples include Global Earth Observation SoS (GEOSS), National Security SoS, Global Telecommunications SoS, North American Power Grid SoS, Global Healthcare SoS, and Transportation Traffic Management SoS.

An SoS also has the characteristic that most of its component systems already exist following years or decades of investment by governments and industry. An SoS does not have the typical top-down development. It is a top-down analysis of the existing SoS performance and the weak links to determine how to use all the existing systems but intervene with minimal cost and disruption to get considerably higher performance with minimal investment. SoSs are a large subject unto themselves. They are rare and very high leverage and a lot more complicated in every dimension than normal large-scale systems.

The critical concept here to the DoD and to major companies who are conceiving smaller SoSs is that a SoS starts by using *existing* system components. This turns the traditional SE approach upside down. The DoD's desire to use COTS is such an application independent of SoSs. One must start with the COTS available and figure out how to use them to enhance operational performance in the larger system rather than starting with a clean sheet of paper and only working downward. I have found this to be a particularly difficult concept to grasp for certain systems engineers. So more training, education, and lessons learned are required in terms of how to do this effectively. One has to do trade studies not just in the technical domain but also simultaneously in the market and business domains. An example is shown below in Fig. 6.22.

Lessons Learned

If an organization performs the above basics of SE and project management, it will have a greater chance of success than if it doesn't, no matter what size or type of project. History

(Continues)

Case Study 6.8.   Cont'd

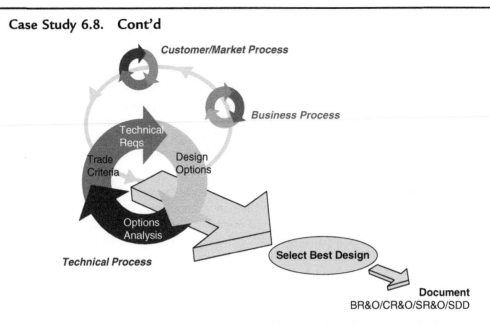

Figure 6.22: Complicated, multidimensional trade space of SoSs.

shows that over 90% of the project failures can be avoided by good SE and project management. The most critical of all these are the top-level understanding of the market, the completion criteria, the requirements, and the architecture including the interfaces. Failures at the top level propagate everywhere and generate "out-of-sequence" work. The cost to fix them can easily be multiple factors (100 to 2000% or more) of the cost of doing it correctly in the beginning. Failures at the top level can also turn the program unstable by extending the schedule, reducing the delivered functionality, missing the market timing, canceling the entire program, or losing the investment. All of these failures can significantly reduce shareholder value either directly through the loss of investment or indirectly through reduction of company brand quality or through loss of market potential.

Final Comments

These comments are the author's view of the state of the practice today. Interactions between "best practices" and the realities of the marketplace force a constant rethinking of where the value is and how much of these best practices to use.

Widely known but not necessarily widely implemented examples of good practices are as follows:

- Have clear project goals and completion criteria.
- Reduce to a minimum confusing RAAs.
- Capture top system requirements, architecture/design, and trade criteria early.
- Use some form of program "agility, stability, and control" such as configuration management and change management.
- Use processes by focusing mainly on key deliverables (the ones that will be put under configuration control).

Personally, I believe that better industry documentation of the SE discipline is necessary to show the value of using best practices. Without this documentation, management will always attempt to cut costs, which usually also means cutting all or parts of SE and of integration. A study needs to be performed and presented that shows the financial and schedule consequences of not implementing the various practices, by practice and by layer, so that management can decide how and where to tailor the practices to the needs and risk tolerance of the project. Then management will know where it is reasonable to save money by taking the risk versus where it is expensive, destructive, and foolish.

References

[1] Barrows J. QML specification to become performance-based, EEE Links 1995;1(2). Available at: nepp.nasa.gov/eeelinks/vol_01/no_02/eee1-2h.html.

[2] Available at: en.wikipedia.org/wiki/Joint_Capabilities_Integration_Development_System.

[3] Available at: www.wbbinc.com/jcids.html#overview.

[4] Available at: akss.dau.mil/dag/DoD5000.asp?view=document.

[5] Government Accounting Office. Defense acquisitions: DOD's requirements determination process has not been effective in prioritizing joint capabilities. GAO-08-1060. Washington, DC: GAO; 2008. Available at: www.gao.gov/products/GAO-08-1060.

[6] National Research Council Air Force Studies Board. Pre-milestone A and early-phase systems engineering: a retrospective review and benefits for future Air Force acquisition. Washington, DC: National Research Council Air Force Studies Board; 2008.

[7] Doran T. IEEE 1220 revision, ISO/IEC 12207 revision, SC7's ITSM development June 13 and 14, 2005. Available at: standards.computer.org/sabminutes/2006Spr/81-IEEE1220_ISOIEC12207-Revision_SC7sITSM&WG25_T.Doran-v.5.ppt.

[8] Available at: www.sepo.nosc.mil/Naval_Systems_Engineering_Guide.pdf.

[9] Available at: webstore.iec.ch/preview/info_isoiec15288%7Bed2.0%7Den.pdf.

[10] Available at: standards.ieee.org/reading/ieee/std_public/description/se/1220-1998_desc.html.

[11] Integrated master plan and integrated master schedule preparation and use guide. Washington, DC; 2005. Available at: www.acq.osd.mil/sse/docs/IMP_IMS_Guide_v9.pdf.

[12] Systems engineering plan preparation guide, Version 2.01, 2008. Available at: acc.dau.mil/CommunityBrowser.aspx?id=19389.

[13] Smith D, Russell S. How to measure the effects of systems engineering on the outcomes of a project. SETE Conference, Canberra, Australia; 2008.

[14] Available at: acc.dau.mil/CommunityBrowser.aspx?id=19397.

[15] Available at: iac.dtic.mil/index.html.

[16] Available at: ammtiac.alionscience.com/services/capabilities.html.

[17] Available at: www.cbrniac.apgea.army.mil/About/Functions/Pages/default.aspx.

[18] Available at: www.cpia.jhu.edu/templates/cpiacTemplate/about/.

[19] Available at: www.thedacs.com/about/.

[20] Available at: iac.dtic.mil/iatac/.

[21] Available at: www.dod-msiac.org/.

[22] Available at: www.theriac.org/.

[23] Available at: www.sensiac.org/external/index.jsf.

[24] Available at: www.bahdayton.com/surviac/.

[25] Available at: wstiac.alionscience.com/.

[26] Available at: www.saviac.org/.

[27] Available at: www.wbdg.org/ccb/NAVFAC/DMMHNAV/engineering_criteria.pdf.

[28] Stamatis DH. Six sigma and beyond: design for six sigma. Boca Raton, FL: CRC Press; 2003.

[29] Available at: www.redstone.army.mil/amrdec/sepd/tdmd/Documents/ADS13FHDBK.pdf.

[30] Freeman H. Software testing. IEEE Instrum Meas Mag 2002;5(3):48–50.

[31] Fowler K. What every engineer should know about developing real-time embedded products. Boca Raton, FL: CRC Press; 2008.

[32] Lansdaal M, Lewis L. Boeing's 777 Systems Integration Lab. IEEE Instrum Meas Mag 2000;3(3):14.

[33] Chan HA, Englert P, editors. Accelerated stress testing handbook: guide for achieving quality products. New York: IEEE Press; 2001.

[34] Available at: www.goes-r.gov/procurement/flight_documents/MIL-STD-461E.pdf.

[35] Available at: acc.dau.mil/CommunityBrowser.aspx?id=30513.

[36] Williams T. EMC for product designers: meeting the European directive. 3rd ed. London: Newnes; 2001.

[37] Montrose MI. Printed circuit board design techniques for EMC compliance: a handbook for designers. 2nd ed. New York: IEEE Press; 2000.

[38] Sanders MS. The Yard: building a destroyer at the Bath Iron Works. New York: HarperCollins Publishers; 1999.

[39] Drees R, Young N. Built-in-test in support system maintenance. IEEE Instrum Meas Mag 2002;5(3):25–29.

[40] Steinmetz M. Built-in-test instrumentation and 21 rules of thumb. IEEE Instrum Meas Mag 2002;5(3):31.

[41] Acquisition Community Connection. Obsolescence management. Available at: acc.dau.mil/CommunityBrowser.aspx?id=32247.

[42] Northrop Grumman. Components obsolescence management, Rolling Meadows, IL: Northrop Grumman, Defensive Systems Division; 2000. Available at: www.bmpcoe.org/bestpractices/internal/ngdsd/ngdsd_7.html.

[43] Bogdanski J, Downey M. Obsolescence management: another perspective. Military Embedded Systems 2009;38–44.

[44] Available at: www.dmea.osd.mil/arms.html.

[45] Hilderman V. DO-178B and DO-254: a unified aerospace-field theory? Military Embedded Systems 2009;16–19.

[46] RTCA. Software considerations in airborne systems and equipment certification. RTCA/DO-178B. Washington, DC: RTCA, Inc.; 1992. (Available for purchase at: www.rtca.org).

[47] Available at: www.do254.com/?p=faq.

[48] RTCA. Design assurance guidance for airborne electronic hardware. RTCA/DO-254. Washington, DC: RTCA, Inc.; 2000. (Available for purchase at: www.rtca.org/downloads/ListofAvailableDocs_April_2009.htm#_Toc228074011).

[49] Available at: en.wikipedia.org/wiki/DO-254.

[50] Torbjörn M, Håkan F. Derating concerns for microprocessors used in safety-critical applications. IEEE A&E Syst Mag 2009; 35–40.

[51] Forrester JW. Counterintuitive behavior of social systems. Technol Rev 1971;73(3):52–68.

[52] Sabbagh K. 21st century jet: the making and marketing of the Boeing 777. New York: Scribner's; 1996.

Index

A

Analysis report, types, 62–63
API, *see* Applications programming interface
Applications programming interface (API), 119, 119*f*
AS9100, 11–12
ATE, *see* Automatic test equipment
Automatic test equipment (ATE), military systems, 534
Autonomy, rule-based autonomy in spacecraft development, 361, 362*t*

B

BIT, *see* Built-in-test
BITE, *see* Built-in-test equipment
Built-in-test (BIT), military systems, 534
Built-in-test equipment (BITE), military systems, 534

C

Capabilities based assessment (CBA), defense acquisition, 463–464
Capability Maturity Model Integration (CMMI)
  definitions, 14
  framework, 15
  ISO 9001 comparison, 14–15, 14*t*
  maturity levels, 13*t*
  process areas, 13*t*, 16*t*
CBA, *see* Capabilities based assessment
CBE, *see* Current best estimate
CCB, *see* Change control board
Change control board (CCB), 188–190, 191
CMMI, *see* Capability Maturity Model Integration
Code inspection, 41
Code of Federal Regulations 820, *see* Quality Systems Regulations
Commercial off-the-shelf components and equipment (COTS)
  military use, 461–462, 541*b*
  spacecraft modules and subsystems, 326–328

Configuration Management Plan, components, 46
Configuration management
  design repository, 43
  file structure, 44–45
  military systems, 498–499
  obsolete documents, 45
  rationale, 42
  records responsibility, 43
  scope, 42
  spacecraft development, 370
  system and location, 531–532
  training, 45–46
  version control, 531–532
COTS, *see* Commercial off-the-shelf components and equipment
Current best estimate (CBE), systems engineering, 353

D

Department of Defense (DoD), *see also* Joint Capabilities Integration Development System; Military projects
  acquisition process, 464–466, 464*f*
  decision-making support systems, 273*f*
Design control, *see* Medical devices
Design description, components, 63–64
Design history file (DHF), medical software, 180–182
Design review, 40
Development plan, components, 57–58
Development process
  example, 29–33, 30*f*, 31*f*, 32*f*
  overview, 27
  processes versus procedures, 27
  spiral model, 28, 29*f*
  V-model, 27–28, 28*f*
DHF, *see* Design history file
DO-178B, 523–525
DO-178C, 525, 538*b*
DO-254, 525–526